AF520193

GENERAL MICROBIOLOGY

GENERAL MICROBIOLOGY

Jamaluddin
Group Co-ordinator Research &
Head, Division of Forest Pathology
Tropical Forest Research Institute
P.O. RFRC, Jabalpur 482 021 (MP)

Naveen Malaviya
Head, Department of Botany,
Govt. P.G. College
Narsinghpur (M.P.)

Published by
SCIENTIFIC PUBLISHERS (INDIA)
Jodhpur –
5 A, New Pali Road
P.O. Box 91
Jodhpur - 342 001 INDIA

Print: 2019

ISBN: 978-81-7233-420-6

Visit the Scientific Publishers (India) website at
http://www.scientificpub.com
Printed in India

CONTENTS

SECTION – V

SECTION – I

CHAPTER - 1

HISTORY OF MICROBIOLOGY

The organisms, which can only be studied with the help of microscope, are known as micro-organisms. The study of microorganisms is called Microbiology. This branch of science deals with the study of structure, nature, physiology, reproduction and economic importance of micro-organisms. The micro-organisms were the first to appear on earth.

Antony Van Leeuwenhoek (1623-1732) Fig. 1A, is said to be the father of the Microbiology and Bacteriology. Leeuwenhoek was a cloth merchant in Dutch. He prepared a lens (microscope, 300X or so), by grinding glass. He observed mobile organisms, which he described as animalcules or little animals. He described it in voluminous letters to the Royal Society of London which were translatated and published in English in the six proceedings of the Royal society. He discovered major classes of bacteria (spheres, rods and spirals), protozoa etc. An abstract of a letter from Mr. Antony Van Leeuwenhoek, dated 17th September 1683, presented in **Philosophical Transactions of the Royal Society of London** — "Though my teeth are kept usually very clear, nevertheless when I view them through a magnifying glass, I find growing between them a little white matter as thick as wetted flower; in this substance though I do not perceive any motion I judged there might probably be living creatures."

Fig. 1A. Antony van Leeuwenhoek (1623-1732)

"Animals, which moved themselves very extravagantly, the biggest sort had the shape of A. Their motion was strong and humble, they darted themselves through the water or spittle, as a Jaet or Pike does through the water. These were generally many in number. The 2nd sort had the shape of B. These spun about like a Top, and took a course sometimes one side, as is shown at G, and b. They were more in number than the first. In the 3rd sort I could not well distinguish the figure which sometimes seemed to be an oval, and other times a circle" Fig IB.

In 1758, **Linnaeus** created the genus *Vermes* which he called as *"Choas"*. A few years latter **Wrisberg** named these organisms *"Infusoria"*. In 1767 Linnaeus in his 11th edition of the system naturae classed these organisms as *Chaos* *infusoria*. **O.F. Muller** in 1773 and 1786 first proposed classification of bacteria. He introduced the terms *Vibrio* and *Monas*. The genus *Bacterium* (Gr. *Bactos* a staff) was first established by **Ehrenberg** in 1829. In 1883 Ehrenberg introduced the terms *Spirillum* and *Spirocheate*. **Nageli** (1857) introduced the term **Schizomycetes**. Nageli noted the relationship of bacteria and fungi.

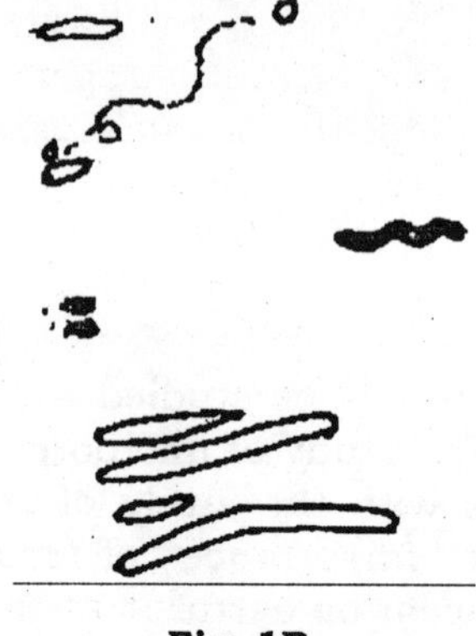

Fig. 1B.

ABIOGENESIS *Vs.* BIOGENESIS

Abiogenesis (spontaneous generation) theory explains the spontaneous formation of living organisms from non-living materials, apart from its application to the evolutionary origin of life. **Aristotle** (384-322 B.C.) taught that animals might originated spontaneously from soil, plants or other unlike animals. **Vergil** (70-19 B.C.) artificially propagated maggots by exposing the meat to warmth and air. The doctorine of abiogenesis persisted upto 17th century and since then it is abandoned.

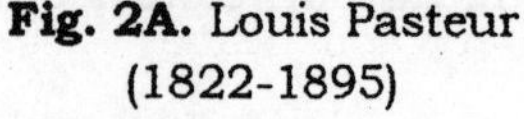

Fig. 2A. Louis Pasteur (1822-1895)

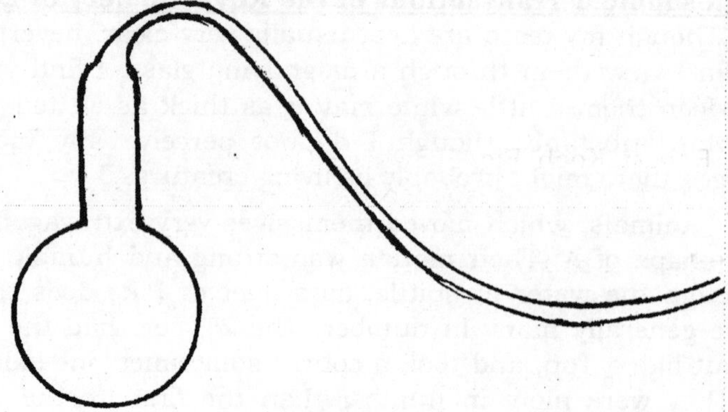

Fig. 2B. Gooseneck flask

Biogenesis (Bios-life+Genesis-source) theory explains that all living organisms are derived from other living things. **Francesco Redi** (1626-1697) an

Italian naturalist, was the first person to oppose Abiogenesis doctorine. He placed pieces of animal flesh in jar and covered it with gauze. Attracted by odour of the meat, flies laid eggs on the covering and the maggots developed from the eggs. **John Turberville Needham** (1748) English biologist and priest, exposed the meat with hot ashes, observed the appearance of organisms not present at the start of the experiment. He concluded that the bacteria originated from the meat. **Lazzaro Spallanzani** (1776) Italian biologist boiled beef broth for an hour and then sealed the flasks, no microbes appeared. His experiment did not disprove spontaneous generation, as air was essential for the production of microbes. **Theoder Schwann** (1837) passed air into the flask containing *via* a red-hot tube, no microbes appeared. **Schroder and Van Dusch** (1850) used cotton plugs, it filtered the air and the microbes were excluded. **Louis Pasteur** (1858-1861) Fig IIA; prepared a flask with a long, narrow gooseneck opening Fig. IIB. The nutrient solutions were heated in the flasks and untreated and unfiltered air was passed through it. The germs settled in the gooseneck and no microbes appeared in the solution. It was in 1864 at Sorbonne when Pasteur delivered a lecture and stated that life is a germ and germ is a life. **John Tyndall** (1877) a British physicist concluded experiment in a specially designed box in which he placed six culture tubes. A beam of light was passed through the tubes. The tubes were filled with nutrient broth in a dust free light and the tubes were boiled. He observed that microbes could not appear in these tubes.

THE GERM THEORY OF DISEASES

Lucretious (967-55 B.C.) wrote a poem saying that there are seeds of disease. In 1546 **Girolamo Fracastoro** a physician, recognized the contagious nature of diseases. **Afanasi Kircher** (1658) said that “worms” cause disease. He recognized that these worms are invisible by naked eyes. **Agostino Bassi** (1835) showed that a fungus caused muscardine disease in silk worms. Louis Pasteur (1822-1895) and Robert Koch (1843-1910) were the pioneers of this field of microbiology. Christian Gram (1884) developed a differential technique of staining bacteria, known as Gram staining.

Robert Koch

Robert Koch Fig. III was a German bacteriologist, in 1876 he worked on anthrax disease. He isolated Bacilli from the blood of cattle that died of anthrax. He prepared a pure culture. Koch formalized a criterion, known as Koch postulates. Koch postulated that:

(i) A specific organism can always be found in association with a given diseases;

(ii) The organisms can be isolated and grown in pure culture in the laboratory.

(iii) The pure culture will produce the disease when inoculated into susceptible animals.

(iv) It is possible to recover the organism in pure culture from the experimentally infected animal.

Fig. 3. Robert Koch (1843-1910).

In 1882 Robert koch isolated the tubercle bacillus (Koch's bacillus) now known as *Mycobacterium tuberculosis*, the causal organism of tuberculosis. In 1890, Koch discovered tuberculin. During 1879 to 1889, the members of German school isolated the Cholera Vibrio, Typhoid Bacillus, Diphtheria, Bacillus, Pneumococcus, Staphylococcus, Streptococcus, Meningococcus, Gonococcus and Tetanus bacillus. For the cultivation of disease-producing bacteria outside the animal body Koch used meat infusions and meat extracts as important constituents of culture media (broth or agar).

Louis Pasteur

Louis Pasteur was a French chemist, made a pioneer work to prove the germ theory of origin of microorganisms. In 1857, Pasteur worked on fermentation and he postulated that fermentation is carried out in living organisms. Later in 1866, developed a technology to kill microbial contaminants of food by heating it gently. This process was named as Pasteurization. In 1860 he observed that when several weeks old cultures of bacteria is used to inoculate the host, no disease symptoms appeared. This observation concluded that the bacteria from old culture loose their ability to produce disease, but these bacteria still retained to protect the host against virulent organisms. This observation led Pasteur to found the basis of immunization. In 1865, Pasteur isolated the parasite causing the silkworm disease pebrine. In 1881, Louis Pasteur discovered vaccine for fowl cholera and anthrax.

Pasteur contribution on fermentation laid the foundations of a new approach in the study of microbiology. In 1857, He discovered that a few microbes take part in breaking down sugar into lactic acid. This proved that fermentation is carried out by microorganisms. Between 1857-1877, Pasteur studied different types of fermentation i.e., some other fermentable materials. Purified and crystallized an animal virus, Poliomyelitis virus causing disease in man and animals.

Howard Taylor Ricketts (1906), in the United States, discovered the bacillary form of rickettsiae. In 1909, he established it as causal agent of Rockey mountains spotted fever.

HISTORY OF MEDICAL MICROBIOLOGY

Medical microbiology deals with the study of causal organism of a disease, their structure, symptoms of disease, diagnostic procedure & control measures.

Girolamo Fracastoro (1530,1546) described the contagious nature of disease. A number of bacteriologists, pathologists, and physicians isolated the causal organisms of different diseases. This proved a landmark in the history of medical microbiology. In 1880, Kark Joseph Eberth isolated typhoid bacteria, the bacilli were named as *Eberthella typhosa.* A French physiologist Charles Louis Alphonse Laveran (1880), discovered malaria parasite. Carl Friedlander (1882), discovered a bacillus and named it as *Friedlander bacillus.* This was later named as *Klebsiella pneumoniae.* In 1879, Albert Neisser discovered gonococcus, the bacteria causing gonorrhea. Daivd Bruce (1903-1906) described that African sleeping sickness was caused by a protozoan, *Trypanosoma gambiense.* He also demonstrated that this disease is transmitted by tsetse fly. Fritz Schaudinn and Erich Hoffman (1905) discovered *Treponema pallidum* the etiological agent of syphilis. Schaudinn also discovered Entamoeba the causal organism of amoebic dysentery.

Fig. 4. Martinus Willem Beijerinck (1851-1931)

In 1798, Edward Jenner used the term "vaccine". Joseph Lister (1867) introduced antiseptic surgery using phenol to soak surgical dressing. Elie Metchnikof (1884) described certain cells of blood and connective tissue, which engulf solid particles. He named such amoeboid cells as "phagocytes". In 1890,

Emil von Behring and Shibasaburo litasato explained the mechanism of immunity in animals to diphtheria and tetanus. The contributions of Louis Pasteur in the field of immunology gave a multiple dimensions of research in this field.

Paul Ehrlich (1908), controlled syphilis by applying arsenical compounds, Salvarsan. He used some chemical compounds to control disease. This aspect of medical microbiology is known as chemotherapy. In the early part of 20th century, first antibiotic was discovered by Emmerich and Low who isolated an anti-microbial substance from bacteria. They named this substance as pyocyanase. Alexander Fleming (1929) isolated penicillin from a fungus *Penicillium notatum.* Rene Dobos (1939) isolated gramicidin and tyrocidine. These active principles were isolated from a bacterium, *Bacillus brevis.* Selman A. Waksman (1944), isolated antibiotics streptomycin from actinomycetes, *Streptomyces griseus*, Some other antibiotics which are of important use in chemotherapy were discovered by Ehrlich and Duggar. In 1947, Ehrlich isolated chloromycetin (chloramphenicol). John Ehrlich, Paul R. Burholder, and David Gottlieb (1947), isolated chloramphenicol while Duggar and co-workers (1948) isolated chlorotetracycline.

HISTORY OF AGRICULTURAL MICROBIOLOGY

Agricultural microbiology deals with microflora of soil. Role conducted by these microorganisms in soil fertility and decomposition of organic matter. Their role in biogeochemical cycle of different inorganic elements as well as their mobilization. It also covers the study of pathogenic microbes responsible for soil borne diseases.

Fig. 5. Sergius Winogradsky (1856-1953)

Boussingault (1837-1838), a French agronomist opined that leguminous plants assimilate nitrogen from atmosphere. Though, Lachman (1858) observed that motile bacteria are associated with nodule formation in leguminous plants but it was Beijrinck (1888) who isolated the root nodule bacterium which symbiotically fixes atmospheric nitrogen. He named this bacterium as *Bacillus radicicola,* it was later named as *Rhizobium.* In the other two decades, Beijerinck and his coworkers isolated a number of bacteria, which were isolated with nitrogen fixation. Beijerinck, Omeliansky and others worked out the process by which ammonia formed in the soil as a result of various biological processes, is oxidized to nitrites by *Nitromonas, Nitrobacter* while the bacteria *Nitrobacter* and *Nitrocystis* convert this nitrite into nitrate. They also worked on transformation of sulphur and carbon in the soil. Winogradsky (1890) Fig V; demonstrated that *Clostridium pasteurianum,* an anaerobic bacteria can fix atmospheric nitrogen non-symbiotically. Beijerinck (1901) discovered *Azotoba-cter* (*A.chroococcum* and *A. agile*). This bacteria is aerobic in nature and is able to fix atmospheric nitrogen non-symbiotically. In 1928, Drewes observed that blue green algae, *Anabaena* and *Nostoc* are able to fix atmospheric nitrogen. This was later confirmed by Allison and Morris (1930). The microorganisms associated with root system form a zone known as rhizosphere. Hiltner (1904) observed a high degree of microbial activity in this zone.

HISTORY OF FOOD MICROBIOLOGY

Food microbiology deals with the study of different microorganisms associated with food, their application in preparation of food and their role in food spoilage. Food borne disease and food preservation i.e., different technique to control microorganisms to avoid food spoilage and to preserve its nutritive values is also studied in this field of microbiology. Microorganisms are used in various fermentations and are consumed as part of the food. Contribution of Louis Pasteur in studying fermentation technique and food preservation (pasteurization) has been discussed earlier.

Coon (1975), Crook (1975), Munro (1976) reviewed the toxic agents naturally present in foods.

Ennis and coworkers (1975 and Albert (1975) estimated the food loss caused by the action of microorganisms and other animals. In 1975, Schweigert estimated food losses from production to consumption to range from 20 to 50%.

The waste product may have food value. Progress is being made in the recovery of wastes and there utilization in preparation of food and feed. Callihan and Dunlap (1971), Cherry *et al.* (1975), Cooper (1976) Toyama (1976) have made important contribution in this field of food microbiology.

HISTORY OF INDUSTRIAL MICROBIOLOGY

This branch of microbiology deals with the study of application of microorganisms in industrial preparation of economically important products like fermented beverages, medicines (antibiotics, vaccines), fine chemicals, enzymes etc. Emil Christian Hansen (1842-1909) was the first to start the study of industrial fermentation. The pioneer studies on fermentation techniques were restricted to some acids and alcohols. Now this technology is applied in the commercial scale production in different industries like foods, chemical processing etc. A number of enzymes have been isolated from bacteria and molds, which are used in these fermentation techniques.

CHAPTER - 2

TAXONOMY OF BACTERIA

The art of biological classification is known as "Taxonomy".

Microorganisms covers a vast group. They are in general haploid showing simple morphological structures and a tendency to reproduce asexually. It is difficult to classify them with only one single criterion. On the basis of nuclear organization the microbes can be basically divided into two types Prokaryotes and Eukaryotes. R.H. Whittaker (1969) classified living organisms in broadly five groups namely Monera, Protista, Fungi, Plantae and Animalia. Monera is a group of prokaryotic micro-organisms including archaebacteria, eubacteria and cyanobacteria. The term Protista was coined by E. Haeckel in 1886. He included both, prokaryotic and eukaryotic micro-organisms in this group. However, now this group includes only eukaryotic organisms like algae, protozoa, slime moulds and water moulds. Fungi is a group of eukaryotic micro-organisms, it includes Zygomycetes, Ascomycetes, Basidiomycetes and Deuteromycetes. Plantae are eukaryotes with bryophytes and tracheophytes. Animalia are also eukaryotes including vertebrate and invertebrate animals.

Microbiologists have formulated certain criteria for presenting a satisfactory classification of micro-organisms. These criteria are based on certain characteristic features such as morphological, cultural, biochemical, serological and molecular characters.

Classification of bacteria is based on certain characters and tests. The microscopic examination covers the size, shape and staining characters of a bacterium. The bacterial cell shows different shapes and aggregation of cells. On the basis of accepting Gram-staining bacteria are divided into two groups. Some other important structural characters are related to capsule, flagella and spores (details are discussed in chapter 3 and 4).

Cultural Characters

Cultural characters include the cultural requirements for multiplication, pattern of growth on solid media. A presumptive identification may be made on the basis of their colony morphology. A Colony may show variation in size and diameter. The morphological characters of colony may be studied under three headings: outline, elevation and translucency. Bacterial colony may be circular, wavy, rhizoidal in outline. It may have flat, raised, convex elevations. Bacterial colony may be transparent, opaque or translucent. A bacterial colony may show variation in size and diameter. However the colonies may show characteristic features related to the colony like outline, elevation and translucency. Bacterium may grow on a simple synthetic media or it may require some specialized media along with some growth factors. The former are known as prototrophs while the later are called as auxotrophs. The physical conditions, pH may show slight variation for the growth of a bacterium on a culture media.

Biochemical characters

Besides the nutritional features the bacteria produces certain enzymes and undergo some metabolic reactions, which are helpful in the identification of a bacterium. The biochemical tests performed are the ability to produce indole, H_2S etc and presence of some enzymes like: catalase, gelatinase, urease, lecithinase, lipase etc.

Serological tests

The serological characters are concerned with the nature of the surface antigens-antibody reactions. Bacteria or its metabolites, acts as an antigen when introduced in the body of an animal (host). The serological tests, includes production of antibodies in an animal host and testing of the antiserum. The antigen-antibody reaction takes place by precipitation or agglutination. Precipitation reaction takes place with soluble antigens. In such type of tests, an antiserum is tested by allowing it to diffuse through an agar gel towards a source of crude extract of bacteria. To perform the test 1.5% agar solution is plated in a perti plate and allowed to solidify. Wells are prepared upon solid surface with the help of a borer. The antigens to be tested are placed inside the wells. Antiserum is also placed in a well near to these antigens containing well. The antigens and antibodies diffuse through the agar and react with each other producing some characteristic types of lines. The way in which these lines appear, reveals if the antigens in adjacent wells have any common determinant groups. In the agglutination test, a drop of the culture of a particular bacterium is mixed on a slide with the anti-serum of an individual infected by it and examined under a microscope. If the corresponding antibody is present, it will form a clump, this confirms that the test bacterium is the same or closely related to the bacterium used as the antigen. Some serological tests are discussed in chapter 21.

Approaches to microbial classification

Artificial approach : This system in based on arbitarily chosen criteria. It is the oldest approach of classification. Based on locomotion and ability to synthesize food, living world was classified into plants and animals. Microbes are placed in both of these groups.

Natural approach : Linnaeus developed natural approach of classification based on biological nature of the objects. However, this classification has not emphasized on classification of microorganisms.

Phylogenetic approach : This approach is based on the evolutionary affinities of a group or an organism. As per this approach, microorganisms are placed in Monera, Protista and Fungi kingdom.

Adanosonian approach (Numerical taxonomy) : Michel Adanson developed a concept of classification. Position of a group or an organism is decided based on the similarities and differences of numerical characteristics among them. The phenotypic characters are being given equal wieghtage. The number of character examined should be as numerous and as varied as possible, to obtain a representative sampling of phenotype. The total number of characters examined and the characters shared between the two organisms gives a numerical relationship among them. Tabulating the characters examined, similarities and differences in the characters and their relationship can be expressed as following.

Number of characters positive in both strains = a

Number of characters positive in strain 1 and negative in strain 2 = b

Number of characters negative in strain 1 and positive in strain 2 = c

Number of character negative in both strains = d

$$\text{Similarity coefficient (Sj)} = \frac{a}{a + b + c}$$

$$\text{Matching coefficient (Ss)} = \frac{a + d}{a + b + c + d}$$

In similarity coefficient, negative characters are ignored while in matching coefficient both positive and negative characters are examined. The data obtained are arranged in similarity matrix.

Macromolecular approach : This approach is based on the studies of genetic material of a microbe. Two types of studies are performed : (i) e.g. could be given analysis of the base composition of DNA (mole % G + C in a DNA). (ii) Nucleic Acid hybridization.

(i) *DNA base composition* : For determining DNA base composition, purified DNA is subjected to increasing temperature and determining the increase in hypochromicity or by centrifugation of the DNAs in cesium chloride density gradients. The temperature at which, hydrogen bonds between the two strands of DNA molecule breaks, but the two strands remain together with each other. This temperature is known as melting temperature of DNA molecule. This technique is used to asses G + C content of a DNA of a species. By heating a DNA sample, gradually the hydrogen bonds breaks and absorbance is increased. This phenomenon reaches a plateau at a temperature at which, the DNA becomes single stranded. Mid point of this rise, the melting temperature (Tm) is a measure of its density. The absorbance is recorded at 260 nm. From melting point, the mole per cent (G+C) may be calculated applying following formula.

$$\text{Mole \% G + C} = \text{Tm X } 63.54/0.47$$

In bacteria the range of G + C content is from about 30 to 70 mole per cent G + C. The relative percentage of guanine and cytosine is expressed as following.

$$\frac{(G + C \times 100)}{(A + T + G + C)}$$

Relative rate of sedimentation in a cesium chloride solution technique may also be applied to determine mole % G+C. This technique is known as bouyant density of DNA. Bouyant density of each DNA molecule depends upon per cent G+C content. DNA preparation is centrifuged with heavy salts solution in an ultracentrifuge. After centrifugation, purified DNA sediments in a region where the density of DNA and salt solution is equal. Per cent G+C content is calculated by applying the emperical formula.

$$P = 1.660 + 0.00098 \text{ (\% G+C) g.cm-3.}$$

The mean G+C content varies little in difficult parts of the genome, the only major exceptions are preparations from organisms that contain two genetic elements of different G+C content. Thus, DNA of mitochondrial or chloroplast origin may differ appreciably in G+C content from the nuclear DNA, in preparations from certain eukaryotic organisms and there is sometimes a marked heterogeneity in the DNA of the bacterium that harbours a plasmid. In such case, the minor constituent may form distinct Satellite band in CsCl gradient; this phenomenon provided one of the clues, that led to the discovery of DNA in mitochondria and chloroplasts.

(ii) *Nucleic acid hybridization* *(DNA-DNA homology or DNA base sequence and RNA sequence).* DNA-DNA homology means the ability of strands of two different source (species or groups) DNA molecule to form a hybrid molecule *in vitro*. The DNA molecule hybridization renders a picture of phylogenetic relations among two groups of living organisms. Single strand of DNA of two different species is isolated and DNA hybrids are prepared. DNA hybrid molecules of two different species of bacteria are prepared by heating a mixture of DNA strand between 10 to 30°C or below their melting temperature. The ability to form DNA hybrids shows the relationship among the two species or groups. RNA sequence demonstrates DNA base pair sequence of a gene. To know the phylogenetic relationship between two groups or species, comparative sequencing of specific RNA (16 S rRNA) is conducted. The degree of relationship is expressed in terms of association coefficient (S), which is characteristic of a pair of organisms.

$$S = \frac{2}{N + N}$$

N and N = Total number of residues represented by oligomers of at least 6 nucleotides in catalogue A and B, respectively.

N = Number of residues in sequences common to two catalogues.

Classification of Prokaryotes

R.G.E. Murray (1968), coined a term "Prokaryotes" and included all the microorganisms having incipient nuclei. David H. Bergey published first edition of his book in 1923. This book is named as Bergey's Manual for Determinative Bacteriology. The 8th edition of this book was published in 1974, has incorporated a descriptive classification of prokaryotes. In this manual, the prokaryotes are divided into two divisions viz., division I consisting of cyanobacteria while the division II covers bacteria. Division II is further divided into 19 parts. Bergey updated his book and later this Manual of Determinative Bacteriology was published as Bergey's Manual of Systematic Bacteriology. There are four volumes of this book in which all bacteria are grouped into 30 sections.

Volume I includes Gram-negative bacteria of general, medicinal and industrial importance;

Volume II includes Gram-positive bacteria other than the filamentous actinomycetes;

Volume III includes archaebacteria, the cyanobacteria along with the remaining Gram-negative bacteria including chemolithotrophs and the

Volume IV includes Gram-positive, filamentous bacteria.

Division- I: Cyanobacteria

This group is discussed in chapter 5.

Division- II:

Bacteria are included in this division. Bacterium is identified by some simple features like cellular organization, plane of division, growth, habit and cell separation. Bacteria have simple cellular organization. Cell wall is rigid or semi-rigid, basically made-up of peptidoglycan. Cells divide by binary fission, sometimes by budding. Some bacteria show locomotion by flagella or by special mechanisms such as gliding, snapping or darting movements. The photosynthetic bacteria are anaerobes, they have bacterial chlorophyll and perform anoxygenic photosynthesis. Chemosynthetic bacteria may be aerobic, anaerobic or facultative.

Key to the Bacteria

Identifying features	Classification
I Phototropic	Part 1
II Chemotropic	
A. Chemolithotropic (Chemoautotrophic)	
1. Derive energy from the oxidation of N, S or Fe compounds; do not produce CH_4 from CO_2	
a cells glide	
aa cells do not glide	Part 3
b cells unsheathed	Part 3

	bb	cells not unsheathed	Part 12
2	Do not oxidise N, S, or Fe compounds; produce CH_4 from CO_2		
B	Chemoorganotrophic (Chemoheterotrophic)		
1	Cells		
2	cells do not glide (exception in part 19)		Part 2
	a	cells filamentous and unsheathed	
	aa	cells not filamentous and unsheathed	Part 3
	b	products of binary fission not equivalent (have appendages other than flagella and pili or reproduce by budding)	Part 4
	bb	not as above	
	c	cells not rigidly bound	
	d	cells spiral shaped, have cell wall	Part 5
	dd	cells not spiral shaped, no cell wall	Part 19
	cc	cells rigidly bound	
	d	Gram -ve	
	e	obligate intracellular parasites	Part 18
	ee	not as above	
	f	curved rods	Part 6
	ff	not curved rods	
	g	rods	
	h	aerobic	Part 7
	hh	Facultatively anaerobic	Part 8
	hhh	Anaerobic	Part 9
	gg	Cocci or coccobacilli	
	h	Aerobic	Parts 10 & 7
	hh	anaerobic	Part 11
	dd	Gram +ve	
	e	Cocci	Part 15
	f	Endospores produced	Part 14
	ff	Endospores not produced	
	ee	Rods or filaments	
	f	Endospores produced	Part 15
	ff	Endospores not produced	
	g	Straight rods	Parts 16 & 17
	gg	Irregular rods (coryneform) or tend to form filaments or filamentous	Part 17

Part I — The phototrophic bacteria

Bacteria included in this part are Gram negative and are of aquatic habit. Cells are spherical, rod shaped or spiral. The diameter varies from 0.3 to 64 μm.

Reproduction takes place by binary fission. DNA base ratios vary from 45-73 moles % G+C. They contain photosynthetic pigments which includes bacterial chlorophyll *a, b, c* and *d.* This part includes two groups the purple bacteria and green bacteria.

Part II — The gliding bacteria

This group includes Myxobacteria. They are Gram negative gliding bacteria. Cells are rod shaped covered by slime measuring more than 1.54 μm in diameter. The vegetative colony is called as swarm or pseudopodium. Bacteria have gliding movement in contact with a solid surface or air-water interphase. Reproduction takes place by binary fission. Cells within the fruiting bodies are dormant and are known as myxospores. DNA base ratios vary from 67-71% G+C. All members are chemoorganotrophs and strictly aerobic. Cells contain carotenoids pigments.

Part III — The sheathed bacteria

Bacteria included in this group are Gram-negative, aerobic, single rod-shaped cells either motile or non-motile. Cells covered with a sheath, which may be attached to sub-merged plants or stones. The motile cells of bacteria have polar or sub-polar flagella. DNA base pair vary from 69.5 to 70 moles % G+C. Sheaths may or may not be encrusted with iron or manganese oxides.

Part IV — Budding and / or appendaged bacteria

Bacteria included in this group are Gram-negative mostly chemohetero-trophic. Cells have excreted appendages and hold fasts. Reproduction takes place by binary fission or budding, resulting in two unequal daughter cells. During budding, cell forms a small tube like outgrowth (prostheca) on the surface.

Part V — Spirochaetes

Bacteria included in this group are unicellular, Gram-negative, aerobic or facultatively anaerobic or anaerobic and chemoheterotrophic. Cells are slender, helically coiled varying from 3 to 500 μm in length. Protoplasmic cylinder present in cells which interwined with one or more axial fibrils. The axial fibrils originate in approximately equal number from the subterminal attachment disc located at either end of the protoplasmic cylinder. Reproduction takes place by transverse fission. Bacteria show motility by three means : (a) by rotation about the long axis of the helix, (b) by flexion of the cells, and (c) by locomotion along a helical or serpentine path. DNA base pair vary from 36 –66 moles % G+C. Some species are pathogenic.

Part VI — Spiral and curved bacteria

Bacteria included in this part are Gram-negative, strictly aerobic or microaerophilic chemoorganotrophs. Cells are curved rod shaped varying 0.2-1.7 μm in length. Cells are lophotrichous or amphitrichous with one or more flagella. Motility is by characteristic cork-screw like motion. DNA base pair vary from 38-65 moles % G+C. Species of *Spirillum* are associated with plants and fix nitrogen.

Part VII — Gram-negative aerobic rods and cocci

This is a large group of bacteria comprising of five families : Pseudomonadaceae, Azotobacteraceae, Rhizobiaceae, Methylomonadaceae and Halobacteriaceae. Bacteria belonging to these families differ in morphology, physiology and metabolic processes.

1. *Family Pseudomonadaceae* — Bacterial cells are straight or curved rods with polar flagella. They are chemoorganotrophs with oxidative metabolism. G+C content vary from 58-70% moles. A few species of *Pseudomonas* and almost all the species of *Xanthomonas* are Phytopathogens. *P. aeruginosa* is pathogenic to animals. Some species of Pseudomonas produce bacteriocins known as *pyocins.*

2. *Family Azotobacteraceae* — Bacterial cells are rod-shaped to oval shaped. Cells are lophotrichous to peritrichous. They are heterotrophic and are capable of fixing nitrogen. There are four genera included in this family. The % G+C vary in each genera. In all, this is 53 –70 mole % G+C.

3. *Family Rhizobiaceae* — Two genera are included in this family *Rhizobium* and *Agrobacterium. Rhizobium* forms root nodules in leguminous plants and fix nitrogen symbiotically. DNA base pair vary 59-65% G+C moles. *Agrobacterium* forms hypertrophy in many plants and is not capable of fixing nitrogen. Bacterial cells are rod-shaped with polar or subpolar flagellum. Cells may be peritrichous with 2-6 flagella.

4. *Family Methylomonadaceae* — includes two genera *Methylomonas* and *Methylococcus.* They utilize one carbon compounds like methane or methanol.

5. *Family Halobacteriaceae* — The family consists of two genera *Halobacterium* and *Halococcus.* They grow on saline medium with NaCl concentration of 2M. Cells are rods or cocci.

Part VIII — Gram-negative facultatively anaerobic rods

This group is divided in two families: Enterobacteriaceae and Vibrionaceae.

1. *Family Enterobacteriaceae* — Bacteria included in this family are found in the human or animal gut. They are Gram Negative rods, chemoorganotrophs, aerobic or facultative anaerobic, motile or non-motile. If motile, the cells are peritrichous. In some bacteria, the cells are capsulated. G+C content varies from 39 to 59 moles per cent. This family is divided into two tribes : Escherichiae and Klebsielleae. Common genera of this family are *Escherichia, Salmonella, Klebsiella, Serratia, Proteus* and *Erwinia.*

2. *Family Vibrionaceae* — Bacteria included in this family are rod shaped and motile. The cells are straight or curved with polar flagella. G+C content varies from 39-63 moles %. Common genera of this family are *Vibrio, Aeromonas* and *Photobacterium.*

Part IX — Gram-negative anaerobic bacteria

Bacteria included in this part are present in natural cavities of man and other animals and intestinal tract of insects. They are chemoorganotrophs, Gram Negative pleomorphic rods. They may be motile or non-motile, if motile, cells are peritrichous. Common genera included in this part are *Desulfovibrio, Butyrivibrio* and *Bacteriodes.*

Part X — Gram-negative cocci and coccobacilli

Bacteria included in this part are parasitic or saprophytic. The cells are coccoid, found in pairs of in a mass. The bacteria are non-motile and aerobic in nature. Few species contain xanthophylls pigments. The common members of this part are *Neisseria* and *Branhamella.*

Part XI — Gram-negative anaerobic cocci

Bacteria included in this part are parasitic. The cells are non-motile, coccoid ranging from 0.3 –0.5 μm to 2.5 μm in diameter. Cells may be found single or in pairs or in a mass. Three genera are included in this part : *Veillonella, Acidamonococcus* and *Megasphaera.*

Part XII — Gram-negative, chemolithotrophic bacteria

This group is divided in two families : Nitrobacteriaceae and Siderocapsaceae.

1. *Family Nitrobacteriaceae* — Bacteria included in this family are commonly known as nitrifying bacteria. The cells are rod-shaped, ellipsoidal, spirillar, spherical or lobular, if flagellate, the flagella may be subpolar or peritrichous. Seven genera are included in this family showing a variable range in G+C content in each genera. In common, the G+C content vary from 47.4 to 61.2 mole per cent. The members of this family are *Nitrosomonas, Nitrosospira, Nitrosococcus, Nitrosolobus, Nitrobacter, Nitrospira* and *Nitrococcus.*

2. *Family Siderocapsaceae* — The bacteria included in this family are encapsulated having thick or thin capsule on their cells. The capsule contains iron or manganese. The cells are spherical ellipsoidal or rod-shaped.

There is a third group in this part, which is not distinguished as family. The members of this group have ability to oxidise sulphur and sulphur compounds. The cells may be rod-shaped, spherical with lobes or round to ovoid. Bacteria may be motile or non-motile, if motile, cells have polar or peritrichous flagella. G+C content varies from 50-68 moles %. The sulphur oxidizing bacteria are *Thiobacillus, Sulfolobus, Thiobacterium, Thiovulum* and *Thiospira.*

Part XIII — Methane producing bacteria

The bacteria included in this part are Gram variable, strict anaerobic and reduce CO_2 to methane. Cells are rods or coccoid, motile or non-motile. There is a single family in this part, Methanobacteriaceae and the common genera included in this family are : *Methanobacterium, Methanococcus* and *Methanosarcina.*

Part XIV — Gram-positive cocci

This group is divided into three families namely : Micrococcaceae, Streptococcaceae and Peptococcaceae.

1. *Family Micrococcaceae* — Bacteria included in this family are coccoid, motile or non-motile, aerobic or facultatively anaerobic and chemoorganotrophs. The cells are found in regular or irregular cluster or packets with diameter varying from 0.5 to 3.5 µm. G+C content varies from 30-75 moles per cent. Common genera included in this family are *Micrococcus, Staphylococcus* and *Planococcus.*

2. *Family Streptococcaceae* — Bacteria included in this family are spherical or ovoid, non-motile and facultatively anaerobic. The cells occur in pairs or chains or tetrads. G+C content varies from 33-44 moles per cent.

3. *Family Peptococcaceae* — Bacteria included in this family are found in mouth, intestine and the respiratory tract in man and other animals. They may be also present in soil. Bacteria are anaerobic and chemoorgnotrophic in nature. Cells vary 0.5-2.5 µm in diameter. The family includes four genera with G+C content varying in each genus. In general the G+C content varies from 28.6-36.7 moles per cent. The four genera of this family are *Peptococcus, Peptostreptococcus, Ruminococcus* and *Sarcina.*

Part XV — Endospore-forming rods and cocci

The bacteria included in this group are rod shaped, Gram-positive, motile or non-motile. If motile, cells have lateral or peritrichous flagells. They show a great variation towards oxygen tolerance, as they may be aerobic, facultative anaerobes, strictly anaerobic or even microaerophilic. They form endospores. There is a single family in this part namely Bacillaceae, including five genera. Each genera show variation in G + C content. In general in G+C content varies from 32-62 moles per cent. Genera *Bacillus, Sporolactobacillus, Clostridium, Desulfotomaculum* and *Sporosarcina* are included in this family.

Part XVI — Gram-positive, asporogenous rod-shaped bacteria

The bacteria included in this part are rod shaped, generally non-motile, anaerobic or facultatively anaerobic and highly fermentative. The cells may be straight or curved occurring singly or in chains. G+C content varies from 34.7-53.4 moles %. There is a single family in this group namely Lactobacillaceae with a single genera *Lactobacillus.* Three genera of uncertain affinities are also included. They are *Listeria, Erysipelothrix* and *Caryophanon.*

Part XVII — Actinomycetes and related organisms

This is a heterogenous group including coryneform, propionibacteria and the actinomycetes.

1. *Family Coryneformaceae* — bears the coryneform bacteria are human pathogens or phytopathogens or even non-pathogenic. Bacteria are Gram-positive, rod shaped, motile or non-motile and aerobic to facultatively anaerobic in nature. Cells may be pleomorphic with irregular outline. Human pathogenic coryneform bacteria have slightly curved cells with club-shaped swellings. G+C content varies in human pathogens and plant pathogens. It is 52-68 moles % in human pathogens while 65-75 moles % in plant pathogens. The common coryneform genera are *Corynebacterium, Arthrobacter, Cellulomonas* and *Kurthia.*

2. *Family Propionibacteriaceae*: This family includes Gram-positive, non-spore forming bacteria. They are pleomorphic rods arranged in a branched or filamentous manner. The bacteria perform fermentation. The end product of fermentation are propionic acid, acetic acid, CO_2 or a mixture of butyric, formic, lactic or other monocarboxylic acids. *Propionibacterium* and *Eubacterium* are included in this family.

The Actinomycetes are described in an order namely Actinomycetales. The bacterial body is made-up of branched filaments forming a mycelium. The members of this group are in general strictly aerobic, however, some members may be anaerobic or facultatively anaerobic. Similarly, in general, the actinomycetes are Gram-positive but this reaction may vary, members of Mycobacteriaceae are acid-fast and those of Nocardiaceae are weakly acid-fast. The whole sugar pattern in actinomycetes varies from galactose, arabinose, xylose and madurose and depending upon the type of sugar pattern, they are classified into four groups A, B, C and D. The cell wall constituent also shows variations. The variety of constituents may be DAP (meso or LL) arabinose, galactose or glycine. On the basis of cell wall constituent, actinomycetes are divided into five groups I, II, III, IV and V.

3. *Family Actinomycetaceae* — Members are diptheroid in shape forming branched filaments during some stages of cultural development. There is no formation of aerial branches and spores. Their basic nature is facultatively anaerobic however, anaerobic and aerobic natured members are also included.

4. *Family Mycobacteriaceae* — *Mycobacterium,* a single member of this family is pathogenic. It is aerobic, acid-fast, nonmotile with slightly curved or straight rods types of cells. It does not form endospores, conidia or capsule.

5. *Family Frankiaceae* — *Frankia*, the only member of this family is found in association with non-leguminous plants. It is capable of fixing nitrogen symbiotically. It has filamentous mycelium.

6. *Family Actinoplanaceae* — Members are found in humus rich soil. They are Gram-positive, chemoorganotrops aerobic. Vegetative body is mycelium which is intramatrical or sometimes aerial. Sporanigospores are found on branched or unbranched hyphae.

7. *Family Dermatophilaceae* — Members of this family are aerobic and pathogenic to man and other mammals. They have mycelial filaments or muriform thalli which divide transversely in two longitudinal planes, and form a mass of motile, coccoid or cuboid cells. There are two genera in this family *Dermatophilus* and *Geodermatophilus.*

8. *Family Nocardiaceae* — *Nocardia* and *Pseudonocardia* are placed in this family. They have rudimentary or extensive mycelium. The principal cell wall carbohydrate is arabinogalactan. It is a co-polymer of arabinose and galactose. Myolic acid, a lipid is esterified to arabinoglactan. Many other complex lipids are found in the wall of these organisms. Excess of lipids in cell wall make it impermeable and do not retain stains. The pathogenic forms are typically acid-fast. Acid-fastness is corelated with the size of the myolic acid. Spore production varies with the genus.

9. *Family Streptomycetaceae* — This family includes four genera, *Streptomyces, Streptoverticillium, Sporichthya* and *Microellobospora.* They are Gram-positive and aerobic in nature. Mycelium is well branched and reproduction takes place by spores or sometimes by growth of mycelial fragments. They do not show any definite type of sugar pattern.

10. *Family Micromonosporaceae* — This family includes six members. They are saprophytic in nature, living in soil, under mesophilic or thermophilic conditions. The members have both aerial and substrate mycelium except in Micromonospora. Sporophores are lacking or are very short. The members differ by the presence or absence of aerial mycelia and other sporulation characteristic.

Part XVIII — Rickettsias

This part is divided in two orders Rickettsiales and order Chlamydiales. The members of this part are Gram negative and parasitic in nature. They are found as parasites in man and other animals. Cells are rod shaped, coccoid or even pleomorphic. They have a typical cell wall composition. Life cycle lacks flagellated structures. They multiply inside the host by binary fission.

Order Rickettsiales is divided into three families, Rickettsiaceae, Bartonellaceae and Anaplasmataceae. The order Chlamydiales has only one family Chlamydeaceae with only one generum, *Chlamydia.*

Part XIX — Mycoplasma

This part is divided in two families each with one genera. Family Mycoplasmatatceae with genus *Mycoplasma* and family Acholeplasmataceae (genus *Acholeplasma*). They are parasitic in nature. The cell is devoid of cell wall with coccoid or peared shaped. They are obligate chemoheterotrophs. The two genera differ in the requirement of cholesterol.

CHAPTER 3

CELLULAR ORGANISATION OF PROKARYOTIC AND EUKARYOTIC CELLS

Cell is defined as structural and functional unit of life. The specific organizational pattern of the cell (structure and metabolism) is determined by the genetic material. All the cells have common macromolecules i.e., DNA, RNA and proteins. The simplest type of microorganisms are unicellular and a complex type of microbe is multicellular. All the cells are surrounded by cytoplasmic membrane. Virus and viroids are considered as acellular because, they do not have cell membrane. The cells are divided into two types, based on the organizational pattern of genetic material, the nucleus. A *prokaryotic* cell lacks nuclear membrane surrounding the genetic material, they never possess membrane bound cell organelles, while a *eukaryotic* cell has well defined nucleus, there is a greater degree of internal structural organization as they contain membrane bound organelles. Bacteria represents a *prokaryotic* cell. It is again divided into *archaebacteria* and *eubacteria*, the basic difference between the two is the cell chemistry. The knowledge about the *archaebacteria* is much confined upto three groups and they are treated separately. Hence, on the basis of cell structure and function, three major groups are recognized viz., archaebacteria, eubacteria and eukaryotes.

THE CELL WALL

Cell wall is present surrounding the cytoplasmic membrane of bacteria and some eukaryotic organisms. It is a rigid structure protecting the cell from osmotic shock. The region between the cytoplasmic membrane and cell wall is known as periplasmic space and is filled with periplasm. The chemical composition of cell wall are different in bacteria and eukaryotic cells. The cell wall of all the prokaryotic cells is basically composed of a macromolecule murein, but it is absent in certain archaebacteria and the *Planctomyces* groups of eubacteria. Murein is also known as peptidoglycan or mucopeptide Fig. 1. It is a heteropolymer of substituted sugars and amino acids. The backbone is made-up of alternating molecules of two amino sugars, N-acetylglucosamine (NAG) and N-acetylmuramic acid (NAM) attached by beta-1,4 linkage. Each chain consists of 10 to 65 disaccharide units. These strands are interlinked by short peptide chains. Both L- and D- amino acids occur in murein. A tetrapeptide chain is attached to N-acetylmurmiac acid. All the tetrapeptide chains are made-up of L-alanine, D-glutamic acid (this may be hydroxylated in some organisms) either L-lysine or diaminopimelic acid and D-alanine. The tetrapeptide are cross linked by one peptide bond present between the carboxyl group of an amino acid in one tetrapeptide chain and the amino group of an amino acid in another tetrapeptide chain. In *Staphylococcus aureus* the cross linkage is present between a tetrapeptide and a pentapeptide (made-up of L-glycine).

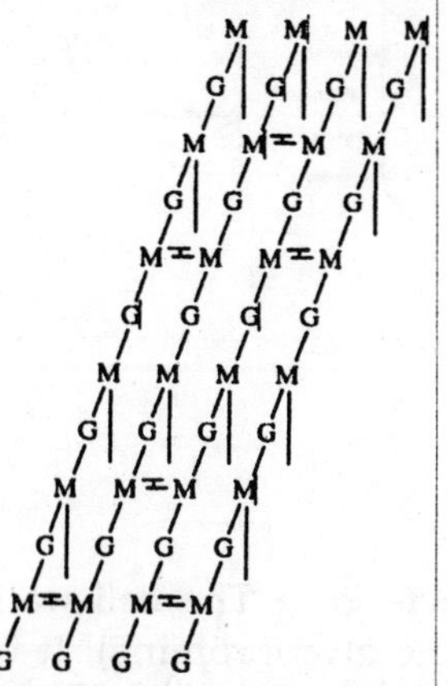

Fig. 1A. Cell wall of Eubacteria
Digramatic representation
G - N-acetyl glucosamine (NAG); M - N-acetyl muramic acid (NAG)

Cell wall of Gram positive and Gram negative bacteria — Eubacteria are divided into Gram positive and Gram negative groups on the basis of cell wall composition and to retain Gram staining. Both groups of bacteria have peptidoglycan in their cell wall. However, the quantity differs. Gram positive bacteria have a thick layer of peptidoglycan comprising about 90% of the cell wall constituent. Beside this major constituent, these bacteria have an acidic polysaccharide, teichoic acid. It is a polysaccharide made-up of glucose, phosphate and alcohol (glycerol or ribitol). In bacilli and staphylococci there are three types of teichoic acids e.g., *S. aureus* have ribitol teichoic acid, *B. subtilis* have glycerol teichoic acid and *B. licheniformis* have glucosylglycerol phosphate teichoic acid. Teichoic acid is present between the gaps of the backbone and cross linkages and it performs the function of a binding substance. Its presence makes easy to diagnose the bacteria serologically. The cell wall of Gram negative bacteria have a very little amount of peptidoglycan i.e., approximately 5 to 10% of the cell wall and completely lack teichoic acid. Their cell wall is more complex containing lipids, polysaccharides and proteins. These components form an outer layer. The lipids contain phospholipids and lipopolysaccharides (LPS). LPS is composed of three regions viz., lipid A, the R core region and the O side chain. Lipid A consists of glucosamine, phosphate and fatty acids. One end of LPS is hydrophilic while the other end is hydrophobic. The hydrophobic end is inserted in the membrane while the hydrophilic end is towards the outer side. The covering layer of lipids is also known as cell envelope.

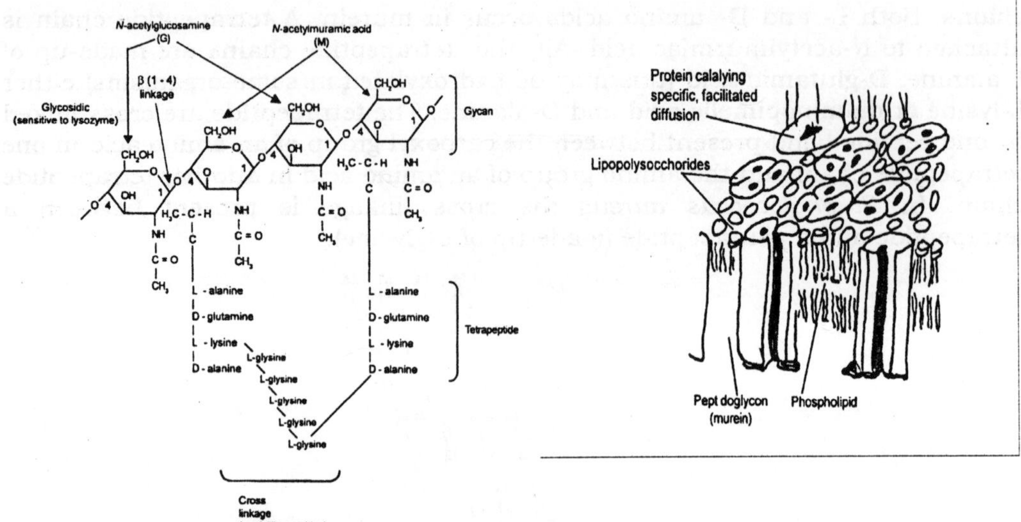

Fig. 1B **Fig. 1C**

Cell wall of Archaebacteria — The cell wall of archaebacteria is made up of proteinaceous subunits (the glycoproteins). It is present in a single layer, and in some more complex forms in Gram negative archaebacteria. Methanogens are also divided in three groups and the cell wall composition in these groups is different. The cell wall of Group I methanogens is made up of pseudomurein. Pseudomurein is a peptidoglycan similar to murein of eubacteria, but it differs due to the presence of N-acetyl-talosaminuronic acid instead of N-acetylmuramic acid and lacks D-amino acids. Pseudomerin, methanochondrotin, or heteropolysaccharide is present in *Methanobacterium* and *Methanobrevibacter*. The cell wall of *Methanococcus* (group II), is made-up of protein with trace of glucosamine. The wall of *Methanosarcina* (group III), is made-up of heteropolysaccharide, the principal constituents of which are galactosamine, neutral; sugars and uronic acids.

Cell wall of eukaryotic organisms — The cell wall of eukaryotic microorganisms is composed of polysaccharides. Cellulose is the major portion of these walls. In algae, the cell wall is composed of cellulose, pectic compounds along with some other major polysaccharides like xylans, mannans and alginic acid. In some algae calcium or silicon forms the test or frustule. Another large group of eukaryotic microbes is fungi having different cell wall from algae. The fungal cell wall is made-up of microfibrils of cellulose (fungal cellulose) and or chitin. Slime molds are pleomorphic. They lack a cell wall but at any stage of their life cycle they have a rigid cell wall.

Functions of the cell wall. The cell wall performs following functions —

(a) It provides a proper shape and rigidity to cell.

(b) It acts as a barrier for diffusion of certain molecules.

(c) It prevents osmotic shocks.

(d) The matrix proteins act as a receptor sites for bacteriophage and bacteriocins.

(e) In certain bacteria, cell wall possess antigenic characteristics.

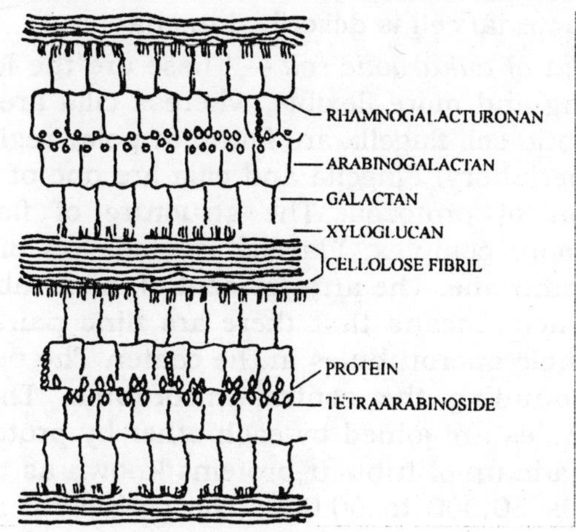

Fig. 2.

CAPSULE AND SLIME LAYERS

Some bacterial cells are surrounded by an amorphous layer made up of polysaccharide and/or proteins. This layer is known as capsule or slime layer. In general a capsule is meant for a structure closely attached to the cell wall and is revealed by negative staining. A slime layer or glycocalyx is similar to capsule but are not tightly bound to the cell. A glycocalyx is a mass of tangled fibers of polysaccharides or branching sugar molecules surrounding an individual cell or a colony of cells. These structures are considered as exopolymers. They show variations in their composition. In *Bacillus anthracis* and *B. megaterium* the exopolymers are peptides made up of a single amino acids glutamic acid. Glutamyl residues are linked through the carboxyl group as in the side chain of peptidoglycans. A majority of bacterial capsule is made up of polysaccharides. The subunits of these exopolysaccharides may be only glucose as in *Acetobacter xylinum, Agrobacterium tumefaciens.* The enteric bacteria have exopolymers made up of glucose, galactose, fucose, glucuronic acid and pyruvic acid. The pneumococcal exopolysaccharides are made up of glucose, rhamnose, glucuronic acid, in some pnuemococci N-acetyl-D-mannosamine, phosphate and ribitol is also present. A bacteria may have antigenic characteristics in their capsule. They are virulent if capsule is present outside their cells. Much more, during cell mediated immune response, the blood cell are unable to engulf the capsulated bacterial cells. The slime layers protects the bacterial cell from dehydration and loss of nutrients. The slime layer may also bind cells together forming multicellular aggregates. The slime layer also helps bacteria to attach to their substratum, which may be rocks in aquatic environment or teeth for bacteria occurring in oral teeth.

FLAGELLA, CILIA AND PILLI

These structures are found emanating from the surface of some bacterial cells. They originate from cell membrane and extend outward through the wall. Flagella, cilia of a bacterial cell is described in chapter 4.

Flagella and cilia of eukaryotic cell — These are the locomotary organs of a cell. Flagella are long and more flexible, whereas cilia are shorter and tough in nature. In a eukaryotic cell flagella are found at polar regions while the cilia are found all over the periphery. Flagella and cilia are one of the taxonomic feature in the classification of protozoa. The structure of flagella and cilia in a eukaryotic cell is more complex. Flagella is made up of pair of microtubules surrounded by a membrane. The arrangement of microtubules is "9 + 2" Fig. III. The 9 + 2 arrangement means that there are nine pairs of peripheral microtubules, and two single microtubules in the center. The peripheral microtubules form a circle surrounding the central micotubule. The central as well as peripheral microtubules are joined by each other by protein microfilament. The microtubules are made up of tubular proteins known as tubulin. The molecular weight of tubulein is 50,000 to 60.000. The peripheral microfilament network contains a protein known as dynein. This system contains enzyme ATPase. ATP is hydrolyzed by enzyme ATPase to provide energy required for the movement of flagella and cilia. The movement of flagella or cilia is due to sliding microtubule mechanism. The peripheral doublet microtubules slide past each other, resulting in bending of the flagella or cilia.

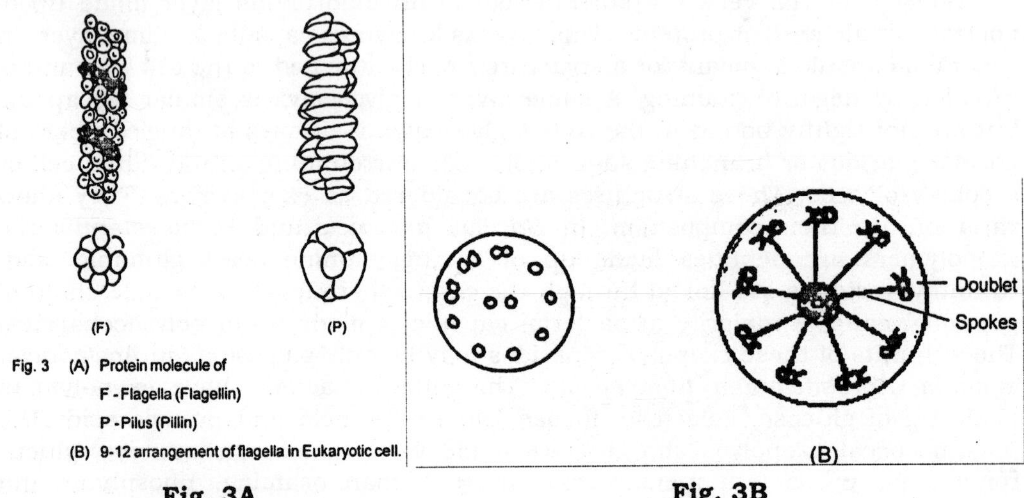

Fig. 3 (A) Protein molecule of
F - Flagella (Flagellin)
P - Pilus (Pillin)
(B) 9-12 arrangement of flagella in Eukaryotic cell.

Fig. 3A **Fig. 3B**

CYTOPLASMIC MEMBRANE

All cells are bounded by cytoplamic membrane. The important characteristic of these membranes is there selective permeability. The cytoplasmic membrane is composed of lipids and proteins Fig. 4. A thin section of cytoplasmic membrane stained with heavy metals when examined in the transmission electron microscope reveal a characteristic triple-layer appearance. The organization of these three layers is the lipid layers present on the either side of centrally located protein layer. The lipids are basically phospholipids are amphoteric as they have both hydrophobic and hydrophilic ends. The hydrophobic ends are placed towards the protein layers and the inserted proteins have hydrophobic amino acid residues buried within the membrane. The lipids of cytoplasmic membrane of archaebacteria always contain branched hydrocarbon chains joined to the polar region by ether bonds. The eukaryotes and the eubacteria contain lipids whose hydrophobic portion is a long, generally unbranched hydrocarbon chain joined in ester linkage to the hydrophilic portion. There are some differences in lipid composition between prokaryotic and eukaryotic cytoplasmic membranes. Sterols, like cholesterol and phospholipid (whose constituent fatty acids are polyunsaturated) are found in eukaryotic cell membrane but generally absent in prokaryotic cells. This gives more rigidity to the cytoplasmic membrane of eukaryotes. Bacterial membrane contains hopanoids or squalene, a biosynthetic precursor of sterols and hopanoids.

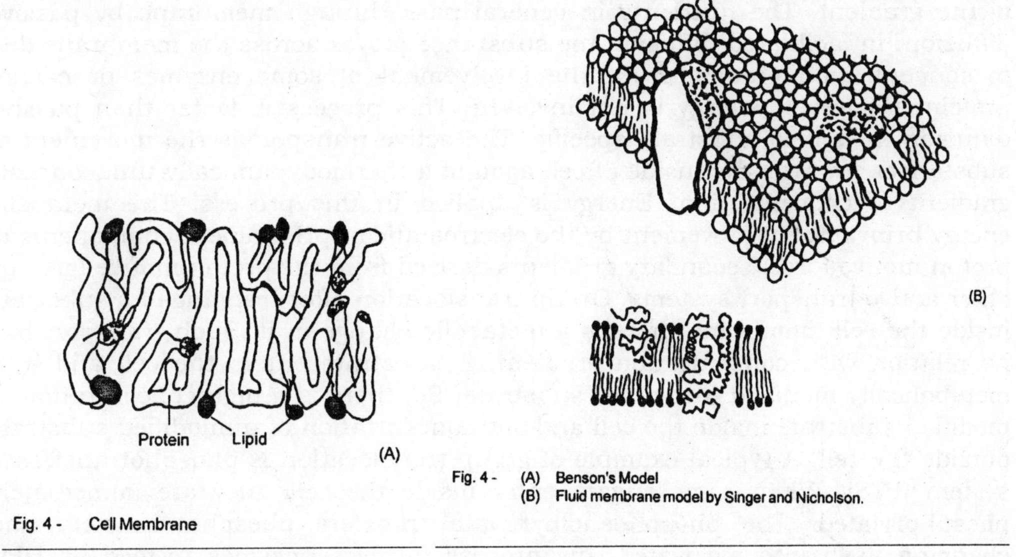

Fig. 4A. Benson Model

Fig. 4B. Fluid membrane model by Singer and Nicholson

A number of models have been proposed to explain structure of cytoplasmic membrane. However, the fluid mosaic model is widely accepted. According to this model, the membrane is a bilipid layer and proteins are distributed in a mosaic pattern. The proteins can easily move in the matrix of lipid. The proteins confined to the membrane surface while others are totally buried in the matrix, they are known as extrinsic and intrinsic proteins, respectively. There are three major categories of proteins associated with membrane. The proteins which, functions in transporting of inorganic and organic nutrients inside the cell are known as the second category of protein is biosynthetic proteins which are responsible of synthesizing lipids of the membrane and the components of cell wall. In some bacteria, there is a third category of proteins which, participate in the generation of ATP and brings about electron transport.

Cell membrane of archaebacteria — The cell membrane of archaebacteria lacks normal bilipid layer. The phosphate groups of lipids are linked by ether bonds. In *Sulfolobus* the cytoplasmic membrane is a monolayer, consisting of a long-chain branched hydrocarbon. Unusual structure of cytoplasmic membrane is also found in other archaebacteria (*Thermoplasma* and *Halobactrium*). These organisms live in extreme habitats and their cell membrane makes them resistant to the environmental conditions.

Functions of cytoplamic membrane — In general cytoplasmic membrane performs following functions :

(a) Diffusion and osmosis, (b) Passive and facilitated diffusion, (c) Active transport, (d) Group translocation, (e) Secretion of extra cellular enzymes, (f) Endocytosis and exocytosis and some other additional functions.

Passive diffusion is movement of substances across the cell membrane due to concentration gradient. The rate of flow is a direct function of the magnitude of the gradient. The nutrients in general pass through membrane by passive diffusion. In facilitated diffusion the substance moves across the membrane due to concentration gradient with the involvement of some enzymes or carrier proteins. However, energy is not involved. This process is faster than passive diffusion. It is also substrate specific. The active transport is the movement of substances towards the inside of cell against a thermodynamically unfavourable gradient of concentration. Energy is applied in this process. The metabolic energy brings about movement by the electrostatic or pH gradient components of proton motive force, secondary gradients derived from the proton motive force by other active transport systems. Group translocation after trapping of a substrate inside the cell membrane there is a metabolic change in it. Such transport has no relation with concentration gradient. A substrate enters the cell and it is metabolically modified into other substrate. So, there is a high concentration of modified substrate inside the cell and low concentration of unmodified substrate outside the cell. A typical example of group translocation is phosphotransferase system (PTS). When some sugars enter inside the cell, they are immediately phosphorylated. The phosphoenolpyruvate transfers phosphate moiety and changing itself into pyruvate. This process involves enzymes termed as HPr, Enzyme I, Enzyme III and Enzyme II.

Cytosis in general means movement of substance inside or outside the cell. The membranes engulf some substrate inside the cell by phagosomes. These structures are formed by the infolding of cytoplasmic membrane and the process is known as phagocytosis. Bacteria intakes small droplet while protozoa and some other cells of metazoa intake solid substrate by phagocytosis. The intake of liquids by eukaryotic cell is known as pinocytosis. Endocytosis is a collective term used for phagocytosis and pinocytosis. Substances synthesized inside the cell are excreted to the exterior by a process known as exocytosis. In eukaryotic cells golgi bodies plays an important role in exocytosis.

In eukayotic cells oxidative phosphorylation takes place in defined cell organelle but in prokaryotes the enzymes associated with this process are localized within the cytoplasmic membrane. Thus, cytoplasmic membrane of bacteria performs this additional function.

THE GENETIC MATERIAL

DNA macromolecule is regarded as genetic material of cell. It controls structural and physiological characteristics. In prokaryotic cell, there is no definite nucleus, while in eukaryotic cell genome is covered by nuclear membrane.

Archebacterial genome In archebacteria DNA molecule wraps around the protein molecule. These basic proteins are known as histone like proteins (Hta). In *Thermoplasma*, the core protein is made up of four molecules of Hla and forty nucleotides surrounds it.

Eubacterial genome Bacterial genome or chromosome consists of a single molecule of DNA. The double helical molecule of DNA is present in the portion cell, is known as nucleiod. Protein covering is absent. Some bacteria may contain extra chromosomal molecule of DNA. It is a circular molecule of DNA with a few specific genes. Details of plasmids are discussed in chapter 17.

Eukaryotic genome In eukaryotic cell, there is a definite nucleus. Nucleus has a covering known as nuclear membrane. Nuclear membrane has very fine pore known as nuclear pore. Nucleus communicates to cytoplasm and endoplasmic reticulum through these pores. Genetic material is present in chromatin, which forms chromosome. DNA molecule is covered by histone protein. Histones are low molecular weight protein with a high content of basic amino acids. DNA molecule forms a definite pattern of coiling. The DNA molecule (165 nucleotides) coils around the histone (9 molecules). This is known as nucleosomes. Nucleosomes are subunits of chromatin. This is a fundamental unit of genetic material. It is complex structure and hence it needs proper processing for giving message for polypeptide synthesis.

RIBOSOMES

Ribosomes are site of protein synthesis. It is made up of ribosomal ribonucleic acid (rRNA) and proteins. Ribosome is expressed in terms of Svedberg units (S). Svedberg unit is a unit of movement in density gradient ultracentrifugation. It depends on both the mass and shape of the ribosome. The ribosomes of prokaryotic and eukaryotic cells show some differences.

Prokaryotic ribosome is a 70 S unit made up of two subunits 30 S and 50 S units. The 30 S subunit contains twenty one proteins and a 16 S rRNA molecule. rRNA of this subunit has about 1,540 nucleotides. The 50 S subunit is made up of thirty four proteins and twenty three fRNA molecules. It contains about 2900 nucleotides. Details of bacterial ribosome is discussed in chapter 4.

Eukaryotic ribosome is a 80 S unit made up of two subunits 40 S and 60 S subunit. The 40 S subunit contains 18 S rRNA. The 60 S subunit has 25-28 rRNA and 5 S rRNA. The ribosomal subunits of eukaryotic cells are synthesized in nucleolus. It is transported to the cytoplasm through nuclear pore. The sequence of nucleotide bases of rRNA provides phylogenetic evidences. Chloroplasts and mitochondria contain 70 S ribosome. It is again an evidence of evolution of eukaryotes from prokaryotes.

The 70 S unit ribosome of bacteria is susceptible for antibiotics inhibiting protein synthesis. This is clinically important.

INTRA CELLULAR ORGANELLES OF EUKARYOTIC CELLS

Eukaryotic cells have unit membrane system. The physiological processes of cell take place in particular organelle. Each organelle is enveloped by unit membrane. The most important organelle is nucleus. It is a control room of all the activities of cell. It contains genetic material, which is already discussed.

MITOCHONDRIA

Mitochondria are spherical or rod shaped organelles Fig 5. They are the sites of ATP synthesis. Mitochondria have double membrane. The structure of these membranes is similar to that of plasma membrane. The center of the mitochondria is a semifluid substance. This is known as matrix. The outer membrane is a covering structure. The inner membrane is folded into convolutions. These folds are extended into the interior of mitochondria inside the matrix. These folds are known as cristae. Cristae contain some enzymes, which carry out respiration and ATP synthesis. Some metabolic steps involved in cellular respiration take place in matrix. The inner matrix space between the invaginated inner membrane of the mitochondria is the site of Kreb's cycle reaction.

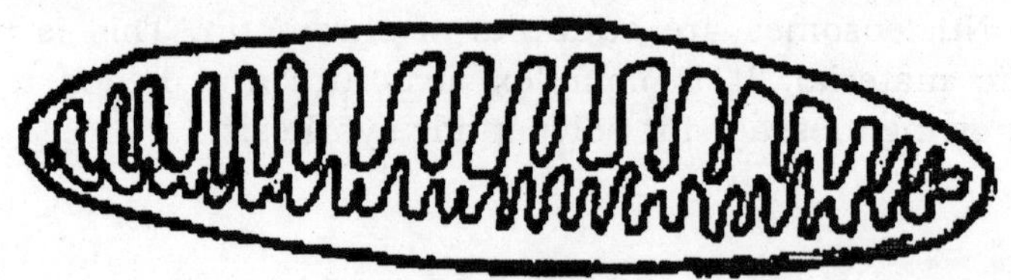

Fig. 5. Mitochondria

Mitochondria contain 70 S ribosome and DNA molecule. DNA is in a circular form and is arranged as in bacteria. Beside these similarities, mitochondrial membrane lacks sterols. The nucleotide arrangement in 16 S rRNA are some of the important characteristics which resembles with prokaryotic cells.

CHLOROPLASTS

Chloroplasts are one form of plastids and the site of photosynthesis. It is a double membrane structure Fig 6. Circular DNA and 70 S ribosomes are present in chloroplasts. It contain chlorophyll and enzymes required for light reaction of photosynthesis. It converts light energy into chemical energy, the ATP. The photosynthetic pigments are on thylakoids. The latter are internal membranous complex system of chloroplasts. Within the chloroplast, the sac-like membranous vesicles (thylakoids) may be stacked to form grana. Grana are densely packed piles of individual thylakoids. Inside the membrane is the interior structure, known as stroma. Enzymes catalyzing carbon-di-oxide's fixation are present in stroma. Thus, chloroplasts contain stacks or columns of grana in the matrix space of the stroma. Algae show variations in the structure of chloroplast. In brown algae, no grana occur and the thylakoid membranes are not stacked.

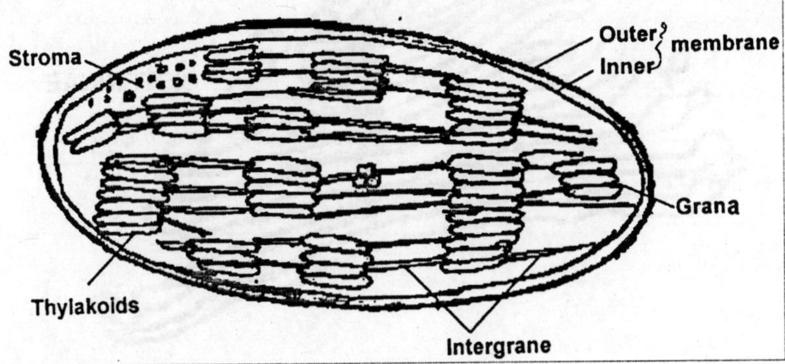

Fig. 6. Chloroplasts

ENDOPLASMIC RETICULUM

Endoplamic reticulum is membranous network of the eukaryotic cells. Endoplasmic reticulum is made up of fluid-filled sacs enclosed by membrane. It forms a connective system with outer nuclear membrane. It may provide a communication network for coordinating the metabolic activities of the cell. Ribosomes may be present on the surface of endoplasmic reticulum. Electron microscopic examination of endoplasmic reticulum shows that it appear rough when ribosome is present, while it is smooth in the absence of ribosomes. Some analogous membrane system is present in prokaryotic cells. However, ribosomes are not found attached to it. The proteins synthesized on these attached ribosomes, are destined to be transported out of the cell or for incorporation into membranes. Endoplasmic reticulum may be the source for some of the membranes, of the other organelles of eukaryotic cells.

GOLGI APPARATUS

Golgi apparatus is flattened membranous sac or stacks of Golgi bodies. Four to eight Golgi bodies assemble to form Golgi apparatus Fig. 7. These membranes are referred as dictyosomes. Golgi apparatus is closely associated with the rough endoplasmic reticulum. The proteins formed at the ribosomes of the rough endoplasmic reticulum are transferred to Golgi apparatus. The transfer takes place by budding and this process is known as blebbing. Golgi bodies are the sites of synthesis of lipoprotein, glycoprotein and some other carbohydrate derivatives. These polysaccharide derivatives are essential for the synthesis of various cell constituents. The proteins and lipids are carried by membrane sacs. These biochemicals reach to Golgi apparatus. After their repackaging, secretory vesicles are formed. These, secretory vesicles move to cytoplamic membrane and releases its contents by exocytosis. Construction of certain structures formed outside the cytoplasm, take place by this method.

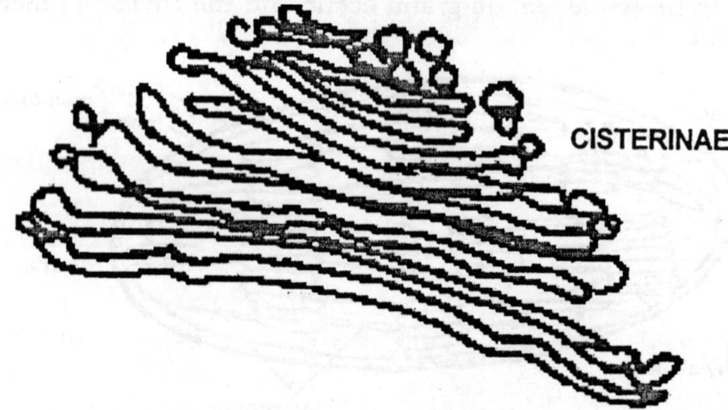

Fig. 7. Golgibodies

VACUOLES

Vacuoles are simple membranous structures performing different functions in eukaryotic cell. The storage vacuole, perform the function of storage of reserve food material. A food vacuole is formed after the cell has engulfed food. These vacuoles unite with lysosomes and thus the food is digested. Some protozoan cells have contractile vacuole. These vacuoles help in balancing the water content inside the cell. Water enters inside the cell by the process of osmosis. Excess water content may be harmful to the cells and cause osmotic shocks. This excess water enters in the contractile vacuole, which after gaining a particular size ejects out water forcefully.

CYTOSKELETON

Cytoskeleton is a network of microtubules, microfilaments and intermediate filaments. These fibrous structures are made up of protein. Microtubules are 25 nm in diameter and it is a hollow structure, made up of tubulin proteins. Microfilaments are of 6 nm in diameter, while intermediate filaments are of intermediate diameter.

Microfilaments are made up of polymerized actin protein with conjugation of another protein, myosine. It is a membrane bound organelle and helps in maintaining the shape of the cell. This system provides all the cytoplasmic organelles to maintain their original location in the cytoplasm. It is a linking system and helps in movement of membrane bound structures.

CHAPTER 4

BACTERIA

Bacteria are very small organisms with 0.5 to 1.0 μm in diameter. This small size provides high surface area/volume ratio. There are three basic cell shapes : the sphere, rods and curved Fig. 1. The spherical bacteria are known as **coccus**, the rods are cylinder with rounded ends are called **bacillus** and the curved rods, spirals or helix are known as **vibrio**. Bacteria showing a variety of shapes are called **pleomorphic**. Bacteria are generally unicellular, they gain multicellular structure by different planes of cell division. The rod divides in one plain. If the bacterial body is made up of a pair of rods, it is **diplobacilli** while if the rods are found in chain it is called as **streptobacilli** e.g., *Bacillus subtilis*. Some chains are known as **trichome** as they show much larger area of contact between the adjacent cells, e.g., *Baggiatoa, Saprospira. Corynebacterium diphtheriae,* show palisade arrangement, the cells are rod shaped and a number of rods are arranged side by side like match-sticks. Coccoid bacteria show a number of types. A **diplococcus**-coccoid cells in pairs, **streptococcus** is a chain of coccoid cells, the **pediococcus** -coccoid cells divide in two planes to give plate like structure, the **sarcina**–coccoid cells divide in three planes to give a cubic structure, the **staphylo**-coccus- the coccoid cells divide in three planes irregularly producing bunches of cocci.

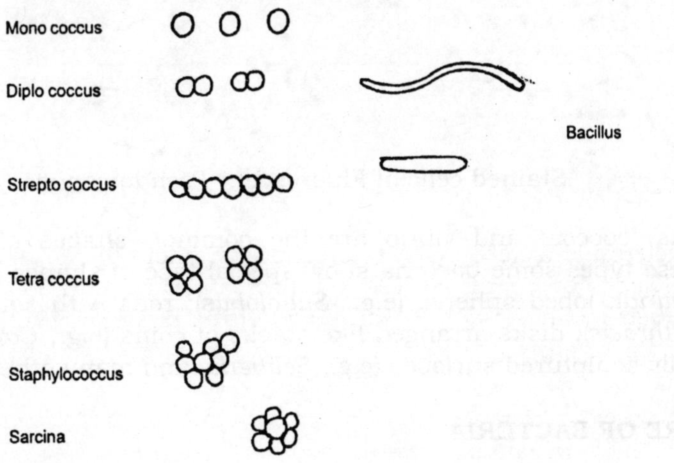

Fig. 1A. Shapes and arrangement of bacteria

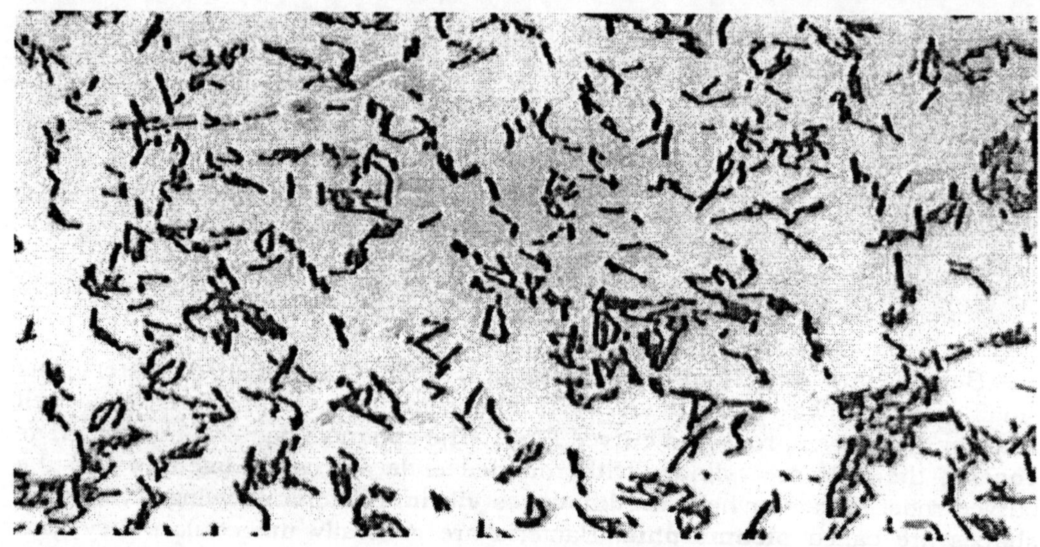

Stained cells of *Bacillus* sp.

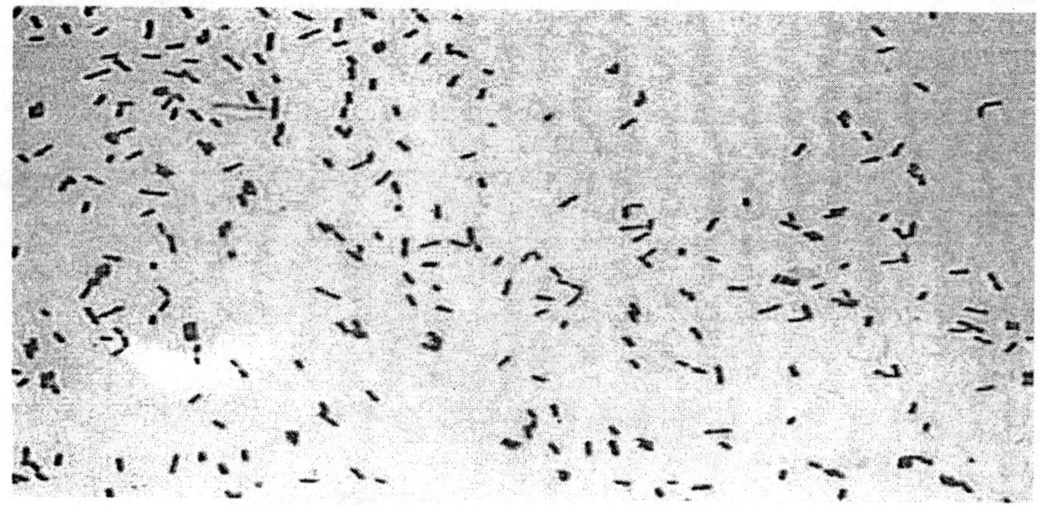

Stained cells of Fluorescent *Pseudomonas*

Bacillus, coccoid and vibrio are the common shapes of bacterial cells, besides these types some bacteria show special type of shapes like pear shaped (e.g., *Pasteuria*); lobed spheres (e.g., *Sulfolobus*); rods with squared ends (e.g., *Bacillus anthracis*); disks arranged like stacks of coins (e.g., *Caryophanon*); rods with helically sculptured surfaces (e.g., *Seliberia*) and many others.

STRUCTURE OF BACTERIA

Bacteria are prokaryotic organisms. There cell structure is very simple. There are some structures present outside the cell wall while others are internal to the cell wall.

Flagella — Flagella are hair like structures with 0.01 to 0.02 nm. in diameter. They are motile organs showing swimming motility. Since, the size of the flagella is below the limit of resolution of light microscope, flagella are not visible in routine stained smears of bacteria. It could be seen only with light microscope if, a special substance called **mordant** is used to build up the diameter of flagellum. A stain is then applied to mordant to colour the thickened flagullum. A flagellum is composed of three parts : a basal body- associated with the cytoplasmic membrane and cell wall, a short hook and a helical filament which is usually several times as long as the cell. The hook and filament are chemically made up of protein. The protein of flagella is known as **flagellin** (mol. wt 30,000 to 40,000 Da). Flagellin is a fibrous protein. Each molecule is 40 Å in diameter. Usually 3-8 longitudinal chains of flagellin molecules run parallel twining to each other in a longitudinal manner. A cross section of the flagellum reveals 8 flagellin molecules around a central spaces. Some bacterial flagella have antigenic property like H-antigen in *Salmonella.* These antigenic properties, resides in flagellin molecule.

Bacterial cells having flagella show following typical arrangements (Fig. 2).

(a) Monotrichous — There is a single flagellum at one pole of cell, e.g., *Pseudomonas aeruginosa.*

(b) Lophotrichous — There are several flagella at one end of cell, e.g., *Pseudomonas fluroesecns.*

(c) Amphitrichous — Cell bears at least one flagellum at each pole e.g., *Aquaspirillum serpens.*

(d) Peritrichous — Their are flagellum all over the surface of cell, e.g., *Salmonella typhi.*

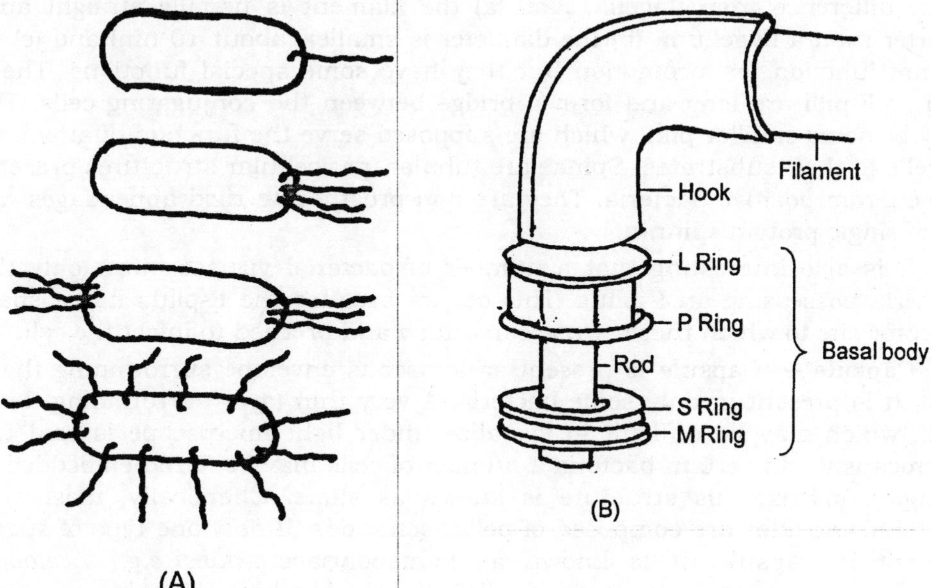

Fig. 2. Flagella: (A) Arrangement of Flagella; (B) Structure of Basal granule

Bacterial cells show propelling movement by rotating their helical flagella. Spirochetes are helical bacteria with no external flagella, yet they exhibit motility by periplasmic flagella. The latter are flagella like structures located within the cell just beneath the outer cell envelope. Periplasmic flagella are also known as axial fibrils or endoflagella. *Cytophaga* show gliding motion, though no organelles are observed showing such motility. *Aquaspirillum magnetotacticum,* show swimming movement in response to the earth's magnetic field. Cells possess magnetosomes, which allows the cell to become oriented as a magnetic dipole.

The mechanism by which the cell provide the energy for the rotation of its flagella is not known, but the fact that bacteria exhibit chemotaxis, that is, they move towards a gradient of various attractants (usually nutrient substrate) demonstrates movement is not entirely random. High speed moving pictures have shown that when the flagellum is rotating counter clockwise, the bacterium will travel in a more or less straight line, but if the direction of flagellar movement is reversed, the organism will tumble aimlessly. When the bacterium is moving-up a gradient of chemotactic attractant, the straight line movement continues longer and tumbling occurs only briefly. However, if the bacterium is moving away from attractant or accidentally towards a repelent, the tumbling occurs frequently until the net movement is properly redirected.

Pilli, Fimbriae and Spinae — Brinton (1959) introduced the term **Pilli** and Duguid *et al.*, (1955) introduced the term **fimbriae**. These terms were used for non-flagellar appendages of cell generally observed in Gram negative rods. They are extremely fine, non-flagellar appendages which look superficially similar to flagella. These structures are 0.5 –20 nm long and 3-25 nm in diameter. The chemical composition of these structures is protein known as **pillin.** There are some difference from flagella, like: (a) the filament is usually straight and is shorter than a flagellum, (b) the diameter is smaller (about 10 nm) and (c) they do not function for locomotion but they have some special functions. The sex pilli or F-pilli are long and form a bridge between the conjugating cells. There may be some smaller pilli, which are supposed serve the function of attachment of cells to their substrates. Spinae are tubular, pericellular structures present in some Gram positive bacteria. They are non-prosthecate rigid appendages made up of single protein spinnin.

It is also interesting that a number of bacterial viruses infect only those bacteria possessing an F-pilus. This occurs because the F-pilus has a specific receptor site to which these viruse can attach and proceed to infect the cell.

Capsule — Capsule is present as a viscous envelope surrounding the cell wall. It is present in only some bacteria. A very thin layer surrounding the cell wall, which may be difficult to visualize under light microscope is said to be microcapsule. In certain bacteria a number of cells may be found embedded in a common matrix, this structure is known as slime. Chemically, most of the bacterial capsules are composed of polysaccharides. If only one type of sugar is present in capsule, it is known as homopolysaccharides e.g., *Acetobacter xylinum,* the capsule is made up of cellulose only, in which the polysaccharide is made up of glucose units linked by 1-4 glycosidic linkage. In contrast to this if the capsule is made up of different types of sugars it is known as heteropolysaccharide. In *Pseudomonas aeruginosa,* The capsule is made up of D-glucose, D-galactose, D-mannose, D-glucuronic acid and L-rhamnose residues. These sugars are synthesized in the cell. It becomes attached to lipid carrier molecule, which acts as a carrier and transports the monmers outside the cell wall where they get polymerized. The capsule of *Bacillus anthracis* is entirely composed of a polymer of glutamic acid while the capsule of Streptococci is made up of proteins consisting of L-amino acids. Capsule performs different functions in different bacteria such as (a) they may help in overcoming dehydration by binding water molecules, (b) they may be antiphagocyctic i.e., they may resist the engulfment of bacterial cells by white blood cells, (c) they may block attachment of bacteriophages to the cells, (d) capsule of *Streptococcus mutans* secretes water insoluble capsular glucan which helps it in attachment of bacterial cells to smooth teeth, (e) the capsule of *Streptococcus pneumoniae* contains toxins which produces the symptoms of disease, (f) some bacteria have capsule made up of electrically charged components such as sugar-uronic acids, this enables the bacterial cell to repel from each other and thus prevents sedimentation.

The presence of a capsule can be of great importance in medical microbiology because, possession of one may tremendously increase the ability of a bacterium to cause disease. This occurs because the capsule acts as a protective layer that resists ingestion by the host's phagocytic cells (cells that engulf and digest foreign material). Furthermore, because the resistance provided by the capsule can be negated in the presence of specific antibodies, immunity to some infectious agents is directed against the organism's capsule.

Sheaths, Prosthecae and Stalks

Some bacterial body is made up of trichomes which are enclosed by a hollow tube called a **sheath**. It is generally present in fresh water and marine water inhabiting bacteria. **Prsothecae** are semirigid extensions of the cell wall and cytoplamic membrane. *Caulobacter* has a single prostheca while *Stella* and *Ancalomicrobium* have many prosthecae. In these bacteria prosthecae increases the surface area of the cells for nutrient absorption. In *Gallionella* or *Planctomyces* there are ribbon like or tubular structures present outside the cells. These structures are non-living and help in attachment of the cells to the surface. They are known as **stalks**.

THE CELL WALL

This is a rigid structure and gives a proper shape to the cell. Chemically the cell wall is made up of peptidoglycan, phospholipids, proteins, lipopolysaccharides, polysaccharides and techoic acids. Peptidoglycan (murein) is a macromolecule. It is a polymer of N-acetylglucosamine, N-acetylmuramic acid, L-alanine, D-alanine, D-glutamate and a diamino acid (LL-or meso-diaminopimelic acid, L-lysine, L-ornithine or L-diaminobutyric acid). Based on the chemical composition of cell wall, bacteria are divided into Gram positive and Gram negative types. The cell wall of Gram negative bacteria is generally thinner (10-15 nm) and is rich in lipid contents. There is an outer membrane surrounding a thin layer of peptidoglycan. This outer membrane anchores the underlying peptidoglycan by means of Braun's lipoprotein. The membrane is a bilayered structure consisting mainly of phospholipids, proteins and lipopolysaccharides (LPS). The space between the outer membrane and cytoplasmic membrane is known as periplasmic space. The cell wall of Gram positive bacteria is thick (20-25 nm). It is made up of a great amount of peptidoglycan along with polysaccharide, techoic acid and a little amount of lipid. The walls of *Mycobacterium* and *Corynebacterium* are rich in lipid contents. (Details of structure of cell walls of Archaeobacteria and Eubacteria is discussed in Chapter 3).

Differences in Cell Walls of Gram Negative and Gram Positive Bacteria

	Gram-negative	Gram-positive
1	Thin (10-15nm), less homogenous, more generally three layered	Thick (20-25 nm), more homogenous, amorphous, single layered.
2	Peptidoglycan 5-15%, Phospholipid-35%, Protein-15%, Lipopolysaccharides-50%.	Peptidoglycan 20-80%, rest is protein, Polysaccharides and teichoic acids.
3	Less rigid, due to plastic nature of lipid protein polysaccharide complex	Much rigid, due to presence of greater amount of mucopeptides.
4	Lipopolysaccharide are the principal surface antigen.	Teichoic acids are the principal surface antigens.

The Cytoplasmic membrane

It surrounds the protoplast and is about 6-8 nm thick. Chemically it is made up of phospholipids (20-30%) and proteins (60-70%). The phospholipids form a bilayer in which proteins are present as integral proteins. There are some loosely attached proteins known as peripheral proteins. The lipid matrix of the membrane has fluidity, allowing the components to move around laterally. Cytoplasmic membrane performs some basic functions, like:

(a) Selective permeability,

(b) site of electron flow,

(c) extra cellular polymer production,

(d) site of chromosome attachment and

(e) ensures aggregation of DNA to Daughter cells during cell division.

The details of structure and functions of cytoplasmic membrane of Archaeobacteria and Eubacteria are discussed in chapter 3.

Protoplast and Sphaeroplast

The cytoplasmic membrane and the cell material enclosed by it are included in protoplast. The cell wall free protoplast may be prepared by treatment with enzyme like lysozyme, or if the cell is cultured in an antibiotic, which inhibits cell wall synthesis. Even in such a treated cell, the flexible outer membrane of the cell wall remains and the treated cell has two membranes, the cytoplasmic membrane of the protoplast plus the outer membrane of the cell wall. Such cell is called a spheroplasts rather than a protoplast.

Membrane Intrusions and Intracellular membrane systems

Bacterial cells lacks unit membrane. Hence, no cell organelles are enclosed in a membrane. The cytoplasmic membrane infoldings may occur which can produce complex internal structures. It increases the surface area of cell. The morphologically and physiologically differentiated into two main groups, chromatophores and mesosomes.

Chromatophores

The photosynthetic bacteria contain membranous structures which bears pigments, alongwith enzymes and the electron transport system. These structures may be found as vesicles, tubes, bundled tubes, stacks, membranous or thylakoids. Chromatophores are able to conduct photophosphorylation but cannot perform dark reaction of photosynthesis. They lack enzymes associated with dark reaction.

Mesosomes

They were earlier known as peripheral body or chondroid Fig. 3. Mesosomes are densely present in Gram positive as compared to Gram negative bacteria. They are complex invagination of cell membrane. The infoldings contain vesicles, tubules or lamellar whorls. Chemically mesosomes are made up of lipid and proteins. Lipid composition is same as that of plasma membrane. However, the proteins composition differs. It was previously believed that mesosomes are site of respiratory activity. However, this concept was rejected as the mesosome do not contain an outer membrane and much more they lack certain enzymes like ATPase, dehydrogenase and cytochromes. It is also said that mesosomes may be sites of some of the wall membrane biosynthetic processes.

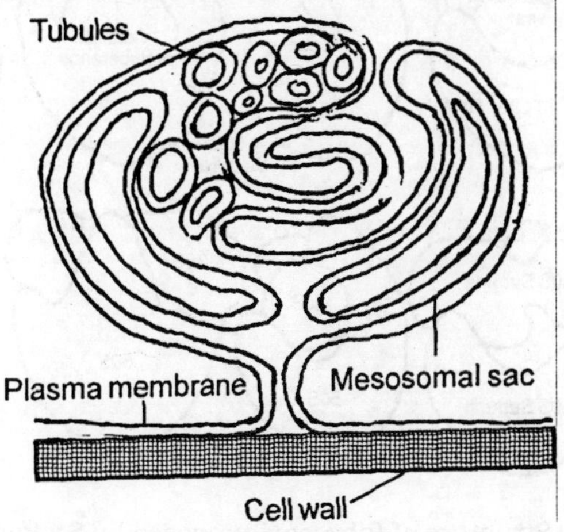

Fig. 3. Mesosome

Ribosomes

A bacterial cell may contain up to 10,000 ribsomes. The ribosomes are designated as 70S consisting of two subunits 50S and 30S Fig. 4. The 50S subunit has a number of shapes like rounded, kidney shaped etc, while the 30S subunit may be ellipsoid, disc shaped etc. The Lake's model describes that 50S subunit is asymmetrical with lateral projections incline at an angle of about 50 from a central protuberance. According to this model, 30S subunit is also asymmetrical with an indentation dividing it into two unequal parts, the upper one-third head and the lower two third body. There is a platform on this lower body. A cleft is present between the platform and the head. It is believed that this cleft is a site of the codon-anticodon interaction and the site of binding for initiation factor. Ribosomes are made up of protein and ribosomal RNA (rRNA). *E. coli* ribsomes contain 53 proteins, 34 in 50S subunit and 21 in 30S subunit. Ribosomal RNA makes two third mass of ribosomes. There are three types of rRNAs (5sRNA, 16sRNA and 23s RNA). The rRNA interacts amongst themselves and other types of cellular RNAs like mRNA and tRNA. During protein synthesis, rRNA binds with other types of RNAs e.g., 5sRNA binds with the loop IV of tRNA (G-T- -C sequence of tRNA). 16s RNA binds with mRNA. A few nucleotides at 3'end of rRNA are complementary to mRNA, which helps in such type of bindings. 16sRNA also binds with tRNA. 23s rRNA binds with 5s rRNA. Ribosomes are sites of protein synthesis.

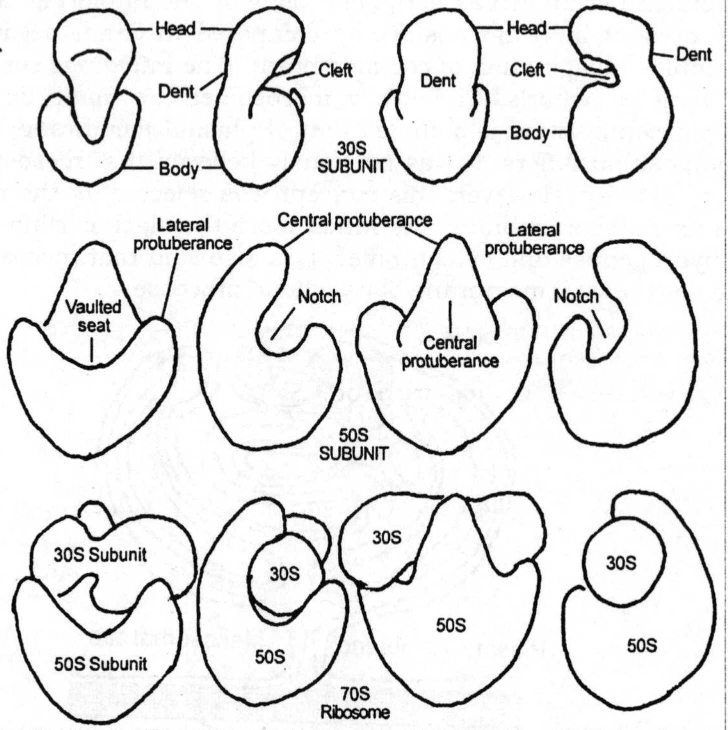

Fig. 4. Structure of Ribosome proposed by Stoffer and Wittmann

Storage granules

Storage granules are molecules of high molecular weight present in a number of bacterial cells. They are osmotically inert. A bacterial cell may contain three types of storage granules viz., polymetaphosphate granules, poly-β-hydoxybutyrate granules and polyglucan granules. *Polymetaphosphate* (PO_3) in are *volutine granules* or *metachromatic granules.* The name volutine is derived for the presence of these granules in *Spirillum volutans.* It is present in corynebacterium and mycobacterium. These granules stain reddish violet with methylene blue or toludine blue. The granules are polymers of inorganic pyrophosphate. The latter are present in a form of linear chains and are incorporated during nucleic acid synthesis. *Poly-β-hydroxybutyrate* are polylactides. They are formed by the condensation of acetylCoA, forming acetoacetyl CoA. Previously they were also named as lipid granules because of their nature of being stained with Sudan Black. They are present in large amount in *Bacillus megatherium* and are the source of Acetyl CoA. Polyglucan granules are storage granules of starch. They are polymers of glucose units with α -1-4 linkage or sometimes α–1-6 linkages. They are stained with iodine. It is present in abundance in clostridia and coliforms.

Nucleoid (Nucleus)

Bacteria is a prokaryotic organism and thus the nucleus is incipient. The nuclear membrane is absent, hence it is known as nucleoid or chromatin body. It consists of single molecule of double stranded DNA. Histone protein is absent thus only double helical DNA molecule is designated as chromosome. The chromosomes of *E. coli* show 12 to 80 super coiled loops. There is present a core of RNA and enzyme RNA polymerase which determines the positions of loops. The DNA molecule may contain some unusual bases like 6-methylaminopurine and 5-methylcytosine.

Plasmids

It is an extra-chromosomal DNA molecule. They are circular, double-stranded DNA molecule. They undergo replication at the time of cell division. Plasmids are discussed in details in chapter 17.

Endospore

Few bacteria e.g., *Bacillus, Clostridium, Sprolatobacillus, Sporosarcina* and *Desulfotomaculum* produces endospores (Fig. 5), under unfavourable conditions. These endospores protect bacteria at high temperature and in chemical disinfectants. The endospores are generally spherical, oval or ellipsoid in shape. The endospores have a covering known as envelope or spore integument. The envelope consists of outermost spore coat, inner to it are layers like outer membrane, the cortex and the inner membrane. Inside this wall is protoplast or core rich in DNA, ribosomes, tRNA, accessory factors and enzymes. The envelope is rich in peptidoglycan. The cortex also contains large amount of dipicolinic acid and calcium ions. When bacterial cell faces adverse conditions, the cell division and growth retards. A cell divides asymmetrically producing forespore within the mother cell. The vegetative cells of *Bacillus subtilis* represents the O stage while the developmental stages of endospore are represented by letters I to VII. In stage I, there is formation of axial filament. The axial filament is formed by spreading of DNA from one end of the cell to the other end. Intracellular salts and storage granules accumulate in the cell. Septum is formed in stage II. A double membrane septum is formed by invagination from one end of the cell. A transverse spore septum is formed which divides the cell asymmetrically. The smaller portion is known as forespore while the bigger part is called mother cell. The nuclear material also divides. In stage III, ingulfment of forespore takes place. The membrane of the larger compartment invaginates and proliferates towards the pole of the cell and engulfs the forespore. In stage IV, cortex is synthesized by deposition of the primordial cell wall and spore-specific peptidoglycan outside the peptidoglycan wall. There is accumulation of calcium and pyridine-2,6 dicarboxylic acid (DPA) in the forespore. The spore gains heat resistant quality by the uptake of Ca^{++} and synthesis of DPA. At stage IV, there is also synthesis of enzymes like adenosine deaminase and ribosidase. Early coat (a proteinaceous covering) is synthesized in stage V. An electron dense covering is synthesized outside the outer membrane. Large amount of cysteine is used in the formation of early coat. The forespore becomes phase white due to continued accumulation of Ca^{++} and DPA. Spore becomes refractile due to maturation in stage VI. The cortex brings about dehydration of mature spore. The contractile cortex theory explains that the wastes are expelled by contraction of the loose peptidoglycan. Ionic cross-linking with calcium ions helps in this process. However, there is an alternate view that the cortex expands within the fixed volume of the coat because of loss of ionic cross link. The spore is liberated by autolysis of mother cell in stage VII.

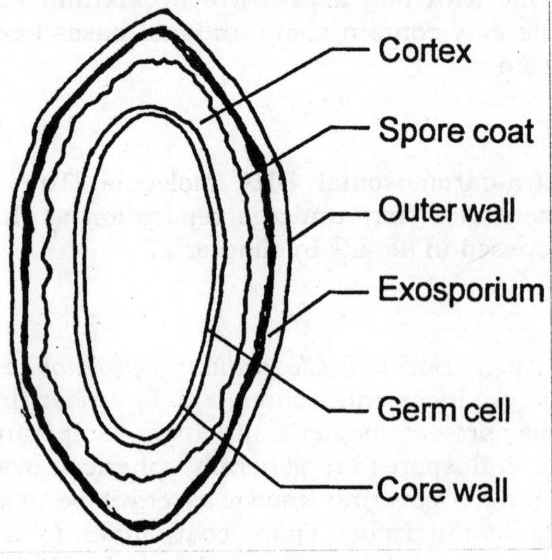

Fig. 5. Endospore

Endospores are dormant structures, which are formed to overcome adverse conditions and they are metabolically inactive. When the favourable conditions appear the endospores germinate. There are three stages of germination : activation, initiation and outgrowth. The endospore becomes active by the influence of traumatic agents like heat, low pH or an SH compound. Temperature is the most effective factor, which favours spore germination. If the spore is kept at 65°C for 15-60 minutes it becomes activated. Endospores of thermophilic species are activated at 105-120°C. Some chemicals like L-alanine, adenosine, glucose and some reducing agents may help in activation. The initiation stage is also known as the germination proper. The (cortex) peptidoglycan is hydrolyzed. Soluble materials are liberated. There is a rapid uptake of water and ions (K^+ and Mg^{++}). Nutrients are also required during germination. The energy stored in phosphoglycerate is transferred to ATP. After germination if the conditions are favourable and the nutrients are sufficiently available outgrowth or postgerminative development takes place. During this stage, the spore wall swells within the spore coat which later splits due to the formation of vegetative cells. The cells become metabolically active. All the enzyme system starts working and DNA synthesis begins after an hour.

Most endospore forming bacteria are inhabitants of the soil, but bacterial endospores exist almost everywhere, including the atmosphere where they ride on invisible particles of dust. The fact that endospores are so hard to destroy is the principal lesson for the lengthy and elaborate sterilization procedures employed in hospitals, canneries and other places where absolute sterilization is required.

REPRODUCTION

The bacteria, reproduces by modes of cell division, i.e., by binary fission. Budding, fragmentation and formation of conidiospores or sporangiospores also brings about reproduction.

In eukaroytic cells, during sexual reproduction there are three important stages viz gamete formation, zygote formation and meiosis. Bacteria shows sexuality as there is formation of incomplete zygote or merozygote. The processes are discussed in chapter 17.

CHAPTER 5

OTHER PROKARYOTES

CYANOBACTERIA

Cyanobacteria (Gr. Kyanas, a dark blue substance) Fig. 1, are prokaryotic microorganisms, in which sexual reproduction does not take place. They occur in fresh water, marine water, in soils and moist rocks. They are also found on damp walls. Cell wall contains diaminopimilic acid. Mucilage envelope is present covering cells or trichomes. Some members are surrounded by their individual mucilagenous sheath, whereas others are surrounded by a common mucilagenous envelope. Aquatic species have gas vacuoles in their cells. Gas vacuoles are a series of gas vesicles surrounded by a protein wall, which is permeable to air but not to water. Flagellate cells absent throughout the life cycle, movement by gliding motion. Cells contain bilo-proteins along with myxo xanthin and myxoxanthophyll. Bilo-proteins are c-phycocyanin and c-phycoerythrin. Pigments are located in peripheral region of the cell. This is known as chromoplasm. They perform oxygenic photosynthesis. Sugars and glycogen are produced as result of photosynthesis. Cyanophycin is main storage product. It is a proteinaceous food material and contains a large amount of arginine. Beside cyanophycin, myxophycean starch is also found as storage products in a cell. Heterocysts are special type of cells present in the vegetative body. These cells are thick walled with homogenous and pale-yellowish content. There are different opinions about the functions of heterocyst. However, it is established that these cells contain enzymes catalyzing nitrogen fixation. Reproduction takes place by the formation of hormogones, endospores, exospores and akinetes.

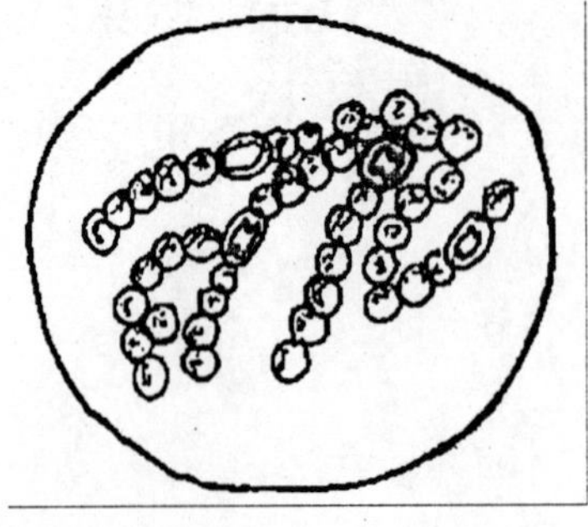

Fig. 1A. Nostoc colony

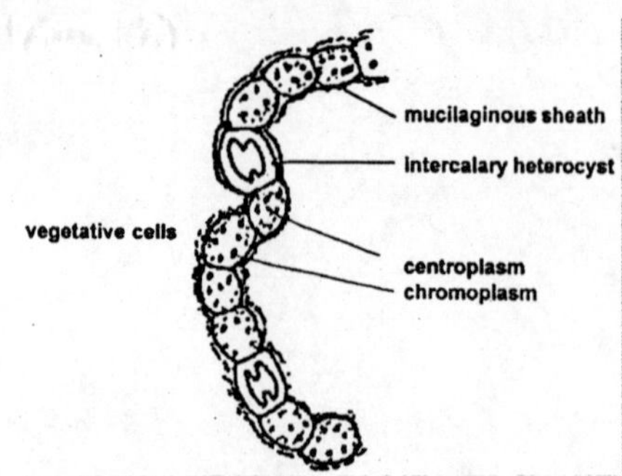

Fig. 1B. Nostoc - Trichome

ACTINOMYCETES

The actinomycetes Fig. 2 are widely distributed in nature. There are numerous oxidative forms of this group of bacteria. They primarily occur in soil. Actinomycetes are Gram positive bacteria. Some of these organisms are known as sporoactinomycetes or euactinomycetes. Structurally they are made up of mycelium. Superficially, their morphology resembles that of the filamentous fungi. The cells are prokaryotic with much smaller diameter. Reproduction takes place by spores. Spores are borne singly or in chain at the tip of the hyphae. Proactinomycetes have less developed mycelium. It may be transitory. In some extreme cases, there may be only a single branched cell. There is diversity in their cell wall composition. The major constituents of cell wall are meso- or LL-DAP (diaminopamilic acid), arabinose, glactose and glycine. About sixty varieties of murein have been described in actinomycetes. Based on the cell wall constituents, members of actinomycetes are divided into five groups. Galactose, arabinose, xylose and madurose form the whole sugar patterns of aerobic actinomycetes. Based on the whole sugar pattern actinomycetes are divided into four groups. In general, the members are non-motile. Some forms may have flagellated spores known as zoospores, which are analogous to the zoospores of algae and fungi. The life cycle is completed in two forms. These phases are regarded as vegetative phase and sporulation phase. The mycelium of vegetative phase may be underground or aerial. If the mycelium is entirely within substratum, it is known as substrate mycelium. It is common in non-sporulating actinomycetes. Most of the sporoactinomycetes have aerial mycelium bearing spores at the tip. A few actinomycetes may develop thallus or sclerotium. The vegetative mycelium grows in length and breadth by developing longitudinal and transverse septa giving an impression of thallus. In sporulating actinomycetes, spores are formed when there is a lack of nutrients. The amount of nutrients varies from center to periphery. Nutrients may be lacking in the center as compared to the periphery of developing colony. Thus, there will be more sporulation in the centre as compared to periphery.

Fig. 2A. Fruiting bodies of actinoplanaceae **Fig. 2B.** Streptomyces

Some actinomycetes are pathogenic in human beings. *Actinomyces israeli* causes actinomycosis, *Mycobacterium* causes tuberculosis. *Streptomyces griseus* produces antibiotics. *Frankia* is able to fix nitrogen and cause nodulation in host plants. Streptomyces produces a gaseous compound called geosmin. This compound gives a typical musty odour to soil.

Actinomycetes are divided into 8 families. The details of classification and characteristics of each family is described in chapter 2.

RICKETTSIAS

Rickettsias are obligate intracellular parasite. They are pathogenic bacteria causing disease in man and other vertebrates. They are transmitted by the bites of arthropod vectors or their excreta. *Coxiella burnetii* is an exception it causes Q-fever. These are Gram-negative, rod-shaped, coccoid or often pleomorphic bacteria. Cells contain proteins, lipids (including phospholipid). DNA is in stable quantity, while RNA is variable. The cell wall is typical. It contains amino acids, polysaccharides and muramic acid. Rickettsias depend upon a host cell for some essential metabolic reactions. Some intracellular metabolic enzymes are present. Rickettsias are capable of carrying on limited metabolic reactions independently. They are capable of utilizing glutamic acid, pyruvate and succinate by their own enzyme system. They are non-motile and non-spore forming. They divide by binary fission.

In Bergey's manual of Determinative Bacteriology, rickettsias are placed in part 18, order Rickettsiales. This order is divided in three families Rickettsieae, Ehrlichieae, Wolbachieae.

MYCOPLASMA

Mycoplasma lacks cell wall. However, there is substantial amount of polysaccharides associated with the cell membrane. They are pathogenic causing disease in plants and animals. It invades the mucous membrane of respiratory and genital tracts, and the synovial membrane of joint capsules. They cause pneumonia or arthitis diseases. *M. pneumoniae* causes primary atypical pneumonia (commonly called "walking pneumonia"). Colonies on solid media have a nippled or "fried egg" appearance. Mycoplasmas lack cell wall, hence they are pleomorphic in shape. They are very small. Cells are 0.1 to 0.25 nm in diameter. Cells are coccoid or pear-shaped, sometimes with filamentous extensions. They may produce filaments, which resembles fungi. They divide by binary fission. Mycoplasmas are aerobic or facultative anaerobic in nature and obligate chemoheterotrophs. Based on the nutritional requirement, two genera are recognised (a) *Mycoplasma* — it requires sterols and cholestrols and (b) *Acholeplasma* — it do not requires cholesterol. *Spiroplasma* and *Ureaplasma* are included in this group. Species of *Spiroplasma* are serious plant pathogens and common parasites of plant feeding insects. The cells are of a tight corkscrew morphology. *Ureaplasma* produces enzyme, which can convert urea to urine. They are pathogens of urinary tract.

Mycoplasmas are cultured on artificial culture media. Sterols (if necessary) and other special nutritional or physical requirement are served while culturing mycoplasma.

CHLAMYDIAS

Chlamydias are basophilic, Gram-negative coccoid, non-motile bacteria. They have typical life cycle. Small cells or elementary bodies range in size from 200 to 350 nm in diameter. "Initial bodies" may have 800 nm in diameter. Chlamydias are obligate intracellular parasites. They are also said to be energy parasites, as they need energy from host cells. This character demarcates them from rickettsias. They are human pathogens and transmitted to humans by interpersonal contact or by air borne respiratory routes. *Chlamydia trachomatis* causes trachoma. *C. psittaci* causes psittacosis and *C. pneumoniae* causes a mild form of pneumonia.

Chlamydias have high lipid content specially phospholipids. The protein content is with eighteen or more amino acids. Life cycle completes with two types of cells, elementary body and initial body. Elementary bodies are small particles having an electron-dense nucleoid. It is phagocytosed by host cells and engulfed in a vacuole. Elementary bodies at this stage are known as initial body. Initial bodies are two to three times large than elementary body. They do not contain electron dense nucleoid material. Initial bodies reproduces by binary fission Fig. 3.

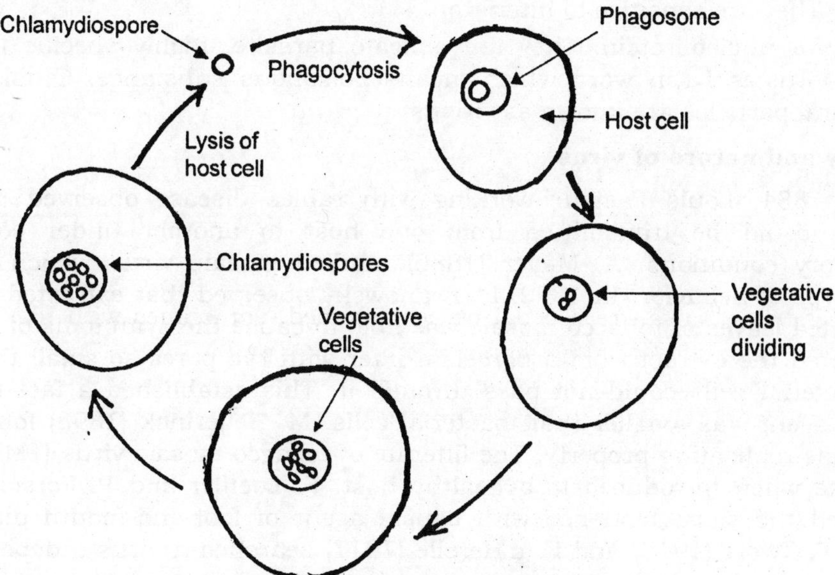

Fig. 3. Life cycle of Chlamydia

Chlamydias are cultured in living media. Cells of animal tissue culture or the yolk sacs of chicken embryos are used as culture media.

CHAPTER 6

VIRUS

Viruses are obligate intracellular parasites. These are unicellular, ultra-microscopic particles containing either RNA or DNA, which reproduce inside living host cells, pass through filters that retain bacteria and are covered by a protein coat.

The general properties of viruses are:

(1) Do not possess cellular organization.

(2) Contain one type of nucleic acid, either RNA or DNA but never both.

(3) Lack enzymes necessary for protein and nucleic acid synthesis and so depend upon synthetic machinery of host cells.

(4) They multiply by complex process at the expense of host energy and not by binary fission.

(5) They are unaffected by antibodies.

(6) They are sensitive to interferon.

Virus is a nucleoprotein. They are obligate parasite, highly specific for their hosts. Virus is Latin word which means poisonous substance. Outside their host, viral particles are known as virions.

History and nature of virus

In 1884, Louis Pasteur working with rabies disease observed that the disease could be transmitted from one host to another under controlled laboratory conditions. A. Mayer (1886) studied mosaic virus, which showed same type of behavior. In 1892, D. Iwanowsky observed that extracted filterate of infected leaves of tobacco mosaic was able to cause the symptoms of disease. He filtered the extract from a porcelain filter with the pores so small that even the bacterial cells could not pass through it. This established a fact that the causal agent was smaller than bacterial cells. M. Beijerinck (1898) found that virus has replicating property. The filterate of tobacco mosaic virus (TMV) could replicate when introduced to a healthy host. F. Loeffler and P. Forsch (1898) observed the same tendency with causal agent of foot-and-mouth disease of cattle. F. Twort (1915) and F. d'Herelle (1917) searched a virus independently. They found that this virus grows on bacteria. This virus was named as bacteriophage (i.e., eaters of bacteria). W. Stanley (1935) crystallized TMV and found that it is composed of proteins. Later other scientist found that RNA is also present in small amount with this protein.

Structure

Viruses show two types of symmetry (a) helical and (b) spherical Fig. 1. There is only one type of nucleic acid present in virus i.e., they contain either DNA or RNA. This is regarded as viral chromosome. Protein subunit covering the viral chromosomes is known as capsid and protein molecules are called capsomers. In helical type of viruses are rod shaped and identical protein capsomer cluster around the viral chromosome. Spherical type virus has polyhedral symmetry. The capsomers are arranged to form a polyhedral shell. It is composed of 20 triangular faces or icosahedron. The capsomers of polyhedral viruses are multimeric proteins composed of five, six or more subunits. In a simple type of polyhedral virus, 12 capsid proteins from aggregates of pentamers. Adenovirus has a complex structure. It has four hexamers on each edge, six hexamers on each face and one pentamer at each vertex. In most of the animal virus unit membrane is present enclosing the viral nucleoprotein. These types of viruses are known as enveloped. A number of plant viruses are naked as they lack unit membrane. Excepting poxvirus, all the enveloped viruses have phopholipid in their envelope, this is derived form their host cell membrane.

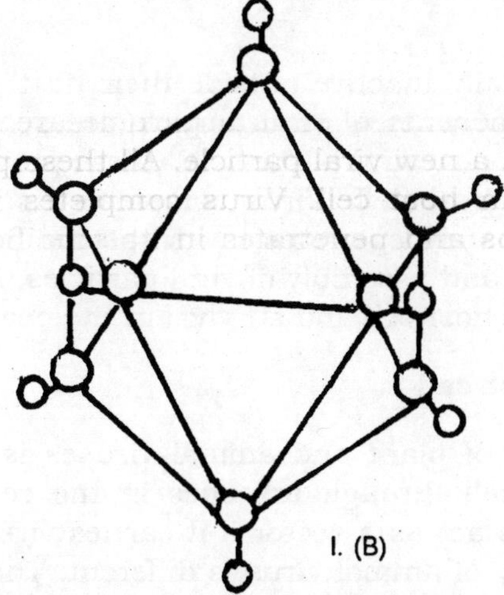

Fig. 1. Isohedral symmetry of Virus

Some plant viruses appear to have a genome of circular ds RNA. But the RNA in animal viruses exist only as linear double stranded or single stranded molecules. Unlike the DNA genome the RNA genome within a viron may exist as segmented genome (divided into several units. For e.g. reo virus contains 10 different segments of ds RNA and influenza virus has 8 separate segments of ssRNA. In some RNA viruses, there are several chromosomes each coding for a separate protein. Such type of genome is known as segmented genome. RNA virus have either plus-strand or minus-strand of RNA. Virions RNA that functions as mRNA is known as plus-strand. In contrast to this viral RNA forms a complementary strand, which acts as mRNA. This type of viral RNA is known as minus-strand.

Classification of virus

In the simplest system of classification, viruses are divided into two groups based on the type of nucleic acid present in them. Thus they may be grouped into ribovira (RNA is present in viral genome) and deoxyribovira (DNA is present as viral genome). The important criteria of classification of viruses are:

(a) size of virions and architecture of their capsid;

(b) presence of envelope (enveloped or naked) and

(c) nucleic acid structures (linear *vs.* circular) and (single stranded *vs.* double stranded).

Plants and animal viruses are classified separately. Animal viruses are classified into family, genera and species. Latin and English names are used for this purpose. Plants viruses are not classified into family, genera and species. They are grouped based on the structure, type of nucleic acid and mode of transmission. A plant virus is grouped along with its prominent virus according to its relevance. If a virus is having characters like tobacco mosaic virus, it will form tobacco group.

Replication of Virus

Virions are biologically inactive outside their host. They do not reproduce, but the molecular components of viral particle are synthesized separately and they assemble to form a new viral particle. All these processes of replication take place intracellularly in host cell. Virus completes its replication cycle in few steps. A virion adsorbs and penetrates in its specific host cell, the next step is intracellular replication and assembly of viral particles, in the last step, the viral particle escape from the host cell and survival in the new environment.

Entry of virus into host cell

The mode of entry of plant and animal viruses is different. Plant viruses enter inside the host cell through breaches in the cell wall. An insect when feeding on plant tissues act as a vector if it carries viruses on its mouth parts. The mechanism of entry of animal virus is different. The molecular components of viral particle adhere on specific receptors of the host cell. This is known as adsorption. Adsorption is followed by penetration of nucleic acid. Penetration of plant virus is a simple process. The mechanism of penetration is different in enveloped and naked animal viruses. The naked viral particles enter the host cell by endocytosis or by direct passage through the cell membrane. Phagolysosome carry out endocytosis. The viral membrane and membrane bounding the phagolysosome fuses with each other.

Uncoating — In all the plant and animal viruses (except for picornaviurses) viral nucleic acid enters in the host cell enclosed in its capsid. After this, the protease enzyme of host cell cleaves the capsid by degrading its proteins. These enzymes are present either on the cell surface or within the lysosomes. Freed DNA from the capsid enters in the cytoplasm or nucleus. This phenomenon is known as uncoating.

Replication of viral genome

Replication of viral DNA and RNA have different mechanism. Most of the viruses have linear or circular DNA molecule. If DNA molecule is linear, it is circularized in the host cell by cohesive ends or terminal redundancy mechanism. In this circular molecule, the replication starts at a specific site and proceeds in both directions around the molecule. Enzyme endonuclease forms a nick in one of the DNA strand. The 3'OH end of the nicked strand acts as a primer. The intact strand act as template and new nucleotides start incoming and joining as per template. The 5' end is thus displaced and later duplicated. The new double stranded DNA molecule thus produced is much longer than the circumference of the viral genome. It is known as concatemers. Later it is cleaved to produce the genomes of progeny virions.

Replication of RNA takes place through double-stranded RNA intermediates. Plus-strand RNA molecule replicates in two steps. In the first step, a minus strand is synthesized. As soon as the plus-strand RNA enters in the host cell, it is translated. As result of translation, a RNA-dependent RNA polymerase enzyme is synthesized. This enzyme is known as replicase. It catalyzes the synthesis of a RNA strand which is complementary to viral chromosome and is regarded as replicative form (RF). In the second step, enzyme replicase synthesizes new copies of plus strand. The replicative form synthesized by enzyme replicase acts as template in this phenomenon. Minus-strand RNA lack ribosome binding sites. Viruses with minus-strand have enzyme replicase in them. This enzyme enters the host cell along with the chromosome. Viral RNA is replicated *via* a double-stranded replicative intermediate.

Assembly

As stated earlier molecular components of virion are replicated and synthesized separately. Assembly of these molecular components is essential so that a mature viral particle is formed. Assembly of helical viruses is simple as compared to spherical viruses. In helical viruses, protein subunits of capsid associate with the viral genome and wind into a helix. In spherical virions first a procapsid is formed. Procapsid is an assembly of capsomeres of icosahedron. It is an empty structure i.e., lacking viral genome. Viral genome enters in this procapsid. It forms a condensed structure and later the procapsid is sealed so that no other large molecule may enter in it.

In enveloped viruses a fragment of host cell membrane wraps around the viral particle and forms envelope, enveloping takes place as the virions passes through host membrane. Virus encoded protein aggregate in the cytoplasmic membrane or inner nuclear membrane.

Escape

RNA viruses are released from the host cell as a part of assembly. DNA viruses migrates to endoplasmic reticulum and ultimately reach to vesicle from where they are released due to exocytosis.

TOBACCO MOSAIC VIRUS

This plant virus has been widely studied. Fig. 2. It has helcal symmetry, the rods measure 18 nm in diameter. The capsid is a hollow cylinder and there are about 2100 capsomers in capsid. Each capsomer is a polypeptide chain consisting of 158 amino acids. It is RNA containing virus.

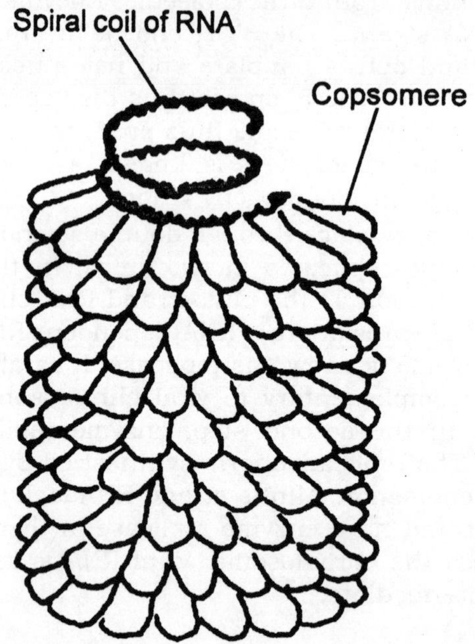

Fig. 2. Tobacco Mosaic Virus — Rod shaped virus with helical symmetry.

Reaction to physical and chemical Agents :

(1) **Heat and Cold** - Viruses are mostly destroyed by heating at 60°C for 30 min except hepatitis virus, adeno associated virus, scrapie virus. Viruses may be preserved by storage at –20°C to –70°C in deep freezer (except polio viruses)

(2) **pH** - They are usually stable at pH 5-9.

(3) **Ether susceptibility** - Arbo, myxo and herpes viruses are destroyed by ether cohereas entero, reo and adeno are resistant to the action of ether.

(4) **Radiation** - UV light, X-rays and heavy particles inactivate viruses.

(5) **Vital dyes** - Toluidene blue neutral red and acridine orange penctrate virus particles. These dyes unite with nucleic acid making viruses susceptible to inactivation by visible light.

(6) **Glycerol** - Viruses remain viable in 50% glycerol where as bacteria die.

(7) **Stabilization by salt** - Magnesium chloride, $MgSO_4$ and Na_2SO_4 stabilize some of the viruses so that they are not inactivated by heating at 25°C in one hour.

(8) **Disinfection** - Leysol, dettol are ineffective against viruses. Higher concentration of chloride may kill viruses. Dilute formaldehydes and beta propiolactone HCL, $KMnO_4$, H_2O_2 are most useful desinfectants against viruses.

BACTERIOPHAGE

Bacteriophage are viruses which eat bacteria. Phage is for eating and as these virus destroy bacteria hence named as bacteriophage. Bacteriophage was discovered independently by F. Twort (1915) and F. d'Herelle (1917).

Bacteriophage show binal structure. Fig. 3. They are made up of polyhdral head and helical tail. Nucleic acid is present in the head. Tail is made up of protein and helps in attachment of virions to their specific host. In some bacteriophage tail has a thick tube, outer tube surrounding the inner simple tube. The outer tube is known as sheath while the inner tube is called as core.

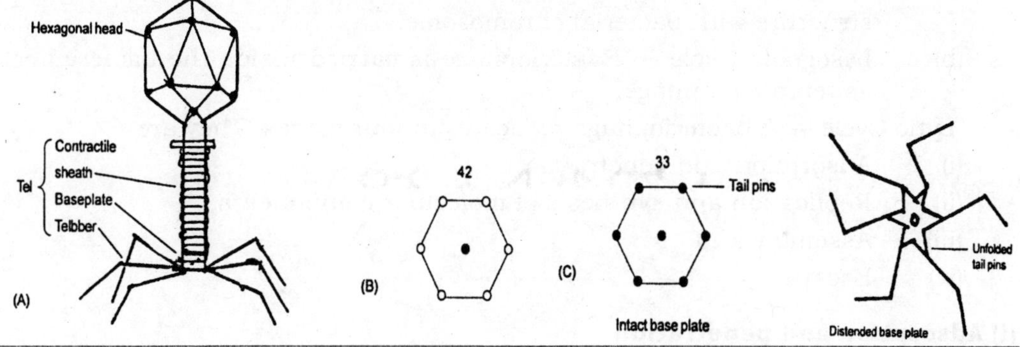

Fig. 3A. Bacteriophage, **3B.** Collar, **3C.** Intact base plate, **3D.** Extend base plate

T-series bacteriophage is an obligate parasite of *E. coli.* Pioneer studies on T-series bacteriophage have enabled to understand the structure of bacteriophage. It contains DNA. In the T-even phages (T2, T4 ...) the sheath contains 24 annual rings. These rings are termed as collar as they join head and tail. A hexagonal base plate is present at the distal end of the tail. At each corners of this base plate, there are present six spike like structures made up of proteins. They are known as tail pin. There are tail fibrils present on the base plate, they actually help in attachment of phage to bacterial cell. Bacteriophage are naked structures.

Classification of Bacteriophage

Bacteriophages are classified in families. Latin names are used in their classification. Some common families of bacterio-phage are as following —

Family Myoviridae : Phage with linear double stranded DNA, e.g., T2, T4, T6 (attacking *E. coli*) and P2 (attacking *Salmonella*)

Family Styloviridae : Phage with linear double stranded DNA, e.g., T5, (attacking *E. coli*)

Family Pedoviridae : Phage with linear double stranded DNA, e.g., T3, T7 (attacking *E.coli*)

Family Microviridae : Phage with circular single stranded DNA e.g., X174 (attacking *E.coli*)

Family Inoviridae : Phage with circular single stranded DNA, e.g., m 13, fd (attacking *E. coli*)

Family Leviviridae : Phage with linear single stranded RNA, e.g., Q, R17 (attacking *E.coli*) MS2

Family Cystoviridae : Phage with linear double stranded RNA.

REPLICATION

Bacteriophages show two types of life cycle —

(a) Lytic Cycle — Bacteriophage lyse host bacterial cell shortly after the infection. Viral genome does not form a stabilized integrated structure with bacterial chromosome.

(b) Lysogenic Cycle — Bacteriophage is carried inside the bacterial cell as temperate phage.

Lytic Cycle — A bacteriophage replicates in four stages. They are:

(i) Adsorption and penetration;

(ii) Replication and synthesis of molecular components;

(iii) Assembly and

(iv) Escape.

(i) Adsorption and penetration

Bacteriophage have binal structure. Fig. 4. They get adsorbed on the receptor of host cell. Teichoic acid of bacterial cell wall, outer membrane proteins, lipopolysaccharides, flagella and pilli acts as receptor for adsorption of phage particles. When a phage particle fixes itself on the bacterial cell surface, its nucleic acid enters inside the host cell through the hollow interior of the tail. Sometimes other molecular components like internal proteins may also enter inside the host cell. The protein capsid remains attached to the cell surface. Helical bacteriophage get adsorbed to a special type of pilli. Nucleic acid as well as capsid protein penetrates inside the cells of the host. T-even phage adsorbs on the cell surface by tail fibers and pins. As a result of contraction of sheath tail, core drives itself through the outer membrane and peptidoglycan layers of bacteria. Following this DNA migrates in the host cell.

(ii) Replication and synthesis of molecular components

R 17, MS2, f2 are the RNA containing bacteriophage. They have only four genes, which code for four types of proteins. The amino acids, ribosomes and nucleoside triphosphates required for viral replication are synthesized by host cells. A number of phages contain unusual base pairs in their nucleic acid. During the synthesis of these unusual base pairs, first the normal nucleoside monophosphate are synthesized. These nucleotides are modified by a viral encoded enzyme. Still another viral enzyme converts these modified nucleioside monophosphates into respective nucleoside triphosphates. In the last stage a viral-encoded polymerase incorporates these nucleotides in viral genome.

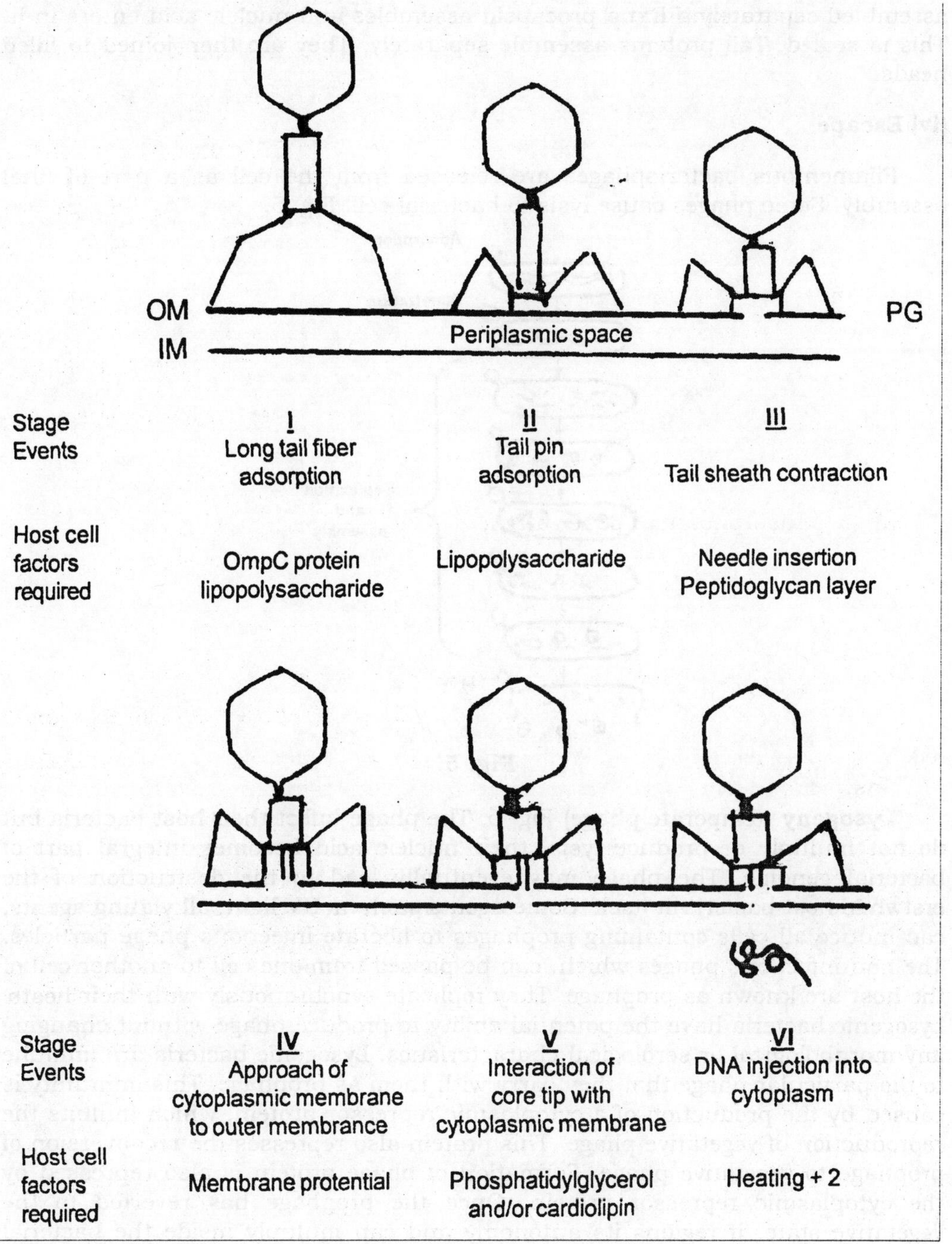

Fig. 4. Steps of penetration of bacteriophage in host cell

(iii) Virion assembly

Most of the bacteriophage have binal structure. After all the molecular components of nucleoprotein are synthesized, icosahedral head and tail are assembled separately. First a procapsid assembles and nucleic acid enters in it. This is sealed. Tail proteins assemble separately. They are then joined to filled heads.

(iv) Escape

Filamentous bacteriophages are released from the cell as a part of final assembly. Some phages cause lysis to bacterial cell. Fig. 5.

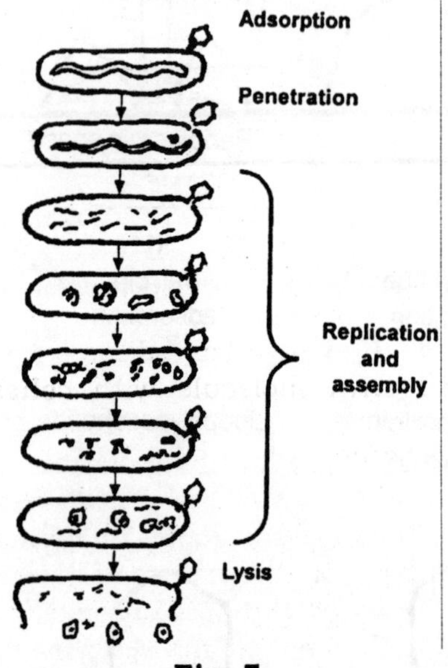

Fig. 5.

Lysogeny (temperate phage) Fig. 6. The phage infect their host bacteria but do not multiply or produce lysis, their nucleic acid becomes integral part of bacterial genome. The phage may eventually lead to the destruction of the erstwhile host bacterium itself. Some agents such as UV light, alkylating agents, can induce all cells containing prophages to liberate infectious phage particles. The non-infectious phages which, can be passed from one cell to another cell of the host are known as prophage. They replicate synchronously with their hosts. Lysogenic bacteria have the potential ability to produce phage without changing any morphological or serological characteristics. Lysogenic bacteria are immune to the particular phage that they carry with them as prophage. This immunity is caused by the production of a cytoplasmic repressor protein, which inhibits the reproduction of vegetative phage. This protein also represses the reconversion of prophage to vegetative phage. Formation of phage protein is also repressed by the cytoplasmic repressor protein. Once the prophage has reverted to the vegetative state, it regains its autonomy and can multiply inside the bacterial cell like a virulent phage.

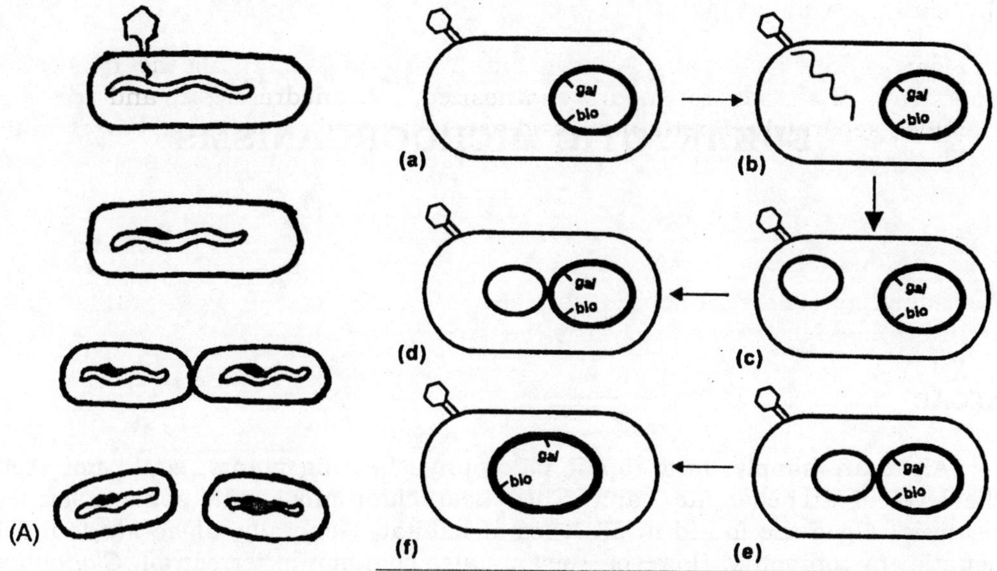

Fig. 6. Lysogeny : A. Mechanism, B. Formation of prophage

Lysogenic phage lambda Phage — Lambda is lysogenic on *E. coli* K 12. It has linear double stranded DNA molecule with cohesive end of about 12 nucleotides. The lambda phage chromosome is about 2% that of the bacterial host chromosome. The phage chromosome firmly attaches to the host chromosome. Phage chromosome is transferred from recipient cell to the donor cell during conjugation. The chromosome mapping and recombination experiments were conducted to study the nature of inserted phage chromosome. The insertion of the phage DNA into the host chromosome apparently proceeds by alignment, breaking, and crosswise rejoining enzyme called lambda integrase. This enzyme has ability to recognise the DNA sequence of bacterial and phage chromosome and brings about close alignment of the two double strands. These nonhomologous DNA strands are broken and then rejoined crosswise. Excision of the phage DNA from the bacterial DNA proceeds by a reversal process of insertion. This leads to the formation of vegetative form of prophage. It multiplies independently and leads to the lysis of the host cell and liberation of lambda phages.

Abnormal Replicative cycles

Abortive infection : sometimes in host cells viral component may be synthesised but maturation or assembly is defective. Either no release occur or the progeny is non infectious.

Defective viruses : They are viruses which are genetically deficient and so capable of producing infectious daughter virion.

Prions : These are composed largely if not entirely of proteins, without any defective nucleic acids. May be prions are responsible for some chronic neurological degenerative diseases of man.

CHAPTER 7

EUKARYOTIC MICROORGANISMS

ALGAE

Algae are simple, autotrophic chlorophyll bearing plants, could not reach the level of archegoniate habit. They bear chlorophyll with some colourless relatives. Algae are found in all types of habitat. Generally, algae are found in aquatic environments. However, they are also common in terrestrial. *Cladophora crispata* grows epizoically on the shells of molluscs. *Stigeoclonium* is endozoic growing in the nose of fishes. *Cephleuros* is a parasitic alga growing on angiospermic plants. Algae are also found as symbionts. Study of algae is termed phycology.

GENERAL FEATURES

Algae show diversity in their range of structures. It may be a single cell or few (but definite in number) cells assemble to form colony. The thallus may be unbranched or branched. Figs. 1, 2, 3.

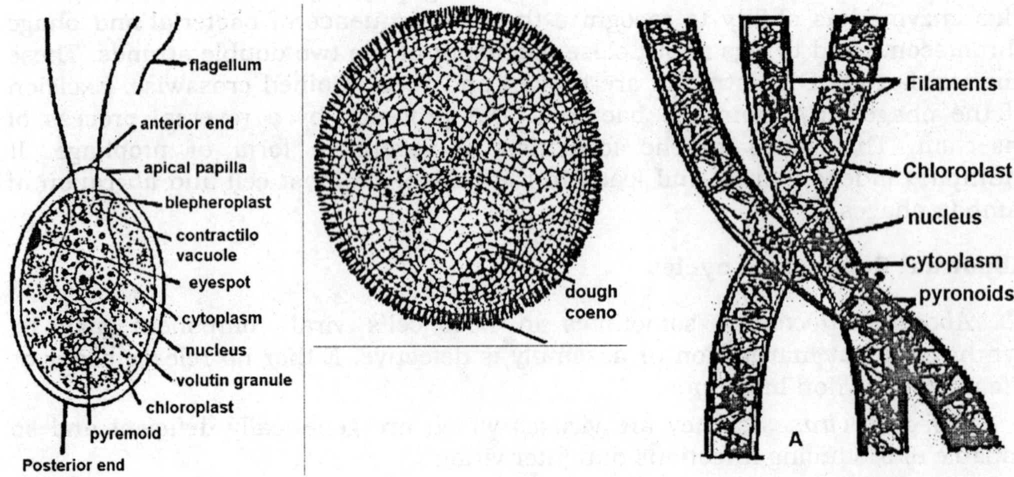

Fig. 1. Chlamydomonas **Fig. 2.** Volvox **Fig. 3.** Spirogyra

Thin cell wall is present. It is bilayered and made up of cellulose along with small quantities of pectin, chitin, algin and fucoidin. In some genera calcium, silica and magnesium carbonate is impregnated in the cell wall. e.g. Diatoms.

In majority of algae, cells are uninucleate. However, a number of algae are multinucleate.

Cells contain different types of photosynthetic pigments. Chloroplast organelles contain chlorophyll, carotene and xanthophyll. Variations in occurence of photosynthetic pigments results in formation of different groups of algae.

Reserve food material is mostly starch. Other types of reserve food materials are fats and oils, laminarin and mannitol.

Flagella wherever present shows 9+2 arrangement. They may be equal or unequal in length. The former is isokontan while the latter is heterokontan.

Flagella are of two types : (a) Acronematic or whiplash type smooth and whip like with stiff base and a flexible upper portion, and (b) pantonematic or tinsel type with lateral rows of fine hairs arranged along the axis of the flagellum giving feathery appearance. Towards the anterior end of motile algae eyespot, a small red or orange body is present.

Asexual reproduction takes place by various types of spores like zoospores, synzoospores, aplanospore, hypnospore, autospore, auxospore, carpospores, tetraspore, cyst etc.

Sexual reproduction may be isogamous, anisogamous or oogamous.

CLASSIFICATION OF ALGAE

Different phycologists have proposed different classification of algae. Algae may be classified in following divisions.

Chlorophycophyta (green algae) — Cell wall made up of cellulose. Photosynthetic pigments mainly chlorophyll a and b. Reserve food material is starch rods. Flagella are 1, 2-8, or many in number. Flagella are apically placed and equal in length.

Chyrsophycophyta (golden and yellow-green algae) — Diatoms are also included in this division. Cell wall is made up of cellulose, silica, calcium carbonate. Photosynthetic pigments mainly chlorophyll a and c, carotenes, fucoxanthin and several other xanthophylls. Storage food material is chrysolaminaran. Flagella are 1-2 in number. Flagella are apically placed and unequal in length.

Cryptophycophyta (cryptomonads) — Cell wall absent. Photosynthetic pigments chlorophyll a and c, carotenes, xanthophylls (alloxanthin, crocoxanthin, monadoxanthin), phycobilins. Storage food material is starch. Flagella are 2 in number. Flagella are subapically placed and unequal in length.

Euglenophycophyta (euglenoids) — Cell wall absent. Photosynthetic pigments chlorophyll a and b, carotenes several xanthophylls. Storage food material is paramylon. Flagella are 1-3 in number. Flagella are apical or subapical in position.

Phaeophycophyta (brown algae) — Cell wall is made up of cellulose, alginic acid, sulfated mucopolysaccharides. Photosynthetic pigments are chlorophyll a and c, carotenes, fucoxanthin and several xanthophylls. Storage food material is laminarin. Flagella are two in number. Flagella are laterally placed and unequal in length.

Pyrrophycophyta (dinoflagellates) — Cell wall is made up of cellulose or it is absent. Photosynthetic pigments are chlorophyll a and c, carotene, several xanthophylls. Storage food product is starch. Flagella are two in number out of which, one is trailing and one is girding.

Rhodophycophyta (red algae) — Cell wall is made up of cellulose, xylans, galactans. Photosynthetic pigments are chlorophyll a, phycocyanin, phycoerythin, carotenes, a number of xanthophylls and in some members chlorophyll d. Reserve food material is floridean starch. Flagella absent.

FUNGI

Alexopoulos and Mims (1979) defined fungi. Fungi are "eukaryotic, spore bearing, achlorophyllous, organisms that may reproduce sexually and asexually, and whose filamentous, branched and somatic structures are typically surrounded by cell walls containing chitin, cellulose or both or these substances with many other complex carbohydrates".

Fungi are ubiquitous organisms found in all kinds of environments. They are found in aquatic as well as terrestrial environments. They may be parasitic or saprophytic in nature. Most of the parasitic forms are pathogenic.

Some fungi are unicellular. In majority of members somatic structure is filamentous body known as hyphae. Hyphae may be unseptate (coenocytic) or septate.

Biotrophs (parasitic fungi) possess haustoria. It is an outgrowth from intercellular hyphae of fungi. It absorbs nutrients from the host cells.

Cell wall is made up of chitin or fungal cellulose along with a small amount of proteins and lipids.

Vegetative reproduction takes place by fragmentation, fission, and by budding. Arthospores, oidia, chlamydospores, gemmae, sclerotia and rhizomorph also contribute in vegetative reproduction.

Asexual reproduction takes place by zoospores, sporangiospores, or conidiospores. Sporangiophores are formed within special sacs called sporangia. When sporangiospores contain flagella termed zoospores and if not they are termed aplanospores. Fig. 4, 5, 6, 7 and 8.

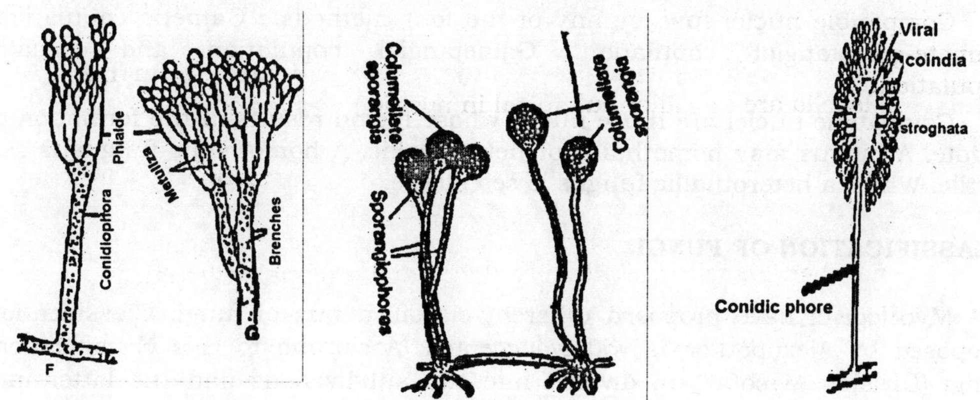

Fig. 4. Penicillium **Fig. 5.** Rhizopus **Fig. 6.** Aspergillus

Oidia, Chlamydospores, blastospores are some of the other asexual spores.

Sexual reproduction completes in three stages : Plasmogamy, Karyogamy and meiosis. In plasmogamy, two cells join and their protoplast fuse resulting in the fusion of two haploid nuclei forming a diploid nucleus called karyogamy. Later the chromosome number reduces again by the process of meiosis.

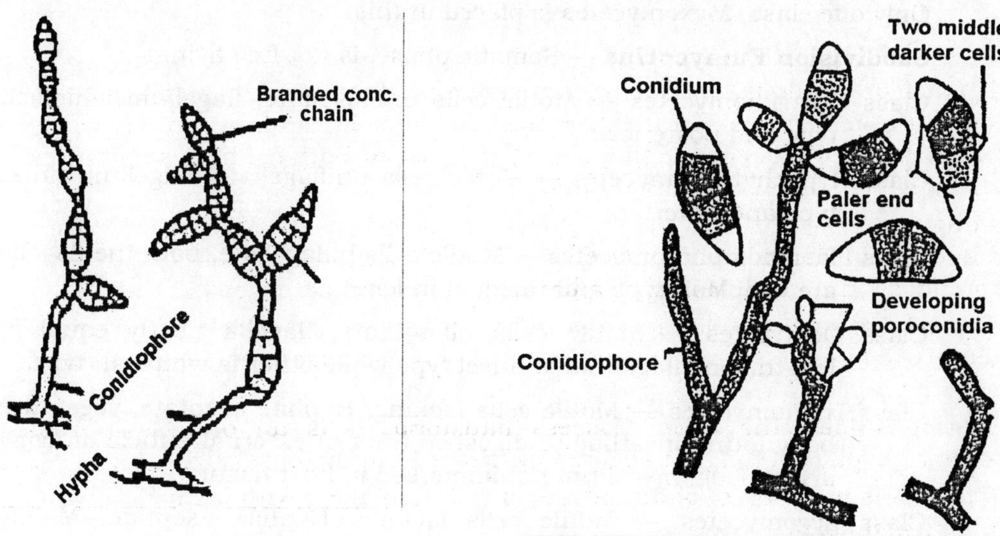

Fig. 7. Alternaria **Fig. 8.** Curularia

Formation of sex organs called gametangia occur. The male gametangium is called antheridium and female gametangium is called oogonium both lead to the formation of gametes.

Sexual spores in smaller numbers are formed such as Ascospores, Oospores, Basidiospores and Zygospores.

Economic importance: Used in industrial fermentation due to their saprophytic characteristics. Although occurrence of fungal diseases not so common yet parasitic fungi cause plants, animals and human diseases.

Compatible nuclei fuse by any of the four methods : Gametic copulation, Gamete-gametangial copulation, Gametangial copulation and somatic copulation.

Compatible nuclei are those nuclei whose fusion results in the formation of zygote. A fungus may homothallic or heterothallic. A homothallic fungus is self fertile. While, a heterothallic fungus is self sterile.

CLASSIFICATION OF FUNGI

Mycologists have proposed different classifications of fungi. Classification proposed by Alexopoulos is widely accepted. According to this classification, fungi (Division Mycota) are divided into two subdivisions and the latter into classes, orders, families etc.

DIVISION MYCOTA

Somatic phase filamentous, unicellular, or plasmoidal. Nuclei with one or more nucleoli and provided with a nuclear membrane;

Subdivision Myxomycota — Somatic phase, is free living plasmodium.

Only one class, Myxomycetes is placed in this.

Subdivision Eumycotina — Somatic phase, is not free living.

Class Chytridiomycetes — Motile cells uniflagellate, flagellum whiplash type and posterior.

Class Hypochytridiomycetes — Motile cells uniflagellate, flagellum tinsel type and anterior.

Class Plasmodiophoromycetes — Motile cells biflagellate, both the flagella are whiplash type and unequal in length.

Class Oomycetes — Motile cells biflagellate, flagella nearly equal in length. One flagellum is tinsel type while other is whiplash type.

Class Trichomycetes — Motile cells lacking. Hyphae aseptate, vegetative body found invading to digestive tract or external cuticle of living arthropods, mycelium not immersed in host tissue.

Class Zygomycetes — Motile cells lacking. Hyphae aseptate. Mainly saprophytic.

Class Ascomycetes — Hyphae septate, sexual reproduction results in formation of ascus and ascospores. Ascospores are endogenous.

Class Basidiomycetes — Hyphae septate, sexual reproduction results in formation of basidium and basidiospores. Basidiospores developed exogenously.

Form-Class Deuteromycetes — Hyphae septate. Sexual reproduction lacking. Parasexual cycle may be present.

PROTOZOA

Protozoan are eukaryotic and simple, unicellular or acellular animal cules, with one or more nuclei. Protozoa are found in aquatic, soil and damp place habits. Generally, they are found as solitary microbes, however, some may form loose colonies. A few protozoan cause disease. The cells are naked or bounded by a pellicle. The cytoplasm in majority of protozoans is differentiated into ectoplasm and endoplasm. There may be a single or many nuclei in a cell. Locomotary organs are absent or are flagella, cilia or pseudopodia. They are mostly aerobic heterotophs. Protozoa found in intestine are anaerobic. Protozoa reproduce asexually by binary fission, multiple fission or by budding. Multiple fission is also known as schizogony. Sexual reproduction takes pace by conjugation or by fusion of gametes (gametocytes). Under adverse conditions, some protozoa produce a protective capsule called a cyst. A cyst also enables a parasite species to survive outside the host. Protozoa may be found as free living forms and others either as symbiotic or parasitic forms. Most of the free living forms require sunlight for photosynthesis while some do not. They are found in variable ranges of temperature and pH from snow of high altitudes to warm waters of hot springs. Optimum temperature between 16-25°C and pH 3.2 to 8.7 respectively.

Sporozoa are parasitic protozoans which result in many diseases.

CLASSIFICATION OF PROTOZOA

Phylum Protozoa is divided into two subphyla : Plasmodroma and Ciliophora.

Subphylum Plasmodroma

Simple and primitive type of protozoan belongs in subphyla. Locomotory organs absent or may be pseudopodia, flagella but not cilia. Cell may have a single or many nuclei. All the nuclei are of same type. Asexual reproduction takes place by binary or multiple fission. Sexual reproduction takes place by fusion of gametes.

Class 1. Sarcodina (Gr. Sarcodes, fleshy) — Amoeboid organisms, having pseudopodia for locomotion. Spore formation and conjugation is not yet observed.

Class 2. Mastigophora (Gr. Mastix, whip: phoros, bearing) — This class is also known as flagellata (*L. flagrum*, a whip). Locomotary organs are flagella. Cells covered by pallicle, which is made up of cellulose, chitin or silica. Asexual reproduction takes place by longitudinal fission. Spore formation or conjugation not yet reported Fig. 9.

Class 3. Opalinata (Protociliata) — This class includes endoparasites in cold-blooded vertebrates. Cilia like flagella present throughout the body. Flagella present in oblique rows. Organisms have saprozoic mode of nutrition. Reproduction takes place by binnary fission or by gametes. Conjugation not yet reported.

Class 4. Sporozoa. (Gr. Spora, seed; zoon, animal) — All the members of this class are parasitic. Locomotary organs are absent. Body covered by thick pellicle.

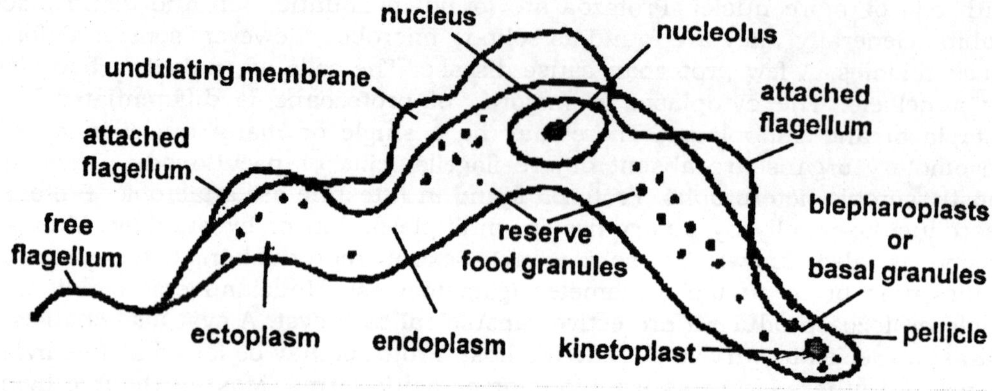

Fig. 9. Trypanosoma

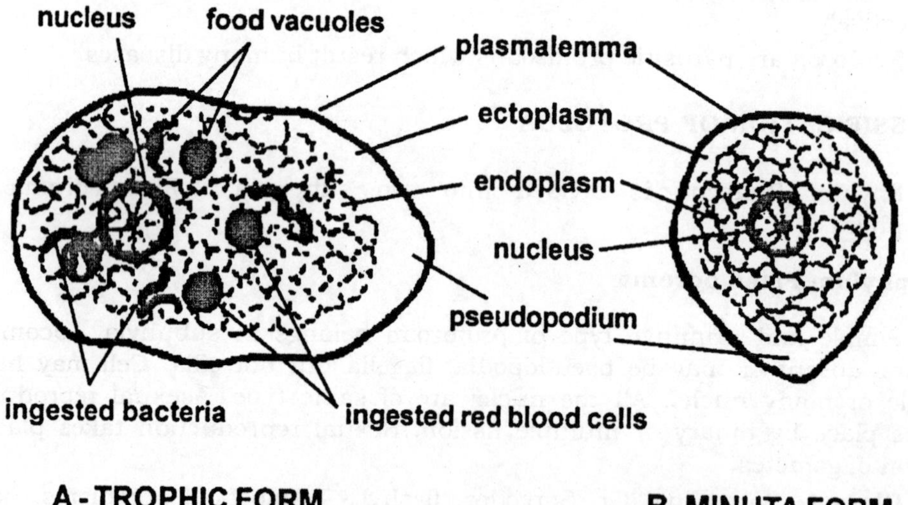

Fig. 10. Entamoeba histolytica

Subphylum Ciliophora (L., cilium, eyelid; ferre, to bear). Fig. 10.

Only one class ciliata or infusoria (L., infusus, croeded) is included in this subphylum. The organisms are considered as complex protozoa having firm pallicle. Locomotary organs are cilia. Nuclei are dimorphic i.e., they are of two types of nuclei. Mode of nutrition is holozoic. Asexual reproduction takes place by binary fission. Sexual reproduction takes place by conjugation and autogamy.

SECTION – II

CHAPTER - 8

BIOCHEMISTRY

BIOCHEMISTRY OF AMINO ACID

Amino acids are monomers of proteins. In 1806, Asparagine was the first amino acid discovered from protein. H. Braconnot (1820) isolated first amino acid Glycine from a protein hydrolyzate of gelatin. In 1935 W.C. Rose isolated Threonine from hydrolyzates of fibrin. It is the 20th amino acid present in protein.

Amino acids are divided into four classes —

1. Nonpolar or hydrophobic R groups

Alanine (Ala); Valine (Val); Leucine (Leu); Isoleucine (Ile); Proline (Pro); Phenylalanine (Phe); Tryptophan (Trp); Methionine (Met).

2. Polar or neutral (uncharged) R groups

Glycine (Gly); Serine (Ser); Threonine (Thr); Cysteine (Cys); Tyrosine (Tyr); Asparagine (Asn); Glutamine (Gln).

3. Positively charged 'R' groups (at pH 6.0)

Lysine (Lys); Arginine (Arg); Histidine (His).

4. Negatively charged 'R' groups (pH 6.0 to 7.0)

Aspartic acid (Asp); Glutamic acid (Glu).

Nutritionally amino acids are of three types-

(a) **Essential amino acids** — These are the ones which are not synthesized by the body (man) and must be taken in diet. They included valine, leucine, isoleucine, phenylalanine, threonine, tryptophan, methionine and lysine.

(b) **Non-essential amino acids** — These can be synthesized by the body (man) and may not be the requisite components of the diet.

(c) **Semi-essential amino acids** — These are growth promoting factors since they are not synthesized in sufficient quantity during growth. They include arginine, and histidine. They become essential in growing children, pregnancy and lactating women.

Table 1.

Name	Symbol	Structural Formula
With Allphatic Side Chains		
Glycine	**Gly (G)**	$H-\overset{H}{\underset{\overset{+}{NH_3}}{CH}}-COO^-$
Alanine	**Ala (A)**	$CH_3-\overset{H}{\underset{\overset{+}{NH_3}}{C}}-COO^-$
Valine	**Val (V)**	$(CH_3)_2CH-\overset{H}{\underset{\overset{+}{NH_3}}{C}}-COO^-$
Leucine	**Leu (L)**	$(CH_3)_2CH-CH_2-\overset{H}{\underset{\overset{+}{NH_3}}{C}}-COO^-$
Isoleucine	**Ile (I)**	$CH_3-CH_2-\underset{CH_3}{CH}-\overset{H}{\underset{NH_3^+}{C}}-COO^-$
With Side Chains Containing Hydroxylic (OH) Groups		
Serine	**Ser (S)**	$HO-CH_2-\overset{H}{\underset{\overset{+}{NH_3}}{C}}-COO^-$
Threonine	**Thr (T)**	$CH_3-\overset{OH}{\underset{H}{C}}-\overset{H}{\underset{\overset{+}{NH_3}}{C}}-COO$

With Side Chains Containing Sulfur Atoms		
Cysteine	**Cys (C)**	$HS-CH_2-CH(NH_3^+)-COO^-$
Methionine	**Met (M)**	$CH_3-S-CH_2-CH_2-CH(NH_3^+)-COO^-$
With Side Chains Containing Acidic Groups or their Amides		
Aspartic acid	**Asp (D)**	$^-O-C(=O)-CH_2-CH(NH_3^+)-COO^-$
Asparagine	**Asn (N)**	$NH_2-C(=O)-CH_2-CH(NH_3^+)-COO^-$
Glutamic acid	**Glu (E)**	$^-O-C(=O)-CH_2-CH_2-CH(NH_3^+)-COO^-$
Glutamine	**Gln (Q)**	$NH_2-C(=O)-CH_2-CH_2-CH(NH_3^+)-COO^-$
With Side Chains Containing Basic Groups		
Arginine	**Arg (R)**	$H_2N-C(=NH_2^+)-NH-CH_2-CH_2-CH_2-CH(NH_3^+)-COO^-$

Lysine	**Lys (K)**	$H_3\overset{+}{N}-CH_2-CH_2-CH_2-CH_2-CH(NH_3^+)-COO^-$
Histidine	**His (H)**	$HC{=}C(-CH_2-CH(NH_3^+)-COO^-)$, ring: $HN^+{=}CH-NH$
Containing Aromatic Rings		
Phenylalanine	**Phe(F)**	$C_6H_5-CH_2-CH(NH_3^+)-COO^-$
Tyrosine	**Tyr (Y)**	$HO-C_6H_4-CH_2-CH(NH_3^+)-COO^-$
Tryptophan	**Trp (W)**	indole $-C-CH_2-CH(NH_3^+)-COO^-$ (ring: $C{=}CH-NH$)
Imino Acids		
Proline	**Pro (P)**	$H_2C-CH_2-CH_2-NH-CH(COO^-)$ (pyrrolidine ring)

Occurrence of Amino Acids

All the standard amino acids occur in almost all proteins. Cereals are rich in acidic amino acids Aspartic acid and Glutamic acid while Collagen is rich in basic amino acids and also proline and hydroxyproline.

Rare Amino Acids

The rare amino acids of protein are derivatives of some standard amino acid *e.g.*, 4-hydroxyproline, a derivative of proline found in abundance in some fibrous protein collagen. Hydroxlysine, the 5-hydroxy derivative of lysine, Sesmosine and isodesmosine occur in fibrous protein elastin. The unusual amino acids N-methyl-lysine, ϵ-N-trimethy-lysine and methyl-histidine are methyl derivatives of standard amino acids. There are no triplet code words for unusual amino acids. They arise by enzymatic modification after their parent amino acids have already been inserted into the polypeptide chain.

Over 300 amino acids are known to occur biologically in free or combined form but, never in proteins. Some non-protein amino acids are important precursors or intermediates in metabolism.

Stereochemistry of Amino Acids

The carbon atom with four different substituents is known as asymmetric carbon. The compounds with such asymmetric carbon show stereoisomerisms (chirality, handedness). In this phenomenon the compounds have optical activity, and are capable of existing in two forms, that are not super imposable mirror images of each other. They exist as dextorotatory (d) or laevorotatory (l) isomers (Fig. 1).

D-Alanine	L-Alanine
COOH	COOH
H—C—NH_2	H_2N—C—H
CH_3	CH_3

Fig. 1. Stereoisomerism in amino acid (Alanine)

Acid Base Properties of Amino Acids

Crystalline amino acids have relatively high melting point usually, above 200°C. They show affinity towards water and are easily soluble in water than in nonpolar solvents. In crystalline form and in neutral aqueous solutions, amino acids occur as dipolar ions or zwitterions, Fig. 2A, rather than as undissociated molecules. The crystalline zwitter ionic amino acid when dissolved in water shows both acid and base properties. Hence, it behaves as amphoteric ions and are called ampholytes. At certain pH amino acids have no net electric charge. This pH is known as isoelectric point. The molecules have no movement towards electric field. Isoelectric pH is symbolized as pHi. The α - carboxyl group of monoamino monocarboxylic acids is a stronger acid than the carboxylic group of comparable aliphatic acids such as acetic acid. While α - amino group of monoamino monocarboxylic acids is a weaker base than the amino group of comparable aliphatic amines.

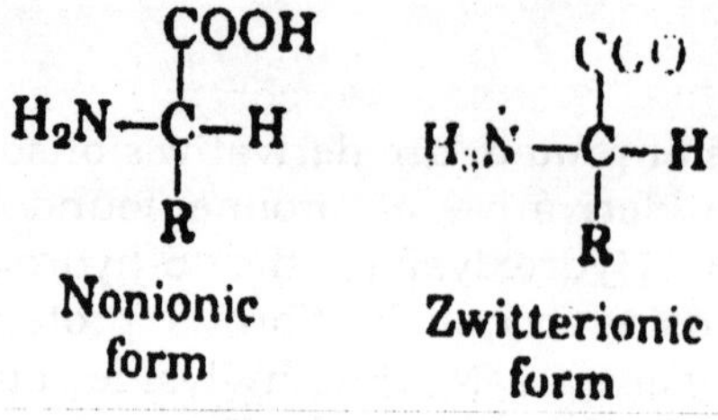

$$\begin{array}{cc} \begin{array}{c} COOH \\ | \\ H_2N-C-H \\ | \\ R \end{array} & \begin{array}{c} COO^- \\ | \\ H_3\overset{+}{N}-C-H \\ | \\ R \end{array} \\ \text{Nonionic form} & \text{Zwitterionic form} \end{array}$$

Fig. 2. Nonionic and Zwitterionic forms of Amino acid

CHEMICAL PROPERTIES

Due to Carboxylic (COOH) Group

Formation of esters — amino acids can form esters with alcohol. Treatment with Na_2CO_3 solution in cold releases the free ester from ester hydrochloride.

Reduction of Amino alcohol — This is achieved in presence of lithium aluminium hydride.

Formation of amines by decarboxylation — Action of specific amino acids decarboxylases dry distillation or heating with $Ba(OH)_2$ or with diphenylamine evolves CO_2 from the –COOH group and changes the amino acid into its amine.

Formation of amides — Anhydrous NH_3 may replace alcohol from its combination with an amino acid in an amino acids ester so that an amide of amino acid and a molecule of free alcohol is produced.

Due to Amino (-NH_2) Group

Salt formation with acids — The basic amino group reacts with mineral acids such as HCl to form salts like hydrochlorides.

Formation of acyl derivatives — Amino group reacts with acyl anhydride or acyl halides such as benzoyl chloride and give acyl amino acids like benzoyl glycine (hippuric acid).

Oxidation — Potassium permanganate or H_2O_2 oxidises the NH_2 group and converts the amino acid into imino acid which reacts with water to form NH_2 and α-keto acid.

Reaction with HNO_2 — Amino acids (expecting proline and hydroxyproline) react with HNO_2 liberating N_2 from NH_2 group. This forms the basis of Van Slyke method for determining –NH_2 group. (Nitrogen).

Reaction with CO_2 — The amino acid anion present in an alkaline solution may react with CO_2 through NH_2 group to form a carbamino acid anion.

Reaction with formaldehyde — Formaldehyde reacts with the $-NH_2$ group to form a methylene compound.

Specific colour reactions — Reaction with Ninhydrin, Millon's test, Sakaguchi test, Hopkin Cole's test amino acids give specific colours.

Due to both NH_2 and COOH Groups

Amino acid form chelated, coordination complex with certain heavy metals and other ions.(Cu^{++}, Co^{++}, Mn^{++} and Ca^{++}).

BIOCHEMISTRY OF PROTEINS

Proteins are highly important biomolecules. The word *Protease* denotes of primary importance. The proteins contain carbon, hydrogen, nitrogen and oxygen, a few contain sulphur. The acid hydrolysis of proteins yields α–amino acid residues which are covalently linked (by peptide bond) to form very long, unbranched chains.

Classification of Proteins

On the basis of their composition, proteins are divided into two major classes : Simple and Conjugated proteins.

Simple proteins : on hydrolysis yield only amino acids and no other major organic or inorganic hydrolysis products. e.g., Globulin, Albumin, Histones, Protamines etc.

Conjugated proteins : Hydrolysis of proteins yields organic or inorganic components in addition to amino acid. The non amino acid portion of a conjugated protein is called the *prosthetic group.* Based on the prosthetic group, the conjugated proteins are classified into:

(a) Glycoprotein : the prosthetic group is carbohydrate. e.g., mucopolysaccharide and mucoprotein.

(b) Lipoprotein : the prosthetic group are lipid substances. These proteins occur mainly in the plasma, brain and eggs. They are also found in bacterial antigens and viruses.

(c) Nucleoprotein : the prosthetic group is nucleic acid, such combination occur both with ribonucleic acid and deoxyribonucleic acid, and the protein portion is usually histone or protamine. e.g., chromosomal protein and viral particles.

(d) Phosphoprotein : the prosthetic is phosphoric acid as organic phosphate but not the phosphate containing substances such as nucleic acids and phospholipids. (1) casein and (2) ovovitellin are two important groups of phsophoprotein found in milk and egg yolk, respectively. They are sparingly soluble in water and very dilute alkali; the phosphoric acid which is esterified through the 'OH' groups of serine and threonine is liberated from organic combination by warming with NaOH and can only be detected by Ammonium molybdate.

(e) Chromoprotein : such proteins contain a coloured non-protein prosthetic group in their molecular structure in addition to the amino acids (1) Hemoproteins : carry heme (heme –red coloured pigment) as a prostheic group; (2) Cytochrome : mitochondrial enzymes of the respiratory chain; (3) Catalase, enzyme that decomposes H_2O_2 to water and O_2; (4) Peroxidase an oxidative enzyme; (5) Flavoprotein: cellular oxidation-reduction protein which has riboflavin a constituent of vitamin B complex as its prosthetic groups This is yellow in colour.

(f) Phosphoprotein : the prosthetic group is metal(s) alone which remain attached to the proteins. e.g., ferritin (metal portion is iron) and carbonic anhydrase (contains zinc).

Derived protein

Protein products formed from the simple and conjugated proteins.

(a) **Primary derived proteins** : Denatured or coagulated proteins are placed in this group, their molecular weight is the same as native protein. They differ in properties like solubility, precipitation and crystallization.

(b) **Secondary derived proteins** : These proteins are formed by the progressive hydrolysis of proteins at their peptide linkages. They represent a great complexity with respect to their size and amino acid composition. They are roughly called proteoses, peptones and peptides according to relative average molecular size.

PROTEIN CONFORMATION

In the native state, protein molecule has a characteristic three-dimensional shape, which is known as protein conformation. Proteins are either fibrous or globular in conformation.

Fibrous proteins have chains usually arranged or coiled along a single dimension (axis) to yield long fibers or sheet. The axial ratio of length : width of a protein molecule is more than 10. Fibrous proteins are physically tough and are insoluble in water or dilute salt and solutions. They are the basic structural elements of in the connective tissue of higher animals. The common examples of fibrous proteins are keratin (of hair, horn, skin, nails and feathers) and collagen (of tendons and bone matrix)

Keratins are rich in cystine residues and thus contain many disulfide cross bridges. α–keratin include the hard, brittle proteins of horns and nails (cystine content upto 22%) while keratin of skin, hair and wool are soft and contain about 10-14% cystine.

The β-Keratin contain no cystine or cysteine. They are rich in amino acids with small side chains particularly glycine, alanine and serine. β–keratin is found in the fibers spun by spider.

Globular proteins have the polypeptide chain tightly folded into compact spherical or globular shapes. The axial ratio of length : width of a protein molecules is less than 10. The common examples of globular proteins are serum albumin and haemoglobin.

Some proteins fall between the fibrous and globular types, resembling fibrous protein in their long rod like structures and the globular protein in their solubility in aqueous salt solution. e.g., myosin and fibrinogen.

The α-Helix and the structure of α-Keratins

Pauling and Korey proposed the model of twisting and coiling the backbone of the polypeptide chain along one axis (Fig. 3). In the α-helix, the backbone is arranged in a helical coil having about 3.6 amino acid residue per turn. The R groups of the amino acids extend outward from the rather tight helix formed by the backbone. A single complete turn of the helix (repeat unit) extends about 0.54 nm (5.4Å) along the long axis. The rise per residue is about 0.15 nm, corresponding to the minor periodicity of 0.15 nm also observed in the differaction patterns. Intrachain hydrogen bonds are found between successive coils of the helix, parallel to the long axis of the helix and extending between the hydrogen atom attached to the electronegative nitrogen of one peptide bond and the carbonyl oxygen of the third amino acid beyond it.

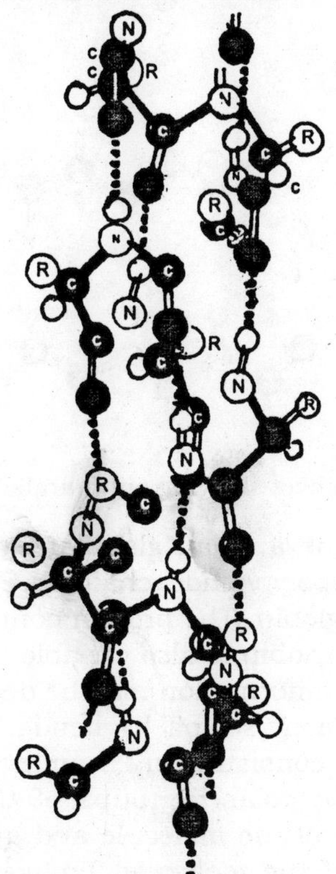

Fig. 3. Hydrogen bonds in a polypeptide chain with α-helical conformation

The β - Keratin : The β-Conformation and the Pleated sheath — When hair or wool is steamed so by the thermal breakage of the intrachain hydrogen bond makes a zigzag conformation of the polypeptide chain. According to Pauling and Korey this is a transition of α-helix into β-helix and is designated as β-conformation (Fig. 4). Side-by-side polypeptide chains in the β-conformation are arranged in pleated sheath, which are cross-linked by interchain hydrogen bonds. The other differences between α -keratin and native β-keratin is in the α-forms all the polypeptide chains are parallel, i.e., run in the same N-terminal to C-terminal direction, whereas in fibroin, the adjacent polypeptide chains are antiparallel i.e., run in opposite directions. α–keratin contains many cystine residues so arranged as to provide interchain –S-S- cross linkages between adjacent polypeptide chains. In contrast, the β-keratins have no –S-S- cross linkages.

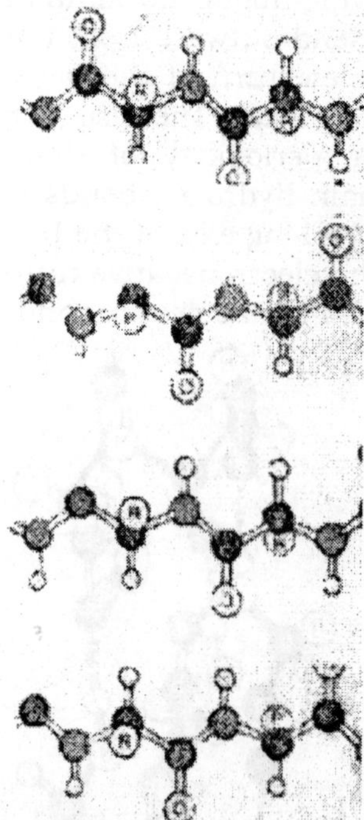

Fig. 4. β - pleated sheet showing anti parallel polypeptide chain

Myoglobin : Myoglobin is a small globular protein. There are 153 amino acids arranged in a single polypeptide chain. J.C. Kendrew performed X-ray studies of sperm whale myoglobin. The protein contains iron-porphyrin or heme group. Like hemoglobin, myoglobin is also capable of reversible oxygenation and deoxygenation. The protein conformation may be described with the backbone of eight relatively straight segment set off by bends. Each straight segment is a length of helix, the longest consisting of 23 amino acids and the shortest only 7; all are right handed. The polar R-groups of the amino acid residues are located on the outer surface of the molecule and are hydrated. The nonpolar R groups are in the interior of the molecule. Proline residues occur only at the bends.

PROTEIN STRUCTURE

Proteins are macromolecules made up of twenty different amino acids attached to each other by peptide bond. Generally proteins have complicated structure having large molecular weight. The simplest protein consists of straight chains of polypeptides but in majority of the proteins there are folds. The connection of protein are due to certain bridges, coiling of chains, folding of such coils and aggregation of such fold entities. The protein structure can be studied at four levels. These levels are referred to as primary, secondary, tertiary and quaternary structures.

Primary structure — The primary structure refers to the sequence of amino acid in the polypeptide chain. The carboxyl group of one amino acid and the amino group of another amino acid link together to form peptide bond (-CO-NH) forming a polypeptide chain. In the straight chains, the left hand side end of the chain bears the amino acid which has a free amino group which is known as N-terminal end, whereas the right hand side end bears an amino acid which has a free amino group which is known as C-terminal end. A protein may have branched polypeptide chains. The branching may be due to disulfide bridges (of sulphur containing amino acids) or may be due to amino acid containing two carboxylic group (glutamic acid and aspartic acid) or may be due to amino acid containing two amino group (lysine, arginine). In certain cyclic proteins the C-terminal and N-terminal do not have free ends. Some of the protein contain more than one chain. These chains may be similar or different in amino acid content and their sequence. If more than one chain is present in a protein, the chains are numbered as alpha (α), Beta (β), gamma (γ) and delta (δ). Insulin contains two chains. The chains differ with one another in the number of amino acids and their sequence. This affects the biological activity of the protein molecule.

Secondary structure — The secondary structure refers to a three dimensional secondary structure by way of folding or coiling consisting of a helically-coiled, zigzag, linear or mixed form. The linkages or bonds involved are hydrogen bonds and disulphide bonds. The disulphide bridges (-S-S-) are formed between two cysteine residues existing in the different chains or in the same chain. Beside disulphide bonds certain non-covalent bonds such as hydrogen bonds are also formed. The hydrogen bonds in secondary structure may form either an α-helix or β-pleated sheet structure. The primary chain acquires a characteristic type of folding or twisting due to the presence of disulphide bridges or hydrogen bonds. The polypeptide chain may acquire helical or other types of secondary structures. The bridges and bonds may be present between two different chains or within the same chain.

Tertiary structure — The tertiary level of organization is acquired generally by globular protein (Fig. 5). The helixes or the twisted structure are further folded to form a globular, ellipsoidal shape. The bonds responsible for interaction between groups of amino acids are : Hydrophobic interactions normally occur between nonpolar side chains of amino acids such as alanine, leucine, methionine, isoleucine and phenylalanine. Hydrogen bonds normally formed by the polar side chains of the amino acids. Ionic or electrostatic interactions are formed between oppositely charged polar side chains of amino acids, such as basic and acidic amino acids. Vander Waal forces occur between non-polar side chains and disulfidė bonds.

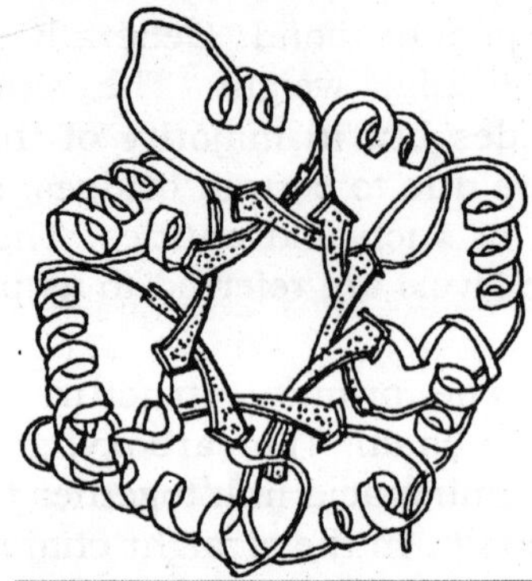

Fig. 5. Tertiary structure built up from four β–α–β motifs

Quaternary structure — Some of the protein molecules are composed by polymerization of the subunits of the secondary or tertiary structures. Two or more peptide chains held together by non-covalent interactions or by covalent cross-links. The assembly is often called as oligomer and each constituent peptide chain is called as monomer or subunit. The polymerized protein show a well defined functional activity. These subunits may be the same or different. The aggregation of the subunits is stabilized by means of the non-covalent forces. A number of enzymes have quaternary structure. It may also make substrate specificity in the enzyme molecule.

Protein Denaturation

When a protein is subjected to some drastic treatment like heating, treatment with acids, alkaline, salts, alcohol, acetone and other organic solvents results in extensive disruption of the secondary, tertiary and whenever applicable quaternary organization of a protein molecule due to cleavage of non-covalent bonds. However, primary structure is not affected. Denaturation results in loss of their natural properties. Denaturation results in decreased solubility, increased viscosity of the solution. Rate of diffusion is decreased. X-ray, ultraviolet light, violent shaking and high pressure may also result in denaturation. Denaturation does not involve hydrolysis. Generally denaturation is an irreversible phenomenon. At a later stage of denaturation, proteins are precipitated and this stage is known as coagulation.

BIOCHEMISTRY OF CARBOHYDRATE

Carbohydrates are defined as a polyhydric alcohols of carbon possessing active aldehydic or ketone groups. They are classified into Monosaccharide, Oligosaccharide and Polysaccharide. The monosaccharides cannot be hydrolysed into simple sugars. Oligosaccharide (Oligo-few) on hydrolysis yield three to six monosaccharides joined by glycosidic bonds. Polysaccharide are sugars with monosaccharides joined by glycosidic linkage, the monomers are joined in long linear or branched chains.

1. MONOSACCHARIDE

The monosaccharides are also called as simple sugars and are expressed by empirical formula $(CH_2O)n$. They are classified into *triose, tetrose, pentose, hexose* and *heptoses.* They contain one asymmetric carbon atoms and thus are Chiral molecules. Suffix *"ose"* denotes that the particular compound belongs to aldo group of carbohydrate while suffix *"ulose"* denotes `keto' sugars.

They can be subdivided into:

(a) depending upon the number of carbon atom they possess, as trioses, tetroses, pentoses, hexoses etc.

(b) depending upon whether aldehyde (-CHO) or ketone (-CO) groups are present as aldoses or ketoses. e.g. —

Erythrose ($C_4H_8O_2$ - tetroses)
Ribose ($C_5H_{10}O_5$ - Pentoses)
Glucose ($C_6H_{12}O_6$ - Hexoses)

In D-glyceraldehyde, the hydroxy group exists on the right side and in the L-glyceraldehyde. The position of the hydroxy group is on left to the asymmetric carbon atom.

In crystalline form, the monosaccharides containing 5 or more carbon atoms exist in the tautomeric rings. The free sugars mostly contain six membered ring in which five members are carbon and the remaining one member is an atom of oxygen. The six-membered ring is known as pyranose ring. (Fig. 6). In five membered ring, four members are carbon and one member is oxygen. The five membered ring is known as furanose ring. Triose, tetrose do not possess ring structure. In the pyranose type of hexoses carbon at '1' position is asymmetric. If the hydroxyl group at position '1' is cis to the hydroxyl group at position 2, it is known as 'α' form and if trans it is known as 'β' form. Isomeric forms of monosaccharides that differ from each other only in configuration about the carbonyl carbon atom are anomers and the carbonyl carbon atom is called anomeric carbon. The formation of pyranose is a special case of a more general type of reaction between and aldehyde and an alcohol to form a *hemi-acetal* which contains an asymmetric carbon atom and, therefore, can exist in two sterio isomeric forms. A *hemiketal* is an analogous product formed by reaction of a ketone with an alcohol.

Anomers and Anomeric carbon : Carbon 1, after cyclization has four different groups attached to it and thus, it becomes now 'asymmetric'. The two cyclic compounds and have different optical rotations, but they will not be same because the compounds as a whole are not mirror images of each other. Compounds related in this way are called "anomers" and carbon 1, after cyclization becomes asymmetric is called now "anomeric carbon atom".

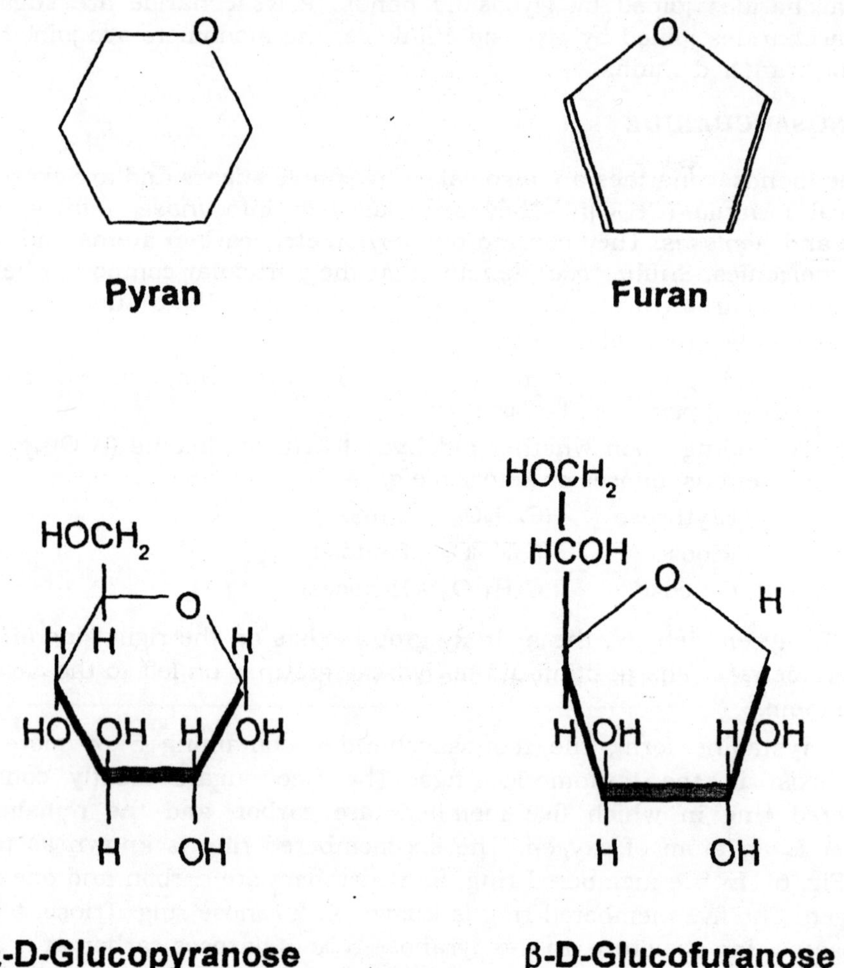

Fig. 6. Pyran and Furan forms of Glucose

D-glucose in solution mostly exists in the pyranose form and it remains in equilibrium with a small amount of *open chain form.* The open chain form is capable of reducing various reagents such as Fehling's solution, Benedicts reagent etc.

Mutarotation — when an aldohexoses is first dissolved in water and the solution is put in optical path so that plane polarized light is passed the initial optical rotation shown by the sugar gradualy changes until a constant fixed rotation characteristic of the sugar is reached. This phenomenon of change of rotation is called as "mutarotation". If D-glycopyranose is dissolved in water, it first shows rotation of +112 but this value gradually falls to +52. This change of rotation is known as mutarotation and this is due to establishment of an equilibrium between α and β forms of the ring structures (Fig. 7).

α-D-Glucopyranose
(α-anomer)

β-D-Glucopyranose
(β-anomer)

Acyclic
aldehyde
form

Fig. 7. Mutarotation of Glucose

Two sugars differing only in the configuration around one specific carbon atom are called *epimers* of each other (Fig. 8).

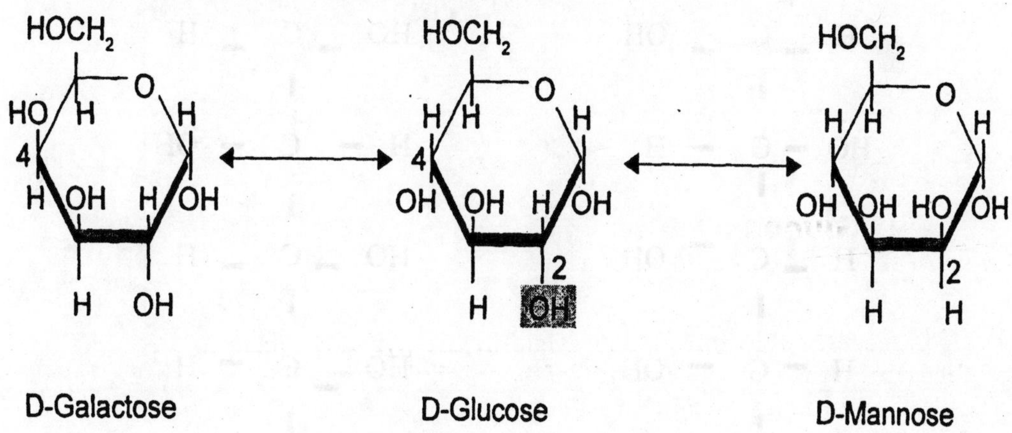

Fig. 8. Epimerisation of Glucose

HAWORTH PROJECTION FORMULA

These formulas indicate the ring forms of monosaccharides. The edges of the ring nearest to the reader is represented in bold lines. The Haworth projections suggest that the five and six membered furanose and pyranose ring are planar, it is somewhat misleading and is not a fact. The pyranose ring exist in two conformations, the *chair* form and the *boat* form.The chair form is much more stable and relatively rigid. In aqueous solutions of hexose chair form is found predominating. The substituent group in the chair form are not geometrically and chemically equivalent.

General properties in reference to glucose

Vant Hoff's Rule 'n' — The number of possible isomers of any given compound depends upon the number of asymmetric carbon atoms the molecule possess. According to the Vant Hoff's rules of 'n'; 2 equals the possible isomers of that compound where n represents the number of asymmetric carbon atom in a compound.

Stereo Isomers — The presence of asymmetric carbon atoms in a compound gives rise to the formation of isomers of that compound. Such compounds which are identical in composition and differs only in spatial configuration are called "stereo-isomers".

Two such isomers of glucose – D- glucose and L-glucose are "mirror" image of each other (Fig. 9).

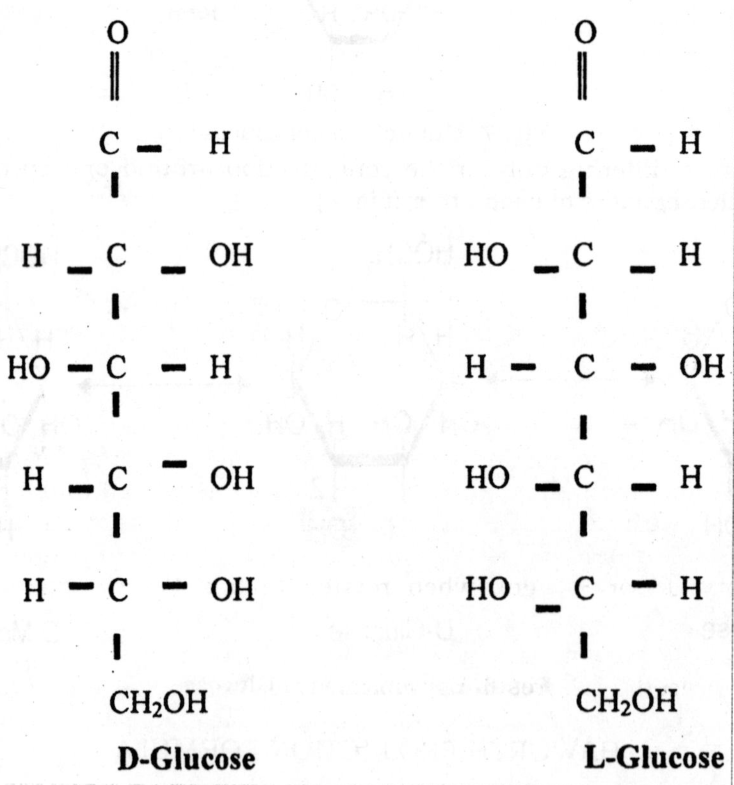

Fig. 9. Stereo isomers of Glucose.

D-series and L-series — The orientation of the H and OH groups around the carbon atom first adjacent to the terminal primary alcohol carbon, e.g. C-atom 5 in glucose determines the series. When the –OH groups on this carbon is on the right, it belongs to D-series, when the OH group on this carbon is on the left it is a member of the L series (Fig. 10).

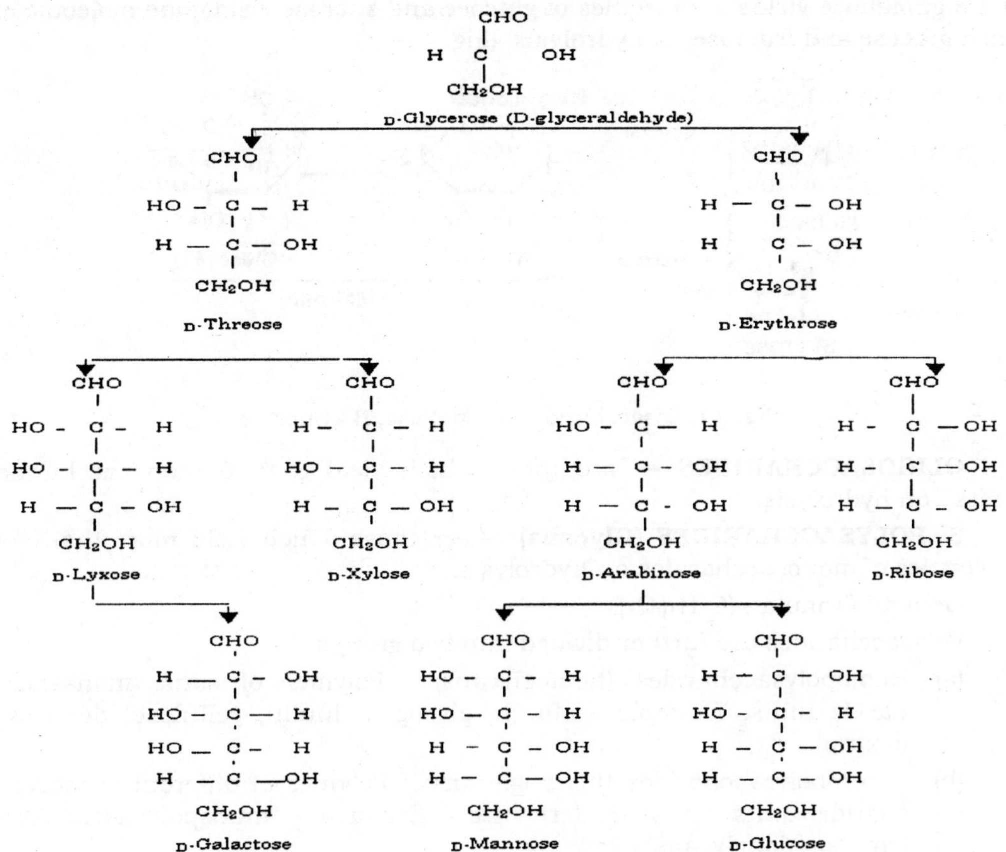

Fig. 10. D-series and L-series in Glucose

Optical Activity — Presence of asymmetric carbon atom also confers optical activity on the compound when a beam of plane polarized light is passed through a solution exhibiting optical activity, it will be rotated to the right or left in accordance with the type of compound, i.e. the "optical isomers" or "enantiomorphs"; when it is rotated to light, the compound is called "dextrorotatory" (d or + sign), when rotated to left, the compound is called "Laevorotatory" (l or – sign).

Racemic — when equal amounts of "dextrorotatory and laevorotatory isomers are present, the resulting mixture has no optical activity, since the activities of each isomer cancels each other. Such a mixture is said to be 'Racemic'.

Resolution — The separation of optically active isomers from a racemic mixture is called resolution.

2. DISACCHARIDES

The disacharides are those sugars which yield two molecules of the same or different molecules of monosaccharides on hydrolysis. General formula Cn(H_2O)

n-1 e.g. Maltose yields 2 molecules of glucose and sucrose yields one molecule of each glucose and fructose on hydrolysis. (Fig. 11)

Fig. 11. Disaccharides A - Maltose, B - Sucrose

OLIGOSACCHARIDES — are those which yield 3 to 6 monosaccharide units on hydrolysis.

3. POLYSACCHARIDES (Glycans) — are those which yield more than six molecules of monosaccharides on hydrolysis.

General formula : $(C_6H_{10}O_5)_n$

Polysaccharides are further divided into two groups :—

(a) Homopolysaccharides (homoglycans) : Polymer of same monosaccharide units. *Example* – starch, glycogen, inulin, cellulose, dextrins, dextran.

(b) Heteropolysaccharides (heteroglycans) : Polymer of different monosaccharide units or their derivatives. *Example* – mucopolysaccharides (glycosaminoglycans)

Starch — It is a common storage polysaccharide of plant cells. Starch occurs in two forms, *amylose* and *amylopectin* (Fig. 12). Amylose consists of long unbranched chains in which all the D-glucose units are bound in α (1-4) linkage Amylose is rather insoluble in water and forms hydrated micelles. Amylopectin is highly branched. There are both α (1-4) and α (1-6) linkages. The backbone glycosidic linkage is α (1-4), but the branch points are (1-6) linkages, it yields colloidal or micellar solutions.

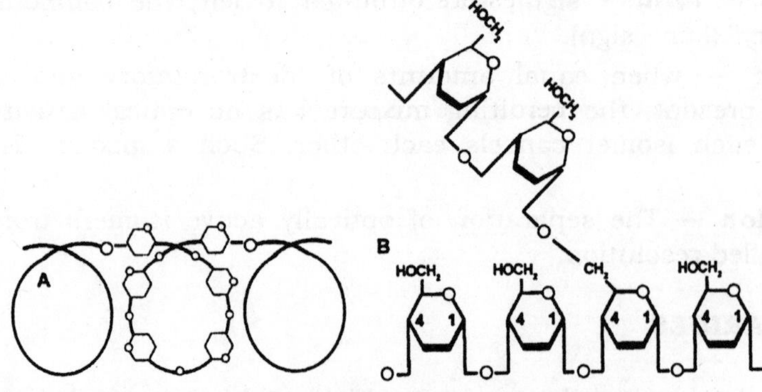

Fig. 12. Starch: A - Amylose, B - Amylopectin

Glycogen —Glycogen is a storage polysaccharide of animal cells and nonchlorophyllous plant cells like yeast and fungi. It is made up of glucose monomers. The monomers are linked by α -(1-4) and α -(1-6), linkages.

Cellulose — Cellulose is a linear polymer and a basic structural polysaccharide of cell wall of plant cell. The monomer D-glucose is linked by β–(1-4) linkage to form the linear polymer. Partial hydrolysis yields disaccharide yields cellobiose while complete hydrolysis yields D-glucose. Cellulose is hydrolyzed by enzyme cellulase. The cellwall polymer is made up of densely packed cellulose fibrils present in regular parallel arrays, often in *cris*-cross layers. These fibrils are cemented together by a matrix of three other polymeric materials: *hemicellulose, pectin* and *extensin*. Hemicellulose is a polymer of D-xylans, polymers of D-Xylose in β-(1-4) linkage with side chains of arabinose and other sugars. Pectin is a polymer of methyl-D-galacturonate. Extension is a complex glycoprotein attached covalently to the cellulose fibrils.

BIOCHEMISTRY OF LIPIDS

LIPIDS

Lipids are Fats and Fat like substance include fats (also oils), wax, phospholipids, cerebrosides, steroids, vitamin A, D, E, K, carotenoids and quinones.

Bloor proposed certain criteria under which lipids are insoluble in water, solubility in one or more organic solvents such as ether, chloroform, benzene, acetone, ethanol is so called fat solvents. Lipids form some relationship to the fatty acids as esters either actual or potential. There is possibility of lipids to be utilized by living organism.

Lipids are classified into simple, compound, derived and miscellaneous lipids.

Simple lipids — are esters of fatty acids with various alcohols. They may be neutral *fats* or *waxes*. Neutral fats or Triacylglycerols (TG) which are triesters of fatty acids with glycerol. Waxes are esters of fatty acids with monohydroxy aliphatic alcohols other than glycerol.

Compound lipids — are esters of fatty acids containing groups other than and in addition to an alcohol and fatty acids. Compound lipids are:-

(a) *Phospholipids* — They are substituted fats, containing in addition to fatty acid and glycerol, a phosphoric acid residue, a nitrogenous base and other substituents. *Examples* : Phosphatidyl choline (lecithin), Phosphatidyl ethanolamine (cephalin) Pohosphatidyl serine, Plasmalogens, Sphingomyelins (Fig. 13 A, B, C).

(b) *Glycolipids* — The lipids containing carbohydrate moiety hence, called glycolipids. They contain a special alcohol called sphingosine and sphingol and nitrogenous base in addition to fatty acids but does not contain phosphoric acid or glycerols.

(c) *Sulfolipids* — The lipids characterized by possessing sulphate groups.

(d) *Aminolipids* — These lipids are also known as proteolipids and contain proteins.

(e) *Lipoprotein* — Lipids as prosthetic group to protein.

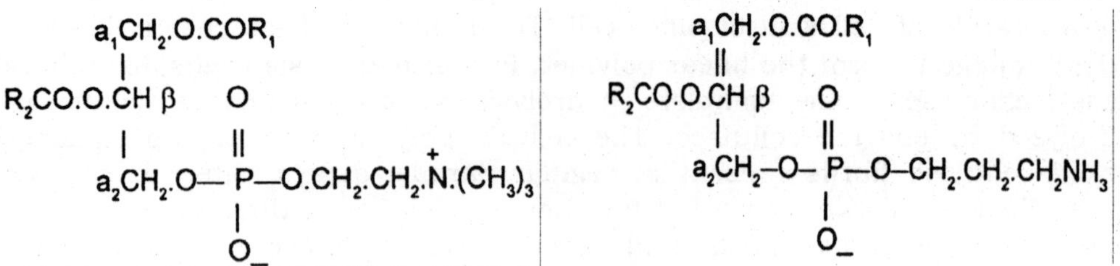

Fig. 13A. General structure of lecithin

Fig. 13B. Structure of Cephalin

$a_1CH_2.O.CH{=}CH.CH_2.R$

$R_2CO.O.CH$ β O

$a_2CH_2.O.P$—O—Base

O_

Fig. 13C. General structure of Plasmalogens

Derived lipids — are derivatives obtained by hydrolysis of those given in group I and II, which still possess the general characteristics of lipids :

(a) Fatty acids may be saturated, unsaturated or cyclic.

(b) Monoglycerides (Monoacyl glycerol) and Diglycerides (Di-acylglycerol).

(c) Alcohols:

  (i) straight chain alcohols are water insoluble alcohols of higher molecular weight obtained on hydrolysis of waxes;

  (ii) Cholesterol and other steroids including Vit. D,

  (iii) Alcohols containing the -ionone ring include Vit. A and certain carotenoids,

  (iv) Glycerol.

Miscellaneous lipids — are (a) Aliphatic hydrocarbons which include iso-octadecane found in liver fat and certain hydrocarbons found in bees wax and plant waxes; (b) Carotenoids; (c) Squalene is a hydrocarbon found in shark and mammalian liver and in human serum and (d) vitamin E and K.

FATTY ACIDS

Fatty acids are organic acid that occurs in a natural triglyceride. It is a monocarboxylic acid ranging in chain length from C4 to about 24 carbon atoms. They are obtained from hydrolysis of fats. Fatty acids may be (a) Saturated Fatty Acids in which there is no double bond and (b) Unsaturated Fatty Acids with one or more double bonds.

Structure and Sources of some Important Fatty Acids

Fatty Acids	Structure	Source
Saturated		
Acetic acid	CH_3COOH	Vinegar
Butyric acid	$CH_3(CH_2)_2COOH$	Butter
Caproic acid	$CH_3(CH_2)_4COOH$	Butter
Palmitic acid	$CH_3(CH_2)_{14}COOH$	Animal/Vegetable
Fats		
Stearic acid	$CH_3(CH_2)_{16}COOH$	— —
Unsaturated		
Oleic acid ^9	$CH_3(CH_2)_7CH{=}CH(CH_2)_7COOH$	— —
Linoleic acid ^9, 12	$CH_3(CH_2)_4CH{=}CHCH_2CH{+}CHCH_2CH{=}CH$ $(CH_2)7COOH$	Linseed oil and fats
Linolenic acid ^9,12,15	$CH_3CH_2CH{=}CHCH_2CH{=}CH{=}CHCH_2CH{=}CH$ $(CH_2)7COOH$ }	— —

Nomenclature of Fatty Acids

Fatty acids are named after the hydrocarbon with the same number of carbon atoms, *OIC* being substituted for the final "c" in the name of hydrocarbon. Thus, Saturated acids end in *ANOIC* e.g., Octanoic acid. Unsaturated acids with double bonds end in *ENOIC* e.g., Octadecenoic acid (Oleic acid). Carbon atoms are numbered from the -COOH carbon (Carbon No.1) The carbon adjacent to –COOH group, i.e. Carbon No. 2 is known as carbon, likewise carbon 3 is carbon and the end $-CH_3$ carbon is known as the carbon.

The number and positions of double bond is also indicated by various conventions. e.g., 9 indicates a double bond between carbon atom 9 and 10 of the fatty acids, for this ^ sign is used e.g., ^9 denotes double bond between 9 and10 carbon. The fatty acid formula is formulated by (a) the number of carbon, (b) the number of double bonds and (c) the position of double bonds e.g., Oleic acid (mol. Formula $C_{17}H_{33}COOH$ has one double bond between C9 and C10 thus

$$\overset{}{CH_3(CH_2)_7\text{-}\overset{9}{CH}{=}\overset{10}{CH}\text{-}(CH_2)_7\text{-}COOH}$$

So, as per above criteria, it is expressed as 18 : 1 : 9 (18 indicates the number of carbon, 1 indicates the number of double bond and 9 indicates the position of the double bond.

Triacylglycerols (Triglycerides)

Acylglycerols or glycerides are fatty acid and ester of alcohol glycerol. In triglycerols all the three hydroxyl groups of glycerols are esterified with fatty acids. If the triglycerols are solid at room temperature they are known as Fats while liquid are known as Oils.

Simple triglycerols — Glycerols esterfied with single kind of fatty acids e.g., tristearoylglycerol (tristearin), tripalmitoylglycerol (tripalmitin), trileoylglycerol (triolein).

Mixed triglycerols — They contain two or more different fatty acid e.g., 1-palmitoyldisteroylglycerol (1-palmitodistearin).

Fatty acids esterified to two of the hydroxyl groups of glycerol and the remaining hydroxyl group is joined in ether linkage with a long alkyl or alkenyl chain are known as *Alkyl Ether Acylglycerols.* If a sugar is present in glycosidic linkage with the unesterified 3-hydroxyl groups of diacylglycerols (galactosyldiacylglycerol) it is known as *Glycosylacylglycerols.*

Phospholipids

Phospholipids are formed when a phosphoric acid, a nitrogenous base is present in addition to fatty acids in the glycerols or sometimes other alcohol molecule. These include various types of Lecithins, Cephalins and Sphingomyelin. *Lecithins* are constituted by glycerol, fatty acids, phosphoric acid and nitrogenous base choline. The alcoholic groups of glycerols are found to be esterified with two fatty acids. The alcoholic group is esterified with phosphorylcholine. *Cephalins* (Phosphatidyl Ethanolamine) are constituted by glycerol, fatty acids, phosphoric acid and a nitrogenous base which may be β-aminoethanol or serine. Both α and β-cephalin are known. *Plasmlogens* are constituted by one molecule each of long chain aliphatic aldehyde, a fatty acid glycerol -PO4 and a nitrogenous base which is usually ethanol amine but may be sometimes choline.

Sphingolipids

The sphingolipids contain a nitrogenous base sphingosine (an alcohol) or its derivative which remains attached to long-chain fatty acid by its amino group and to the phosphoryl-choine by its alcoholic group (Fig. 14). *Sphingomyelins* contain phosphorylethanolamine or phosphorylcholine as their polar head group esterified to the 1-hydroxyl group of ceramide (Fig. 15). *Cerebrosides* are also known as the glycolipids These lipids contain a nitrogenous base sphingosine, a long chain fatty acid and sugar usually galactose.

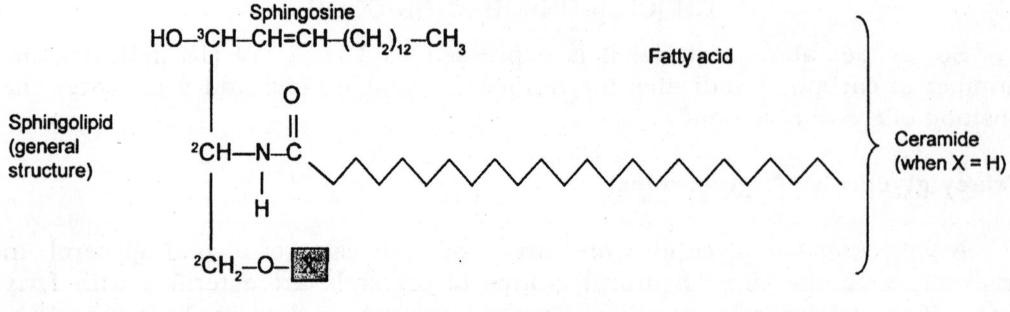

Sphingosine
$HO{-}^{3}CH{-}CH{=}CH{-}(CH_2)_{12}{-}CH_3$
Fatty acid
Sphingolipid (general structure)
$^{2}CH{-}N(H){-}C(=O)$
Ceramide (when X = H)
$^{2}CH_2{-}O{-}X$

Fig. 14. Sphingolipids - General structure of Sphingolipids

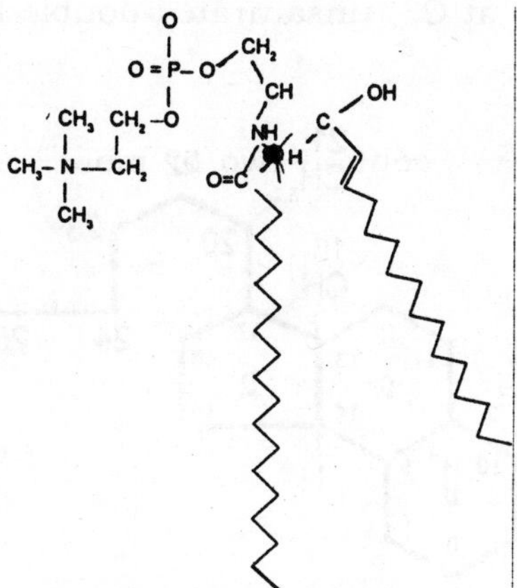

Fig. 15. Sphingomylin

WAXES

Waxes are water insoluble, solid esters of higher fatty acids with long chain monohydroxylic fatty alcohols or with sterols. Waxes are found as protective coating. The major components of bees wax are palmitic acid esters of long chain fatty alcohols with 26 to 34 carbon atoms. Lanolin or wool fat is a mixture of fatty acid esters of the sterols lanosterol and agnosterols.

NON-SAPONIFIABLE LIPIDS

The non-saponifiable lipids contain no fatty acids. A cell contains a small amount of nonsaponifiable lipids. They include many substances having profound biological activity like vitamin, hormones and other highly specialized fat soluble biomolecules. There are two major classes of non-saponifiable lipids, the *terpenes* and the *steroids.* Terpenes are constructed of multiples of the five carbon hydrocarbon isoprene (2-methyl-1,3-butadiene). Terpenes containing two isoprene units are called monoterpenes, those containing three isoprene units are called sesquiterpenes and those containing four, six and eight units are called diterpenes, triterpenes and tetraterpenes, respectively. Terpenes may be either linear or cyclic molecules.

STEROIDS AND STEROLS

Steroids are derivative of saturated tetracyclic hydrocarbon (Fig. 16). All of the steroids have a similar cyclic nucleus resembling "phenanthrene" (ring A, B and C) to which a "cyclopentane" (ring D) is attached. It is designated as "Cyclopentano-perhydro-phenanthrene nucleus". Methyl side chains occur typically at positions 10 and 13 (constituting carbon atom 19 and 18, respectively). A side chain at position 17 is usual. If the compound has one or more OH groups and no carbonyl or carboxyl group, it is called a "sterol" and the name terminates in OL. *Cholesterol* (Fig. 17) has molecular formula

$C_{27}H_{45}OH$ with -OH group at C3, unsaturated double bond between C5 and C6 positions. It

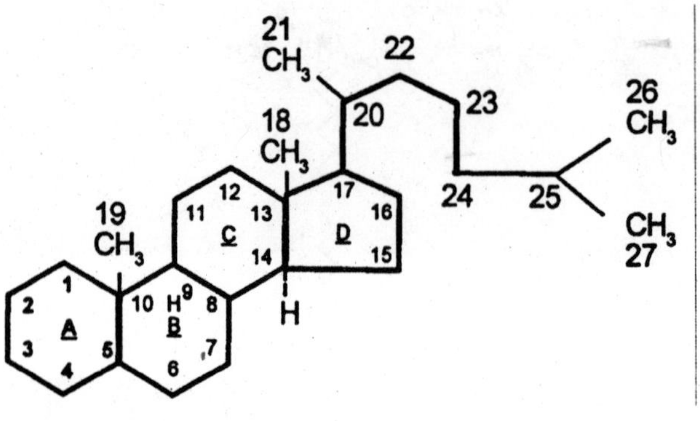

Fig. 16. General structure of cholesterol

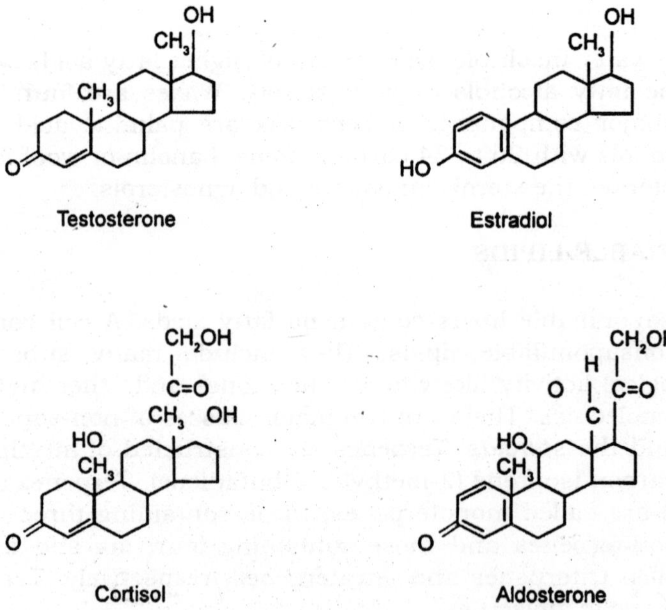

Testosterone

Estradiol

Cortisol

Aldosterone

Fig. 17. Steroids derived from cholesterol

has two CH_3 group at C10 and C13 positions. It has eight carbons side chain attached to C17. *Dehydrocholesterol* differs from cholesterol only in having a second double bond, between C7 and C8. *Ergosterol* is a plant sterol. It was first isolated from ergot (a fungus of rye) and later from yeast and certain mushroom. Structurally this sterol has the same nucleus as 7-dehydro-cholesterol but differs slightly in its side chains. Ergosterol is changed to vitamin D2 by the opening of the ring B of the sterol. Hence, ergosterol is also known as Pro-vitamin-D2. Stigma sterol and Sitosterol are plant sterol occurring in higher plants. They have no nutritional value for human being.

CHAPTER 9

ENZYMES

DEFINITION

Enzymes are found as colloids in a living cell, they are protein in nature, having very high catalytic activity.

E. Buchner (1897) extracted enzymes from yeast and hence the name enzyme (en = in + zyme = yeast). In 1935, J.J. Berzelius published the general theory of chemical catalysis and included diastase in it. J.B. Sumner (1926) isolated urease from jack bean and J. Nouthrop (1930-36) crystallized enzyme pepsin, trypsin and chymotrypsin.

Enzymes are protein molecules. Certain enzymes require a non proteinaceous component which may be organic compound or inorganic metallic ions. Such enzyme molecules with protein and non-protein components are known as holoenzyme and its protein part as apoenzyme. Some enzymes require an additional chemical component called a cofactor. The cofactor may be either one or more inorganic ions such as Fe^{2+}, Mg^{2+}, Mn^{2+} or Zn^{2+} or a complex organic or metalloorganic molecule called a coenzyme. Some enzyme or metal ion that is covalently bound to the enzyme protein is called a prosthetic group. It is essential for the functioning of enzyme. Many times these terms (coenzyme and prosthetic group) are used as synonym. However, these terms are differentiated on the basis of the binding tendency with the apoenzyme. The tightly bound groups that cannot be removed (e.g., by dialysis) are recognized as "prosthetic groups" and those which dissociate easily are "coenzymes". Most of the coenzymes have close relationship with vitamins.

NOMENCLATURE AND CLASSIFICATION OF ENZYMES

In the past there was no system of naming enzymes and the name did not give much idea about the type of reaction and the type of substrate acted upon e.g., pepsin, trypsin, ptyalin etc. However, still these names are of common use. Later on a system was developed to name the enzyme and the suffix 'ase' was used either after the name of the substrate (e.g., enzyme maltase, sucrase, proteinase and lipase acting upon maltose, sucrose protein and lipid, respectively) or after the name of the type of reaction (e.g., dehydrogenase, hydrolyase etc. conducting oxidation reaction and hydrolysis reaction, respectively).

In 1961, International unit of biochemists have adopted a new system of classification and nomenclature of enzymes. Dixon and Webb (1964); Karlson (1965) have outlined the details or rules for systematic and trivial nomenclature of enzymes.The name of enzyme consist of two parts. The first part denotes the name of the substrate and second part denotes the type of reaction catalyzed. Enzymes were divided into following six major classes.

1. **Oxidoreductases** : Enzymes catalyzing oxidation reduction are included in this class, e.g., cytochrome oxidase, alcohol dehydrogenases, catalase peroxidase etc.
2. **Transferases :** Enzymes responsible for transferring of a group other than hydrogen (e.g., amino, ethyl, formyl, acyl, phosphate, glyoxyl, thiol, ketol aldol etc.) are included in this class.
3. **Hydrolases** : Enzymes responsible for hydrolysis of substrate (e.g., ester, peptide, glycosidase, lipases, phosphatases, deaminases nucleases) are included in this class.
4. **Lysases** : Enzymes responsible for removal of groups from the substrate by the mechanism other than hydrolysis leaving double bonds in the substrate (e.g., fumarases, aldolase, decarboxylase) are included in this class.
5. **Isomerases** : Enzymes responsible for isomerism like substrate is converted into its positional or optical isomer by intramolecular rearrangement of various atoms or groups (e.g., aconitase, phosphohexoisomerase, triose phosphate isomerase) are included in this class.
6. **Lignases (Synthetases)** : Enzymes which catalyze linking together of compound utilizing energy made available by simultaneous breakdown of ATP or by similar high energy bond compound (e.g., Succinate thiokinase, Glutmine synthetase, Acetyl CoA carboxylase) are included in this class.

According to this system each enzyme is given a systematic code number which denotes class, subclass, sub-subclass and the name of the enzyme e.g., an enzyme hexokinase is denoted by code number 2.7.1.1, where 2 denotes class transferse, 7 denotes subclass transferase of phosphate, 1 denotes sub-subclass alcohol phosphate as acceptor and the last digit 1 denotes the name of the enzyme i.e. Hexokinase.

MECHANISM OF ENZYMATIC REACTION

Enzymes are highly specific in their catalytic activity. Two theories have been proposed to explain the catalytic activity of an enzyme. These theories are known as Rigid model and Flexible model.

Rigid Model : In 1894, Emil Fischer proposed a Lock and Key theory which is also known as rigid template model. According to this model enzyme combines with substrate to form enzyme substrate complex. This complex breaks down to give the product and enzyme.

$$E + S \rightarrow ES \rightarrow E + P.$$

Michaelis and Menton (1913) modified this theory and proposed that enzyme has some special points on its surface. These sites were called as active sites and they in turn have special groups that bind the substrate molecule. The substrate binds with this site and the site undergoes chemical changes to form enzyme substrate complex (Fig. 1). The number of substrate molecule changed per minute by a molecule of the enzyme is called turnover number.

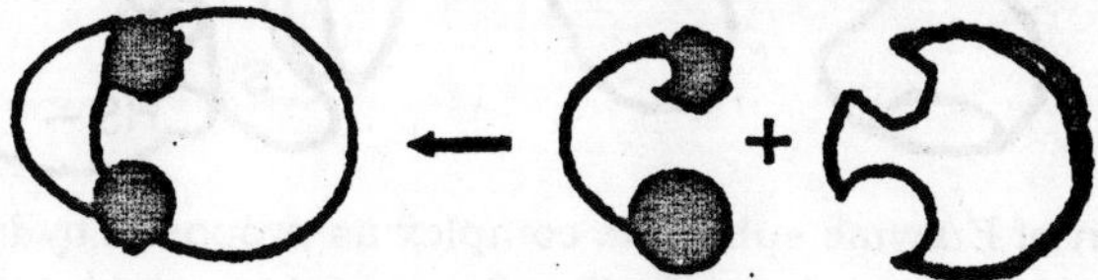

Fig. 1. Formation of Enzyme substrate complex (ES) and later ES complex into enzyme and product as proposed by Fischer.

Flexible Model or Induced Fit Modes : This model was proposed by Koshland. This model explains that the substrate induces a conformational change in enzyme (Fig. 2). This change occurs in the amino acid residue or in the prosthetic group of the holoenzyme. The conformational changes takes place in its correct special orientation for substrate binding, catalysis or both simultaneously other amino acid may become buried in the interior of the molecule. To explain the mechanism Koshland distinguished three types of amino acid residues in an enzyme. These three residues are —

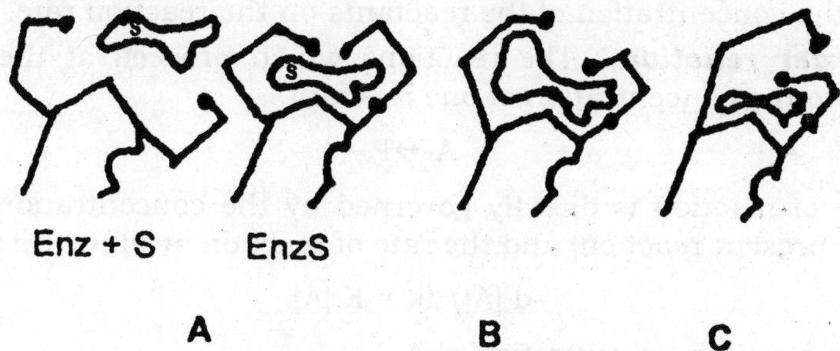

Fig. 2. Conformation changes in enzyme

(a) *Contact residue* : It includes both the catalytic and specific residue and designated to the amino acid residue with one bond distance (0.2 nm) of the substrate or other ligand concerned. The ligands are described as any small molecule bound to the enzyme by non covalent forces (activators, substrate and inhibitor).

(b) *Specificity residue* : The amino acid residue which takes part for binding the substrate with enzyme. It helps in subsequent catalytic process.

(c) *Catalytic residue* : An amino acid residue directly involved in covalent bond chain during enzyme action.

Koshland explained his model by enzyme Chymotrypsin (Serine protease). This enzyme has catalytic property on His-57 and Ser-195 present near each other in peptide bond. Beside these active site amino acid residue a covalent intermediate is formed involving OH group of serine 195 (Fig. 3).

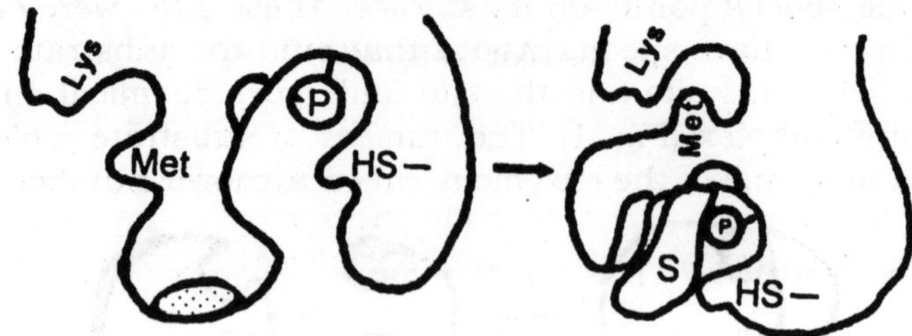

Fig. 3. Formation of Enzyme substrate complex as proposed by induced fit model (enzyme-Chymotrypsin)

CHEMICAL KINETICS

Chemical reaction may be classified on the basis of the number of molecules that take part in a chemical reaction to form its product. If only a single molecule takes part in the reaction and it is converted into product the reaction is known as Monomolecular, like wise the reaction may be Bimolecular and Trimolecular when two and three molecules take part in chemical reactions, respectively. On the kinetic basis, chemical reactions are classified into Zero, First, Second and Third order reaction. These reaction types are based on the influence of the concentration of the reactants on the reaction rate.

First order reaction : The reactions which proceed at the rate exactly proportional to the concentration of one reactant.

$$A \rightarrow P$$

The rate of reaction is directly governed by the concentration of substrate (i.e., A in the present reaction) and the rate of reaction at time 't' is presented as

$$-d[A]/dt = K[A].$$

where [A] is molecular concentration of A,

K is rate constant or specific rate reaction.

As the reaction proceeds the concentration of A decreases and hence the value of $-d[A]/dt$ also decreases. First order constant have the dimensions of reciprocal time.

Second order reaction : The reactions in which rate of reaction is proportional to the product of concentration of two reactants or the second power of a single reactant

$$A + B \rightarrow P$$

The rate of reaction may be presented as

$$-d[A]/dt. -d[B]/dt \text{ or } +d[P]/dt.$$

The second order rate equation is given by

$$-d[A]/dt = K[A][B].$$

If the substrate of the reaction is similar i.e., the reaction has form 2A →P, the second order rate equation is

$$- d[A]/dt = K[A]\ [A]$$

$$= K[A2]^2$$

The rate constant of second order reaction have the dimension 1/ (conc. X time) or M.

Third order reaction : The reaction in which rate is proportional to the product of three substrate concentration. In general such type of reactions are rare.

MICHAELIS-MENTEN EQUATION

In 1913, L. Michaelis and M.L. Menten proposed a general theory of enzyme action and kinetics. According to this theory the enzymatic reaction proceeds with the formation of an intermediate enzyme substrate complex. First the enzyme combines with the substrate to form an intermediate enzyme substrate complex. This forms product and enzyme is regenerated. These reactions are assumed to be reversible. Thus, the rate constant for the forward and reversible direction have positive and negative subscript, respectively. The reaction kinetics may be deduced as follows.

$$E + S \underset{k2}{\overset{k1}{—}} ES \rightarrow P + E \qquad (1)$$

The dissociation constant (Km) of the enzyme substrate complex is equal to the

$$K_M = \frac{[E]\ [S]}{[ES]} \qquad (2)$$

Where, [E], [S] and [ES] represents the molar concentration of E, S and ES, respectively.

If, the initial total concentration of enzyme is represented as [Eo], the concentration of free enzyme is [E] and the concentration of bound enzyme is [ES]. The equation 2 may be represented as following.

$$K_M = \frac{[Eo - ES]\ [S]}{[ES]} \qquad (3)$$

The rate of forward reaction will be

$$= [Eo - E.S]\ [S]\ K_1$$

The rate of backward reaction will be

$$= [E.S]\ K_2$$

The rate of dissociation of ES (i.e., formation of product) is equivalent to

$$= [\ E.S\ ]\ K_3$$

When the reactions are in equilibrium

$$[Eo -ES]\ \ [S]\ \ K1 = [E.S]\ K2 + [E.S]\ K3. = [E.S]\ [K2 + K3].$$

$$\frac{[Eo - E.S.]\,[S]}{[E.S.]} = \frac{K2 + K3}{K1} = Km$$

The constant Km is called as the Michaelis - Menten constant.

The Michaelis Menten equation expresses the mathematical relationship between the initial rate of an enzyme catalyzed reaction, the concentration of substance and certain characteristic of enzyme (Fig. 4 A and B).

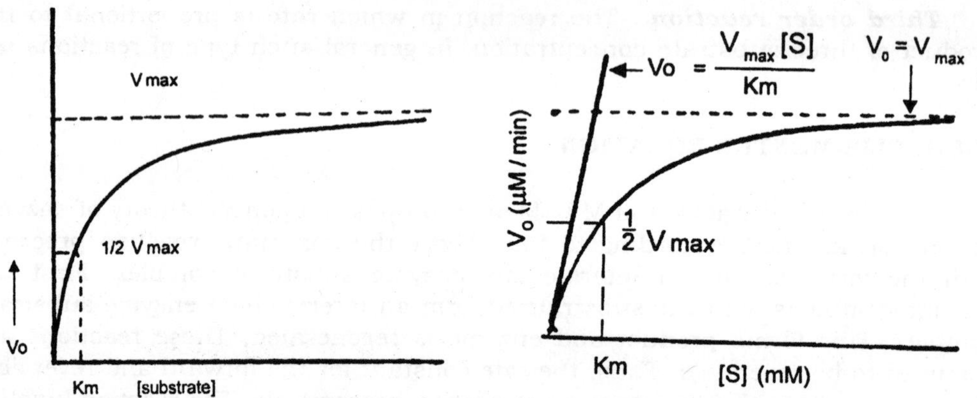

Fig. 4A. Effect of substrate concentration on enzyme reaction, 1/2 V max express Km (Michaelis-Menton constant)

Fig. 4B. The Kinetic parameter of enzymatic reaction

The substrate concentration that produces half maximal velocity the Km value or the Michaelis constant. It may be determined experimentally by graphing Vo as the function of [S]. Km has the dimension of molar concentration of S.

When S is approximately equal to Km, almost all enzymes have Km values. It is related to the physiological concentration of their substrate. At this stage Vo is directly related to the change in S and the enzyme is working at ½ maximal velocity. The Michealis-Menten expression will be

$$Vo = \frac{Vmax\,[S]}{Km + [S].}$$

This expression describes the behavior of many enzymes and as the substrate concentration will vary, it will directly effect the velocity of reaction. This effect and the relationship between [S] and Km at any concentration may be illustrated by evaluating Michaelis-Menten equation.

Lineweaver – Burk plot or Double–reciprocal plot

Km may be estimated using either the slope and y intercept or the negative x intercept. Since [S] is expressed in molarity or moles per litre, Velocity, v1, may be expressed in any units, since Km is independent of [Enz]. The algebrabical transformation of Michaelis-Menton constant can be expressed as following:

$$\frac{1}{Vo} = \frac{Km + [S]}{Vmax\ [S]}$$

or
$$\frac{1}{Vo} = \frac{Km}{Vmax\ [S]} + \frac{[s]}{Vmax\ [s]}$$

or
$$\frac{1}{Vo} = \frac{Km}{Vmax} + \frac{1}{[s]} + \frac{[1]}{Vmax}$$

This expression is known as Lineweaver – Burk plot equation (Fig. 5).

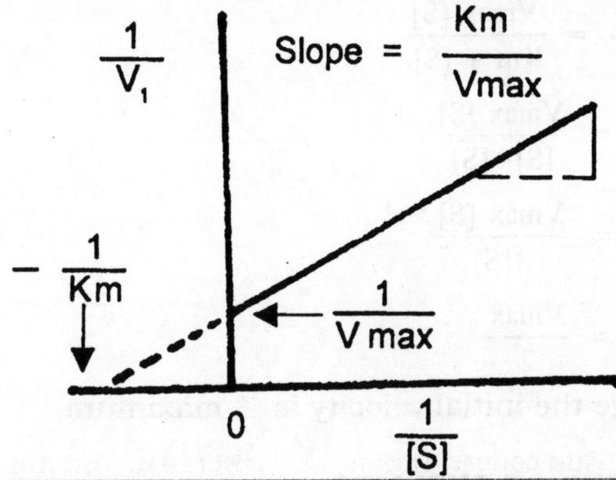

Fig. 5. Lineswaver - Burk plot of 1/V *vs.* 1/(S) used to evaluate Km and Vmax

The relationship between S and Km is discussed in following heads:

When the S is very much less than Km : If we add S to Km in the equation, it will make a very little change in the value and thus S can be dropped from the denominator. Since Vmax and Km are both constant, it may be replaced by a new constant K. By making such changes the expression will be presented as following :

$$Vo = \frac{Vmax\ [S]}{Km + [S].}$$

$$Vo = \frac{Vmax\ [S]}{Km}$$

$$= Vmax\ [S]$$

$$Vo = \frac{Km}{K[S]}$$

where K is a new constant equaling Vmax / Km because both Vmax and Km are constant for a particular enzyme.

Thus, when the substrate concentration is below the required amount to produce ½ maximal velocity, [Km value], the initial velocity depends upon the substrate concentration.

When S is much more the Km : At this stage if we add Km to S it will make a very little change in the value of denominator and the term Km may be dropped. The equation may be expressed as following —

$$Vo = \frac{Vmax\ [S]}{Km + [S].}$$

$$Vo = \frac{Vmax\ [S]}{[S]}$$

$$Vo = Vmax$$

Thus, the initial velocity is maximal.

When S = Km : The equation may be expressed as following :

$$Vo = \frac{Vmax\ [S]}{Km + [S].}$$

$$= \frac{Vmax\ [S]}{[S]+[S]}$$

$$= \frac{Vmax\ [S]}{2[S]}$$

$$= \frac{Vmax}{2}$$

Thus, at this stage the initial velocity is ½ maximum.

GENERAL PROPERTIES OF ENZYMES

Enzymes are protein molecule present in colloidal form in a cell. It exhibits all the general properties of a colloid. Some of the important general properties of enzymes are discussed here.

1. **Substrate specificity** : Enzymes are highly specific for their substrate they act on their specific substrate only.
2. **Effect of pH** : Every enzyme is optimally active at a specific pH. The enzyme activity can be measured as maximum at their specific pH. There may be a slight variation in the activity of enzyme at pH *in vivo* and *in vitro* conditions. In general, for all enzymes the pH ranges from pH between 5.0 to 9.0. However, pepsin shows its activity outside this range.
3. **Effect of temperature** : Each enzyme has its own liking of temperature and the temperature co-efficient (Q_{10}) varies from one enzyme to another. As the enzymes are protein molecule they are denatured at higher temperature and become inactive. This temperature may vary from 55 to 60°C. Enzymes of thermophilic bacteria are active upto 85°C.
4. **Effect of substrate concentration** : At a constant enzyme concentration an increase in substrate concentration increases the rate of enzymatic reaction. This increase is observed upto a particular concentration of substrate and above this concentration the rate may be inhibited. At the optimal concentration substrate fully occupies the catalytic site of enzyme.
5. **Denaturation** : When the protein molecule of enzyme undergoes conformational changes, it becomes denatured. Such denaturation may

be brought about by acid, high salt concentration, heavy metal salts, alkaloid reagent and even by UV light. Denatured enzyme shows loss of catalytic activity.

6. **Activators and Inhibitors** : A number of molecules or chemicals act as activator and increases enzyme activity enzyme looses its catalytic activity. Likewise some chemicals inhibit enzyme activity, they are known as inhibitors.

ENZYME INHIBITION

Some chemicals inhibit enzymatic activity. They are also known as negative modifiers. The inhibitors may be of physiological or unphysiological in nature. In general the substances which effect the chemical kinetics of enzyme activity and may be analyzed in terms of Michaelis - Menten equation are regarded as inhibitors. There are three types of inhibitors, viz., *Competitive, Uncompetitive* and *Non-competitive.*

Competitive Inhibitors : Substances which behave like substrate analog are regarded as competitive inhibitors. They react with enzymes at the catalytic site in a similar manner as the substrate reacts. During the reaction when both the substrate and inhibitor are present there is a competition between substrate and inhibitor to form either enzyme-substrate [ES] complex or enzyme-inhibitor [EI] complex. The quantity of formation of ES or EI depends upon the relative affinity of substrate or inhibitor with enzyme. The concentration of substrate and inhibitor also effects the formation of ES or EI. The inhibition constant Ki {the dissociation constant of enzyme inhibitor complex} may be formalized as following.

$$Ki = \frac{[E][I]}{[EI]}$$

Ki may be comparable to Ks, the dissociation constant of E.S complex.

An enzyme of TCA, Succinate dehydrogenase is inhibited competitively by malonate or other dicarboxylic anions.

Kinetic studies of the competitive inhibition have revealed that the apparent Km value on addition of I is changed. It becomes greater the Km value obtained without I. For quantitative analysis, the effect of varying substrate concentration [S] on the initial velocity (Vo) is determined with a constant concentration of inhibitor. This experiment is repeated at different concentration of inhibitor. In each experiment there is a fixed concentration of inhibitor with varying substrate concentrations. A plot of [1/Vo] *vs.* [1/S] is prepared for each concentration of inhibitor. These plots give a straight lines intersecting at a common intercept on 1/Vo axis. The apparent Km value for the substrate will be greater than the true Km value by the increase in the intercepts on the 1/S axis. (Fig. 6).

$$E + S + I \begin{cases} \nearrow ES \\ \searrow EI \end{cases}$$

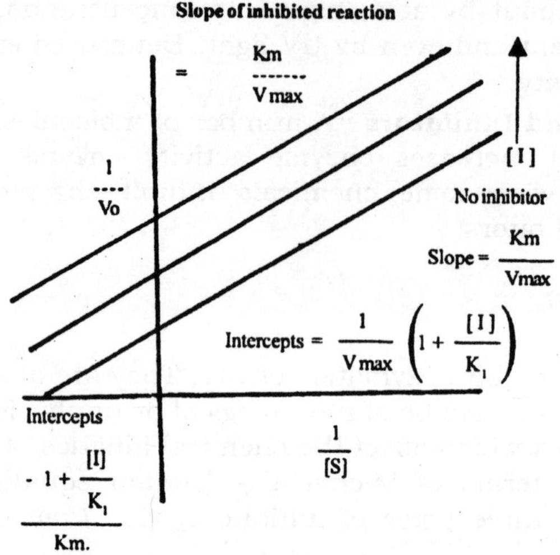

Fig. 6. Lineweaver - Burk plot to express competitive inhibition in enzymatic reaction

The relationship of the slope of the plot of uninhibited reaction is Km / Vmax and the slope of inhibited reaction can be expressed as following.

$$Km / Vmax (1 + [I] / [KI])$$

The slope is increased by the factor

$$1 + [I] / KI$$

Uncompetitive Inhibition : The inhibitor does not directly combines with enzyme but it combines with enzyme-substrate complex to forms enzyme-substrate-inhibitor complex. This results in an inactive complex. This type of inhibition may be explained as following

$$E + S \rightarrow ES + I \rightarrow ESI.$$

The inhibitor constant is thus –

$$Ki = \frac{[ES][I]}{[ESI]}$$

Such type of inhibition is common in two substrate reactions.

The plot of 1/Vo *vs.* 1/S at a constant concentration of inhibitor expresses that the slope of the plot remains constant at increasing concentration of inhibitor but Vmax decreases.

Non-competitive Inhibition : (Fig. 7). The non-competitive inhibitor binds with the enzyme and deforms the later. The binding is at any other surface than catalytic site. Due to such deformation rate of formation of enzyme-substrate complex is inhibited. The deformation may go upto the limit that even the increase in substrate concentration will not be able to reverse the enzyme.

When the enzyme is deformed, it is not reversed by increasing the substrate concentration. In non-competitive inhibition the reaction with inhibitor yields two inactive forms, EI and ESI

$$E + I \rightarrow EI; \quad ES + I \rightarrow ESI$$

Slope inhibited reaction

$$= \frac{Km}{Vmax}\left(1 + \frac{[I]}{K_i}\right)$$

[I]

No inhibitor

$$\text{Slope} = \frac{Km}{Vmax}$$

$$\text{Intercepts} = \frac{1}{Vmax}\left(1 + \frac{[I]}{K_i}\right)$$

$$\frac{-1}{Km} \qquad \frac{1}{[S]}$$

Fig. 7. Lineweaver - Burk plot to express reversible-noncompetitive inhibition

The two inhibitor constants

$$KI = \frac{[E][I]}{[EI]}$$

$$KIESI = \frac{[ES]\,[I]}{[ESI]}$$

ALLOSTERIC ENZYME

J. Monad, J.P. Changeux and F. Jacob coined the term allosteric enzyme. Allosteric means another surface. These enzymes have a surface other than catalytic site at which the modulator or effector bind reversibly and non-covalently. These are the regulatory enzymes. Allosteric enzyme are oligomeric with much larger molecular weight. They are made up of two or more polypeptide chains. They fail to confirm Michaelis-Menten equation.

Allosteric enzymes may show homotropic or heterotropic control. It is based on the nature of the modulator. In homotropic control the substrate functions as modulator, while in heterotropic control any substance other than substrate acts as modulator.

Most of the allosteric enzymes exhibit a typical dependence of the initial reaction velocity on substrate concentration. In a biochemical pathway more than one enzyme take part and thus it is a multienzyme system. The end product of the pathway influences the activity of enzyme catalyzing the first reaction of the pathway and thus regulates the pathway. The enzyme whose activity is controlled is known as allosteric enzyme. A simple example of such allosteric enzyme is L-threonine dehydratase. L-threonine is converted into L-isoleucine in a series of five biochemical reactions. L-isoleucin inhibits the activity of L-threonine dehydratase. This type of inhibition is also known as end-product inhibition or feed-back inhibition.

Kinetics of Allosteric Enzymes : Allosteric enzymes do not follow Michaelis-Menten equation. Variation in the concentration of modulator greatly alters the behaviour of these enzymes. Only a nine time fold increases in [S] is required to raise the activity from 10 to 90% maximal. Allosteric enzyme shows a sigmoid curve relating initial velocity to substrate concentration. Such sigmoid relationship is an example of positive cooperativity (Fig. 8). First the substrate binds with the catalytic site and it enhances the binding of subsequent molecules at the other site, which shows the sigmoid curve. It is important to note that all the allosteric enzyme do not show sigmoid curve and much more the enzymes showing sigmoid curve may not be necessarily allosteric in nature.

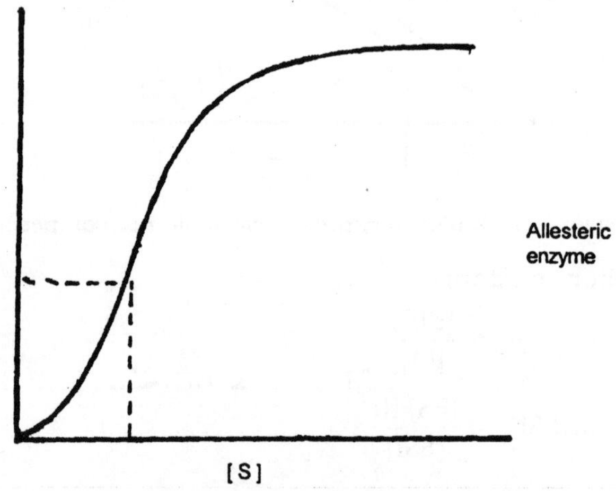

Fig. 8. Sigmoid Curve

ENZYME IMMOBILIZATION

Enzyme immobilization is a technique in which an enzyme is confined in a phase through which substrate and products moves freely in and out. It is economic for expensive enzymes where it is not to be purified after each enzymatic reaction. The support material used for enzyme immobilization is known as the Carrier Matrices. The latter are usually inert polymers or inorganic material. An ideal carrier matrices holds following properties :

(a) inertness; (b) physical strength under the applicable pH; (c) regenerability after the reaction; (d) enhancement of enzyme specificity; (e) reduced microbial contamination; (f) non-specific adsorption and (g) reduction in product inhibition.

Methods of Enzyme Immobilization

There are five different techniques for enzyme immobilization —

1. Adsorption of the enzyme on a carrier surface;
2. Covalent coupling of an enzyme to a carrier surface;
3. Crosslinking between enzyme molecules (copolymerization);
4. Entrapment of an enzyme in a matrix; and

5. Encapsulation or confinement of an enzyme solution in a membrane structure.

1. Adsorption : It is the most economical and a simple process in which enzyme molecule adheres on the surface of carrier matrices (Fig. 9). The nature of bond between the enzyme protein and the carrier matrices depends upon the nature of the protein surface and the matrices. The bonds involved in adsorption may be ionic, hydrogen, co-ordinate-covalent, covalent or hydrophobic. There may be combination of any two of these bonds. These bonds are very strong and help in tightly binding the enzyme protein to the carrier matrices. However, it may be weakened during use by many factors e.g., addition of substrate, pH or ionic strength. If the enzyme is immobilized externally the carrier particle size must be very small in order to achieve appreciable surface by bonding. These particle may have diameter which range from 500 Å to 1 mm. Carrier matrices commonly used in adsorption are ion exchange matrices, porous carbon, clays, hydrous metal oxides, glasses and polymeric aromatic resins. Out of these ion-exchange matrices are costly but they can be readily regenerated by a simple operation, like washing the adsorbed enzymes with concentrated salt solutions.

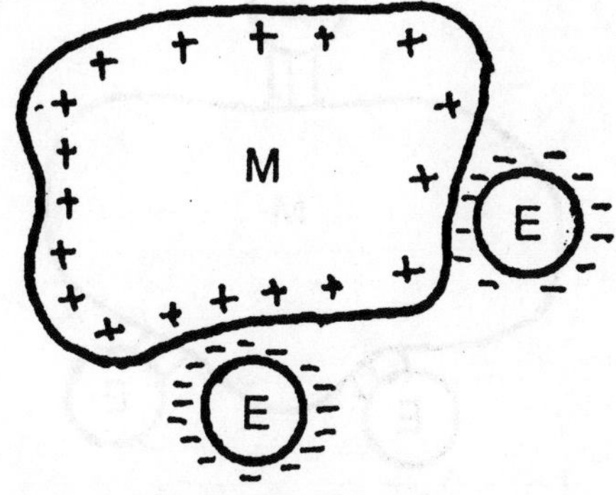

Fig. 9. Adsorption

There are **four** methods of enzyme immobilization by adsorption. They are described as following —

(i) The Static process In this process the carrier is kept in contact with enzyme solution. Agitation or stirring is not performed for adsorption of enzyme to be immobilized. This process is not adopted due to its inefficiency.

(ii) The Dynamic process In this process carrier is kept in enzyme solution. They are mixed with agitation or stirring in shaker. This procedure is adopted for laboratory preparations of immobilized enzymes.

(iii) The Reactor loading process In this process the carrier is placed into the reactor and enzyme solution is transferred to it. The enzyme is circulated so as to make the carrier in a dynamic state or both the enzyme solution and carrier are agitated. This method is common for industrial purposes.

(iv) The electrodeposition process In this process carrier is placed proximal to one of the electrodes in an enzyme bath. Under electric current enzyme migrates to the carrier and is deposited upon the surface.

Advantages

(i) Little or no damage to the bio catalyst.
(ii) Simple, cheap and quick to obtain immobilization.
(iii) No chemical changes required for the support of enzymes.
(iv) Easily reversed to allow regeneration with fresh enzymes.

2. Covalent binding : In this technique enzyme molecule is allowed to bind with carrier matrices by covalent bonds (Fig. 10). The strength of binding is very strong and hence this method is useful under a broad spectrum of pH, ionic strengths and uncontrolled variable conditions. The side chains of amino acids take part in covalent bond formation. The specific groups like amine, hydroxyl, tyrosyl etc. present on the surface of the enzyme molecule directs covalent attachment, e.g., carbohydrate carrier is generally through a hydroxyl group. The cyanogens bromide coupling enzyme to carbohydrate involve the attachment of cyano group to two adjacent hydroxyls on the surface according to the following manner.

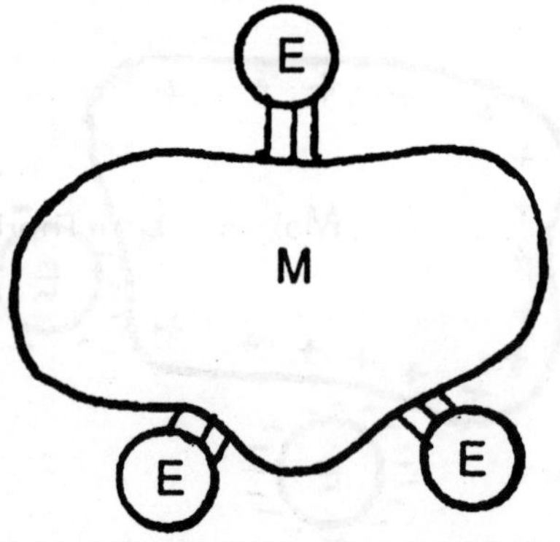

Fig. 10. Covalent Binding

One of the earliest procedures employed for attachment of carbohydrate was that of azide coupling. Incarodiamide procedure the attachment is through carboxylic group, which may be formed on cellulose in a manner similar to the first reaction under azide coupling.

$$\begin{matrix} -OH \\ -OH \end{matrix} + BrCN \rightarrow \begin{matrix} -O \\ -O \end{matrix}\!\!>C = NH \xrightarrow{NH_2-E} \begin{matrix} -O-\overset{\overset{O}{\|}}{C}-NH-E \\ -OH \end{matrix}$$

Protein and amine bearing carriers are also used for enzyme immobilization. Difunctional reagents have been employed in order to bind selectively to tyrosyl group. Diazdobenzidines and its derivatives are utilized in this application. Glutaraldehyde forms a Schiff's base with amine group. Isocyantes form substituted urea bond with amine under alkaline conditions.

Silane { γ-amino propyl trimethoxy silane } is utilized as inorganic carrier for co-valent enzyme immobilization. Silane is attached inorganic functional group to the inorganic surface leaving the organic moiety like amine. This amine takes part in covalent enzyme immobilization.

```
   |                           |          |
   O                           O          O
   |                           |          |
—O—Si—OH                 —O—Si—O—Si
   |                           |          |
   O      EtO)3 Si(CH2)3NH2    O          O      + EtOH
   |                           |          |
 O—Si·  OH                —O—Si—O—Si(CH2)3NH
   |                           |          |
   O                           O          O
```

Agrose, cellulose, polyacrylamide, saffrose are most commonly used matrices. Agrose beads are highly hydrophilic and is generally inert to microbial attack. It is activated by the treatment of CNBr. Enzyme molecule binds to these activated groups. As cyanogens bromidé is highly toxic hence it is not used commercially. However, it is used for research purposes. Saffrose is commercially used. It is activated with chloroformates, carbodiimides, glutaraldehyde or other compounds.

In covalent immobilization technique the active site of enzyme is utilized for immobilization. Hence, there is a loss in enzyme activity.

3. Cross linking between enzyme molecules or copolymerization : This type of immobilisation (Fig. 11) is support free and involves joining the cells (or the enzymes) to each other to form a large three dimensional complex structure and can be achieved by chemical or physical methods. Chemical methods of cross linking normally involve covalent bond formation between the cells by means of a bi – or multifunctional reagent e.g., glutaraldehyde and toluene diisocyanate. Physical crosslinking of cell by flocculation is well known in biotech industry and lead to high monomers are insoluble preparations used in continuous reactor. Enzymes immobilized by copolymerization amongst the enzyme molecules or with other preparations are gummy materials or gels and difficult to operate. Copolymerization of enzyme in a membrane configuration is utilized in a spiral-wound reactor module. The later is used as a packed-bed reactor.

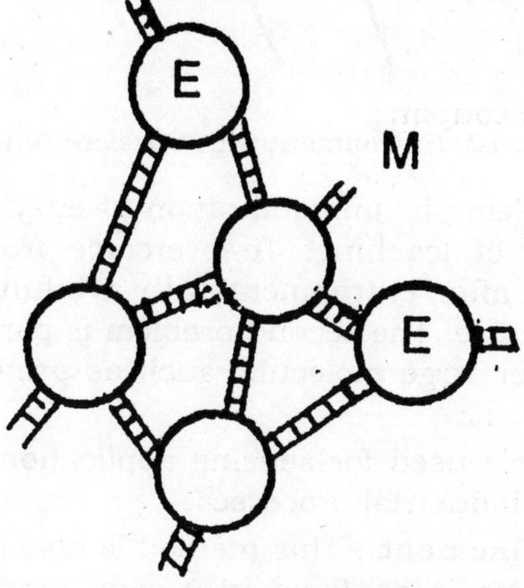

Fig. 11. Cross Linking

In 1960 Bar-Eli and Katchalski immobilized trypsin by copolymerizing it with ethylene and maleic anhydride.

4. Entrapment : In 1963, Bernfeld and Wan made a pioneer work on this technique. The carrier matrix used was polyacrylamide. In 1965, Bauman and coworkers used starch for immobilization of cholinesterase. Vieth and Venkatasubramanian (1976) used collagen as matrix for enzyme immobilization and silicone rubber was used as matrix by Pennington and coworkers.

In this technique enzyme molecules are held or entrapped within suitable gels or fibers (Fig. 12). It is different from crosslinking of enzymes and copolymerization of enzymes in that the enzyme in not chemically part of the matrix. The binding between the enzyme and matrix may be of non-covalent or covalent type. A non-covalent type of binding may be compared as putting the enzyme in a matrix as an animal or bird is caged in a net. For covalent binding the enzyme is usually treated with a suitable reagent. Acryloyl chloride is used to prepare lysine residues for binding by forming acryloyl amides. Following this it is co-polymerized and cross linked with acrylamide and bis acrylamide to form a gel containing the entrapped enzyme. This entrapped enzyme may be used to form small beads or a film on a solid support. For the entrapment of enzyme in cellulose acetate fiber, an emulsion of enzyme and cellulose acetate is prepared in methylene chloride. This emulsion is extruded as fibers into a solution of an aqueous precipitant. A common method of entrapment is by using calcium aligenate as matrix.

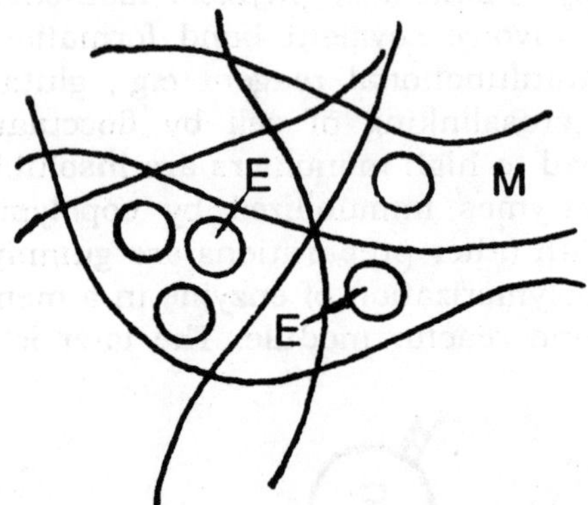

Fig. 12. Entrapment with Covalent binding

There are few problems in immobilization of enzymes by entrapment. The major concern is that of leaching. To overcome from this problem simple crosslinking of enzyme after entrapment with a bifunctional reagent such as glutaraldehyde is applicable. The second problem is pore diffusion limitations. If the substrate is a rather large molecule, such as protein, it may be restricted from entry into the pore.

Entrapment is widely used for sensing applications, it has not achieved a great deal of success in industrial processes.

5. Membrane confinement : This method is also known as encapsulation. In this technique enzyme is confined in a semi permeable membrane in an aqueous solution. This method is helpful for a free movement of substrate and product in either direction while the enzyme remains confined in the membrane. In 1964, Chang made a pioneer work in this technique but was not successful for industrial purposes. Later in 1977, Chang developed a few more methods out of which encapsulation in membranes of cellulose nitrate and nylon are widely used. Membrane confinement can be distinguished from entrapment by the fact that a solution of enzyme is separated from the bulk of solution by membrane encapsulation while entrapment involves a lattice immobilized enzyme. The general techniques of membrane confinement are as following :—

(i) A semipermiable membrane is placed in a vessel, separating it into two chambers. One chamber contains enzyme while the other contains substrate and product.

(ii) Enzyme is placed in hollow space of fiber membrane units. These membranes are submerged in substrate. This technique is useful for only those substrates, which are smaller than the enzyme molecules.

(iii) Enzymes may be packed in micro capsules. These are prepared by polymerization reactions. 1,6 diaminohexane is a common chemical used in such type of polymerization.

The encapsulation technique developed by Chang (1977) is as following:

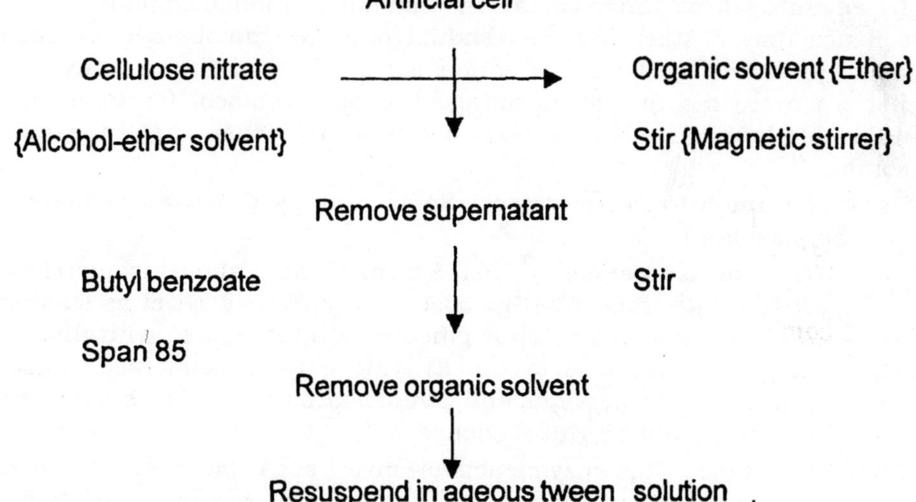

This technique was adopted when a combination of urease and ammonia adsorbent was encapsulated and administered to gastro-intestinal tract. Blood urea levels were significantly reduced utilizing this procedure.

Ehler *et al.,* (1973) proposed another technique for encapsulating enzyme in which RBCs are used as encapsulating membrane. The cells are swollen in hypotonic solution, which results in formation of holes in the membrane. The size of holes is sufficiently large to permit the enzyme molecules to enter into it. These swollen cells are loaded with enzyme solution so that the enzyme molecule enters into it. The enzyme containing cells are then kept in isotonic solution, which reseals the holes.

Effects of enzyme immobilization

There is a change in the kinetic behaviour of immobilized enzymes. Different enzymes show different effects to the same immobilization protocol. Enzyme kinetics may be effected as the immobilization procedure may bring about change in conformation of enzyme molecules or due to the influence of matrix. Immobilization may increase or decrease enzyme stability. Immobilized enzymes may be more prone to inactivation by higher temperature or pH. In contrast, binding of an enzyme molecule at several points without creating any strain in enzyme molecule may lead to substantial stabilization.

Advantages of immobilization : It is important that to recover an enzyme after the reaction. Generally the costlier enzyme are immobilized. Immobilization helps in repeated use of enzymes. As the substrate and product can freely move in the immobilized system, it is easy to free the product from the enzyme. Thus, the cost of down streaming process of the product is saved. Another important advantage of immobilized enzyme is that the enzyme can be used in non-aqueous system. Other advantages of immobilization are that immobilized enzymes may be used in continues process. Enzymes may be used at much higher concentration. Thermostability of some enzymes is also increased. Recovery of enzyme may also reduce effluent handling problems.

Disadvantages of immobilization : Immobilization techniques are costly affair hence only costlier enzymes should only be immobilized for economic purposes. In some cases immobilization effects the stability or activity of the enzyme. To overcome this disadvantage a proper protocol for immobilization should be developed. Immobilization is not applicable when one of the substrate is insoluble.

Uses of immobilized enzymes. Some examples of uses of immobilized enzymes are as following :

(i) **Glucose isomerase** : This enzyme bring about isomerization of glucose into fructose. D-glucose is only 70% as sweet as sucrose and less soluble in water. Thus glucose syrup is treated with this enzyme to produce fructose syrup. Enzyme glucose isomerase is obtained from *Bacillus coagulans* and *Streptomyces* spp. It is immobilized with cross linking with glutaraldehyde.

(ii) **Invertase** : This enzyme acts on invert sugar (sucrose). It is obtained from autolysis of yeast cells. It is immobilized by adsorption on to bone charcoal and sterilizing with cross linking. One kg of immobilized granular enzyme produces 16 tons of inverted sugar.

(iii) **Lactase** : This enzyme acts upon lactose sugar. It is applied in processing of milk and sweet whey. The enzyme is obtained from yeast (*Kluyveromyces lactis*). It is immobilized in cellulose tri acetate fibers.

(iv) **Raffinase** : This enzyme acts on raffinose sugar. This enzyme is used to remove raffinose and stachyose from soyabean milk as these sugars cause flatulence. The enzyme is obtained from *Mortierella vinacea*.

(v) **Penicillin amidases** : This enzyme is also known as penicillin acylases and is used in the production of penicillin and cephalosporin. It is obtained from *E. coli,* and is immobilized in a variety of ways. The general method of immobilization is by covalent linkage with cyanogens bromide.

EXCLUSION CHROMATOGRAPHY OR GEL FILTRATION

The principle of this technique is based on the separation of the sample component according to molecular size, hence also known as molecular size chromatography. The basic procedure is that a mixture of enzyme is dissolved in a suitable buffer and is allowed to flow by gravity down a column packed with beads of an inert, highly hydrated polymeric material. The solid support material or gel is a polysaccharide derivative. It may be Sephadex, Agarose or Polyacrylamide derivative-Bio-Gel. The column can be prepared with different degrees of internal porosity. In molecular sieve chromatography separation are carried out on natural and synthetic zeolites {meta alumino silicates} called molecular sieve. In gel chromatography the granulated or beaded gel material is known as packing material. There are stationary phase and mobile phase. The stationary phase is the solutes distributed throughout the entire gel phase while the liquid flowing through the bead is mobile phase. The mixture of protein contains protein molecule of different size and thus their movement in the column takes place. In the column, proteins of different molecular size penetrate to the internal course of the beads to different degree and they travel down the column at different traits. Very large protein molecules are unable to enter the pores of the beads and are called as excluded molecules. They represent the excluded volume. The later is defined as the volume of aqueous phase outside the beads. The protein molecules entering the pores of the packing material may be small as well as large in size. The small proteins molecules can enter the pores of the beaks freely and are retained by the column. The large protein molecules cannot enter the hydrated polymer particle and thus they pass through the column rapidly. Protein molecules between these sizes will exclude from the beads to a degree that depends on their size. The term exclusion chromatography is used due to this distinction of flow of protein molecule through the column.

Technique : The apparatus used for exclusion chromatography is simple made up of a straight tube and a bead support at the bottom. The bead support allows only the mobile phase to pass through without disturbing the packing material. The total volume of column packed with a gel is as following :

$$Vt = Vg + Vl + Vo \qquad (1)$$

Where, Vg = volume occupied by the solid matrix of the gel

Vl = volume of solvent held in the pores or interstices

Vo = free volume outside the gel particles.

Molecules, which are neither too large nor too small i.e., of intermediate size may penetrate with a fraction Kd of the interstiantly held solvent.

The exclusion volume (Ve) can be represented as—

$$Ve = Vo + Kd \cdot Vl. \quad (2)$$

The behaviour of a gel column which restrict to all solvent can be described as Kd = 0 and thus Ve = Vo for protein molecules which are too large to enter the pores. The equation for protein molecules that can penetrate the pores without any obstacle is —

$$Kd = 1 \text{ and } Ve = Vo + Vl \quad (3)$$

Application of exclusion chromatography : This type of chromatography is widely used for mixtures of sugars, polypeptides, liquids, polythenes, butyl rubbers etc.

For the separation of protein molecules Sephadex is generally used as packing material. Different types of Sephadex is used in the separation of proteins of different molecular weight. Some examples are as following :

Sephadex types	Molecular weight of proteins, which can be separated
G-25	3500 – 4500
G-50	8000 – 10,000
G-75	40,000-50,000
G-100	1,00,000
G-200	2,00,000.

For most of the naturally occurring low molecular molecules, Sephadex G-25 and G-50 are used as packing material. It forms a tight linking with small pore size. Sephadex G-25 is used to separate salts and amino acids from protein mixture. Sephadex G-75 is used in the fractionation and purification of proteins, polysaccharides and nucleic acids.

ENZYME LINKED IMMUNO SORBENT ASSAY (ELISA)

This is a latest test in diagnosis of disease (antigen and haptans). In this test antigen antibody reaction is detected keeping one of the components quantified by enzyme labelled counterpart and subsequent demonstration of fixed enzyme by its substrate. In ELISA, estimation of different antigens is made by allowing them to react with appropriate antibody to which an appropriate enzyme molecule has been coupled so that the enzyme activity would help in assess that how much of the antigen concentration is there and hence the antigen concentration. It is a simple technique utilizing only a micro liter of test reagent. This technique is widely used for the detection of hormones, toxins and virus. Two types of procedures are used viz., competitive saturation analysis and non-competitive saturation analysis.

1. Competitive Saturation Analysis : The principle of this procedure is same as RIA {Radio Immune Assay}. The antigens labelled with enzyme and unlabelled antigens are incubated with the limited amount of antibodies.

(a) *Equilibrium Technique*: In this technique all the reagents are incubated together to achieve equilibrium.

(b) *Sequential Saturation or Non Equilibrium Technique*: In this technique unlabelled antigen is incubated with antibodies prior to the addition of labelled antigens.

2. Non-Competitive Saturation Analysis. In this technique labelled antigens behave differently than the unlabelled antigens. Binding of labelled antigen to antibody results in steric hindrance to enzyme activity. The enzyme activity is blocked or reduced. Labelled antibody antigen complex so called homogenous EIA.

GENERAL PROCEDURE FOR ELISA TEST

To set an ELISA test there are few steps. A test needs preparation of microtiter plates, preparation of sera, addition of conjugate and substrate and finally the reading of results.

For ELISA test preparation and washing of microtitre plates is important. For the preparation of microtiter plates polystyrine microtiter plates are used. The antigens are diluted in carbonate bicarbonate buffer pH 9.6 optimal dilution of each antigen is established by using various dilutions in a chequer board titration method. 200 ml of antigen solution is dispensed into each well. The plates are sealed and incubated at 4°C overnight.

For washing of microtiter plates first the antigen solution is removed from the wells of microtiter plates by inverting the plates. The wells are then filled with washing buffer and incubated at room temperature for 5 minutes. The buffer is removed from the wells of the plate and these are then filled with fresh buffer and incubated for 5 minutes. This process is repeated for three times.

All sera are diluted in PBS containing 0.5% BSA. The sera is added in the well after proper washing of the antigen senitised plated. Plates are washed with washing buffer and 200 ml of optimal dilution of test serum is then dispensed into each well of microtiter plate. The covered plates are incubated at 37°C for one hour in a moist chamber. After incubation plates are again washed and conjugate is added. The conjugate is Anti Human IgG Goat tagged with Horse. Radish peroxidase is diluted in PBS-T with 0.5% BSA. The dilution, which can be detected optically is gained by performing a chequer board titration using 100 mg/ml human IgG. 200 ml of diluted conjugate is added to each well of the plate and these are incubated at 37°C for one hour in a humid chamber. When the incubation period is completed the plates are again washed for three times and dried. This step is followed by addition of substrate O-phenylene diamine in phsophate citrate buffer pH 5.0 with H_2O_3. After this the plates are incubated for 30 minutes in dark. To stop the enzyme substrate reaction, 50 ml of 2.5 M H_2SO_4 is added in each well of the plate. At the end of the test, colour developed in each well is recorded by a photocolorimeter (a ELISA reader).

METHODS FOR ASSAY OF ANTIBODY

There are following methods for the detection of anitibody :

1. **Indirect ELISA method** : In this method a single conjugated enzyme linked antiglobulin is used to measure all antibody of a species. Sensitized wells are prepared by lining the wells of microtitre plates with antigen. A serial dilution of patients serum is poured in these wells. To these wells enzyme labelled anti-globulins is added and after a fixed

time enzyme activity is stopped. Colour intensity is recorded with the help of a photocolorimeter. If the reaction is not stopped, kinetic measurement of substrate breakdown can be made.

2. **Assay of specific IgM Antibody** : In this method, solid phase is coated with anti-IgM. This enables to fix IgM of patient serum. After the treatment of patient serum enzyme labelled antigen or unlabelled antigen is treated with it. It is followed by enzyme linked antibodies to that antigen. The fixed enzyme is assayed by its activity. Measurement of enzyme activity is carried out with the help of photocolorimeter. This method is useful for impure antigen.
3. **Sandwich method** : In this method first known antigen is coated on the solid phase. This is followed by treatment with patient serum and after this enzyme labelled antiglobulin is treated and in the last again treated with substrate. This method is useful to detect antigen and antibody.

METHOD FOR ASSAY OF ANTIGEN

In this method, solid phase is treated with specific antibodies. The excess antibody is washed. After this it is treated with sample, which may contain antigens. It is followed by enzyme linked specific antibody or specific antibody and the corresponding enzyme linked antiglobulin. This method is modified in which a simultaneous test is performed in which test sample is first mixed with known amount of specific antibody. This will reduce the amount of antigen fixed. It will ultimately result in the colour production by substrate. The concentration of antigen in the test sample is estimated by the difference in the colour production between the test sample and test sample treated with antibody.

CHAPTER 10

PHOTOSYNTHESIS

Photosynthesis is a process by which, plants and certain bacteria convert radiant energy (in the form of light) into metabolic energy and reducing power. Photosynthesis is defined as the conversion of light energy to biochemically utilisable energy (ATP) and reducing power ($NADPH+H^+$) in the cells of phototropic organisms. Phosphorylation and photosynthetic pyridine nucleotide reduction are the elementary processes of photosynthesis.

The photosynthesis in microorganisms may be oxygenic or anoxygenic. In oxygenic photosynthesis, microbes use water and carbon di oxide as raw material. Oxygen is evolved as a result photolysis of water, hence it is named oxygenic photosynthesis. The cyanobacteria perform oxygenic photosynthesis and possess a pigment system similar in basic respect to that of photosynthetic eucaryotes. Their pigment system includes chlorophyll a, carotenoids and phycobilin. The cyanobacteria utilise water as hydrogen donor and evolve oxygen in the light. There is a group of photosynthetic eubacteria which do not use water as hydrogen donor. Oxygen is not evolved, hence this type of photosynthesis is known as anoxygenic photosynthesis. These bacteria utilize H_2S, H_2 or organic compounds as hydrogen donor. The photosynthetic eubacteria are Gram-negative to include two distinct groups purple bacteria and the green sulphur bacteria.

W. Engelmann (1885) was first to suggest that the purple bacteria might perform photosynthesis. He discovered that these bacteria are phototactic and their growth is favoured by light. S.Winogradsly showed that some purple bacteria can oxidize H_2S to sulphate. In 1930, C.B. van Niel first recognized and defined the various metabolic versions of anoxygenic photosynthesis. He demonstrated the characteristic mode of energy yielding metabolism in both purple and green bacteria.

Order Rhodospirillales

Order Rhodospirlillales (the photosynthetic bacteria) is divided into two classes Rhodospirillineae (purple bacteria) and Chlorobiineae (green bacteria).

Rhodospirillineae In all the members of Rhodospirillineae, the whole photosynthetic apparatus is present on the intracytoplasmic membranes. They all contain bacteriochlorophyll a (BChl a) and fix carbon di oxide *via* the ribulose-di-phosphate cycle. They utilize organic compounds as hydrogen donor. This class is divided into two families Chromatiaceae formerly Thiorhodaceae (purple sulphur) utilize hydrogen sulphide as hydrogen donor and Rhodospirillaceae formerly Athiorhodaceae (purple non-sulphur bacteria) do not utilize hydrogen sulphide as hydrogen donor.

Family Chromatiaceae (Fig. 1) *Chromatium, Thiospirillum, Thiocyctis* and *Thiocapsa* are included in this family. *Chromatium okeniiis* 5 μm in diameter and 20 μm in length, contains intracellular highly refractile sulphur granule. *C. warmingii* has smaller dimensions with polar location of its sulphur globules. The *Thiocapsa* are also spherical but non-motile. Several species included in this family contain gas vacuole. The important characteristic of all these bacteria is that they have intermediary deposition of intracellular sulphur on the outside of the cell, where it is further oxidized extracellularly. The deposition of sulphur is due to oxidation of H_2S.

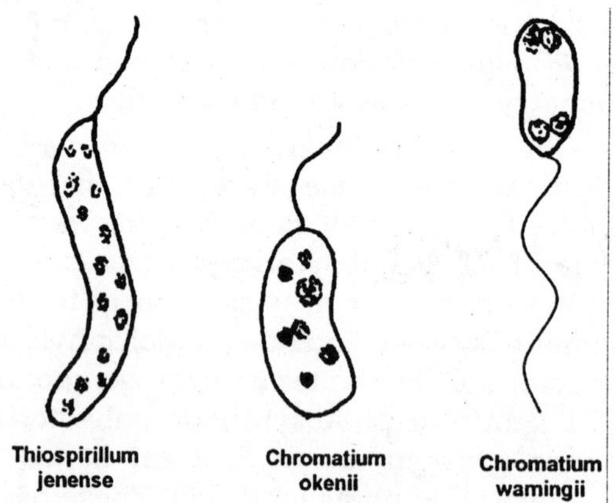

Fig. 1. Members of Chromatiaceae

Family Rhodospirillaceae (Fig. 2) *Rhodopseudomonas, Rhodospirillum, Rhodomicrobium* and *Rhodocyclus* are included in this family. The thylakoids have different shapes. *Rhodomicrobium vannielii* multiplies by budding. *Rhodocyclus purpureus* is non-motile and its cells are semicircular. Growth of most non-sulphur purple bacteria is inhibited by hydrogen sulphide. However, some species can tolerate or even use it as hydrogen donor for the fixation of CO_2.

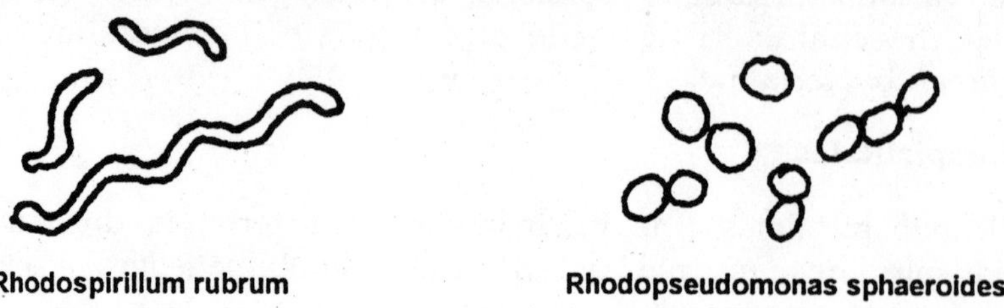

Fig. 2. Members of Rhodospirillaceae

Class Chlorobiineae Bacteria included in this class possesses pigmented organelles, the so-called chlorosomes. They contain small amount of BChl *a* and

BChl *c, d,* or *e* as antenna pigments. BChl *a* is directly connected to the photosynthetic reaction center and is localized in the cytoplasmic membrane. The difference in Chlorobiineae and Rhodospirillineae is by the absence of the enzyme ribulose-di-phosphate carboxylase and hence unable to fix carbon-di-oxide.

Chlorobniaceae (Fig. 3) *Chlorobium* and *Chlorochromatium* are included in this family. *Chlorobium vibirioforme* and *C. limicola* are green in colour while *Chlorobium phaeobacterroides* is brown in colour. *Chlorochromatium aggregatum* is a symbiotic association of two bacterial species : a colourless, chemo-organotrophic rod, motile by a long polar flagellum has attached to its outer surface a green, phototropic bacteria ectosymbionts.

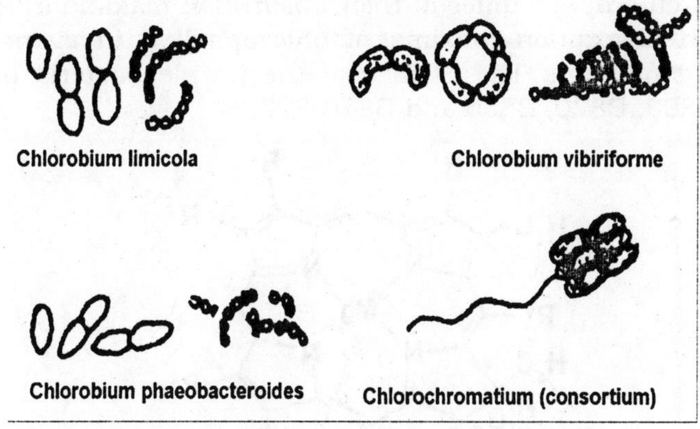

Fig. 3. Members of Chlorobniaceae

Chloroflexaceae e.g., *Chloroflexus* is a filamentous gliding bacteria. It contains bacteriochlorophyll *c* and *a.* It possesses chlorosomes corresponding to those of the green sulphur bacteria. *Chloroflexus* differs from *Chlorobium* species, by its aerobic nature and its heterotrophic growth in complex media in the dark as well as in the light. The Chloroflexaceae are more nearly photoheterotrophs.

Table 1. Characteristics of the major groups of Photosynthetic bacteria

Characteristic	Bacterial Group			
Pigments	Bacterio-chlorophylls a plus c, d, or e	Bacterio-chlorophylls a or b	Bact. Chl. a & b	Chlorophyll a plus phycobilin proteins
Photosynthetic electron donors	H_2, H_2S, S	H_2, H_2S, S	Usually organic molecules	H_2O
Sulphur deposition	Outside the cell	Inside the cell	—	—
Nature of photosynthesis	Anoxygenic	Anoxygenic	Anoxygenic	oxygenic

PHOTOSYNTHETIC PIGMENTS AND PHOTOSYNTHETIC APPARATUS

Basic molecular organisation of the photochemical apparatus is the same in all photosynthetic bacteria. It consists of a reaction center, the antenna pigments with an electron transport system.

Chlorophyll The major group of photosynthetic pigment is chlorophyll (Fig. 4). The chlorophylls are cyclic tetrapyrroles within which a magnesium ion is chelated. The chlorophylls of purple and green bacteria are known as Bacteriochlorophylls. Pheophytins are chlorophylls without a chelated magnesium atom. The various chlorophyll types (Table 2) differ from each other chiefly by the presence or absence of a double bond between carbon atoms 3 and 4 and by the substituents on the porphyrin skeleton. Due to these differences, the different types of chlorophyll differ in their absorption maxima in the near infrared region. Main absorption maxima of chlorophyll a in cyanobacteria, lies between 680 and 685 nm. The BChl *a* of the purple bacteria occurs in four spectral forms B800, B820, B850 and B870-890.

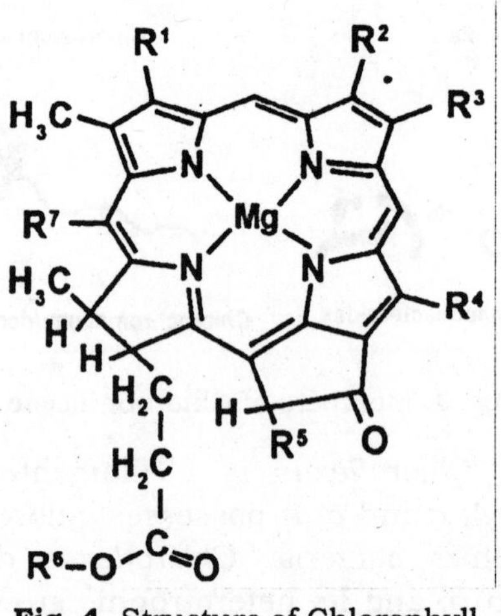

Fig. 4. Structure of Chlorophyll

Table 2. Relation between Chlorophyll *a*, Bacteriochlorophyll *a* and Bacteriochlorophyll *b*.

Pigment	R 1	R 2	R 3	R 4	R 5	R 6	R 7
Chlorophyll a	$-CH=CH_2$	$-CH_3$	$-CH_2-CH_3$	$-CH_3$	$O=C(OCH_3)-$	Phytol	-H
Bacterio-chlorophyll a	$O=C(OCH_3)-$	$-CH_3$	$-CH_2-CH_3$	$-CH_3$	$O=C(OCH_3)-$	Phytol or geranyl-geranol	-H
Bacterio-chlorophyll b	$O=C(OCH_3)-$	$-CH_3$	$=C(OCH_3)H$	$-CH_3$	$O=C(OCH_3)-$	Phytol	-H

Both chlorophyll *a* and the bacteriochlorophyll *b* are synthesized from 5-aminolevulinate (ALA) by substantially the same biochemical pathways. In purple and green bacteria, ALA is a product of a condensation between glycine and succinyl-CoA, whereas in cyanobacteria and the chloroplasts of eukaryotes, it is formed from L-glutamic acid.

The phycobiliproteins are the major components of antenna pigments of cyanobacteria. They are water-soluble proteins that contain covalently bound linear tetrapyrroles (bilins) as chromophores. The cyanobacteria contains two blue pigments, allophycocyanin and phycocyanin, which have absorption maxima at relatively long wavelengths. Phycobiliproteins can constitute about 60 per cent of the cell's soluble proteins.

Carotenoids are the so-called accessory pigments of photosynthetic organisms. The absorption maxima is 400-550 nm. In purple bacterium the carotenoids have a tetrapenoids (C40) with tertiary hydroxy or methoxy groups as the basic structure. The deep colour may be due to the presence of oxo or aldehyde group. Carotenoids with aromatic rings derived from carotenoids are also found. Carotenoids perform two functions in photosynthesis. First they may serve as light-harvesting pigments and belong to the antenna pigments, which channels energy to chlorophyll. The second function is, their indispensable role as quenchers of chlorophyll-catalysed photoxidation, thus they shield the chlorophyll and protect it from photoxidation.

The photosynthetic pigments in purple bacteria are bound to intracytoplasmic membranes. The membranes or thylakoids originate as vesicular or tubular invaginations of the cytoplasmic membrane into the cell interior and remain connected. The vesicular membrane fragments that can be isolated from broken cells by differential centrifugation are called 'chromatophores'. In Chlorobiineae the photosynthetic pigments are bound to two separate cell structures : the antenna pigments of the reaction center are in the cytoplasmic membrane.

THE LIGHT HARVESTING MECHANISM

Radiant energy is always transferred in discrete packets known as photons. The energy content of which is inversely related to wavelength. The electro-magnetic radiation between 200–1200 nm is capable of mediating biological reactions. Wavelengths longer than 1200 nm in the infrared region have energy content so small that the absorbed energy is immediately converted into heat. While, the wavelengths less than 200 nm are termed ionizing radiation and their energy content is so high that molecules in their path are immediately ionized.

In Eukaryotic cells and cyanobacteria, there are two kinds of antennas associated with two different photosystems. Photosystem I absorbs longer wavelength light (>-680 nm) and funnels the energy to a special chlorophyll. This molecule is called P_{700}. The term P_{700} signifies that the molecule most efficiently absorbs light at a wave length of 700 nm. Photosystem II traps light at shorter wavelengths (<=680 nm) and transfers its energy to the special Chl P_{680} (Table 3).

Table 3. Relationship between photosynthetic organisms

Property	Eukaryotes	Cyanobacteria	Green & Purple Bacteria
Pigment	Chl. a	Chl. a	Bacteriochlorophyll
Photosynthesis II	+nt	+nt	-nt
Electron donors	H_2O	H_2O	H_2, H_2S, S
O_2 production pattern	oxygenic	oxygenic	Anoxygenic
Primary products of energy conservation	ATP^+ NADPH	ATP^+ NADPH	ATP
Carbon source	CO_2	CO_2	Organic acid

Properties of microbial photosynthetic systems

The photochemical reaction center contains a chlorophyll molecule. After absorbing light, it is oxidized by donating an electron to a carrier molecule. The chlorophyll molecule of the reaction center differs from the other chlorophyll molecules (of antenna pigments) in two respects.

1. They are associated with certain protein that interact with them in a manner that decreases the energy required to raise them to the activated state.
2. They are in close proximity with carrier molecules that can accept an electron from them when they are activated.

The energy required to activate a molecule of chlorophyll of the reaction center is designated as P. In purple bacteria the reaction center chlorophyll is activated by photons of wavelength 870 nm and hence, designated as P_{870}.

The photo activated chlorophyll molecule of reaction center donates its electron, it moves into electron transport chain, located within the photosynthetic membranes. The chain is composed of carrier molecules cytochromes, quinones and iron-sulphur centers. As electron flows through the chain, a proton motive force is generated. This released energy is partially used to synthesize ATP by a membrane located ATP phosphohydrolase.

MECHANISM OF PHOTOSYNTHESIS

Almost all the photosynthetic bacteria can fix carbon-di-oxide via the ribulose-di-phosphate pathway. They use $NADH+H^+$ for the reduction of 3-phosphoglycerate. Ferredoxin or NADP-dependent reductive carboxylations are used for assimilation of carbon. The anaerobic phototrophs are dependent on an external hydrogen donor. They can use gaseous hydrogen or hydrogen sulphide, elemental sulphur, thiosulphate, organic acids, alcohols, sugars and even some aromatic compounds. The purple and the green sulphur bacteria and a few non-sulphur purple bacteria oxidise hydrogen sulphide to sulphate. The quantum yield is same as in oxygenic photosynthesis.

The following formula gives a general idea of photosynthesis in purple bacteria.

$$CO_2 + 2H_2S \longrightarrow \lt CH_2O \gt + H_2O + 2S$$
$$2CO_2 + H_2S + 2H_2O \longrightarrow 2 \lt CH_2O \gt + H_2SO_4$$

ANOXYGENIC PHOTOSYNTHESIS

Only one cyclic electron transport system is found in anoxygenic photosynthesis (Fig. 5 A & B). There is no evolution of oxygen i.e., water does not acts as hydrogen donor.

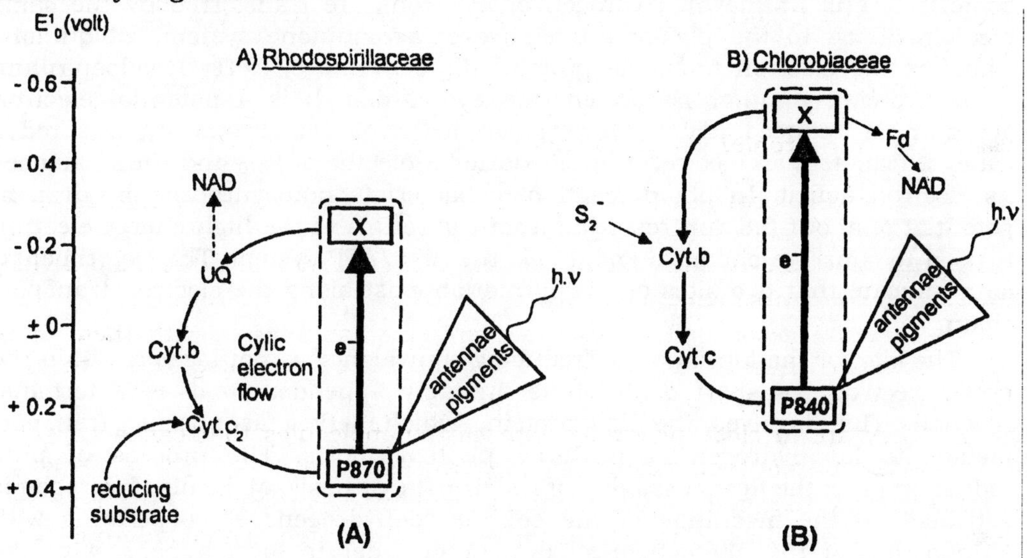

Fig. 5 A & B. Photosynthetic electron transport system in A - Rhodospirillaceae, B. Chlorobiaceae

In purple bacteria photosynthetic pigments absorb energy. The reaction center is P_{870}. The redox potential is in the region +450 to +490 mV. Isolated reaction center consists of a protein complex containing BChl a, Bacterio-phaeophytin, carotenoid, ubiquinone and iron sulphur proteins. A complex of ubiquinone and Fe-S protein probably functions as the primary electron acceptor. The redox potential of the complex is – 100mV. The electron ejected from the pigment travels to ubiquinone, cytochrome b and C_2 and goes back to possibly to Fe-S protein. Hence, the electron required for the reduction of NAD comes from the cyclic electron flow. It is transferred through an energy dependent reverse electron-transport flow to NAD. The electron deficiency is met out by external electron donors : hydrogen sulphide, sulphur or thiosulphate. The electron transport leads to the creation of a proton gradient. Infact, cells respond to illumination by extrusion of protons. In chromatophores illumination causes transport of protons into vesicles. In the vesicles, the electrons are transferred to the outside and the protons to the inside. This proton potential is the force driving photosynthetic phosphorylation.

In green bacteria the redox potential of primary electron acceptor of the light reaction has a potential of about – 500mV. The nature and arrangement of the redox catalysts that, participate in this "cyclic electron transport" are only partly known. The primary electron acceptor "X" has still not been identified. It is because of its strong negative potential (about –0.4 volts). It is assumed it could be ferredoxin. A corresponding compound has been extracted and purified from cells of *Chromatium*, it is firmly to the thylakoid structure. The participation of a specific flavoprotein, which would be located between "X" of ferredoxin and ubiquinone, is yet not clear. Ubiqunones are chemically related to the plastoquinone of the chloroplasts. The structure of ubiquinone in phototropic bacteria is still unknown. Hydrogen or electrons are transferred by the same mechanism as in the plastoquinone/hydroplastoquinone system. Ubiquinone transfers the electron to "heme protein" (Cytochrome c; RPH= Rhodospirillum heme protein). Another hemin enzyme cytochrome b is a potential electron acceptor for reduced RHP. The cycle is closed by cytochrome c_2; this redox catalyst transfers an electron to the oxidized molecule of P_{870} and thus balances its electron deficit. In photo reaction of bacterial photosynthesis the drop in potential of about 0.8 volt resulting from the return of the high energy electron is used to conserve chemical energy *via* the ADP/ATP system. The experimental data indicate that two sites of ATP formation exist along the electron transport chain.

The site for the formation of "reducing equivalents" is not located within the cyclic electron transport chain. It is probably appended to it as a terminal reaction. This involves the flavoprotein, which serves as specific transport metabolite for hydrogen in bacterial photosynthesis. The process of phto reduction gives the first possibility of $NADH+H^+$ synthesis. Molecular hydrogen is activated by the mediation of the enzyme "dehydrogenase" cooperating with ferredoxin and the photochemical system in order to form a bond with the coenzyme. The second possible formation of $NADH+H^+$ apparently takes place outside the photochemical reaction complex and utilizes exclusively organic substrates via an anaerobic reaction cycle largely corresponding to the "citric acid cycle".

The conversion of CO_2 to carbohydrate in phototropic bacteria is apparently brought about by the same or a very similar reaction mechanism as the one in green algae or higher plants, i.e., by Calvin cycle.

OXYGENIC PHOTOSYNTHESIS

The microorganisms performing oxygenic photosynthesis use water as hydrogen donor. The process of photosynthesis takes place in the thylakoids. The thylakoid membranes contain chlorophyll, carotenoids, electron carriers and enzymes. The antenna pigments harvest the light and channel it to the reaction center chlorophyll (carotenoid- carotenoid, chlorophyll + carotenoid –chlorophyll + carotenoid). The photochemical reaction is completed in two pigment systems (Fig. 6). The reactions are arranged in a series. The pigment system with reaction center P_{700}, is known as photosystem I. The second system has reaction center P_{680}, and is known as photosystem II. Both the photosystems contain special Chl a molecule of absorption maxima 700 nm and 680 nm, respectively.

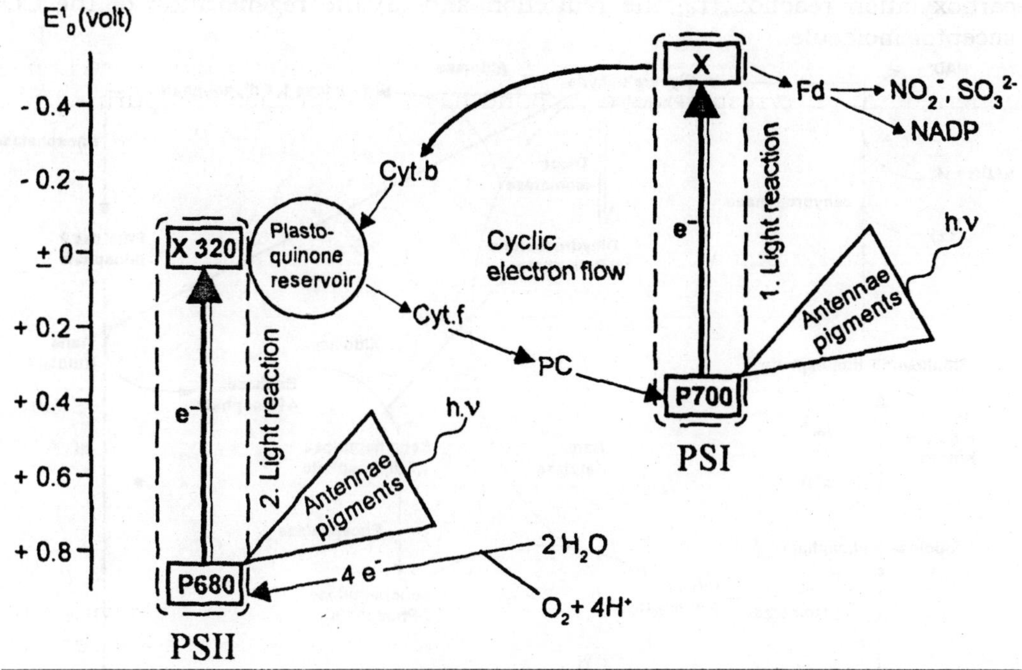

Fig. 6. 'Z' Scheme

Photosystem I exhibit cyclic transport system. The emission of an electron in the reaction center produces an electron deficit. The ejected electron is accepted by an iron-sulphur protein (X). This has more negative redox potential than –420 mV, possibly –530 mV. This acceptor transfers the electron to ferredoxin, and from ferredoxin (reduced), the reducing power can be transferred to $NADP^+$ or other acceptors. Alternatively, the electron can pass from "X" *via* plastoquinone, cytochrome, and plastocyanine, in a cyclic electron flow back to the reaction center chlorophyll and thus fill the electron gap.

In case of photosystem II, the activation of Chl *a* molecule of the reaction center causes the emission of an electron : this is accepted by a specific platoquinone molecule (X 320), which is thereby reduced to a semiquinone. Water acts as the electron donor for photosystem II. The splitting of water molecule is carried out with the participation of manganese. The two pigment systems are connected by an electron transport chain. The electron is transferred from reduced platoquinone to cytochrome f, which is a membrane bound type cytochrome and then to plastocyanine, which is a soluble copper containing protein. This electron fills the electron gap of Chl *a* of photosystem I.

CARBON-DI-OXIDE FIXATION

The photosynthetic eubacteria fix carbon-di-oxide *via* the ribulose-di-phosphate cycle (Calvin-Bassham cycle) (Fig. 7). In this cycle, carbon di oxide is reduced to the level of carbohydrates. Three stages can be distinguished : (1) the carboxylation reaction, (2) the reduction and (3) the regeneration of the CO_2 acceptor molecule.

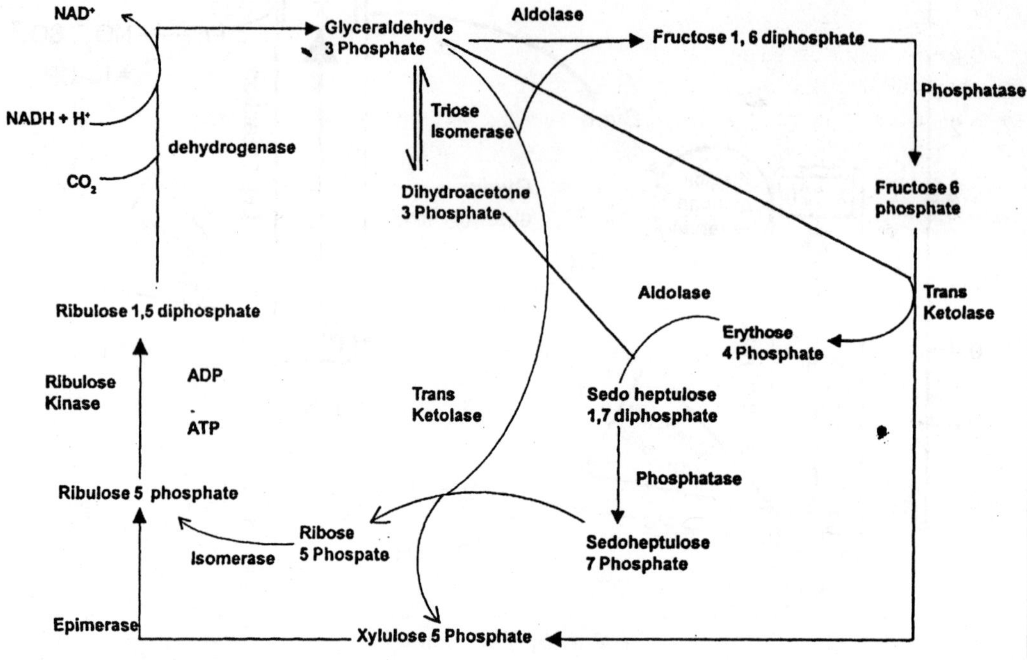

Fig. 7.

Enzyme ribulose-di-phosphate carboxylase fixes CO_2 to ribulose di phosphate, which is converted into two molecules of 3-phosphoglycerate.

Enzyme triose phosphate isomerase creates equilibrium with glyceraldehydes-3-phosphate and dihydroyacetone-3-phosphate. These two sugars combine to form fructose, 1-6-di phosphate. This reaction is catalyzed by enzyme fructose, 1-6 diphosphate aldolase.

The reactions, which regenerate CO_2 acceptor is initiated by transketolase. This enzyme acts upon fructose-6-phosphate and triose phosphate to form erythose, 4-phosphate and xylulose, 5-phosphate. the tetrose phosphate (erythose, 4-phosphate) produced by the transketolase reaction is then converted in an aldolase reaction with dihydroxyacetone phosphate, to yield sedoheptulose, 1-7-di-phosphate. This is dephosphorylated in the 1- position by enzyme phosphatase to produce sedoheptulose 7-phosphate. From sedoheptulose, 7-phosphate, transketolase then transfers a glycol group to glyceraldehydes, 3-phosphate. Two molecules of pentose phosphate are produced. The pentose-phosphate molecules are ribose, 5-phosphate and xylulose, 5-phosphate. Again, enzymatic equilibirium is established among these pentose phosphate molecules and another pentose phosphate molecule, ribulose, 5-phosphate. The phosphorylation of ribulose, 5-phosphate with ATP forms ribulose-1, 5-di-phosphate by enzyme phosphoribulokinase. This is the last reaction of Calvin cycle.

CHEMOSYNTHESIS

Chemosynthesis (chemoautotrophic or chemolithotrophic) micro-organisms use the energy derived from the oxidation of inorganic compounds to supply energy needed for the synthesis of ATP. Chemolithotrophic micro-organisms couple oxidation of an inorganic compound with the reduction of a suitable coenzyme. Chemoorganotrophs use the energy obtained from oxidation of organic compounds. Chemolithotrophic bacteria play important role in biogeochemical cycles. Their metabolic activities form critical links in the biogeochemical cycling of various elements.

Chemolithotrophic bacteria

1. *Nitrosomonas, Nitrosococcus, Nitrosocystis, Nitrosogloea* and *Nitrosospira* (the nitroso group) oxidize ammonia.
2. *Nitrobacter* and *Nitrocystis* (the nitro group) oxidize nitrite.
3. *Hydrogenomonas* (the hydrogen bacteria) oxidize hydrogen.
4. *Thiobacillus* and *Ferrobacillus* (the ferrous iron bacteria) oxidize iron.
5. *Pseudomonas methanica* and *Methanomonas methanooxidans* (the methane oxidizing bacteria) oxidizes methane.
6. *Thiobacillus* (the sulphur-oxidizing bacteria) oxidizes sulphur.

In general terms, the chemosynthetic bacteria reduces inorganic sulphur compounds in various oxidation states are anaerobic in nature. While, those bacteria, which reduce nitrogen compounds in various oxidation states are facultative anaerobes, and those that oxidize reduced sulphur and nitrogen compounds are obligate aerobes.

SULPHUR OXIDATION

Hydrogen sulphide, elemental sulphur, and thiosulfate are oxidized by chemolithotrophs. Oxidation yields electrons that are fed into an electron transport chain. *Thiobacillus thiooxidans* oxidizes reduced sulphur with the formation of sulphates, while *Thiobacillus ferrooxidans* can oxidize reduced sulphur as well as reduced iron. The elemental sulphur thus formed is deposited in the bacterial cells. It is deposited either as sulphur granule inside the cell or as elemental sulphur outside the cell.

NITRIFICATION

The nitrifying bacteria oxidize ammonium to nitrite (e.g., *Nitrosomonas*) or nitrite to nitrate (*Nitrobacter*). A detail of nitrification is discussed in chapter 25.

OXIDATION OF FERROUS IRON

Thiobacillus ferrooxidans oxidize ferrous to ferric iron. *Sulfolobus acidocaldarius* is a thermophilic bacterium. It has the ability to oxidize ferrous iron. *Gallionella ferruginea* and *Leptothrix ochracea* are found in draining pipes and mountain streams associated with flakes and thick coverings of iron oxides.

OXIDATION OF MOLECULAR HYDROGEN

Hydrogen is produced during anaerobic decomposition of organic matter in sediments and anaerobic microhabitats in soil. It is produced aerobically in well-aerated ecosystems. *Pseudomonas, Alcaligenes, Aquaspirillum* are Gram-negative bacteria, and *Bacillus, Mycobacterium* are Gram-positive bacteria oxidizing hydrogen aerobically. In aerobic hydrogen oxidation, oxygen is the ultimate electron acceptor. Enzyme hydrogenases take part in this process. This enzyme is found in two types. A soluble NAD^+ reducing hydrogenase and a membrane bound hydrogenase. Only a few bacteria possess both types of hydrogenases.

CARBON MONOXIDE UTILIZING BACTERIA

Pseudomonas carboxidovorans oxidize carbon monoxide into carbon-dioxide. There are some other bacteria also which utilize carbon monoxide as their sole energy source i.e., electron donor. Mechanism of metabolism of carbon monoxide is as following equation.

$$7\,CO + 2\tfrac{1}{2}\,O_2 + H_2O \longrightarrow \langle CH_2O \rangle + 6\,CO_2.$$

CHAPTER 11

RESPIRATION

Respiration is a catabolic process. In this process, there is breakdown of inorganic or organic compounds with the supply of energy. The intermediates are precursors for biosynthesis of cell material. Micro-organisms are aerobic or anaerobic in nature and hence, the catabolic process may take place in either of the following ways :

In aerobic or oxybiontic respiration the chemical energy-yielding reactions which requires molecular oxygen as the terminal electron acceptor.

Anaerobic or anoxybiontic respiration the chemical energy-yielding reactions in which inorganic compounds other than oxygen act as the terminal electron acceptor.

Fermentation : In fermentation chemical energy-yielding reactions requires organic compounds as electron acceptor.

Glucose is the main energy yielding carbon source in most of the micro-organisms. Glucose is converted into two molecules of pyruvate. Different pathways have been discovered by various biochemists and they are named after the name of the discoverer or according to their main components. These pathways are as following.

1. Embden-Meyerhof –Paranas (EMP) pathway;
2. Warburg-Dickens or hexose monophosphate (HMP) pathway;
3. Enter-doudoroff (ED) pathway;
4. Phosphoketolase pathway.

The first two pathways occur in many organisms including both eukaryotes and prokaryotes. In certain prokaryotes third pathway occurs instead of EMP pathway. *E. coli* metabolizes glucose *via* the EMP pathway and also have the capacity to synthesize enzymes of the ED pathway. These enzymes are induced by gluconic acid.

GLYCOLYSIS

The biochemical pathway of breakdown of hexose sugar into pyruvic acid is known as glycolysis (Gr. Glyco = sweet: lysis = splitting) or 'glycolytic' pathway. Embden Meyerhof and Paranas discovered the biochemical reactions of this pathway, hence the name EMP pathway. The term 'glycolysis' is generally restricted to mammalian usage.

The EMP pathway is a widely found in most of the groups of bacteria. Members of family Enertobacteriaceae, Lactobacillaceae and also saccharolytic clostridia perform this pathway. Breakdown of glucose to pyruvate takes place in ten steps. The overall reaction may be expressed in a single formula :

$$\text{Glucose} + 2\text{ATP} + 2\text{NAD}^+ \longrightarrow 2 \text{ pyruvate} + 2 \text{ ATP} + 2 \text{ NADH+H}^+$$

In this pathway one molecule of hexose sugar (glucose/fructose) is catalysed into two molecules of pyruvic acid. It completes in ten biochemical reactions. Out of these reactions the first five reactions form preparatory phase where the hexose sugar is phosphorylated with ATP. The remaining five reactions form pyroff phase where energy is released and is utilized in phosphorylation of ADP to ATP. All the intermediates of this pathway are phosphorylated compounds. These intermediates are biologically important as (i) they are negatively charged and hence they are impermeable through cell membrane, (ii) The intermediates can very easily form enzyme-substrate complex, (iii) The phosphorylated intermediates pass their phosphate moiety to ADP to give molecule of ATP.

The detailed reactions of the pathway are as following (Fig. 1) :

1. **Phosphorylation of Glucose :** In the first reaction glucose is converted into glucose 6 phosphate with the help of enzyme hexokinase. This enzyme is known as kinase as it catalyzes the transfer of the terminal phosphate group from ATP to the substrate glucose (hexose sugar). Mg ions act as cofactor. ATP supplies phosphate moiety.

$$\text{Glucose} + \text{ATP} \xrightarrow{\text{Hexokinase; Mg++}} \text{Glucose 6 phosphate} + \text{ADP}$$

In certain micro-organism glucose is replaced by D-Mannose, D-fructose or D-glucosamine and inosine triphosphate (ITP) and deoxy-ATP may act as phosphate donor.

2. **Conversion of Glucose 6 phosphate to Fructose 6 phosphate** :- Glucose 6 phosphate is a phosphorylated aldo sugar. It undergoes reversible isomerization and forms a phosphorylated Keto sugar, fructose 6 phosphate. Enzyme phosphohexose isomerase catalyzes this reaction. This enzyme is also known as phosphoglucose isomerase.

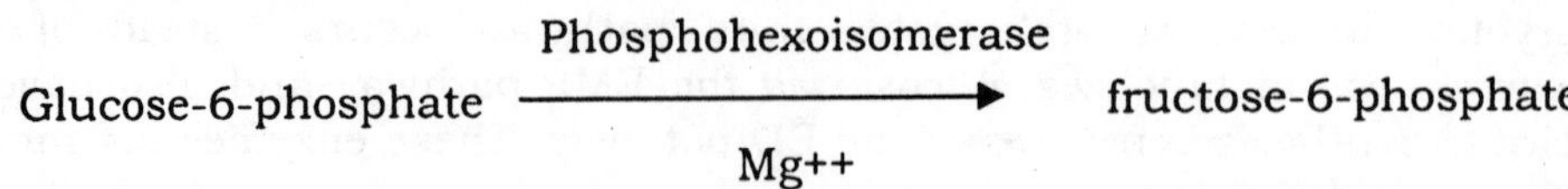

(3) **Phosphorylation of fructose-6-phosphate** : A phosphate moiety of ATP is transferred to fructose-6-phosphate and yields fructose-1,6-diphosphate. This reaction is catalyzed by enzyme phosphofructokinase.

$$\text{Fructose-6-phosphate} \xrightarrow{\text{phosphofructokinase}} \text{Fructose-1,6-diphosphate}$$

In some micro-organism phosphate moiety of pyrophosphate is added to fructose-6-phosphate.

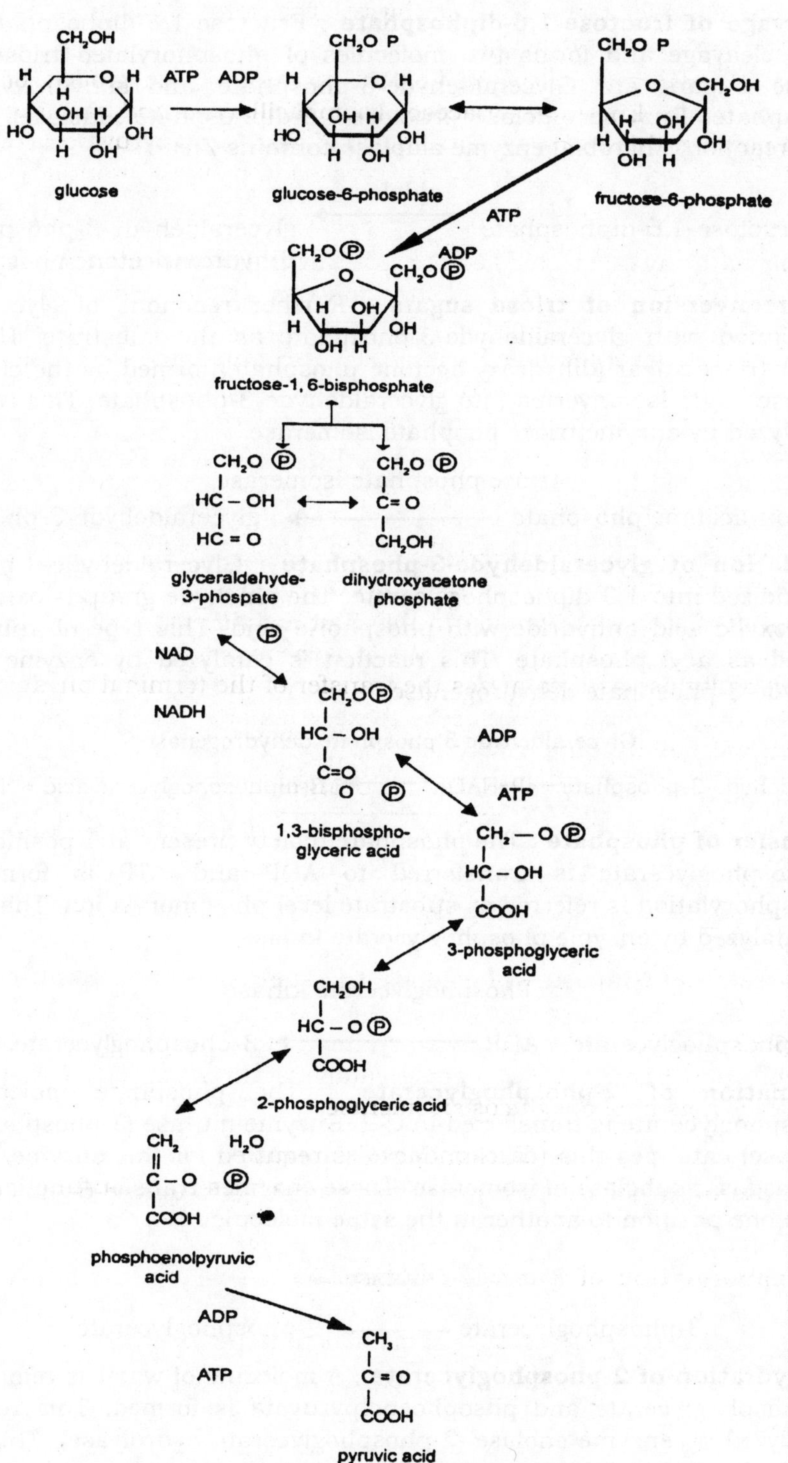

Fig. 1. Glycolysis (EMP Pathway)

(4) **Cleavage of fructose-1,6-diphosphate** : Fructose-1,6-diphosphate undergoes cleavage and forms two molecules of phosphorylated triose sugars. These sugars are Glyceraldehyde-3-phosphate and dihydroxyace-tone-phosphate. Enzyme aldolase (fructose-1,6-diphosphate aldolase) catalyzes this reaction. Microbial enzyme aldolase contains Zn++.

$$\text{Fructose-1,6-diphosphate} \xrightarrow{\text{aldolase}} \text{glyceraldehyde-3-phosphate} + \text{dihydroxyacetone phosphate}$$

(5) **Interconversion of triose sugars** : Further reactions of glycolysis are continued with glyceraldehyde-3-phosphate as the substrate. Hence the other triose sugar (dihydroxy-acetone phosphate) formed by the cleavage of hexose sugar is converted into glyceraldehyde-3-phosphate. This reaction is catalyzed by enzyme triose phophate isomerase.

$$\text{Dihydroxyacetone phosphate} \xrightarrow{\text{Triose phosphate isomerase}} \text{glyceraldehyde-3-phosphate}$$

(6) **Oxidation of glyceraldehyde-3-phosphate** : Glyceraldehyde-3-phosphate is oxidized into 1,3-diphosphoglycerate. The aldehyde group is oxidized to a carboxylic acid anhydride with phosphoric acid. This type of anhydride is called as acyl phosphate. This reaction is catalyzed by enzyme glyceraldehyde-3-phosphate dehydrogenase.

$$\text{Glyceraldehyde-3-phosphate} + \text{iP} + \text{NAD}^{+} \xrightarrow{\text{Glyceraldehyde 3 phosphate dehydrogenase}} \text{1,3-diphosphoglyceric acid} + \text{NADH} + \text{H}^{+}$$

(7) **Transfer of phosphate** : The phosphate moiety present at 1 position of 1,3-diphosphoglycerate is transferred to ADP and ATP is formed. The phosphorylation is referred as substrate level phosphorylation. This reaction is catalyzed by enzyme phosphoglycerate kinase.

$$\text{1,3-diphosphoglycerate} + \text{ADP} \xrightarrow{\text{Phosphoglycerate kinase}} \text{3-phosphoglycerate} + \text{2ATP}$$

(8) **Formation of 2-phosphoglycerate** : The phosphate moiety of 3-phosphoglycerate is transferred to C-2. Enzyme mutase (3-phosphoglycerate mutase) catalyzes this reaction. Mg++ is required for this enzyme. Enzyme mutase is a subclass of isomerase. These enzymes transfer functional group from one position to another in the same molecule.

$$\text{3-phosphoglycerate} \xrightarrow{\text{mutase}} \text{2-phosphoglycerate}$$

(9) **Dehydration of 2-phosphoglycerate** : A molecule of water is removed from 2-phosphoglycerate and phsophoenolpyruvate is formed. This reaction is catalyzed by enzyme enolase (2-phosphoglycerate hydrolyase). This enzyme is a hydratase.

$$2\text{-phosphoglycerate} \xrightarrow{\text{enolase}} 2 \text{ phosphoenolipyruvate} + 2H_2O$$

(10) **Synthesis of pyruvate** : Phosphate moiety of 2 phosphoenol pyruvate is transferred to ADP and the later is converted into ATP. Pyruvate is synthesized in this reaction. This reaction is catalyzed by enzyme pyruvate kinase. This enzyme needs Mg++ and K+ for its function. Mn++ may replace Mg++. 2 phosphoenolipyruvate+2ADP

$$2 \text{ phosphoenol pyruvate} + \text{ADP} \xrightarrow{\text{Pyruvate kinase}} \text{Pyruvate} + \text{ATP}$$

Table 1. Energetics of Glycolysis (per glucose / Fructose molecule).

Reaction	Enzyme	ATP/NADH+H⁺ Consumed
1. Glucose → Glucose-6-phosphate Fructose → Fructose-6-phosphate	Hexokinase	1 ATP Consumed
2. Fructose-6-phosphate → Fructose-1, 6 diphosphate	Phospho-fructokinase	1 ATP Consumed
3. Glyceralddehyde –3-phosphate → 3 phospho-glyceric acid	Glyceralddehyde -3-phosphate dehydrogenase	2 NADH+H⁺ produced
4. 1, 3 diphosphoglyceric acid → 3 phospho-glyceric acid	Phospho glycerate kinase	2ATP
5. Phosphoenol pyruvate → Pyruvic acid	Pyruvate kinase	2ATP

FORMATION OF ACETYL CoA

Enzyme pyruvate dehydrogenase complex catalyzes the oxidative decarboxylation of pyruvate. A vitamin B (CoA) takes part in this reaction and acetyl CoA is synthesized. CO_2 is released in this process.Acetyl CoA is biologically significant intermediate and takes part in a number of biosynthetic pathways. In aerobic respiration acetyl CoA enters in mitochondria where it is completely oxidized *via* Kreb's Cycle.

Pyruvate dehydrogenase is a multienzyme complex of three enzymes. These enzymes are (a) pyruvate dehydrogenase, (b) dihydrolipoamide transacetylase and (c) dihydrolipoamide dehydrogenase. The mechanism of formation of acetyl-CoA from pyruvate is as following :-

$$\text{Pyruvate} + \text{TPP} \xrightarrow{\text{Pyruvate dehydrogenase}} \text{Hydroxyethyl-TPP} + \text{CO2}$$

$$\text{Hydroxyethyl-TPP} + \text{lipoate} \xrightarrow{\text{dyhydrolipoamide transacetylase}} \text{acetyl-dihydrolipoate} + \text{TPP}$$

$$\text{Acetyl-dihydrolipoate} + \text{CoA} \xrightarrow{\text{Dihydrolipoamide transacetylase}} \text{acetyl-CoA} + \text{dihydrolipoate}$$

$$\text{Dihydrolipoate} + \text{NAD}^+ \xrightarrow{\text{Dihydrolipoamide dehydrogenase}} \text{lipoate} + \text{NADH+H}^+$$

KREB'S CYCLE

In 1937, Hans Kreb proposed a cycle for catabolic oxidation of pyruvic acid and all the intermediates of this cycle were acids. He studied the effect of the anions of various organic acids on the rate of oxygen consumption during pyruvate oxidation. In his studies he used suspensions of minced pigeon-breast muscle. The first intermediate of this cycle is citric acid a tricarboxylic acid, hence this cycle is also known as citric acid cycle or TCA. In 1948, Eugene Kennedy and Albert Lehninger observed that isolated mitochondria could also show oxidative catabolism of pyruvic acid. A number of prokaryotes contain the enzymes of TCA in their cytosol. Their plasma membrane plays a role analogous to that of the inner mitochondrial membrane in ATP synthesis. In anaerobic micro-organisms some reactions of TCA occur in a linear biosynthetic processes. These reactions do not serve as the source of energy but the intermediates act as precursors of other biosynthetic pathways. These organisms are capable of forming malate, fumarate, succinate and succinyl-CoA from oxaloacetate.

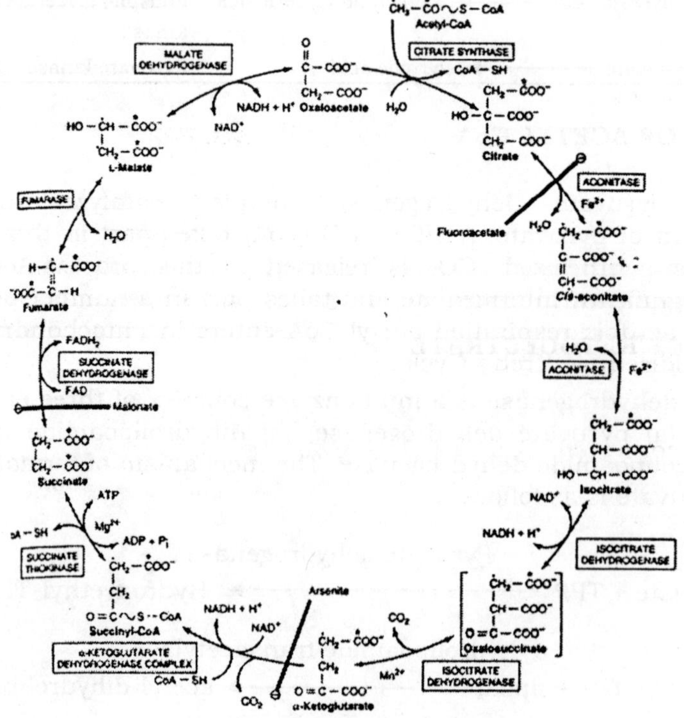

Fig. 2. Kreb's Cycle

Pyruvic acid formed by the catabolism of glucose molecule in the biochemical pathway of glycolysis is further activated and is converted into acetyl CoA. Acetyl CoA (CH3-CO~S-CoA) is an ester of coenzyme A. CoA contains pantothenic-acid, which is a vitamin. The later is an important metabolite and has significant role in many biological processes. It acts as a bridge between glycolysis and Kreb's cycle. The biochemical reactions of TCA are as following (Fig. 2).

Citrate and isocitrate synthesis

The first intermediate of TCA is citric acid. Acetyl CoA condenses with a dicarboxylic acid (oxaloacetic acid) forming a tricarboxyilic acid (citric acid) and releases CoA due to hydrolysis of thioester bond of CoA. The acetyl moiety of acetyl CoA and oxaloaectic acid combines and produces citric acid. This reaction is catalysed by enzyme citrate synthetase. In this reaction the methyl carbon of the acetyl group is joined to the carbonyl group of oxaloacetate. Citroyl-CoA is a transient intermediate. This enzyme is inhibited by ATP, NADH and fatty acetyl CoA esters. However, there significance is not much important. Citrate thus formed is converted into isocitrate. In this reaction first a water molecule is released from citric acid to form an intermediate aconitic acid, and again a water molecule is added to it forming isocitrate. The position of H and OH is changed. Enzyme aconitase (aconitate hydratase) catalyzes this reaction. This enzyme contains iron-sulphur center. This center acts both in binding of the substrate at the active site and in catalysis of the addition or removal of H_2O. This reaction is inhibited by fluroacetate.

$$\text{Acetyl CoA} + \text{oxaloacetate} + H_2O \xrightarrow{\text{Citrate synthetase}} \text{citrate} + \text{CoA}$$

$$\text{Citrate} \xrightarrow{\text{Aconitase}} \text{cis aconitic acid} + H_2O \xrightarrow{\text{aconitase}} \text{isocitrate}$$

FORMATION OF KETOGLUTRATE

Isocitrate is oxidized into oxalosuccinate releasing two hydrogen ($2H+2e^-$). The reducing equivalent is received by NAD^+ or $NADP^+$ and is converted into $NADH+H^+$ or $NADPH+H^+$. It is followed by oxidative decarboxylation and α Ketoglutrate is synthesized. This reaction is catalyzed by enzyme isocitrate dehydrogenase complex.

$$\text{Isocitrate} + NAD^+ \xrightarrow{\text{Isocitrate dehydrogenase}} \text{Oxalosuccinate} \longrightarrow \alpha\ \text{Ketoglutrate} + CO_2 + NADH+H^+$$

Isocitrate dehydrogenase occurs in two forms (isozymes). These two enzymes require NAD^+ and $NADP^+$, respectively.

FORMATION OF SUCCINYL CoA

Succinyl CoA is formed by oxidation of α ketoglutrate. CoA takes part in this reaction and NADH+H+ is formed due to oxidation reduction reaction. This reaction is catalyzed by Succinate thiokinase (succinyl-CoA synthetase). It includes three enzymes, analogous to E1, E2 and E3 of the pyruvate dehydrogenase. The mechanism of this reaction is similar to that of formation of acetyl CoA from pyruvic acid.

$$\alpha\text{ Ketoglutrate} + \text{CoA} + \text{NAD}^+ \xrightarrow{\alpha\text{ Ketoglutrate dehydrogenase}} \text{Succinyl CoA} + CO_2 + \text{NADH} + H^+$$

Deacylation of succinyl CoA

Succinyl CoA releases CoA forming succinic acid. A molecule of GTP is produced in this reaction. Enzyme Succinate synthetase catalyzes this reaction.

In eukaryotes, this enzyme is tightly bound to the inner membrane of mitochondria while in prokaryotes, it is bound to the plasma membrane. This enzyme is inhibited by malonate.

$$\text{Succinyl CoA} + \text{GDP} + \text{Pi} \xrightarrow{\text{Succinate synthetase}} \text{Succinate} + \text{CoA} + \text{GTP}$$

Oxidation of succinate to fumerate

Succinate is oxidized to fumarate. A flavoprotein enzyme, succinate dehydrogenase catalyzes this reaction. FAD is converted into $FADH_2$ by accepting reducing equivalent released from succinate.

$$\text{Succinate} + \text{FAD} \xrightarrow{\text{Succinate dehydrogenase}} \text{Fumarate} + FADH_2$$

Hydration of Fumarate

A molecule of water is added to fumaric acid and produce malate. This reaction is catalyzed by enzyme fumarate hydratase.

$$\text{Fumarate} + H_2O \xrightarrow{\text{Fumarate hydratase}} \text{Malate}$$

Oxidation of malate

Malate is oxidized into oxaloacetic acid and thus the later is recovered. NAD^+ receives the reducing equivalents forming NADH+H+. This reaction is catalyzed by enzyme malate dehydrogenase.

$$\text{Malate} + \text{NAD}^+ \xrightarrow{\text{Malate dehydrogenase}} \text{Oxaloacetate} + \text{NADH} + H^+$$

It is important to note that four components of vitamin B complex take part in Kreb's cycle. They are (a) riboflavin, (b) niacin, (c) thiamin and (D) pantothenic acid.

ENERGETICS OF KREB'S CYCLE

Table 2. Energetics of Kreb's Cycle (per glucose molecule)

Reaction	Enzyme	NADH+H$^+$ /$FADH_2$/GTP produced
1. Isocitrate dehydrogenase → oxalosuccinate	Isocitrate dehydrogenase	2 NADH+H$^+$
2. α-ketoglutarate → succinyl CoA	α-ketoglutarate dehydrogenase	2 NADH+H$^+$
3. Succinyl CoA → succinate	Succinate thiokinase	2 GTP
4. Succinate → fumarate	Succinate dehydrogenase	2 $FADH_2$
5. L. Malate → Oxaloaxcetate	Malate dehydrogenase	2 NADH+H$^+$

COMPONENTS OF ELECTRON TRANSPORT SYSTEM (ETS)

The enzymes associated with ETS are grouped into four complex systems. They are known as (1) Pyridine linked dehydrogenase, (2) Flavin linked dehydrogenase, (3) Iron sulphur protein and (4) Cytochrome.

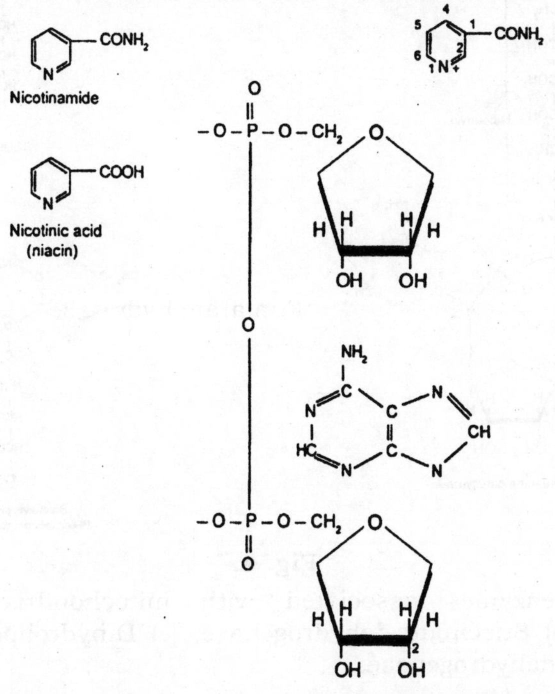

Fig. 3. Nicotinamide, Adenine dinucleotide

Pyridine linked dehydrogenase : These enzymes contain nicotinamide rings resembling pyridine as coenzyme. One of the soluble vitamin niacin provides the nicotinamide moiety for the synthesis of the pyridine nucleotides. NAD (nicotinamide adenine dinucleotide) and NADP (nicotinamide adenine dinucleotide phosphate) are composed of two nucleotides joined through their phosphate groups by a phosphoric acid anhydride bond (Fig. 3). When NAD^+ or $NADP^+$ comes in contact with reduced substrate the reducing equivalent is transferred to it and it is converted into $NADH+H^+$ or $NADPH+H^+$. The reducing equivalent is accepted as hydride ion. It attaches at 4 end of nicotinamide ring (Fig. 6 A). The redox potential of this enzyme is –320 mV.

Flavin linked dehydrogenase : These enzymes are known as flavoprotein. They contain coenzyme flavin adenine dinucleiotide (FAD) or flavin mononucleotide (FMN) (Fig. 4). FAD and FMN are formed in the body from the vitamin riboflavin. The coenzyme is tightly bound by covalent bond to protein and never leaves it during the reaction. Isoalloxazine ring of the riboflavin moiety accepts reducing equivalents. After reduction, the coenzyme is converted into $FADH_2$ and $FMNH_2$, respectively. Flavoprotein enzymes can receive either one or two reducing equivalents (Fig. 6 B). FAD or FMN when receives only one reducing equivalent the semiquinone form of isoalloxzine ring is produced. The redox potential is 0.0 mV.

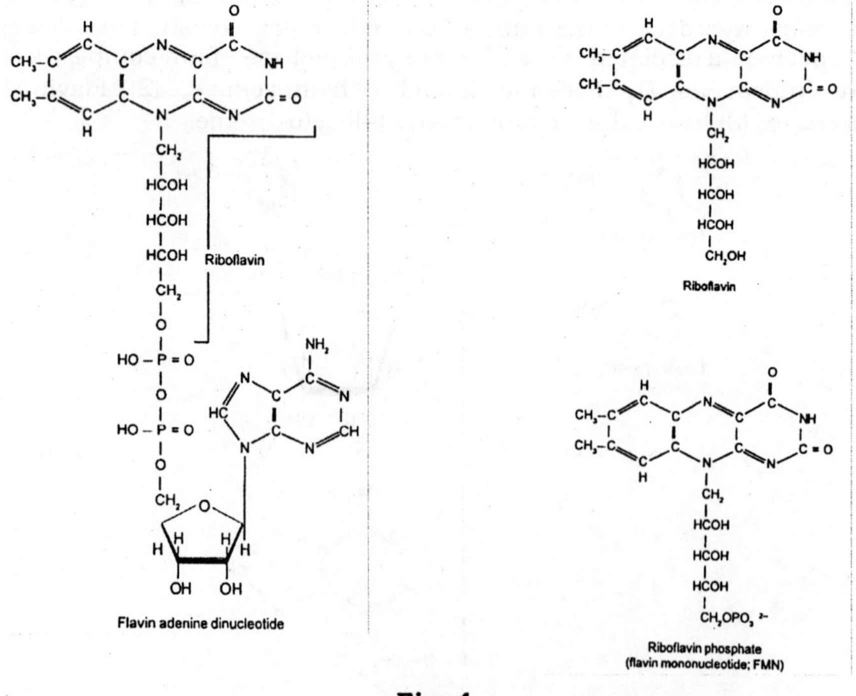

Fig. 4.

The flavin enzymes associated with mitochondria are (a) NADH dehydrogenase, (b) Succinate dehydrogenase, (c) Dihydrolipoyl dehydrogenase and (d)Acetyl CoA dehydrogenase.

Iron sulphur proteins (FeS) : These enzymes have equal amount of iron and acid labile sulphur. Ferredoxin of mitochondria is an iron sulphur protein. Ferric ion takes part in redox reactions. They are associated with the flavoprotein (metalloflavoproteins) and with cytochrome b. Redox potential of these enzymes is +120mV.

Cytochromes : Cytochromes are made up of tetrapyrrole ring, the porphyrin (Fig. 5). They have Fe nucleus. Five types of cytochromes are present in mitochondria. They are named as Cyt b, Cyt c, Cyt c1, Cyt a and Cyt a3. The redox potential varies from +260 mV to +290 mV.

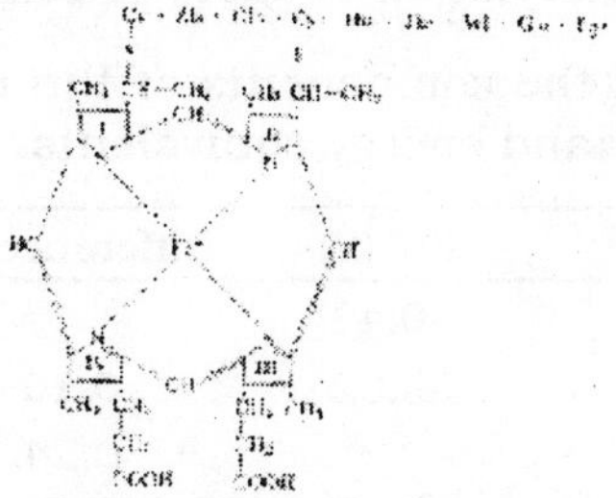

Fig. 5. Cytochrome C-peptide

Ubiquinone : Ubiquinone or Coenzyme Q also takes part in ETS and helps in transforming hydrogen of the substrate to molecular oxygen. This enzyme is a constituent of the mitochondrial lipids. It is very similar to vitamin K and vitamin E. It has isoprenoid side chain in their coenzyme. Ubiquinone exists in mitochondria in the oxidized quinone form under aerobic conditions and in the reduced quinol form under anaerobic conditions. Redox potential is + 810 mV (Fig. 6C).

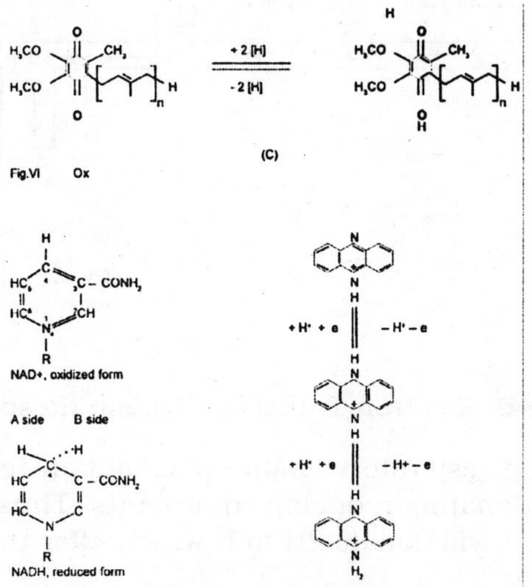

Fig. 6. Oxidized and reduced forms of some components of ETS
(A) NAD+ - NADH; (B) Oxidized and reduced from riboflavin; (C) Ubiquinone-Ubihydroquinone.

RESPIRATORY CHAIN

A respiratory chain is an electron transport chain (Fig. 7 AB). When a pair of electrons or hydrogen atoms (which contains electrons) from an oxidizable substrate is coupled with the reduction of an ultimate electron acceptor, such as oxygen, there is a large free energy change (ΔG^o). The Flow of electrons through the transport chain allows a stepwise release of this energy, some of which is conserved in the form of ATP at several steps in the chain. At these specific steps the difference in Eo values is great enough to permit sufficient energy to be liberated for oxidative phosphorylation to occur (Table 3).

Table 3. Redox potentials of the components of the respiratory chain and their potential differences and energy equivalents

Respiratory chain component	Eo' (V)	Difference (V)	Go (kJ/mol)	(kcal/mol)
Hydrogen	-0.42			
NAD	-0.32	0.10	19.3	461
		0.24	46.4	11.1
Flavoprotein	-0.08	0.04	7.7	1.84
Cytochrome b	-0.04	0.31	59.8	14.3
Cytochrome c	0.27	0.02	3.8	0.92
Cytochrome a	0.29	0.52	100.4	24.0
Oxygen	0.81			

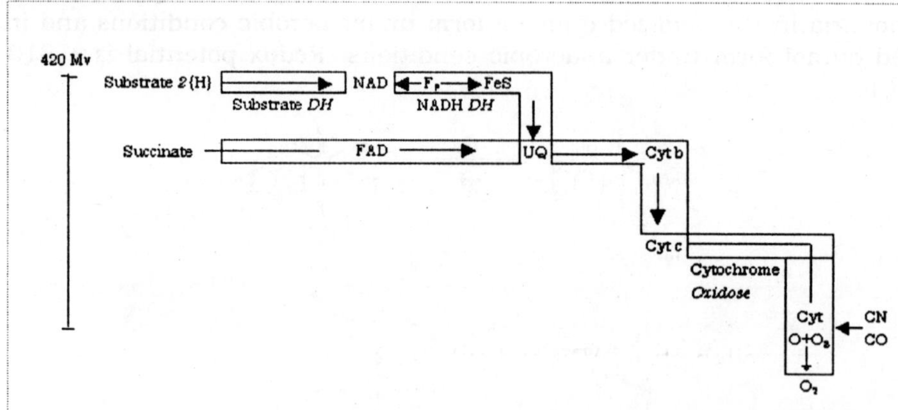

Fig. 7 A and B. Electron Transport System (Respiratory chain)

All the reactions in respiratory chain are oxidation-reduction reactions. The substrate is oxidized donating reducing equivalents. Thus, in a general reaction if AH_2 is oxidized to A it will donate 2H to B which after the reaction is converted into BH_2.

AH_2 B

A BH_2

The above example is a four component system. It consist of two redox couples AH_2/A and BH_2/B. 2H donated from A is accepted by B. 2H is transferred as $2H^+ + 2e^-$ or as hydride ions ($2H^+ + 2e^-$). This movement is governed by the tendency of donor (a reduced component) to acceptor (an oxidized component). It can be quantiated by reference to a standard redox couple and is known as redox potential. The later determined under standard conditions of temperature (25°C), concentration ($[AH_2]=[A]$) and pH (7.0 unless stated otherwise). It is expressed by the symbol E and is measured in millivolts. Under non-standard conditions, redox potential is expressed by the symbol Eh. $NADH+H^+/NAD^+$ is the first redox component of the respiratory chain. The reducing equivalents are ultimately accepted by molecular oxygen. The redox potential of $NADH^+H^+/NAD$ is –320 mV, thus the reducing equivalent travels to increasing positivity and reaches the ultimate acceptor whose redox potential is +820 mV. The standard free energy change (G′o) of this reaction is related to the difference in the standard redox potential of the oxidizing and reducing couples (Eo′). It is expressed by a simple equation

$$\Delta G'_o = -n \times F \times E'o$$

$$= -n \times 96.5 \times E'o \text{ (KJ/mol)}$$

where n is the number of electrons transferred and F is the Faraday constant. In the respiratory chain there is a yield of –219KJ mol^{-1} amount of free energy.

OXIDATIVE PHOSPHORYLATION

In 1929, Fiske and Subbarow isolated ATP from extracts of mammalian muscles. It plays an important role in energy cycle.

Synthesis of ATP from ADP and inorganic phosphate is known as phosphorylation. It has been observed in intact mitochondria that when substrates are oxidized *via* an NAD-linked dehydrogenase, approximately 3 molecules of inorganic phosphate are incorporated into 3 molecules of ADP to form 3 molecules of ATP per ½ molecule of O_2 consumed. The P:O ratio = 3 and if the substrate is oxidized *via* a flavoprotein-linked dehydrogenase, only two molecules of ATP are formed and P : O=2.

Chance and Williams have described following five conditions that control the rate of respiration in mitochondria:

State 1 availability of ADP and substrate.

State 2 availability of substrate only.

State 3 The capacity of the respiratory chain itself, when all substrates and components are present in saturating amounts.

State 4 availability of ADP only.

State 5 availability of oxygen only.

In a resting state, mitochondrion is in state 4 and when respiration increases, it undergoes in state 3 or state 5.

There are three hypotheses to explain the mechanism of ATP synthesis during ETS. They are known as chemical, chemiosmotic and conformational hypothesis. The chemical and chemiosmotic theories are Chemiosmotic theories are more accepted on account of the coupling of oxidation and phosphorylation.

Chemical hypothesis : It postulates direct chemical coupling at all stages of the process. This theory was proposed by Slates in 1953. The basic principle is the substrate level phosphorylation in which oxidation-reduction leads to ATP synthesis via the sequential formation of nonphosphorylated intermediates. The chemical hypothesis may be expressed as following:-

$$AH_2 + B + I.OH \longrightarrow A{\sim}I + BH_2 + OH^-$$

$$A{\sim}I + X.H \longrightarrow X{\sim}I + A + H^+$$

$$X{\sim}I + P.OH \longrightarrow X{\sim}P + I.OH$$

$$X{\sim}P + ADP \longrightarrow ATP + X.H$$

$$AH_2 + B + ADP + P.OH \longrightarrow A + BH_2 + ATP + H_2O$$

AH_2 and B are respiratory carriers and are the site of oxidative phosphorylation. I and X are coupling factor ligands of unknown identity but which must be capable of forming anhydride bond with high free energy of hydrolysis.

Chemiosmotic theory :- Mitchel proposed this theory in 1961. According to this theory, in respiratory chain oxidation of components is coupled to the translocation of hydrogen ions (proton's, H^+) from the inside to the outside of the inner mitochondrial membrane (Fig. 8). This electron transfer produces an electrochemical potential difference of asymmetric distribution of the hydrogen ions is used to drive the mechanism responsible for the formation of ATP. Respiratory complexes I, III and IV acts as a proton pump. Later on Mitchel and Jennifer Moyle expanded this concept and explained it in four main postulates:-

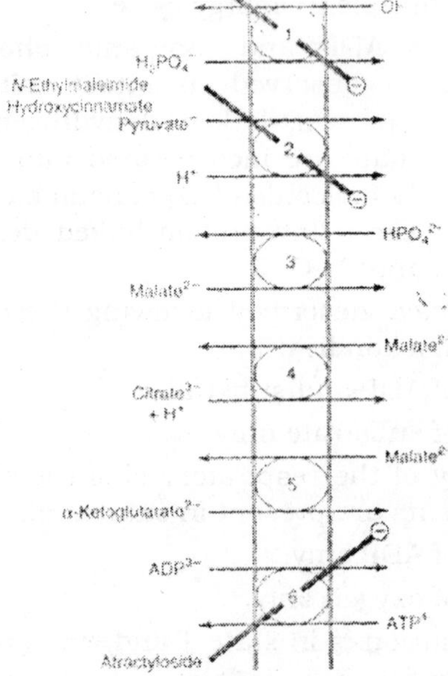

Fig. 8. Transport systems in the inner mitochondrial membrane

1. phosphate transporter, 2. pyruvate symport, 3. dicarboxylate transporter, 4. tricarboxylate transporter, 5. α-ketoglutarate transporter, 6. adenine nucleotide transporter, N-Ethylmaleimide, hydroxycinnamate, and atractyloside inhibit - the indicated systems

1. An anisotropic (direction oriented), proton translocating redox systems.
2. An ion impermeable coupling membrane.
3. Specific exchange-diffusion systems and
4. An anisotropic, proton-translocating, reversible ATPase.

The rate of respiration is controlled by the effective proton conductance of the coupling membrane. F1 particles are present on the inner membrane facing towards the matrix of mitochondria. These are proteinaceous phosphorylating bodies and contain ATP synthase. Here synthase is used as these enzymes catalyze condensation reactions in which no nucleoside triphosphate is required as an energy source. These subunits are attached to a membrane protein, known as Fo. The subscript letter o in Fo denotes that it is the portion of the ATP synthetase that confers sensitivity to an inhibitor, oligomycin. Protons pass through the Fo-F1 complex, and form ATP from ADP and Pi. The electro-chemicalpotential difference is used to drive a membrane-located ATP synthase. It forms ATP with ADP and Pi. In the controlled state the inward flow of protons *via* the ATP synthase is restricted by the lack of ADP. Since the permiability of the membrane to protons is low, a very high proton-motive-force (pmf) is generated (Fig. 9). This exerts back pressure on the redox systems and the respiration is maintained at a low rate.

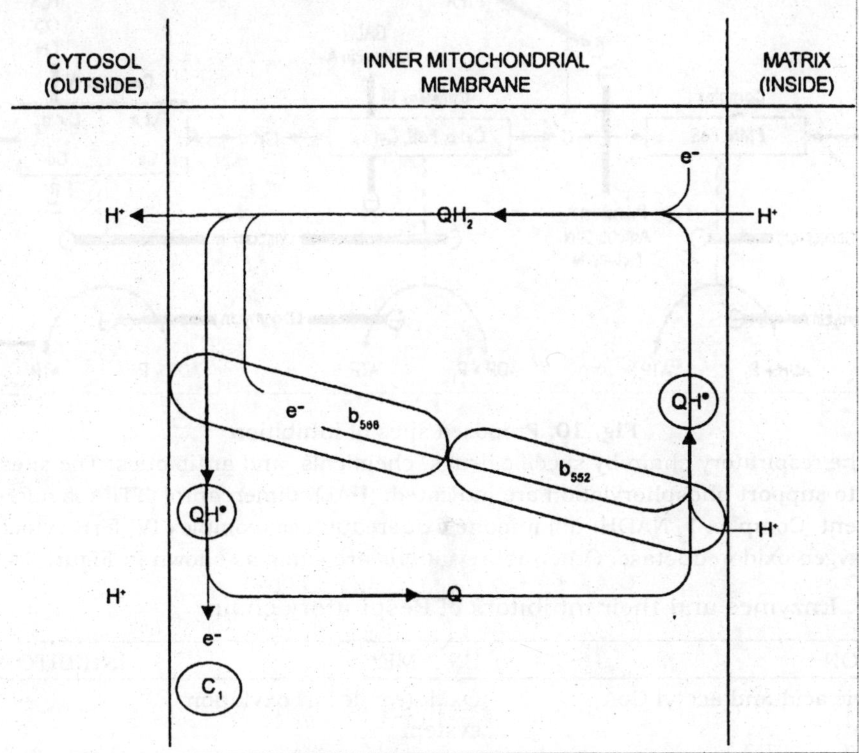

Fig. 9. Proton motive `Q'cycle

Conformational hypothesis : Boyer in 1964, proposed this hypothesis. It is a variation of chemical hypothesis. According to this hypothesis, energy released by respiration is conserved initially bringing about conformational changes in certain protein components of the coupling membrane and subsequently as non-phosphorylated covalent compounds. Now this theory has made some changes dropping the idea of covalent and phosphorylated intermediates. It is suggested that the conformational changes in complexes I, III and IV constitute the primary energy conserving event.

INHIBITORS OF RESPIRATORY CHAIN

Some chemicals acts as a poison for respiratory chain (Fig. 10). Their inhibition may be —

(i) on respiratory chain proper,
(ii) of oxidative phosphorylation, and
(iii) as uncoupler of oxidative phosphorylation.

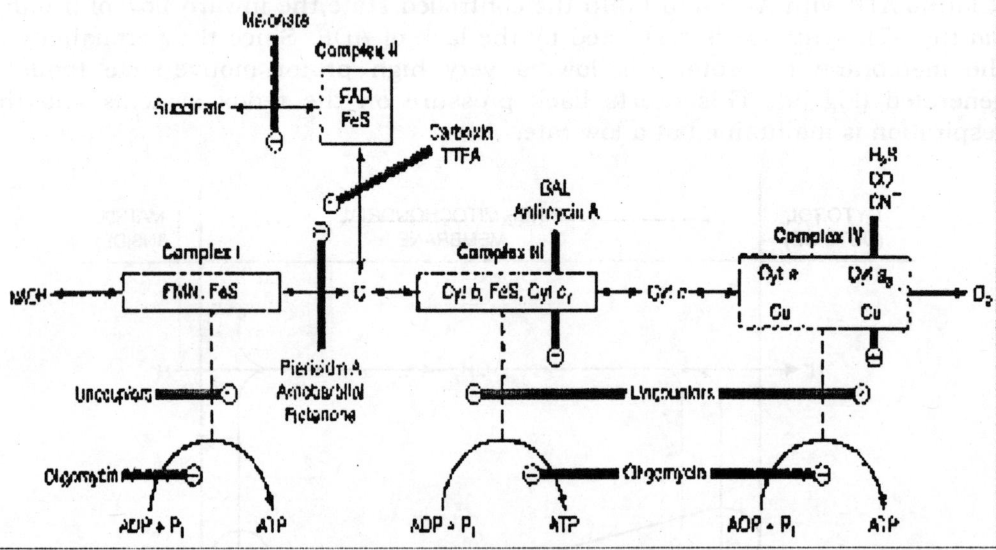

Fig. 10. Proposed sites of inhibition (-) of the respiratory chain by specific drugs, chemicals, and antibiotics. The sites that appear to support phosphorylation are indicated. (BAL), dimercaprol. TTFA is a Fe-chelating agent. Complex 1, NADH: ubiquinone oxidoreductase; complex IV, ferrocytochrome c:oxygen oxidoreductase. Other abbreviations are same as shown in Figure 14-4.

Table 4. Enzymes and their inhibitors of Respiratory chain

REACTION	ENZYME*	INHIBITOR
1. Pyruvic acid and acetyl CoA	Oxidative decarboxylation system	—
2. Acetyl CoA and oxaloacetic acid	Condensing enzyme	—
3. Citric to cisaconitic acid	Aconitase	Fluorocitrate
4. Cisaconitic to isocitric acid	Aconitase	—

5. Isocitric to oxalosuccinic acid	Isocitric dehydrogenase; $NADP^+$	Anaerobiosis
6. Oxalosuccunic to α-ketoglutaric acid	Oxalosuccunic dehydrogenase; Mn^{++}	—
7. α-ketoglutaric to succinic acid	Oxidative decarboxylation system	Arsenite
8. Succinic to fumaric acid	Succinic dehydrogenase	Malonate
9. Fumaric to malic acid	Fumarase	—
10. Malic to oxaloacetic acid	Malic dehydrogenase; NAD^+	—

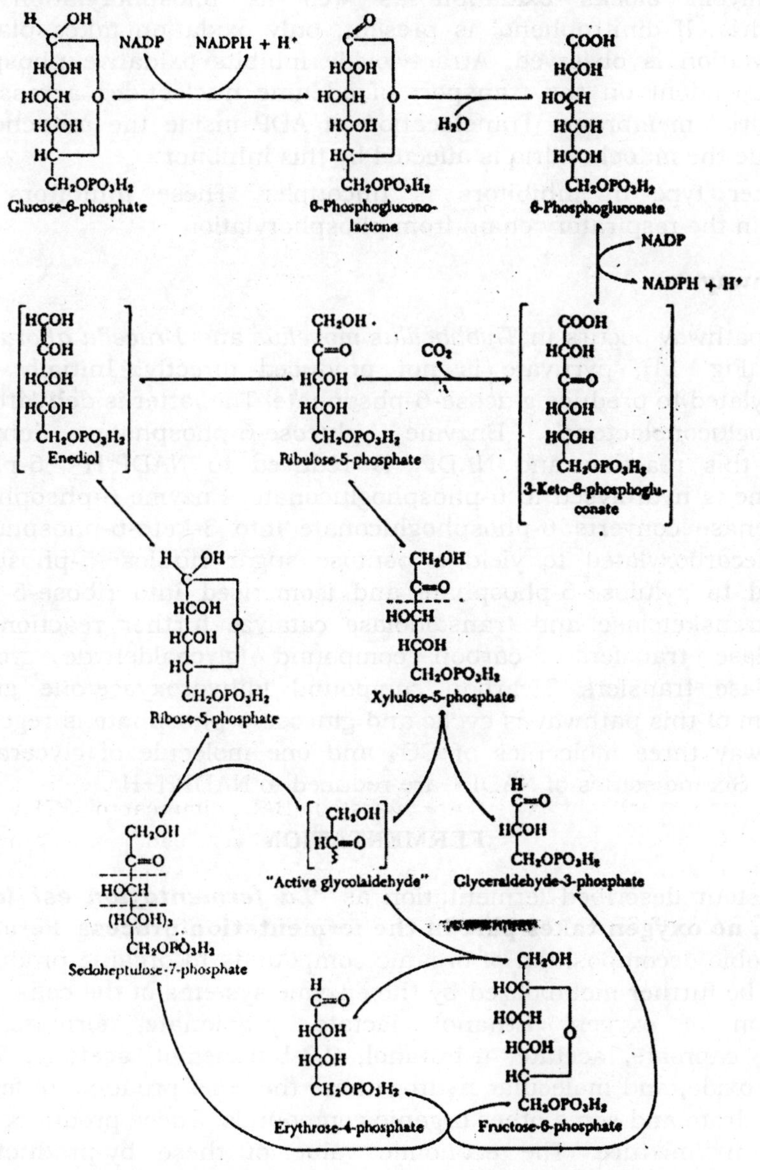

Fig. 11. Pentose-phosphate shunt pathway

Barbturates, piericidin A, and rotenone blocks respiratory chain at three sites, respectively. These inhibitors prevent the oxidation of substrates that communicate directly with the respiratory chain *via* an NAD-linked dehydrogenase by blocking the transfer from FeS to Q.

Inhibitors Dimercaprol and antimycin A act between cytochrome B and cytochrome *c*. H_2S, carbon Monoxide and cyanide inhibit cytochrome oxidase. These are treated as poison. Carboxin and TTFA inhibit transfer of reducing equivalents from succinate dehydrogenase to Q. Malonate inhibits succinate dehydrogenase. It acts like a competitive inhibitor of succinate dehydrogenase.

Oligomycin blocks oxidation as well as phosphorylation in intact mitochondria. If dinitrophenol is present, only oxidation takes place and no phosphorylation is observed. Atractyloside inhibits oxidative phosphorylation that is dependent on the transport of adenine nucleotides across the inner mitochondrial membrane. Translocation of ADP inside the mitochondria and ATP outside the mitochondria is affected by this inhibitor.

Another type of inhibitors is uncoupler. These inhibitors dissociate oxidation in the respiratory chain from phosphorylation.

HMP Pathway

This pathway occurs in *Thiobacillus movellus* and *Brucella aboratus*. In this pathway (Fig. 11), pyruvate is not produced directly. Initially glucose is phosphorylated to produce glucose-6-phsophate. The latter is dehydrogenated to 6-phosphogluconolactone. Enzyme glucose-6-phosphate dehydrogenase catalyzes this reaction and $NADP^+$ is reduced to $NADP^+H^+$. 6-phosphogluconolactone is hydrolysed to 6-phosphogluconate. Enzyme 6-phsophogluconate dehydrogenase converts 6-phosphogluconate into 3-keto-6-phosphogluconate. This is decarboxylated to yield a pentose sugar ribulose-5-phosphate It is epimerized to xylulose-5-phosphate and isomerised into ribose-5-phosphate. Enzyme transketolase and transaldolase catalyze further reactions. Enzyme transketolase transfers 2-carbon compound (glycoaldehyde group) while transaldolase transfers 3-carbon compound (dihydroxyacetone group). The mechanism of this pathway is cyclic and glucose-6-phosphate is regenerated. In this pathway three molecules of CO_2 and one molecule of glyceraldehyde is produced. Six molecules of NADP+ are reduced to $NADPH+H^+$.

FERMENTATION

L. Pasteur described fermentation as ***"La fermentation est la vie sans l'air"*** **i.e., no oxygen takes part in the fermentation process**. Fermentation is the anaerobic decomposition of organic compounds to organic products, which could not be further metabolized by the enzyme systems of the cells without the intervention of oxygen. Ethanol, lactate, propionate, formate, butyrate, succinate, caproate, acetate, n-butanol, 2,3-butanediol, acetone, 2-propanol, carbon di oxide, and molecular hydrogen are the chief products of fermentation of carbohydrate and some other organic compounds. These products are formed singly or in mixture. The economic value of these by-products led the development of industrial microbiology. Fermentation is functional in anaerobic micro-organisms. There is no oxidative phosphorylation. The only pathway common with respiration is glycolysis, and hence ATP synthesized is only in this pathway. In glycolysis there is a net gain of two ATP molecules that may be said to be the actual energy gain. Reoxidation of NDAH+H^+ produced in glycolysis depends on the reduction of pyruvate molecule. There are different types of anaerobic micro-organisms showing different types of fermentation and their end products. Hence depending upon the type of fermentation, pyruvate is utilized in different ways and reoxidation of NADH+H^+ to NAD is also different.

Alcoholic Fermentation pathway: In general ethanolic fermentation is common. A molecule of glucose is converted into two molecules of ethanol. The process can is expressed by following simple equation:

$$\text{Glucose} + 2\text{ADP} + 2\text{Pi} \longrightarrow 2 \text{ ethanol} + 2\text{ATP} + 2CO_2$$

This is common in most of the micro-organisms. Yeast specially, *Saccharomyces cerevisae* is commercially used for alcoholic fermentation. In yeast fermentation of glucose to ethanol and carbon di oxide takes place in following steps:

(a) Glucose is converted into pyruvate by EMP pathway.

(b) Pyruvate is decarboxylated to acetaldehyde. Enzyme pyruvate decarboxylase catalyzes this reaction.

(c) Acetaldehyde is reduced to ethanol. Enzyme alcohol dehydrogenase catalyzes this reaction and NADH$^+$$H^+$ acts as a hydrogen donor.

Ethanol formation by bacteriaum : Few bacteria produce ethanol. *Sarcina ventriculi* is a rod shaped bacterium. It produces ethanol by the same mechanism like that of yeast. *Zymomonas mobilis* uptakes glucose. The later is metabolized by 2-keto-3deoxy-6-phosphogluconate pathway and pyruvate is formed. Enzyme pyruvate decarboxylase splits pyruvate into acetaldehyde and carbon di oxide. Acetaldehyde is reduced to ethanol. Ethanol is one of the products of fermentation of heterofermentor lactic acid bacteria, Closridia and Enterobacteriaceae.

Lactic Acid Fermentation pathway : Lactic acid fermentation is carried out by bacteria grouped in Lactobacillaceae, they are commonly known as lactic acid bacteria. They are Gram-negaitve, cocci or rods and non-sporulating. Depending upon the type of end product lactic acid bacteria are divided into two types (a) homolactic and (b) heterolactic. *Streptococcus, Pediococcus* and few species of Lactobacillus are homolactic bacteria. Lactic acid is the only end product of fermentation. The overall process can be expressed by following equation:

$$\text{Glucose} + 2\text{ADP} + 2\text{Pi} \longrightarrow 2 \text{ Lactic acid} + 2\text{ATP}$$

Homolactic acid fermentation is important in dairy industry.

Leuconostoc and few species of *Lactobacillus* are heterolactic. Ethanol and CO_2 is also formed along with lactic acid as the end products of fermentation. In this type of fermentation phosphoketolase type of pathway of glycolysis is adopted. The overall process can be expressed by following equation:

$$\text{Glucose} + \text{ADP Pi} \longrightarrow \text{Lactic acid} + \text{ethanol} + CO_2 + \text{ATP}$$

Heterolactic acid fermentation is important for the production of sauerkraut.

Enzyme Lactate dehydrogenase catalyzes formation of lactate from pyruvate. This enzyme oxidizes NADH+H$^+$ to NAD+ and pyruvate is reduced to lactate.

$$\text{Pyruvate + NADH+H}^+ \longrightarrow \text{lactate + NAD+.}$$

The amount of lactate synthesized depends upon the conditions in which micro-organisms are growing and the type of micro-organism.

Propionic acid fermentation pathway This type of fermentation takes place in *Propionibacterium.* It is rod shaped and Gram positive. It has a habitat in rumen and intestine of ruminants (cattle and sheep). Beside metabolizing carbohydrate and forming propionic acid, this bacterium is able to carry out fermentation of lactic acid. Lactic acid bacteria forms lactic acid from milk and in late fermentation, propionic bacteria converts this lactic acid into propionic acid. *Veillonella alcalescens, Clostridium propionicum, Selenomonas* and *Microspora* also carry out propionic acid fermentation. Mechanism of propionate formed by lactate can be expressed by following simple formula:

$$\text{Lactate} \longrightarrow \text{Propionate + Acetate + } CO_2 + H_2O$$

Propionic acid is formed from pyruvate of lactate by methylmalonyl-CoA pathway. The sequential intermediates of formation of propionate from pyruvate are as following :

Pyruvate → oxaloacetate → malate → fumarate → succinate → succinyl CoA → methylmalonyl CoA → propionyl-CoA → Propionate.

In *Clostridium propionicum,* propionate is formed as per following intermediates :

Lactyl-CoA → Acryloyl-CoA → Propionyl-CoA → Propionate.

Acetic acid fermentation pathway *Acetomonas* and *Acetobacter* carry out acetic acid fermentation. Frateur (1950) included all the acetic acid bacteria in one genus *Acetobacter* and divided this genus into four groups *peroxydans, oxydans, mesooxydans* and *suboxydans.* DeLey (1961) included *peroxydans, oxydans* and *mesooxydans* in *Acetobacter* and *suboxydans* in *Gluconobacter.*

Acetate formation in Acetobacter peroxydans takes place in following pathway

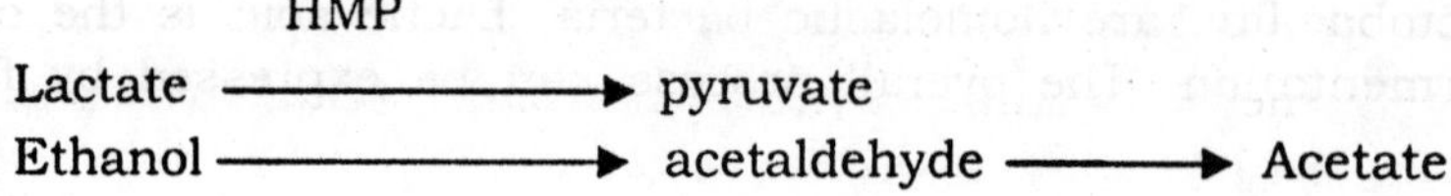

CHAPTER 12

AMINO ACID AND LIPID METABOLISM

BIOSYNTHESIS OF AMINO ACIDS

In the synthesis of amino acids, inorganic ammonium nitrogen is incorporated to organic acid *via* two pathways. In both the pathways ammonium ions are assimilated to form the amino acid L-glutamate. The ammonium moiety can be transferred to other organic acids to form other types of amino acids.

Reductive deamination

In reductive deamination, ammonium ion combine directly with α-keto carboxylic acids to form amino acids. An intermediate of Kreb's cycle, α-ketoglutarate accepts an ammonium ion to form L-glutamic acid. This reaction is catalyzed by enzyme glutamate dehydrogenase and NADPH+H⁺ acts hydrogen donor.

$$\alpha\text{ -ketoglutaric acid} + NH_4^+ \xrightarrow[\text{NADPH+H+} \quad \text{NADP+}]{\text{Glutamate dehydrogenase}} \text{L-glutamic acid}$$

Glutamate synthetase pathway

This pathway is also known as glutamine synthetase pathway. It takes place when concentration of ammonium is low. In the first step of this pathway, L-glutamate reacts with ammonium ions to form the amino amide L-glutamine. In this reaction energy is supplied by the hydrolysis of ATP. L-glutamine reacts with α-ketoglutarate to form two molecules of L-glutamate. This reaction is catalyzed by enzyme glutamine synthetase and glutamate synthetase. NADPH+H⁺ acts as reducing power.

$$\text{Glutamate} + NH_4 \xrightarrow[\text{ATP} \quad \text{ADP}]{} \text{Glutamine} + H_2O$$

$$\text{Glutamine} + \alpha\text{-ketoglutarate} \longrightarrow \text{2 Glutamate.}$$

After the synthesis of Glutamate, the process of transamination yields six essential amino acids are synthesized. These amino acids act as precursors of six families of amino acids, which form all the twenty amino acids required for the protein synthesis.

CATABOLISM OF AMINO ACIDS

The disintegration of protein yields amino acid. The catabolism of amino acids may be discussed in following heads as far as ammonia group is concerned:

(1) Transamination,

(2) Deamination (a) Oxidative deamination and
(b) Nonoxidative deamination.

(3) Transdeamination.

1. TRANSAMINATION

Transamination was first observed by Braunstein and Ritzmann in muscle. Lichstein and Cohen (1944) demonstrated this reaction in bacteria. They observed that suspension of *E. coli* and number of other organism catalyzes the reaction.

l(+) glutamic acid + oxaloacetic acid — α-ketoglutaric acid + l(-)aspartate

Transamination is a reversible reaction in which alpha amino group of one amino acid is transferred to alpha carbon atom of another keto acid, resulting in the formation of new amino acid. The general process of transamination may be represented as follows —

Amino acid + Organic acid ⟶ New amino acid + New keto acid.
(donor) (recipient)

Glutamic pyruvic transaminase
Glutamic acid + Pyruvic acid ⟶ Ketoglutaric acid + Alanine

NH_2 is intramolecularly transferred. NH_3 is not formed in transamination reactions. These reactions are catalyzed by enzyme Transaminases (amino transferases). The coenzyme required for this reaction is pyridoxal phosphate (PLP), the coenzyme form of vitamin B6.

Details of mechanism are that the amino acid reacts with enzyme bound pyridoxal phosphate to form an enzyme bound complex of the Schiff-base type which then yields keto acid and pyridoxamine- phosphate. Pyridoxal phosphate is an aldehyde form while the pyridoxamine phosphate is its aminated form. The pyridoxamine phosphate then reacts with a second keto acid to produce a similar enzyme bound Schiff-base complex. The later breaks down forming a second (new) amino acid and regenerates the pyridoxal phosphate (Fig. 1).

There are some limitations to the reaction of transamination. Such as (i) Group transaminases and specific transaminases and (ii) Alanine transaminasse (or Alanine amino-transferase).

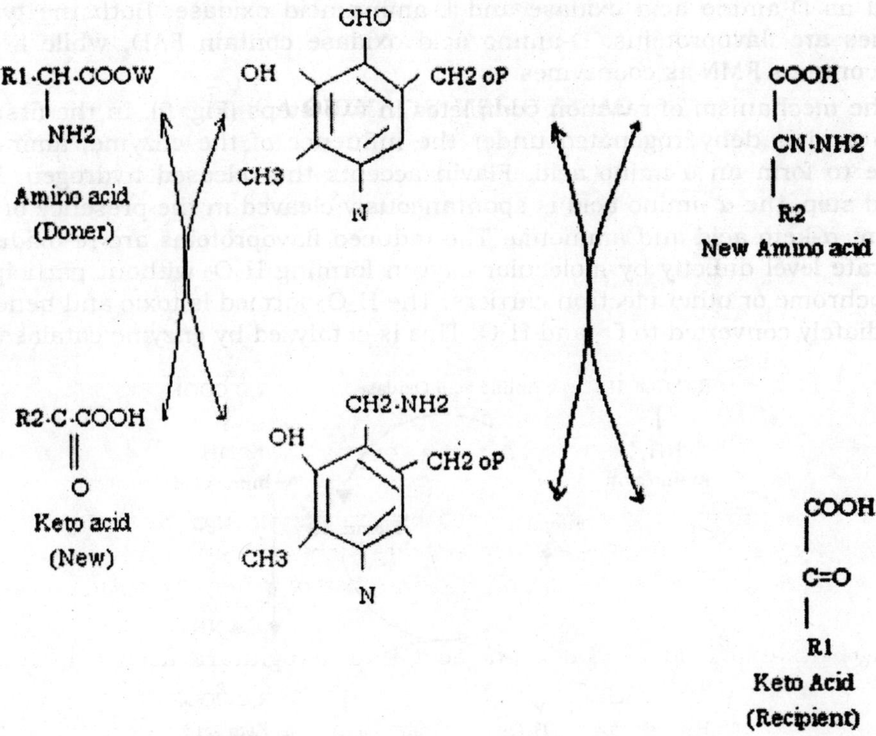

Fig. 1.

Group Transaminases and Specific reactions : Most of the amino acids may act as donor I and Keto acids acts as recipient. Recipient keto acids are the intermediates of TCA. They may be α-keto-(oxo) glutrate or oxalo-acetate or pyruvate. The new amino acids formed by these keto acids are glutamic acid, aspartic acid and alanine. The enzyme transaminases, which catalyze the conversion of these keto acids into their respective amino acids, are known as group transaminases.

There are few amino acids, which do not take part in transamination e.g., lysine, Threonine and proline. Transamination is not restricted to α $-H_2$ groups. The $-NH_2$ groups of orthine is readily transmitted forming glutamate-γ-semialdehyde.

2. DEAMINATION

As the name suggests, deamination means loss of amino groups from amino acids. It is of two types, nonoxidative deamination and oxidative amination.

(a) Oxidative Deamination

Enzyme amino acid oxidase deaminase catalyzes the deamination of amino acid. Depending upon the type of amino acid (D or L amino acid) the enzyme is named as D-amino acid oxidase and L-amino acid oxidase. Both the types of enzymes are flavoproteins. D-amino acid oxidase contain FAD, while L-amino acids contains FMN as coenzymes.

The mechanism of reaction completes in two steps (Fig. 2). In the first step, amino acid is dehydrogenated under the influence of the enzyme, amino acid oxidise to form an α-imino acid. Flavin accepts the released hydrogen. In the second step, the α -imino acid is spontaneously cleaved in the presence of water to form α-keto acid and ammonia. The reduced flavoproteins are re-oxidized at substrate level directly by molecular oxygen forming H_2O_2 without participation of cytochrome or other electron carriers. The H_2O_2 formed is toxic and hence it is immediately converted to O_2 and H_2O. This is catalyzed by enzyme catalase.

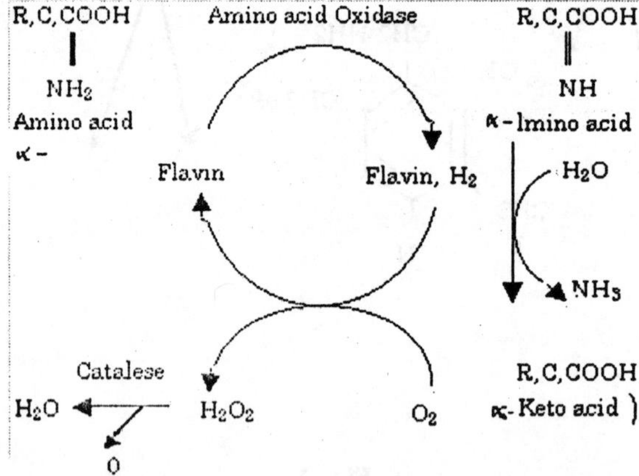

Fig. 2. Mechanism of Oxidative deamination of Amino acids

Non-Oxidative Deamination

Some amino acids are not oxidized while deamination. Ammonia is formed in this reaction. The hydroxy amino acids (serine, threionine) are deaminated by specific enzymes called amino acid dehydrases, which requires pyridoxal-P (B6-P) as coenzyme.

Serine + H_2O ⟶ Imino acid

Imino acid + H_2O ⟶ Pyruvic acid + NH_3

Deamination of Histidine Histidine is non-oxidatively deaminated. Ammonia and uronic acid are the products. This reaction is catalyzed by enzyme histidase.

$$\text{Histidine} \xrightarrow{\text{Histidase}} NH_3 + \text{Uronic acid}$$

Clostridium tetanomorphum converts histidine to butyrate, acetate, CO_2, ammonia and formamide.

Deamination of cysteine In the deamination of cysteine, first H_2S is removed and imino acid is formed. This is known as desulfhydration. This reaction is catalyzed by enzyme desulfhydrase. Imino acid thus formed is spontaneously hydrolyzed releasing ammonia and pyruvic acid is formed.

$$\text{L-cysteine} \xrightarrow{\text{desulfhydrase}} \text{imino acid} + H_2S$$

$$\text{Imino acid} + H_2O \longrightarrow \text{Pyruvic acid} + NH_3$$

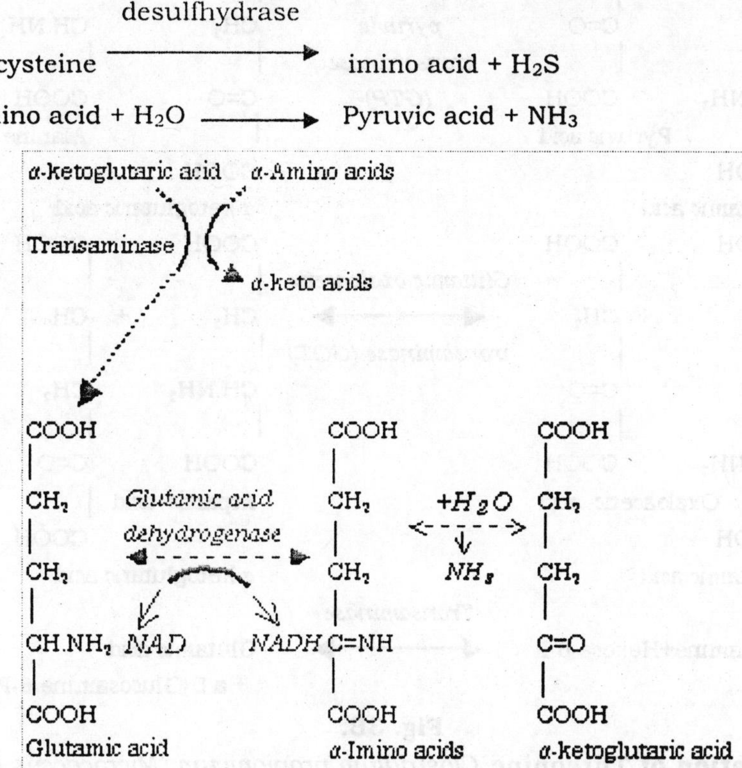

Fig. 3A. Trans-deamination

3. TRANSDEAMINATION

Deamination of L-Glutamic acid — L-glutamic acid is deaminated by a specific enzyme L-glutamate dehydrogenase (Fig. 3 A, B). This enzyme is made up of four polypeptide chains, each containing a Zn ion. This enzyme requires NAD^+ or $NADP^+$ as coenzyme. In the reaction, first α-imino glutaric acid is formed. A molecule of water adds to this imino acid and α-ketoglutrate is formed, ammonia is liberated.

$$\text{L-glutamate} \xrightarrow[\text{NAD}^+/\text{NADP}^+ \;\rightarrow\; \text{NADH+H}^+/\text{NADPH+H}^+]{\text{L-glutamate dehydrogenase}} \alpha\text{-imino glutaric acid}$$

$$\text{-imino glutaric acid} + H_2O \longrightarrow \alpha\text{-keto glutaric acid} + NH_3$$

L-histidine degradation has been reported from studies with *Esherichia coli.*

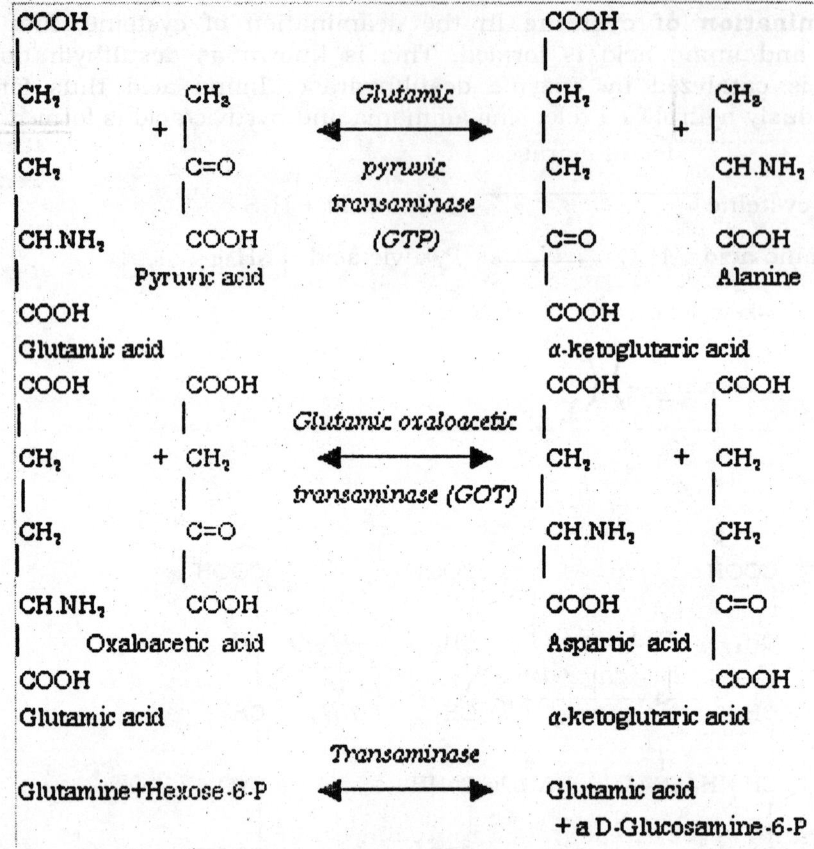

Fig. 3B.

Deamination of Threonine *Clostridium propionicum, Micrococcus aerogenes* utilize threonine. The latter is deaminated into α-oxobutyrate. Enzyme threonine dehydratase catalyzes this reaction.

Deamination of Glycine Glycine metabolism occurs in the spices of *Diplococcus* and *Micrococcus* (*Diplococcus glycinophilus, Micrococcus anaerobius* and *M. variable*). Ammonia, CO_2, and acetic acid are produced by anaerobic deamination of glycine

$$\text{Glycine} + H_2O \longrightarrow NH_3 + CO_2 + \text{acetic acid.}$$

PATHWAY OF AMINO ACID DEGRADATION

There are 20 amino acids in proteins, with a variety of carbon skeletons. Correspondingly, there are 20 different catabolic pathways for amino acids degradation. The 20 catabolic pathways converge to form only five products, all of which enter the TCA (Fig. 4). From here the carbons can be diverted to gluconeogenesis or ketogenesis, or they can be completely oxidized to CO_2 and H_2O. All or part of the carbon skeletons of ten of the amino acids are ultimately broken down to yield acetyl – CoA. Five amino acids are converted into α–ketoglutarate, four into succinyl CoA; two into fumarate and two into oxalate.

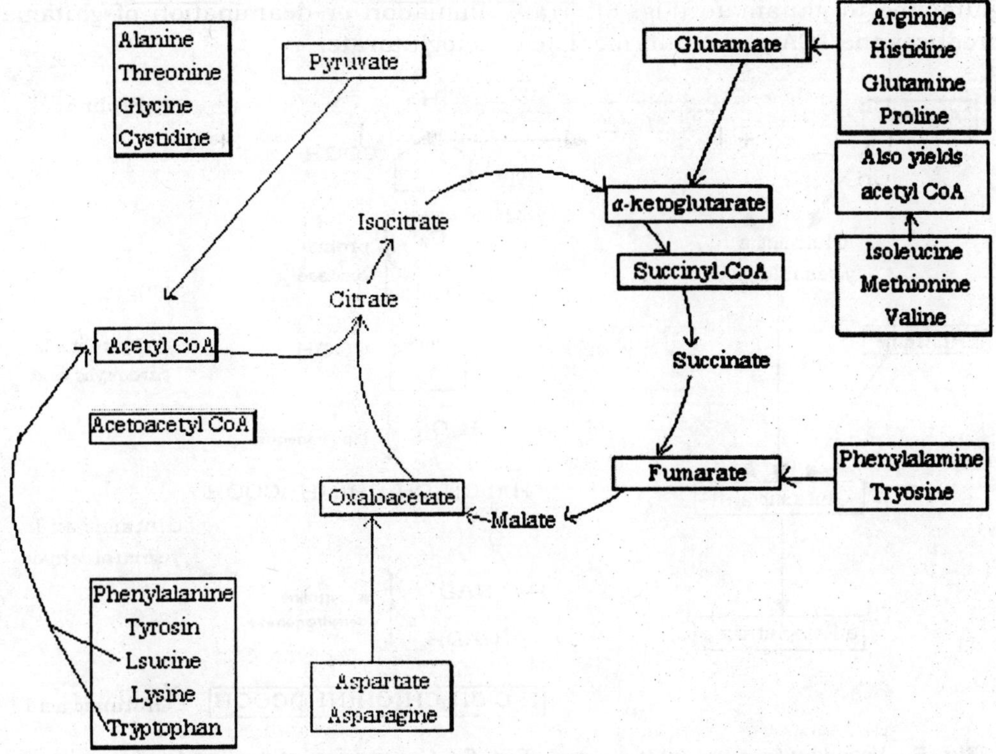

Fig. 4. Pathway of oxidation of Amino acids *via* TCA

1. Ten amino acids are degraded to Acetyl CoA

The carbon skeletons of ten amino acids yield acetyl-CoA, which enters the TCA cycle directly. Five of the ten are degraded to acetyl CoA *via* pyruvate. The other five are converted into acetyl CoA and or acetoacetyl-CoA, which is then cleaved to form acetyl CoA. The five amino acids entering *via* pyruvate are alanine, glycine, serine, cysteine and tryptophan. Portions of the carbon skeleton of six amino acids- tryptophan, lysine, phenylalanine, tyrosine, leucine and isoleucine yield acetyl CoA and/or acetoacetyl CoA; the latter is then converted into acetyl CoA.

2. Five amino acids are converted in α- ketoglutarate

The carbon skeletons of five amino acids (argine, histidine, glutamate, glutamine and praline) enter the TCA cycle *via* α-ketogluterate (Fig. 5). Proline, glutamate and glutamine have five carbon skeletons. The cyclic structure of proline is opened by oxidation of the carbon most distant from the carboxyl group A to create an Schiff base and hydrolysis of Schiff base to a linear semialdehyde (glutamate semialdehyde). This is further oxidized at the same carbon to produce glutamate. The action of glutamine or any of several reactions in which glutamine donates its amide nitrogen to some acceptor, converts glutamine to glutamate (Fig. 6). Trans-amination or deamination of glutamate produces the TCA cycle intermediate α–ketoglutarate.

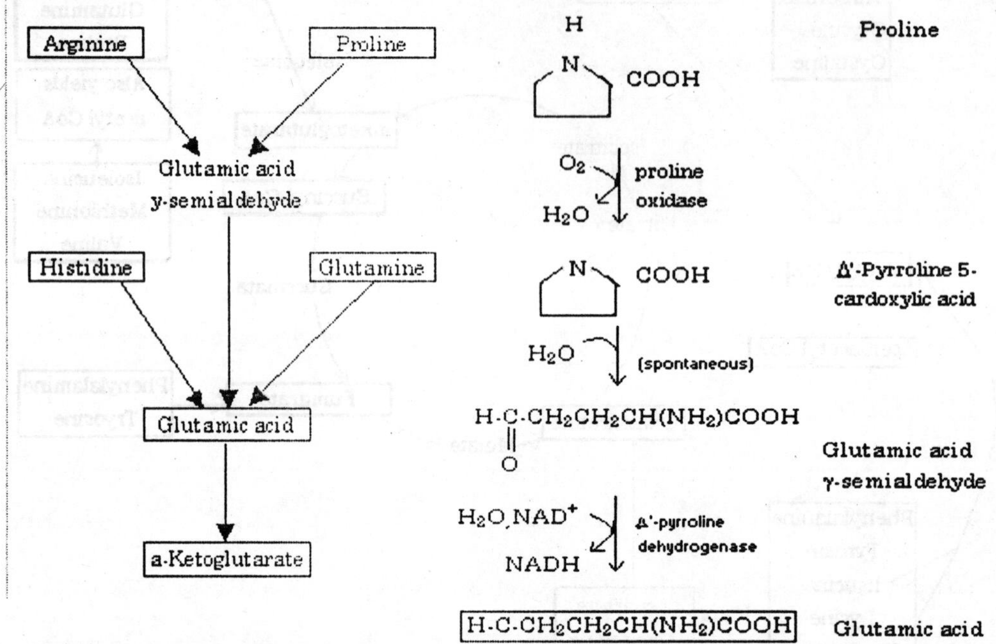

Fig. 5 – Pathway to α-ketogluterate **Fig. 6** – Conversion of proline to glutamic acid

Arginine and histidine contain five adjacent carbons and a sixth carbon attached through a nitrogen atom. The catabolic conversion of these two amino acids to glutamate is therefore slightly more complex than the path from proline or glutamine to glutamate. Arginine is converted to the five-carbon skeleton of ornithine in the urea cycle and the ornithine is transaminated to glutamate semialdehyde. The conversion of histidine to the five-carbon glutamate occurs in 4 multistep pathways the extra carbon is removed in a step that employs tetrahydrofolate as a cofactor.

3. Four amino acids are converted into succinyl CoA

The carbon skeletons of methionine, isoleucine, threonine and valine are degraded by pathway that yield succinyl CoA an intermediate of TCA cycle (Fig. 7).

4. Asparagine and Asparate are degraded to oxaloacetate

The carbon skeletons of asparagines and asparate ultimately enter the TCA cycle via oxaloacetate. The enzyme asparaginase catalyzes the hydrolysis of asparagine to yield asparate, which undergoes a transmission reaction with ketoglutarate to yield glutamate and oxaloacetate. The latter enters the TCA cycle.

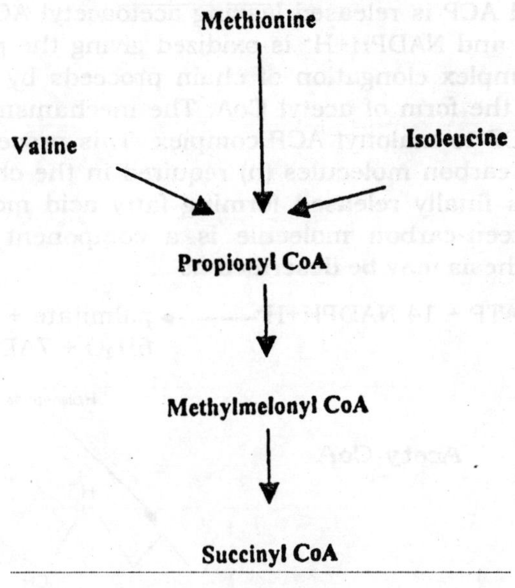

Fig. 7.

5. Some amino acids can be converted to glucose; others to ketone

Some carbon atoms from six of the amino acids (those that are degraded to acetoacetyl CoA /or acetyl CoA, tryptophan, lysine, phenylalanine, tyrosine, leucine and isoleucine) can yield ketones in the liver by conversion of acetoacetyl CoA into acetone and b hydroxybuterate. These amino acids are ketogenic.

The amino acid that can be converted into pyruvate, α-ketoglutarate, succinyl – CoA, fumarate and oxaloacetate can be converted into glucose and glycogen. They are called glucogenic amino acids.

LIPID METABOLISM

Lipids are essential structural component of a cell. In some cells it is a stored food. Glycerol molecules are formed from triglycerides of fatty acid with the help of enzyme lipase. The glycerol molecule is metabolized into dehydroxyacetone phosphate, which is isomerised into 3-phosphoglycerate. The latter enters the metabolic pathway to yield energy to the growing cell.

FATTY ACID SYNTHESIS

The synthesis of fatty acid is derived by addition of carbon in CoA form (acetyl CoA and Malonyl CoA. The precursor of fatty acid is malonyl CoA (a three carbon CoA). This is synthesized by addition of CO_2 to acetyl CoA (Fig. 8 A,B). ATP supplies energy in this reaction. A protein known as acyl carrier protein (ACP) takes part in further reaction. One molecule of CO_2 is released and malonyl CoA-ACP complex is formed. Two carbons (acetyl CoA) are added to it. Acetyl CoA is incorporated as acetyl CoA-ACP complex. In the next step, a molecule of CO_2 and ACP is released leaving acetoacetyl ACP complex. A water molecule is released and NADPH+H^+ is oxidized giving the product butyryl ACP complex. To this complex elongation of chain proceeds by the addition of two carbon at a time in the form of acetyl CoA. The mechanism is same as that of addition of acetyl-ACP to malonyl ACP complex. This process may be repeated upto the number of carbon molecules (n) required in the chain of fatty acid. In the last step ACP is finally released forming fatty acid molecule. A saturated fatty acid with sixteen-carbon molecule is a component of phospholipid of membranes. Its synthesis may be described as

$$8 \text{ Acetyl CoA} + 7 \text{ ATP} + 14 \text{ NADPH}+H^+ \longrightarrow \text{palmitate} + 14 \text{ NADP} + * \text{ CoA} + 6H_2O + 7ADP + 7Pi$$

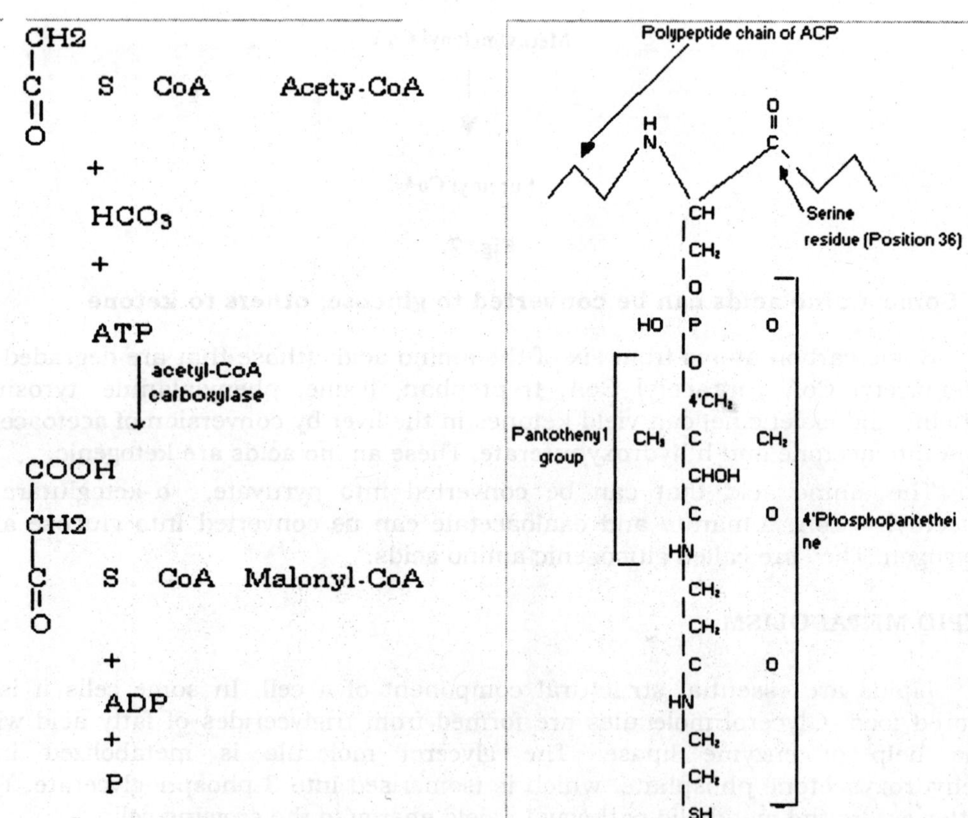

Fig. 8 A. Carboxylation of Acetyl CoA

Fig. 8B. Structure of ACP

SYNTHESIS OF POLY- β-HYDROXYBUTYRIC ACID

Poly-β-hydroxybutyric acid is a common stored food of many bacteria. The synthesis of this storage product takes place in the same pattern as that of synthesis of fatty acid. Acetoacetyl CoA formed during the pathway is reduced to β-hydrobutyryl CoA. NADH+H^+ acts as hydrogen donor. In this pathway NADH+ H^+ takes part instead of NADPH+H^+. The chain undergoes elongation by the repetitive sequential addition of acetyl CoA. In the last step CoA is removed and poly- β-hydroxybutyric acid is synthesized.

BIOSYNTHESIS OF PHOSPHOLIPIDS

Phospholipids are important components of cellular membrane of both prokaryotic and eukaryotic organisms. It is synthesized by the addition of fatty acids to glycerol phosphate. The glycerol 3-phosphate reacts with acetyl CoA to form phosphatidate. The latter is a common intermediary metabolite in the synthesis of phospholipids and triglycerides. The fatty acid-acetyl carrier protein molecule reacts with the phosphatidate to form elongated fatty acid.

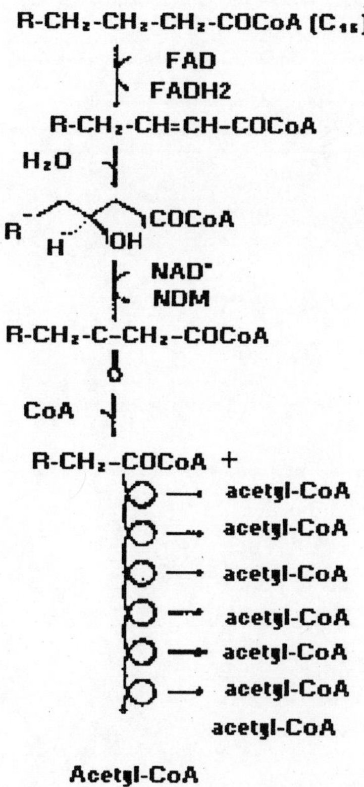

Fig. 9. β-oxidation of Fatty acid

CATABOLISM OF FATTY ACID

In the catabolic pathway of digestion of fatty acid portion of lipid, two carbon molecules are removed at a time in the form of acetyl CoA. This type of breakdown of fatty acid is known as β-oxidation. In the first step CoA reacts with fatty acid to from a fatty acid-CoA molecule. ATP supplies energy in this reaction. Fatty acetyl CoA is oxidized and a molecule of fatty acid is synthesized, with the chain two carbon shorter than that of the substrate chain (Fig. 9). The release of acetyl CoA is coupled with the formation of $FADH_2$ and one molecule of $NADH^+H^+$. In every step of breakdown a molecule of CoA is released with the formation of a molecule of $FADH_2$ and a molecule of $NADH+H^+$. The acetyl CoA thus formed enters in to the Kreb's cycle and yield ATP as a result of oxidative phosphorylation.

CHAPTER 13

NUTRITION AND GROWTH

NUTRITION

Nutrients are the substances required by an organism to synthesize its cell materials. Microorganisms show diversity in their nutritional requirements. A living cell contains about 80 to 90% water of the total weight of the cell. Hydrogen and oxygen are derived from water and are available to a cell as per its need. Other matters found in a cell are carbon, nitrogen, phosphorus and sulphur. Some metallic ions also have significant role in cell physiology. The metals present in a cell are potassium, magnesium, calcium, iron, manganese, cobalt, copper, molybdenum and zinc. These are also termed as trace elements. These metallic ions are cofactors of a number of enzymes catalyzing different biochemical cycle in a cell.

Carbon source

Carbon is a basic constituent of all organic compounds in a cell. Photosynthetic microorganisms utilize CO_2 and reduce it to carbohydrate. The energy required for this purpose is gained from sunlight or released from oxidation of reduced inorganic compounds. Major groups of microorganisms utilize organic nutrients. These nutrients have dual role. They act as nutrients (carbon source) as well as a source of energy. Many microorganisms can use a single organic compound to supply completely both these nutritional needs. While some microbes require more than one carbon source for their growth. The additional organic nutrients are known as growth factors. They act as a precursor of biochemical pathways. Cells are unable to synthesize these growth factors. Hence, they are supplied with nutrient media.

Nitrogen, Sulphur and Phosphorus source

Nitrogen is a constituent of proteins, nucleic acids and coenzymes. Sulphur is present as sulphydryl group in amino acids (cysteine and methionine) and is a constituent of some coenzyme (CoA, carboxylase). Phosphorus is a constituent of nucleic acids, nucleotides, teichoc acid, phospholipids, and coenzymes. Most of the bacteria and fungi assimilate these two elements in the oxidized inorganic state, as nitrates and sulphates. The biosynthetic utilization of these elements involves a preliminary reduction. However, some microbes are not able to reduce these anions. Reduced nitrogen is supplied in forms like ammonium salts. Sulphur is not supplied in reduced forms. An important source of phosphorus is the phosphate ion.

Oxygen

Microorganisms show different behaviour for affinity towards oxygen requirement. The capacity of microbes to grow in the presence of oxygen and to utilize it depends on the possession of a cytochromic oxidase system. Based on the requirement of oxygen by microorganism, the latter are divided into following categories.

- Obligate aerobes Microorganisms require molecular oxygen for their life.
- Facultative aerobes Microorganisms are able to use oxygen when available but can also grow in absence of oxygen.
- Obligate anaerobes Microorganisms are not able to use molecular oxygen, and if oxygen is present it is harmful for their growth.
- Aerotolerant anaerobes Microorganisms cannot use oxygen for growth, but they tolerate it fairly well.
- Microaerophilic Microorganisms grow in a very low concentration of oxygen.

It is interesting that how oxygen is toxic for some microorganisms. Oxygen is both beneficial and poisonous to living organisms. It is beneficial due to its strong acidifier nature, which makes it a terminal electron acceptor for energy yielding process known as respiration. However, O_2 is also a toxic substance. Aerobic and facultative organisms have protective mechanisms but microaerophytes and anaerobes are deficient in these mechanisms. They are restricted to habitats where little or no oxygen is present.

Damage due to toxic derivatives of oxygen

Various cellular enzymes catalyze chemical reactions involving molecular oxygen. Some of these reactions can result in addition of a single electron to an oxygen molecule, thereby producing a super oxide radical (O_2):

$$O_2 + e^- \rightarrow O2^-$$

This superoxide radical can cause inactivation of vital cell components. Superoxide becomes more harmful through the production of more toxic substances such as hydrogen peroxide (H_2O_2) and hydroxyl (OH).

$$2O2^- + 2H^- \rightarrow O_2 + H_2\ O_2$$

$$O_2^- + H_2O_2 \rightarrow O_2 + OH^-\ OH^-$$

Hydroxyl radicals can cause damage to every kind of molecule of living cell while hydrogen peroxide that is a powerful oxidizing agent and is highly toxic to many kinds of cells.

Another toxic derivative of oxygen is an energized form known as singlet oxygen $(\Delta G)O_2$, that is produced in biological systems by certain photochemical reactions.

Aerobes and aerotolerant anaerobes are protected from these products by the presence of enzymes super oxide dismutase, catalase and peroxidase, former eliminates superoxide radicals by greatly increasing rate of reaction and H_2O_2 is dissipated by latter two enzymes. All strict anaerobes lack super oxide dismutases, catalase and peroxidase.

Singlet oxygen Molecular oxygen is boosted into a higher-energy state and is extremely reactive. It is present in phagocyte cells.

Superoxide free radicals (O_2^-)

Microorganisms have enzymes, which can react with O_2. The oxidation of flavoproteins by O_2 invariably results in the formation of H_2O_2. With H_2O_2 Superoxide free radicals (O_2^-) are formed in small amounts. These products are toxic to a cell. Enzyme superoxide dismutase is produced in aerobic, facultative anaerobe and aerotolerant anaerobes. This enzyme converts the super oxide free radical into molecular oxygen (O_2) and hydrogen peroxide (H_2O_2).

$$2O_2^- + 2H^+ \rightarrow O_2 + H_2O_2$$

Hydrogen peroxide contains peroxide ion. This is also toxic to a cell. Enzyme peroxidase catalyzes the oxidation of organic compounds by H_2O_2. It is reduced to water.

Some bacteria employ a catalase-like reaction dependent on high intracellular concentrations of Mn^{+2}. It also reduces hydrogen peroxide into water.

Strict anaerobes lack both superoxide dismutase and catalase.

NUTRITIONAL CATEGORIES

In the old concept of nutrition, microorganisms were classified into two categories (I) Autotrophs microbes utilising inorganic nutrients and capable of synthesizing their own food. They use CO_2 as the main source of carbon and simple inorganic salts.

(II) Heterotrophs microbes utilising organic nutrients and deriving their food from autotrophs or lower categories of heterotrophs.

In the modern concept of nutrition, classification of microorganisms is based on two parameters, i.e., nature of energy source (either light or chemical energy) and the nature of the principal carbon source (either inorganic or organic carbon source). Based on above two criteria microorganisms are classified in following categories.

(i) Phototrophs microbes utilising light as energy source.

Phototrophs are again divided into two classes.

(a) Photoautotrophs microbes using light as energy source and CO_2 as the principal carbon source. Photosynthetic microorganisms are included in this category.

(b) Chemoautotrophs microbes using light as energy source and organic compounds as the principal carbon source, e.g., some green and purple bacteria.

(ii) Chemotrophs microbes using chemical energy source.

Chemotrophs are again divided into two classes.

(a) Chemoautotrophs microbes using a chemical energy source and CO_2 as the principal carbon source. The chemical energy is gained by oxidation of reduced inorganic compounds, like iron, sulphur, ammonia and nitrite. These organisms are also known as chemolithotrophs (Gr. Lithos rock). It is because of their nature to grow in strictly mineral media in the absence of light e.g., *Thiobacillus.*

(b) Chemoheterotrophs microbes using a chemical energy source and an organic matter as the principal carbon source. Protozoa, fungi and majority of bacteria are included in this category.

Chemoheterotrophs are divided in two groups osmotrophs and phagotrophs. Osmotrophs include fungi which take nutrients in dissolved form. Phagotrophs e.g., protozoa can take up solid food particles by the mechanism called phagocytosis.

GROWTH

In simple terms growth may be considered as increase in the number of cells or increase in all chemical components of a cell. Sometimes growth is confused in the increase in biomass. Sometimes this is measured as growth because increase in stored food will increase the biomass. As it is difficult to define growth, a concept of balanced growth has been developed. In a state of balanced growth, all the components are increased. There should be a correlation between the increase in biomass and increase in number of cells.

CONTINUOUS CULTURES OF MICROORGANISMS

This method of culture is being used for both research and industrial purposes. In a bacterial culture, there is mixture of three developmental stages. A few cells might be active young cells, some in an optimum metabolic stage and a few in their dying stage. In these extreme parameters of growth, there is a stage when the cells are in state of balanced growth. This is the logarithmic phase. It is important for metabolic studies of an organism and its use for industrial purposes. It is possible to maintain a culture in which microorganisms are in a constant state of growth (logarithmic phase). This is known as continuous culture. Microbial populations can be maintained in a state of exponential growth over a long period of time by using a system there is a continuous addition and simultaneous removal of culture medium. The growth chamber is connected to a reservoir of sterile medium. The flow of medium is maintained constant and overflow is removed by siphon (Fig. 1). The excess of old cells also overflows with the medium through siphon. Thus, the rate of growing cells in the medium is equal to the rate of loss of cells with the medium through the overflow. Continuous cultivation is possible in a chemostat or a turbidostat. Chemostat and turbidostat are usually operated at different dilution rates. The dilution rate is the ratio of inflowing amount of nutrient medium per hour to the volume of the culture.

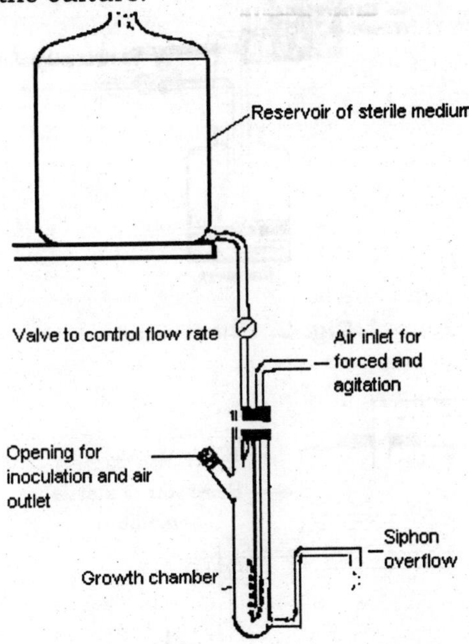

Fig. 1. Continuous culture system

Chemostat is a continuous cultivator system (Fig. 2). It is operated with continuous addition of sterilized medium from a reservoir and simultaneous removal of culture medium from the system. This system depends upon the fact that the concentration of an essential nutrient (substrate) within the culture vessel will control the growth rate of the cells. The effluent removed includes accumulated waste products and older and dead cells. Maximum homogeneity is maintained. Growth is controlled by way of limiting one of the nutritional factors. The growth-limiting nutrient is added into the medium at a concentration below the required amount.

Turbidostat has an optical sensing device (Fig 3). Continuous growth is monitored by turbidity. Here a photoelectric device continuously monitors the cell density within the true vessel and controls the dilution rate to maintain the cell density at a constant value. If density becomes too the dilution rate is increased and if the density becomes too low the dilution rate is increased. A photoelectric cell monitors cell density in the vessel. The electrical signal from the photoelectric cell regulates the flow rate of the nutrient solution in such a way that turbidity remains constant.

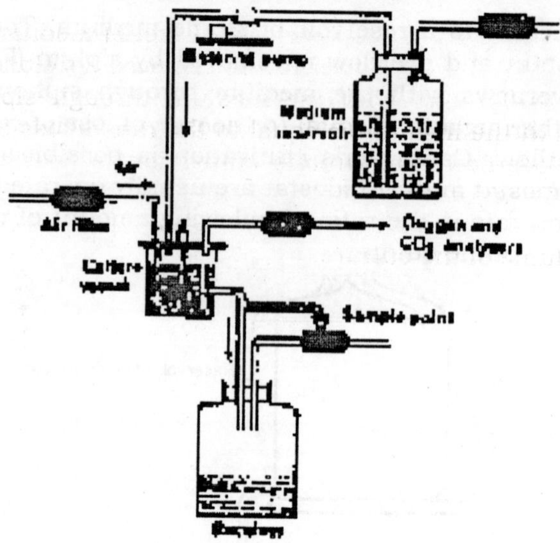

Fig. 2. Chemostat

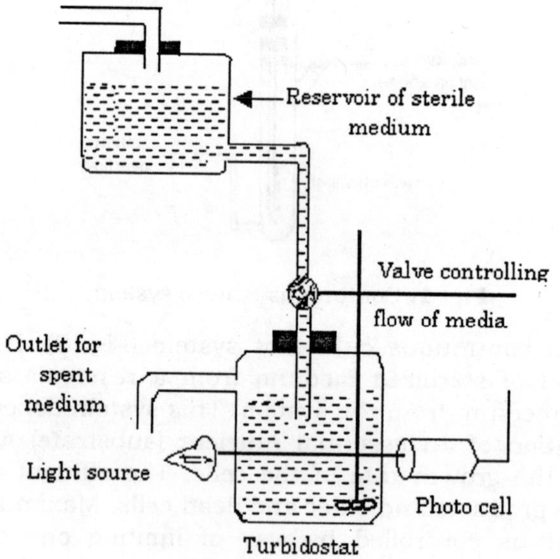

Fig. 3. Turbidostat

SYNCHRONOUS CULTURES

A culture in which all the cells are at the same stage of the cell cycle is known as synchronous culture. To maintain such culture, important is the selection of cells from a population. These cells should be of same size and age so that they grow uniformly. Few techniques have been developed to select such cells.

Helmstetter-Cummings technique - Certain bacteria stick tightly to cellulose nitrate membrane (Fig. 4). A culture containing cells of different size and age is passed through a membrane filter. The filter is then inverted and fresh medium is allowed to flow through it so that the loosely associated cells are washed away. The cells adhered to the membrane are newly formed by cell division. These cells are collected with effluent stream. Cells thus collected are at the same stage of the cell cycle.

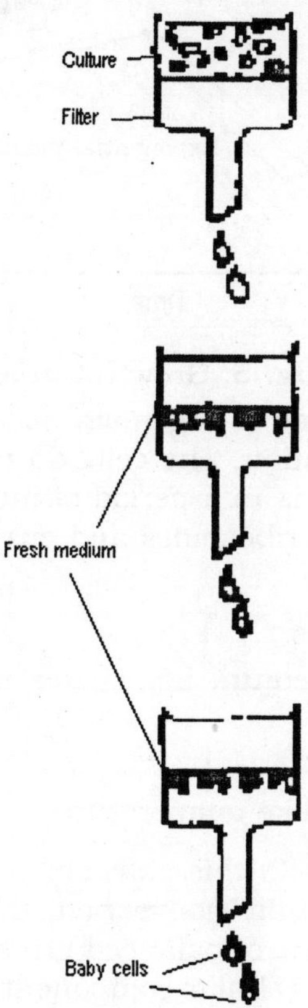

Fig. 4. Helmstetter-Cummings technique of Synchronous culture

THE GROWTH CURVE

When any medium is inoculated with micro-organisms, they start growing. This process continues until the nutritional components start finishing or there is accumulation of toxic products. In a closed system (vessel) if microorganism is inoculated and the medium is kept constant. No additional medium is added periodically and the metabolic products of growth are not removed during growth, this system is known as static culture or batch culture. During growth cells attain different size and age or in other words they are at different stages of cell cycle. In a batch culture, there are different phases of growth of a micro-organism. The sequences of growth phase are known as growth curve (Fig. 5). To obtain a growth curve, the logarithms of the viable cells are plotted against time.

Few bacterial cells are inoculated into a broth and the population of bacteria is counted at a definite interval. The sequences of phases are (i) lag phase; (ii) log (exponential) phase; (iii) stationary phase and (iv) death phase.

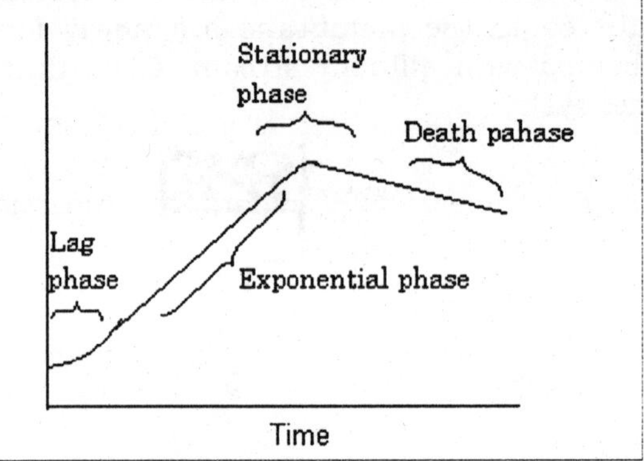

Fig. 5. Growth Curve

Lag phase – When a micro-organism is inoculated in a medium, it undergoes a process of adaptation. The cells do not increase in number but in size. The microbial population is in a period of intense metabolic activity. There is increased synthesis of RNA, ribosomes and enzymes. Length of the lag phase depends upon:

(a) types of inoculated cells
(b) constituents of the medium i.e., better medium makes the lag phase shorter.
(c) size of the inoculum
(d) environmental factors like temperature.

Log (exponential) phase – In this phase of growth the cells start dividing in a constant rate and after each division period, the total population is doubled. The micro-organism is in a state of balanced growth. The increase of biomass is proportional to the increase of cellular components. The growth is optimum due to availability of proper nutrition and other conditions required for it. The rates of exponential growth vary among organisms. During the log phase of growth, microorganisms are particularly sensitive to adverse conditions. Radiation and other antimicrobial agents directly inhibit the growth of the microorganism.

Stationary phase – In this stage, the availability of nutrients starts declining and toxic metabolites start accumulating. Hence, there is no further increase in cell density. The biomass present in this phase is the maximum of the crop. The metabolic activities of individual surviving cells also slow at this stage.

Death phase – The death phase is a linear decrease in number of viable cells with time. Half the surviving cells die in each successive equal time interval. Factors responsible for death phase are:

(a) nutritional exhaustion
(b) toxic accumulation

(c) autolytic enzymes — Death results from a number of factors : an important one is depletion of the cellular reserves of energy. The cellular death is accompanied by autolysis due to cellular enzymes.

ENVIRONMENTAL FACTORS EFFECTING GROWTH

In a culture, growth of a micro-organism is controlled by the availability of nutrients and some other environmental factors. These factors include osmoregulation (osmotic tolerance), temperature, pH, and oxygen relation.

Osmoregulation and osmotic tolerance

Solutes play a vital role in a cell. They maintain the concentration of metabolites inside the cell. Osmoregulation controls the transport of these solutes. The solute concentration of the medium directly regulates growth of a microorganism. Microorganism that can grow in solutions of high osmolarity are known as osmophiles. Marine environments are highly saline. Based on the ability to tolerate salt concentration (particularly sodium chloride) microbes are divided into four categories.

- Nonhalophiles Microbes growing at NaCl concentration from 0.0 to 4% g/ml. e.g., *Escherichia coli.*
- Marine forms Microbes which grow at NaCl concentration from 0.1 to 5.0% g/ml. e.g., *Pseudomonas marina.*
- Moderate halophiles Microbes which grow at NaCl concentration from 2.3 to 20.5% g/ml. e.g., *Vibrio costicolus.*
- Extreme halophiles Microbes growing at NaCl concentration from 12 to 36 % g/ml. e.g., *Halobacterium salinarium.*

Osmolarity inside the cell is always maintained higher to that of the medium. The concentration of potassium ions inside the cell is maintained higher than that of the concentration of sodium ions. In *E.coli,* the intracellular K^+ concentration increases progressively with increasing osmolarity of the growth medium. Diamine putrescine maintains intracellular ionic strength. The concentration of intracellular putrescine varies inversely with the osmolarity of the medium. When the osmolarity of the medium is increased, putrescine is excreted outside the cell and uptake of K^+ increases.

Temperature

Temperature directly affects the growth of a micro-organism as all the biological activities are controlled by enzymes. These enzymes are sensitive to temperature. The lipid composition of almost all organisms alters with the growth temperature. The range of temperature that a micro-organism can tolerate is designated as cardinal points. The cardinal point includes minimum, optimum and maximum values of temperature in which a micro-organism can grow. Microbes are classified into three categories on the basis of their liking for temperature and the cardinal points of their growth.

- Psychrophiles (Gr. Psychros = cold) These micro-organisms can grow on freezing temperature. However, microbes growing upto 20°C are also included in this category.
- Mesophiles (Gr. Meso = middle) This category includes saprophytes and parasites that can grow from 20-35°C. Animal parasites and commensals growing upto 45°C are also included in this category.
- Thermophiles (Gr. Thermos = heat) Microbes growing above 45°C or at higher temperatures are known as thermophiles. A microbe which is heat resistant but unable to grow at higher temperature is known as thermoduric e.g., a bacterium that can survive during pasteurization but fails to grow at higher temperature is a thermoduric.

Hydrogen ion concentration (pH)

All the biochemical reactions are specific for a particular pH. Like temperature microbes also show affinity towards pH. Based on the optimum pH requirement for their growth, microbes are classified into three categories.

- Acidophiles microbes requiring pH of 2 – 3.
- Alkalophiles microbes, which can grow between 7 – 12 pH.
- Neutrophiles microbes prefer neutral pH for their growth.

SECTION – III

CHAPTER - 14

FUNDAMENTALS OF GENETICS

The concept of genetic material to be inherited in generation to generation was first proposed by Mendel. As the science developed, new and new concepts started pouring.

GENOTYPE AND PHENOTYPE

Genotype: This term was proposed by Johannson in 1909. It is the sum of total genetic constitution of an individual. It means the genes (alleles) contained in the chromosomes. The phenomenon connected with or controlled by the genotype is known as *genotypic.* The manifestation by gene action of specific genotypic activity is known as *genotypic expression.* The share of one particular genotype in the next generation is determined by the viability and reproductive success of its carrier. The genotypic distance between two individuals for a given locus is the probability that the genotype of one individual is not the same as that of other. If a specific locus primarily responsible for the appearance of a particular meiotic character is studied, the rest of the genotype is referred to as the "*residual genotype* or *background genotype.*" An interaction between the gene and their environment is known as genotype-environment interaction. The interaction is generally regarded as part of the environmental variance. In animal viruses, after mixed infection with genetically different types of a particular virus, the enclosure of two or more different genomes within the virus envelope occurs. Such genotypically mixed virions are different from heterozygous bacteriophages in which heterologous DNA is part of a single viral DNA molecule.

Phenotype: This term was proposed by Johannson in 1909. It refers to the observable structural and functional properties of an organism, produced by the interaction between the organism's genetic potential and the environment. Phenotype can be applied either to the totality of expressions of the genotype or to only a part, i.e., to particular characters, traits or phenes. The manifestation via gene action of a particular gene present in effective dosage in an individual's genotype is known as *phenotypic expression.* The delayed phenotypic manifestation of a gene mutation in bacteria is known as *phenotypic lag* e.g., a mutation from prototrophy to auxotrophy. The delay may last several generations. The ability of certain genotypes to respond to varying environments by producing radically different phenotypes is known as *phenotypic plasticity.* The non-genetic restoration, partial or complete of the functional activity of mutated genes without directly substituting the required function itself is known as *phenotypic reversion.* The mutation may be missense and nonsense. Phenotypic reversion may be due to miscoding or mistranslation. A number of agents may bring about phenotypic reversion and result in nonheritable phenotypic changes. Lewis and John in 1963 proposed a term *endophenotypic* which are those adapted components of the phenotype which are not directly related to the environment and the survival of the individual. The adaptive components, which affect an individual's ability to compete and to survive, are known as *exophenotypic.* Heredity and environment overlap extensively in the shaping of the phenotype. Phenotypic similarities between individuals may occur, when they possess the same or different genotypes, and phenotypic differences between individuals may or may not be accompanied by genotypic differences. Environmental differences can cause the same genotype to produce a variety of phenotypes. Novick and Szilard (1951), Hershey, Rossel, Chase and Forman (1951) proposed a term *phenotypic mixing.* It is the production in the same host cell of virus particles with discrepant genome and virion specificities due the incorporation into the same capsid or envelope of proteins made under the direction of different viral genomes. The genomes in question produce different structural components of the protein "coats". The components of the coats are drawn, without regard to the gentoype of the particle, from a pool of mutant and wild-type protein molecules and thus mixed coats may be produced giving rise to phenotypic mixing. In bacteriophages, phenotypic mixing has been observed from stability at high temperature, serological properties, the properties of particles to attach to different strains of bacteria, as well as other features depending on the protein moiety of mature phages.

The genotype of an organism is its genetic makeup. It is the sum total of the genes, its entire DNA. It can said to be the genetic constitution with respect to the alleles of the genetic loci. The portion or frequency of any particular genotype among the individuals of a population is known as genotype frequency. Phenotype is the actual, expressed properties, such as the organism's ability to perform a particular chemical reaction. In simple terms phenotype is the manifestation of genotype. It is the structural and functional properties of an organism. Heredity and environment overlap extensively in the shaping of the phenotype. Environmental differences can cause the same genotype to produce a variety of phenotypes. This phenotypic range of reaction of a particular genotype is referred to as its reaction norms.

NUCLEIC ACID

DNA-The Genetic material — Some experiments have confirmed that DNA is the genetic material. It is mainly found in chromosome in nucleus. The other nucleic acid RNA is found in cytoplasm. O.T. Avery, C.M. Macleod and M.McCarty (1944) first demonstrated that DNA is the genetic material. They performed an experiment with bacterium *Diplococcus pneumoniae* and observed the transforming principle is DNA. The process of transformation is discussed in chapter 4 of this section. The nonvirulent strains of this bacterium were converted into virulent strains by receiving transformation principle from virulent cells. This principle was purified and was found as a molecule of DNA.

A.D. Hershey and M. Chase (1952) produced another evidence to prove that DNA is the genetic material. They experimented with bacteriophage T2. During replication of this bacteriophage only DNA entered inside the cells of *E. coli*, while the protein capsid remained outside the cell. The genetic information necessary for viral reproduction was present in DNA. The basis for the Hershey-Chase experiment is that DNA contains phosphorus but no sulphur, while protein contains sulphur but no phosphorus. They performed their experiment with radioactive isotope ^{32}P and ^{35}S. Bacteriophage with such labelled phosphorus and sulphur were the tools of experiment. During replication of such bacteriophage, it was found that DNA containing 32_P was present inside the cells. As the progeny viruses are produced inside the cell. Hence, it could be well understood that the genetic information directing the synthesis of both the DNA molecules and the protein coats of progeny viruses must be present in the parental DNA. The progeny viruses were shown to contain some of the ^{32}P, but none of the ^{35}S of the parental phage.

RNA as genetic material — Viruses contains only one type of nucleic acid i.e., either DNA or RNA. Some viruses contain only RNA as their genetic material.

NUCLEIC ACID STRUCTURE AND REPLICATION

F. Miescher (1869) isolated nucleic acid from nucleus of the cell. It contained phosphorus. He termed it as "nuclein". He observed that in a number of nucleic acid he studied, the number of purine nucleotides is always equal to number of pyrimidine nucleotides (A+G = C+T). Moreover, the number of A residues are always equals the number of T residues and likewise, number of C residues is equal to number of C residues. Franklin and Wilkins studied the X-ray diffraction patterns of fibers of DNA; it clearly showed that DNA is made up of more than one helical polynucleotide chain. The bases are stacked with a spacing of around 3Å.

BIOCHEMISTRY OF NUCLEIC ACID

Nucleic acid is made up of nitrogenous base, pentose sugar and phosphoric acid moiety. Nitrogenous base forms nucleoside by forming α-N-glycosidic bond with pentose sugar (cyclized sugar). By the addition of one or more phosphoryl groups esterified to hydroxyl groups of the sugar, nucleotides are formed. Sugar moiety is present in the furanose ring form. The difference between ribose and deoxyribose sugar is that in the latter, oxygen is absent at 2´ position (Fig. 1).

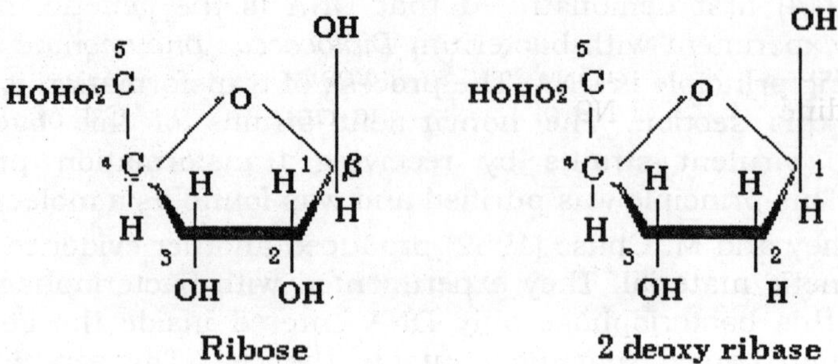

Fig. 1. Structure of Pentose sugars of Nucleic acid

Nitrogenous base

Purine and pyrimidine are the nitrogenous bases of nucleic acid (Fig. 2). These bases are heterocyclic compounds having both carbon and noncarbon atoms. These are planer molecules. Pyrimidine (cytosine, thymine and uracil) is six-atom and purine (adenine and guanine) is nine-atom compound. The nitrogenous base undergoes keto-enol and amine-imine tautomerism. These functional groups exist as tautomeric mixtures of amino-imino and oxo-hydroxy pairs. Some minor or unusual bases are present both, in proakryotes and eukaryotes. These bases are functionally and physiologically important. 5-Methylcytosine and 5-Hydroxymethylcytosine are uncommon naturally occurring pyrimidines, while; Dimethylaminoadenine and 7-methylguanine are uncommon naturally occurring purines.

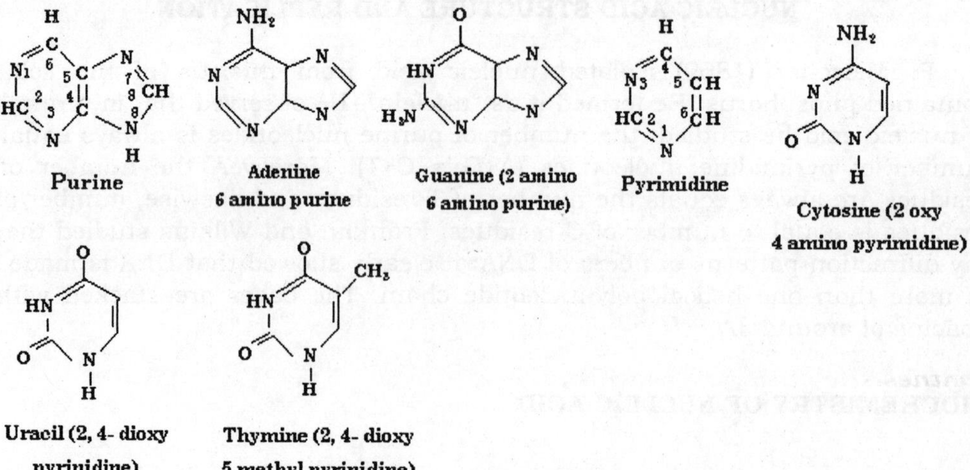

Fig. 2. Structures of Nitrogenous bases of Nucleic acid

Nucleoside — A nucleoside is made up of nitrogenous base linked covalently to a pentose sugar by N-C glycosidic bond (Fig. 3), which is in the (β)-configuration. The nucleosides of the five nitrogenous bases are named as

Adenosine (nucleoside of adenine base), Guanosine (nucleoside of guanine base), Thymidine (nucleoside of thymine base), Cytidine (nucleoside of cytosine base) and Uridine (nucleoside of uracil base). The glycosidic linkage is present between N1 of the pyrimidine ring and N9 of the purine ring and the C'1 of the pentose sugar. One unusual nucleoside is pseudouridine, which differs from uridine in that the glycosidic bond is between C of uracil and the C1 of the ribose.

Purine nucleoside **Pyrimidine nucleoside**

Fig. 3. Structure of Nucleosides

Nucleotide — A nucleotide is the monomeric unit, which is a building block of the polynucleotide chain. It is made up of nitrogenous base, pentose sugar and phosphoric acid (Fig. 4). It is also referred as a phosphate ester of the N-glycoside of a nitrogenous base.

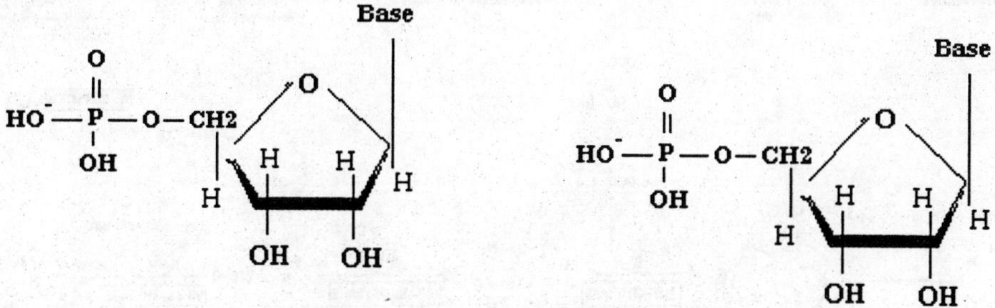

Fig. 4. Structures of Nucleotides

Biosynthesis of purine nucleotides

Synthesis of purine nucleotides takes place by three processes (1) synthesis from amphibolic intermediates (synthesis de novo), (2) phosphribosylation of purines, and (3) phosphorylation of purines (Fig. 5A).

Adenosine mono phosphate and Guanosine monophosphate are synthesized by a common parent nucleotide, inosine monophosphate (IMP). While, the latter is synthesized from the amphibolic intermediate α-D-ribose-5-phosphate (I) in a linear sequence of 11 reactions. Intermediates are 5-phsophoribosyl-1-pyrophosphate (PRPP) (II), 5-phospho-β-D-ribosylamine (III), Glycinamide ribosyl-5-

phosphate (IV), formylglycinamide ribosyl-5-phsophate (V), formylglycinamidine ribosyl-5-phosphate (VI), aminoimidazole ribosyl-5-phosphate (VII), aminoimidazole carboxylate ribosyl-5-phosphate (VIII), succinyl carboxamide ribosyl-5-phosphate (IX), aminoimidazole carboxamide ribosyl-5-phosphate (X) and formimidoimidazole carboxamide ribosyl-5-phosphate (XI). EnzymeIMP cyclohydrolase catalyzes the closure of ring (XI) intermediate and forms inosine monophosphate, the first purine nucleotide.

Conversion of IMP to AMP and GMP

After the formation of IMP, the pathway branches and two short reaction sequences lead to the formation of AMP and GMP (Fig. 5 B). Aspartate adds to IMP and forms adenylosuccinate. Enzyme adenylosuccinate synthase catalyzes this reaction. This reaction is superficially similar to reaction 8. GTP is required. Release of fumarate forming adenosine 5'-monophosphate (AMP) is catalyzed by adenylosuccinase, the same enzyme that catalyzes reaction 9. IMP is oxidized by NAD+ with the help of enzyme dehydrogenase and xanthosine monophosphate (XMP) is formed. 6-oxo group of XMP is transamidated by the amide nitrogen of glutamine. Transamidation proceeds by analogy with reaction 5.

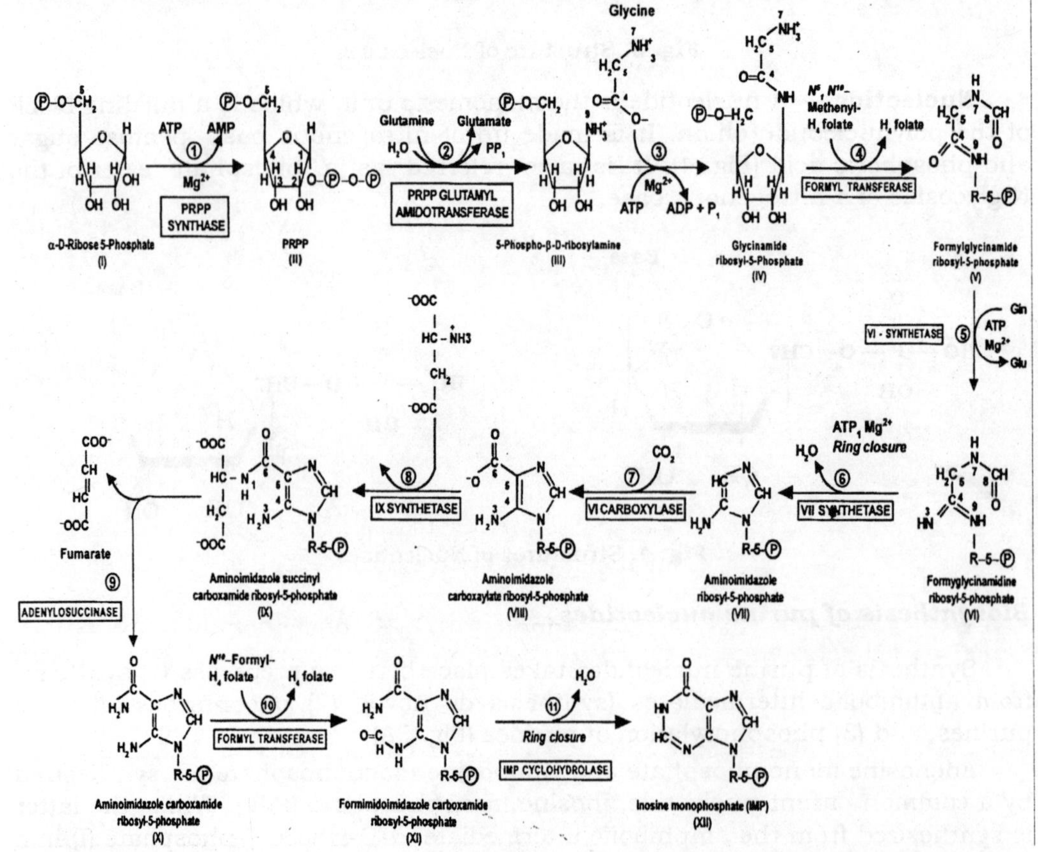

Fig. 5. The pathway of de novo purine biosynthesis from ribose 5-phosphate and ATP

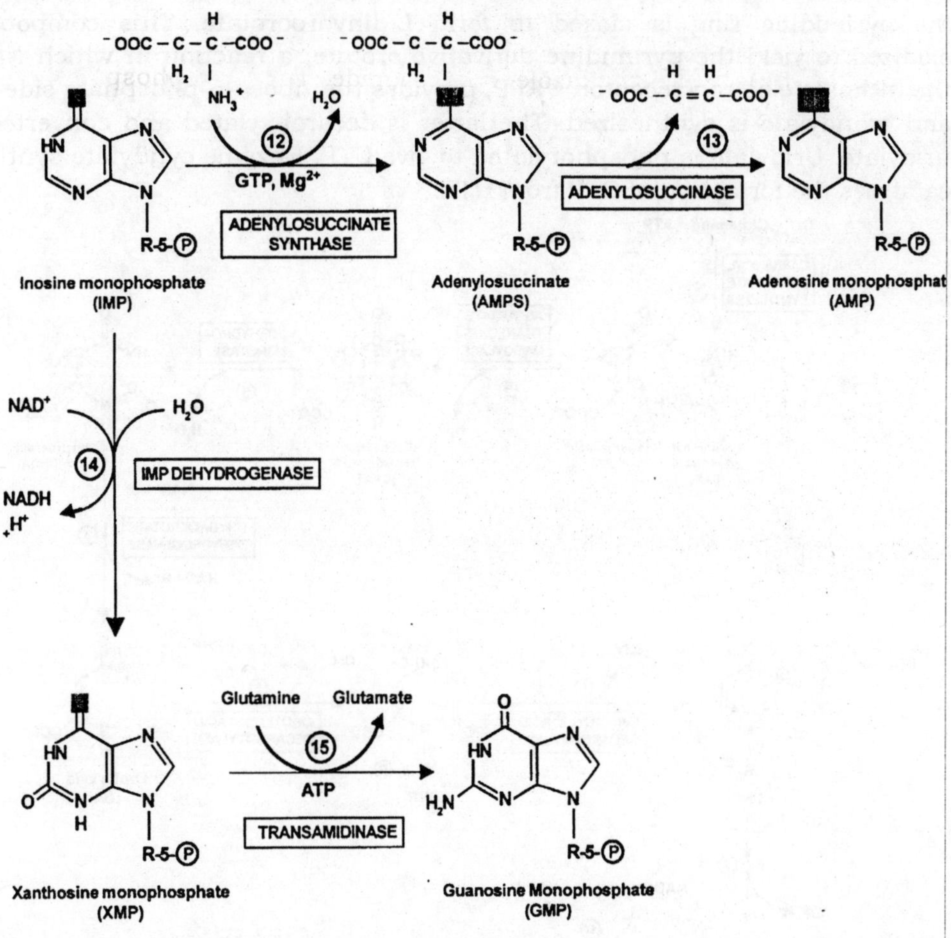

Fig. 5B. Conversion of IMP to AMP and GMP

Synthesis of di- and tri-phosphate nucleotide

Enzyme nucleoside monophosphate kinase catalyzes formation of diphosphate nucleoside from their monophosphate nucleotides. The phosphate moiety comes from ATP. GDP is again phosphorylated by ATP with the help of enzyme, nucleoside diphosphate kinase. ADP is phosphorylated to ATP by oxidative phosphorylation and secondarily by reactions of glycolysis and the citric acid cycle.

Biosynthesis of pyrimidine nucleotides

During biosynthesis of pyrimidine nucleotide (Fig. 6), first the six carbon member pyrimidine ring is made and then it is attached to ribose-5-phosphate. Carbamoyl phosphate is needed in this process. It reacts with aspartate to yield N-carbamoylaspartate. This reaction is catalyzed by enzyme aspartate transcarbamoylase. Enzyme dihydroorotase removes water from this intermediate and the pyrimidine ring is closed to form L-dihydroorotate. This compound is oxidized to yield the pyrimidine derivative orotate, a reaction in which NAD^+ is the ultimate electron receptor. PRPP, provides the ribose-5-phosphate side chain and orotidylate is synthesized. The latter is decarobxylated and converted into uridylate. Uridylate is phosphorylated to give UTP. Enzyme cytidylate synthetase catalyzes the formation of CTP from UTP.

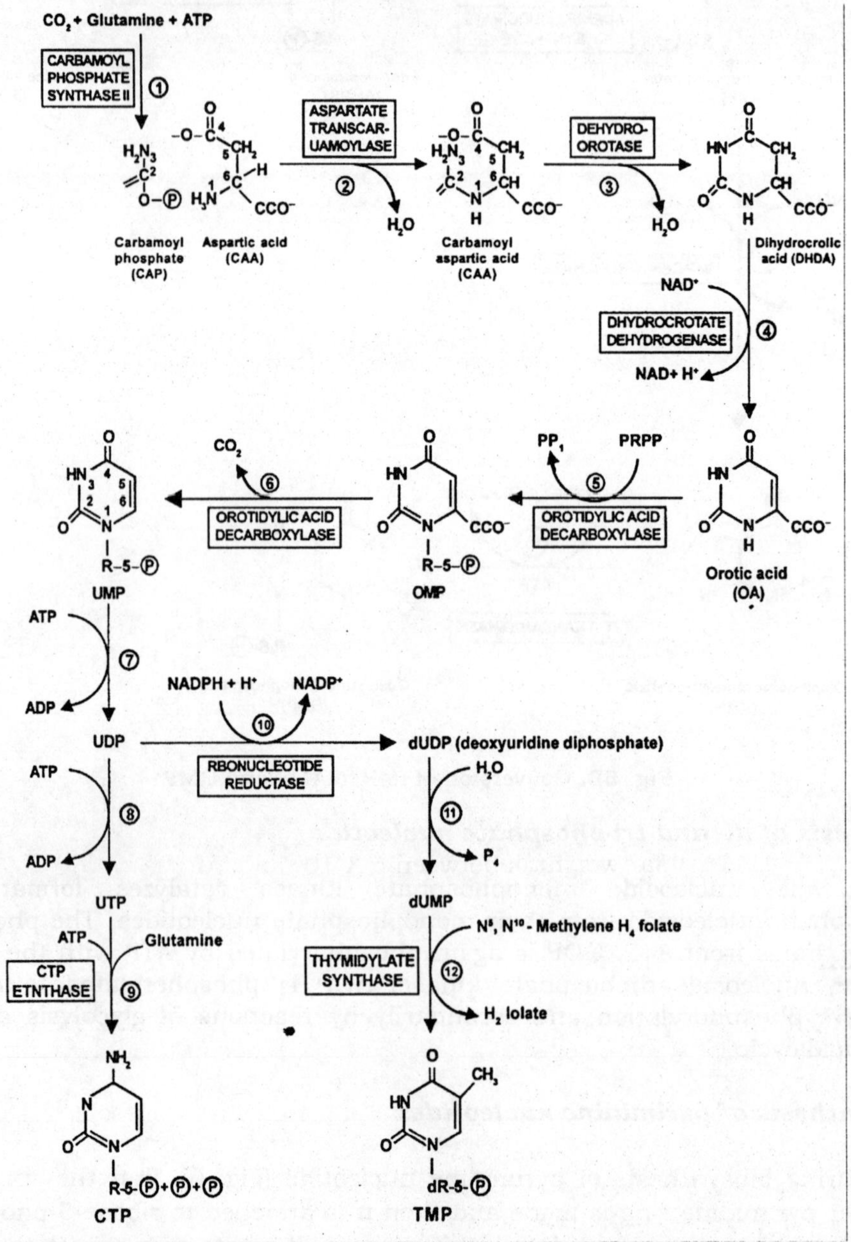

Fig. 6. The biosynthetic pathway for pyrimidine nucleotides

STRUCTURE OF DNA

J.D. Watson and F.H.C. Crick (1953) proposed a double helix model of DNA structure (Fig. 7).

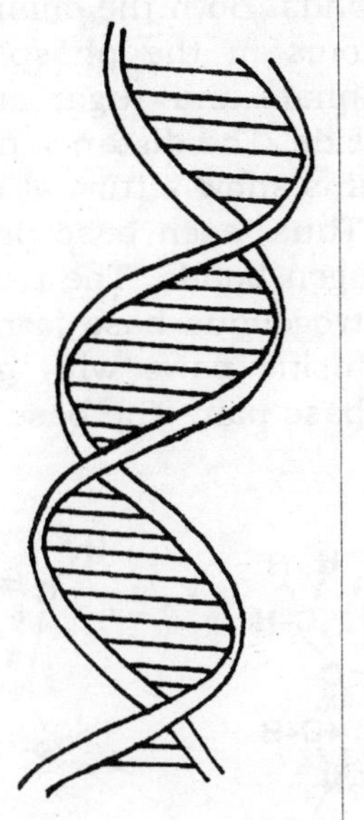

Fig. 7. Double Helical structure of DNA

EVIDENCE OF FIBROUS NATURE OF DNA

DNA molecule is made up of long and unbranched chain. The backbone is made up of alternatively arranged deoxyribose sugar and phosphate groups. Phosphodiester bonds join these in 3´ —> 5´ direction. Nitrogenous base is attached to sugar. The physicochemical analysis involving sedimen-tation, diffusion, light scattering, and viscosity measurement techniques have indicated that DNA is a very asymmetrical molecule with 20Å in diameter and thousands Å long. The molecular weight is between 5 X 10^6 and 10^7.

EVIDENCE FOR THE EXISTENCE OF TWO CHEMICAL CHAINS IN THE FIBER

X-ray crystallographic studies of sodium salt of DNA, shows that the fibers are highly birefringent, having marked ultraviolet and infrared dichroism. Sodium salt of DNA shows two structures. Structure A is crystalline with 75 percent relative humidity and contains approximately 30 per cent water. At higher humidity the fibers take up more water, increase in length by about 30 percent. It is paracrystalline. Structure B has a very strong 3.4 Å reflexion on the meridium. Structure A is shortens by about 30 per cent. It is about 2.5 per cent Å apart axially. All these evidences show that there are two polynucleotide chains.

THE PROPOSED STRUCTURE OF DNA

DNA molecule is a three dimensional crystal. The two-polynucleotide chains are complementary to each other and are coiled round a common axis and joined together by hydrogen bonds. Both the chains follow right-handed helices, but the sequences of the atoms in the phosphate-sugar backbones run in opposite directions. The phosphate and sugar are on the outside of the helix whilst the bases are on the inside. The distance of a phosphorus atom from the fiber axis is 10 Å. The two chains show a turn at each 34 Å distance. There are ten base pairs in each turn. Thus, each base pair has distance of 3.4 Å. The base pairs are joined by hydrogen bonds. The pairing is very specific, with one purine and one pyrimidine nitrogenous base forming the pair. Adenine always pairs with thymine while cytosine pairs with guanine (Fig. 8). There is no restriction to the neighboring base pair. The base pair are flat and are arranged one above the other.

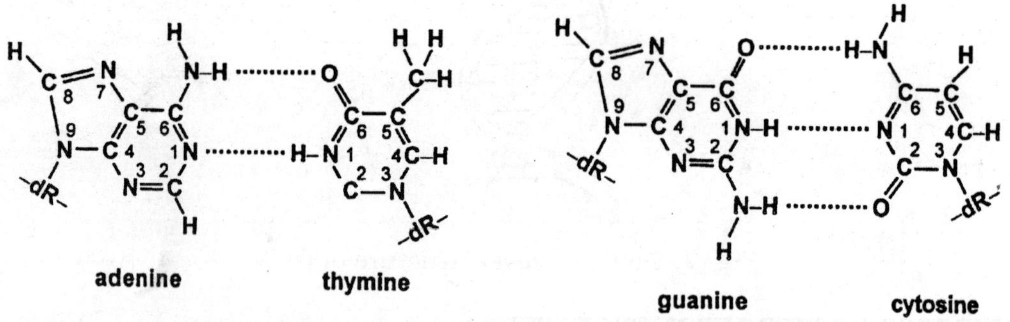

adenine thymine guanine cytosine

Fig. 8. The proposed structure of DNA

REPLICATION OF DNA

DNA molecule has the capability to replicate and produce daughter DNA molecule having the same base pair sequence as that of parent DNA molecule. Three possible modes of replications are being proposed. These are named as dispersive, conservative and semiconservative methods.

Dispersive method — The parental DNA is broken down into smaller fragments. These are the precursors, which complete the remaining portion of the DNA molecule.

Conservative method — The two strands of DNA molecule remains joined to each other while, they give programmes formation to daughter DNA molecule.

Semi-conservative method — The two strands of DNA molecule separate from each other and each strand give template to the complementary strand. Now it is well established that DNA molecule replicates by semi-conservative method.

M.S. Meselson and F.W. Stahl (1958) experimentally proved the semi conservative replication of DNA in *E. coli* (Fig. 9). The bacterium was grown in a culture containing ^{15}N in nitrogen source for several generations. It was later transferred to a culture medium containing ^{14}N nitrogen source. The cells developed in these culture media contained isotopic nitrogen in the nitrogenous base of DNA. Cells grown in ^{15}N will have a greater density (weight per unit volume) than the DNA of cells grown on medium containing ^{14}N. These molecules can be separated by equilibrium density-gradient centrifugation. When a heavy salt solution such as 6 M CsCl is centrifuged at vary high speeds (30,000 -50,000 revolutions per minute) for 48-72 hours, an equilibrium density gradient is formed. The centrifugal force sediments the salt toward the bottom of the tube. Some salt molecules moved towards the top of the tube i.e., towards the low salt concentration due to diffusion. After a definite period, equilibrium is established between sedimentation and diffusion. At this stage linear gradient with increasing density, exist from the top to bottom of the tube. DNA molecule will settle where the density of the salt solution is equal to its won density. A DNA molecule with mixed chain i.e., one heavy chain (^{15}N) and one light chain (^{14}N) are present in the solution and two bands will appear in CsCl tube. This confirms the fact that DNA replicates in a semiconservative method (Fig. 10).

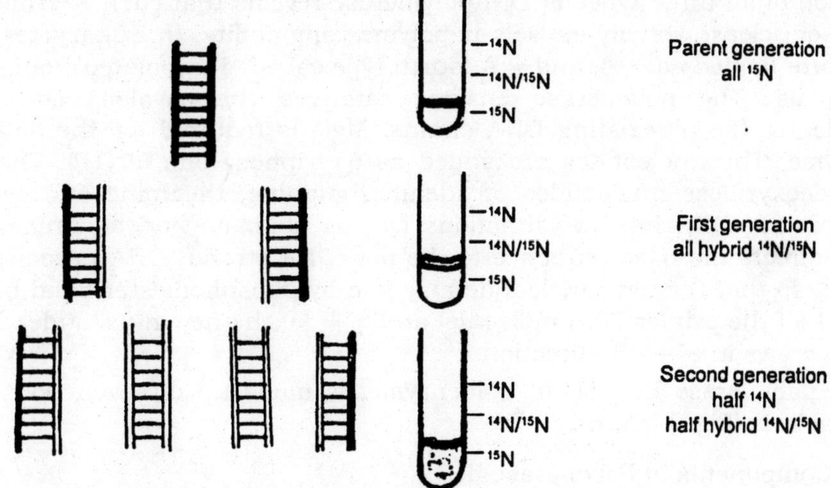

Fig. 9. Meselson and Stahl experiment to demonstrate semi-conservative replication of DNA moelcule

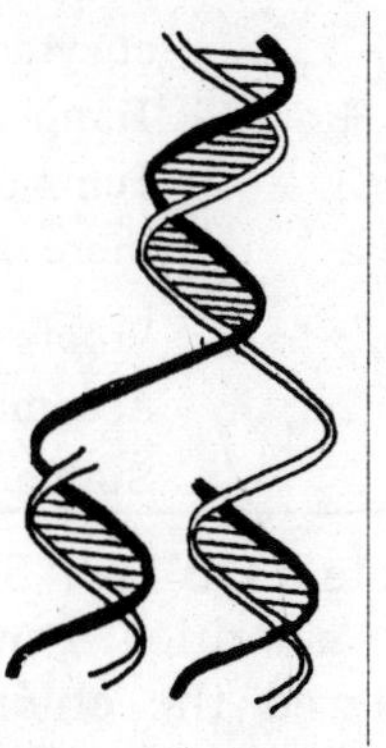

Fig. 10. Semi-conservative replication of DNA molecule

DNA replication is a complex process and requires some enzymes, protein factors to complete this process. The enzymes are dimeric polymerase, a primosome, and a helicase. They collective entity forms a ribosome size and is known as replisome.

DNA POLYMERASE

Arthur Kornberg and his coworkers (1957) isolated an enzyme from *E.coli* and named it as DNA polymerase or "Kornberg enzyme". Later other types of DNA polymerase enzymes have also been isolated. In *E.coli* and *B.subtilis* three types of DNA polymerase enzymes have been identified and studied. These enzymes are named as DNA polymerse I, II and III. Previously, DNA polymerase I was thought to be the major enzyme that joins together deoxyribonucleotides. Now it is established that it is actually DNA polymerase III, the polymerizing enzyme. DNA polymerase I is largely used for filling in gaps between the small Okazaki fragments. Function of DNA polymerase is still not clear. Extensive purification of all three types of DNApolymerase reveals that each enzyme has a 3'—>5' exonuclease activity as well as polymerizing ability. In Eukaryotes, these enzymes are named as α, β, and γ. A fourth type called δ is isolated from tissues of mammals. The polymerase enzyme catalyzes the covalent addition of nucleotides to the preexisting DNA chains. Mg^{++} is required for the activity of this enzyme. The nucleotides are added as 5´ triphosphate (dNTP). The dNTP denoted deoxyribose nucleotides of adenine, guanine, thyamine and cytosine. The preexisting DNA has two functions i.e., as template and as primer. DNA template directs the base sequence to the new DNA strand. DNA primer gives a free 3´OH, so that the new nucleotide may join by phosphodiester bond between the 3´-OH (of the primer DNA molecule) and 5´-P (of the new nucleotide). Hence, the chain grows in 5´—> 3´ direction.

DNA polymerase III (pol III) holoenzyme, is made up of seven polypeptide chains Table 1, (Fig. 11 A, B).

Table 1. Components of Polymerase III

Sub-unit	Gene	Mol. Wt KD	No. of subunits/ mol. of enzyme	Function
α	PolC (dnaE)	130	2	DNA synthesis (dnaE), part of the core enzyme
β	dnaN	37	4	Template association
γ	dnaX	52	2	Template binding, increased processivity
δ	holA	32	2	Increased processivity
ε	DnaQ(mutD)	25	2	Proof reading (mutD), core subunit
θ	holE	10	2	Assembly
τ	dnaX*	71	2	Stable template binding dimmer formation

The exonuclease enzyme acts in 3' —> 5' direction and cuts nucleotides from the ends. It carries out a critical "proof reading" or "editing". If a mismatched nucleotide has entered the chain, this enzyme will clip off the unpaired nucleotide.

DNA gyrase or topiomerases catalyze the formation of negative supercoils in DNA are essential for replication, and are believed to play key role in the unwinding process. DNA molecule is supercoiled under normal cellular conditions. At the time of replication, unwinding is essential so that the two strands separate.

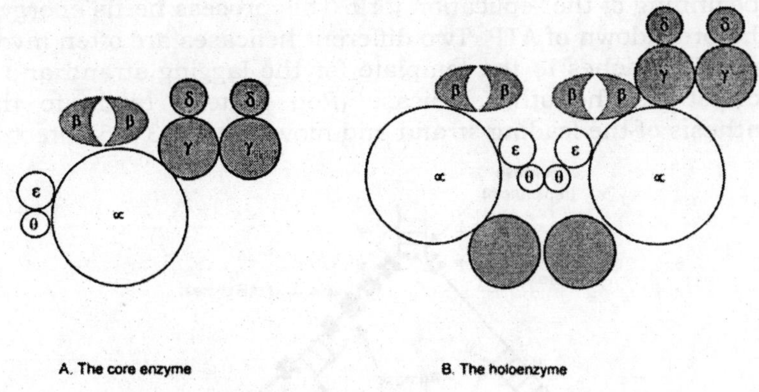

Fig. 11A. Structure of DNA polymerase III

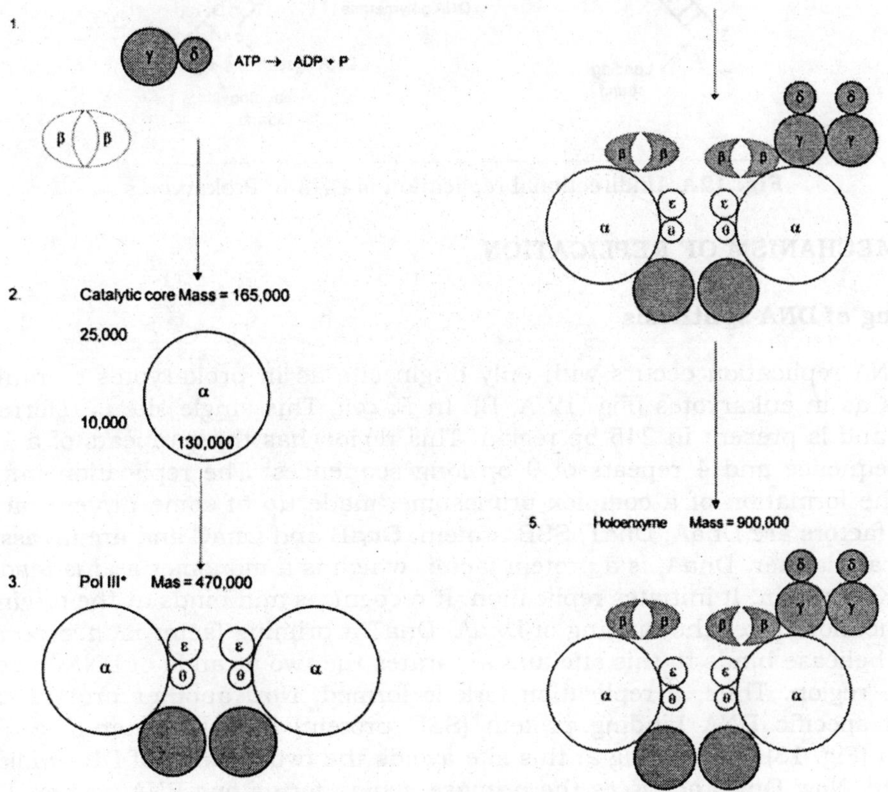

Fig. 11B. Assembly of DNA polymerase III

DNA primase is a specific RNA polymerase. It synthesizes RNA primer needed for the initiation of replication of DNA. Primase activity requires at least six other proteins and forms a complex primosome. In *E.coli*, DNA primase is made up of a single polypeptide chain. It is present in 50 to 100 copies per cell.

DNA helicases separate the two strands of DNA and form single stranded DNA at the beginning of the replication fork. This process needs energy, which is derived by the breakdown of ATP. Two different helicases are often involved. One (helicase II or III) attaches to the template for the lagging strand and moves in the 5'->3' direction. The other helicase (Rep protein) binds to the strand directing synthesis of the leading strand and moves in the 3'->5' direction.

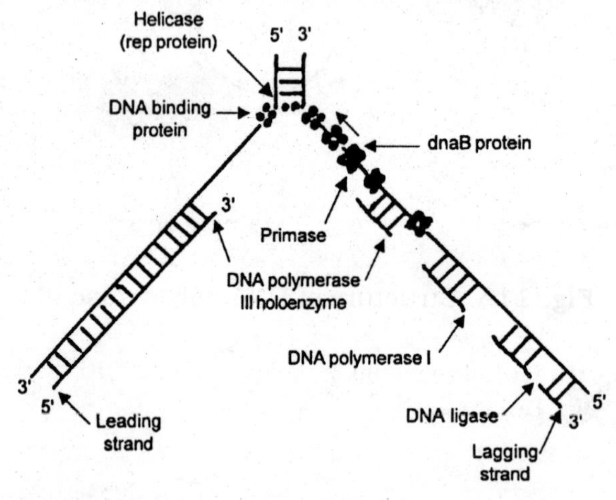

Fig. 12A. Undirectional replication of DNA in Prokaryotes

THE MECHANISM OF REPLICATION

Priming of DNA synthesis

DNA replication occurs with only origin site as in prokaryotes or multiple origins as in eukaryotes (Fig. 12 A, B). In *E. coli*, This single site is referred as 'OriC' and is present in 245 bp region. This region has three repeats of a 13 bp long sequence and 4 repeats of 9 bp long sequences. The replication initiates with the formation of a complex primosome, made up of some protein factors. These factors are DnaA, DnaT, SSB protein, DnaB and DnaC and are involved in a typical manner. DnaA, is a protein factor, which is a monomer and is made up of 82 KD protein. It initiates replication. It recognizes and binds at the origin site of replication. After the binding of DnaA, DnaT a priming factor is incorporated. Later, helicase binds to this site and separates the two strands of DNA molecule in this region. Thus, a replication fork is formed. Now another protein single strand specific DNA binding protein (SSB protein) is a tetramer of a 74KD protein (Fig. 13). Its binding at this site avoids the two strands of DNA molecule to unite. Now DnaB activates the primase, which forms one RNA primer. DnaB is a hexamer and to each monomer chain, six molecules of another protein

DnaC become attached. At this stage, primer is initiated. RNA primer is complementary to 3'-end of template. Once the primer is synthesized enzyme DNA polymerase becomes functional and attaches new nucleotides by phosphodiester bonds.

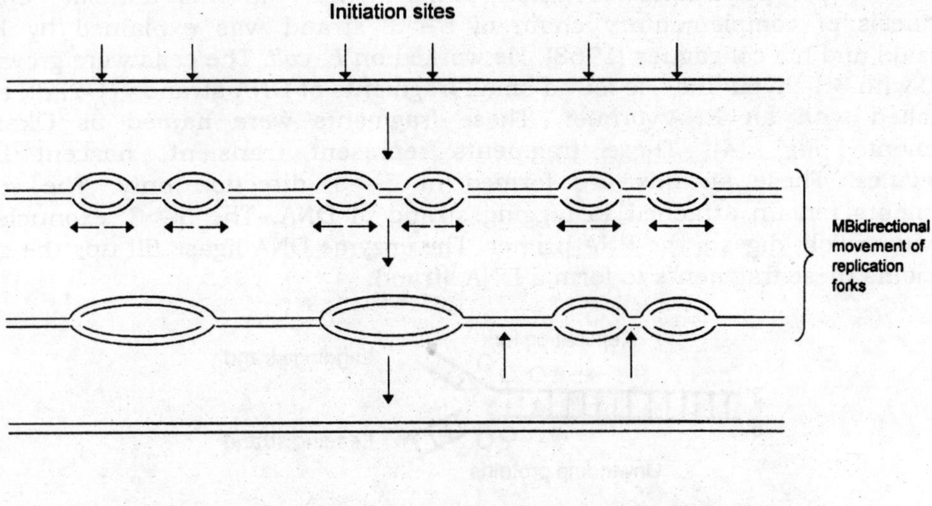

Fig. 12B. Multidirectional replication of DNA in Eukaryotes

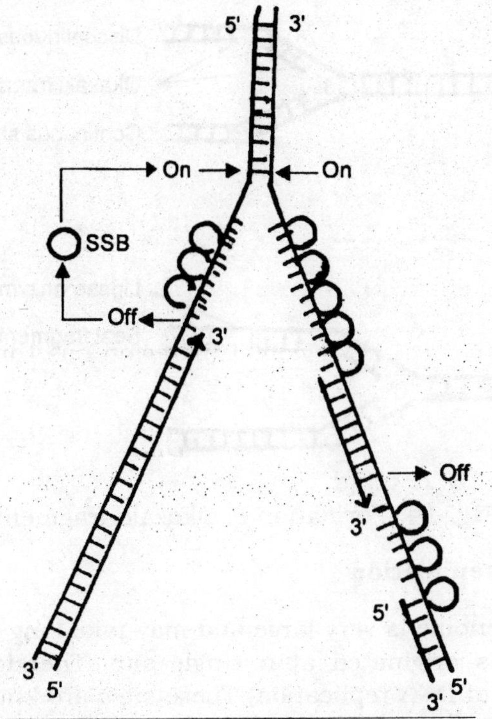

Fig. 13. Binding of SSB protein in replicating DNA molecule

The leading and lagging strands

The 3'->5' strand acts as a template strand. This is referred as leading strand while the other as lagging strand. The DNA polymerase is able to synthesize the polynucleotide chain complementary to this template chain. Synthesis of complementary chain of 5'->3' strand was explained by Reiji Okazaki and his colleagues (1968). He worked on *E. coli.* The cells were grown at 20 C with 3H-thymidine. He found small fragments of DNA strands (1-2kd), each attached with an RNA primer. These fragments were named as Okazaki fragments (Fig. 14). These fragments represent transient, nascent DNA molecules. These strands are formed in 5'->3' direction only. The small fragments remain attached to lagging strand of DNA. The 5'->3' exonuclease activity of PolI, digests the RNA primer. The enzyme DNA ligase fill ups the gaps and joins these fragments to form a DNA strand.

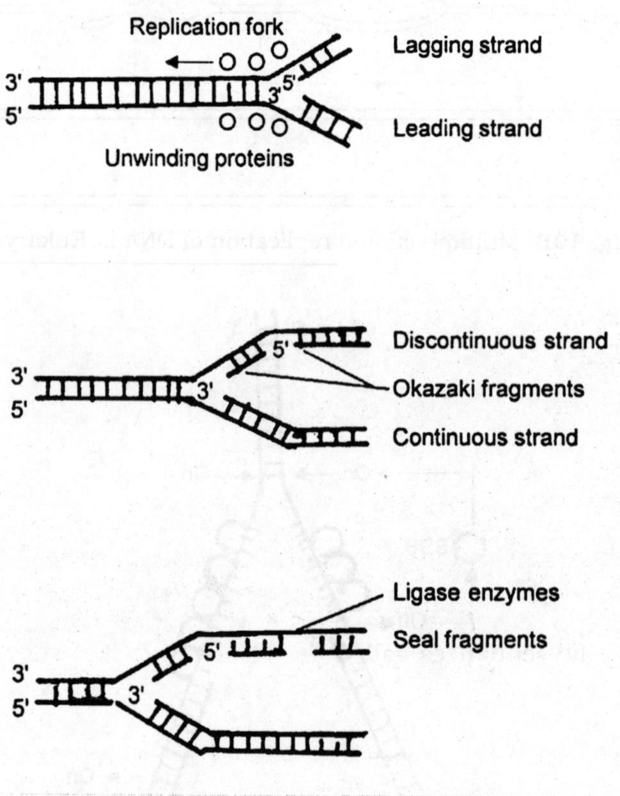

Fig. 14. Formation of Okazaki fragments

Multiple origins of replication

In eukaryotes genome is very large and may take long time for replication of DNA if replication is originated at a single site. Therefore, these cells have multiple origin sites of DNA replication. These sites are known as replicons. The priming of DNA takes place on each site and the replication proceeds in both the directions i.e., for leading and lagging chain (Fig. 12 B). The points at which these chains meet, the nicks are sealed by enzyme DNA ligase.

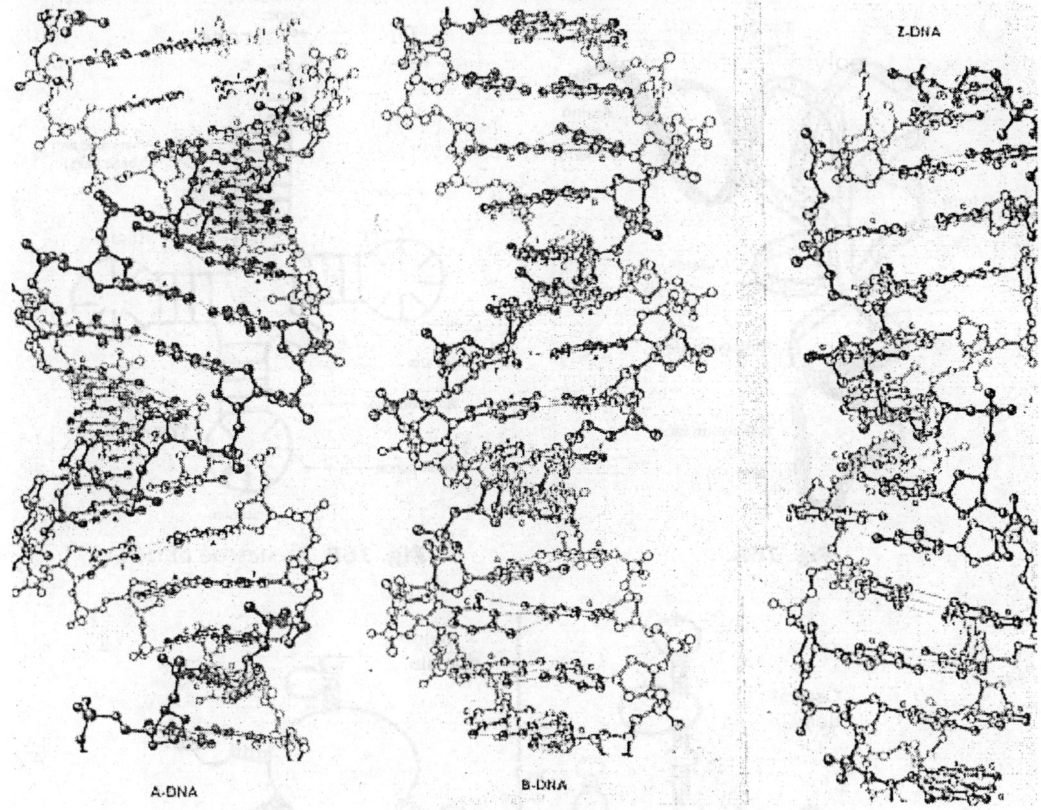

Fig. 15. Z-DNA

Termination of DNA replication

The *E. coli* chromosome carries a large termination zone, diametrically opposite from Oric, which blocks the progress of replication fortis (of bidirectional replication), meeting at this region. *E. coli* chromosomes and several plasmids carry specific sequences called ter sites where TBPC ter bindig protein or 'Tus protein' binds. In the termination zone of *E. coli* there are 3 ter sites (ter A, ter D and ter E) for counter-clockwise fork and 3 ter sites (ter B, ter C and ter F) for clockwise fork.These six sites are arranged in overlapping manner leaving no replication free gap other chromosome. TBP-ter complexes formed at 'ter'sites stalls the replication forby inhibiting the DNA helicase or DNA B when this termination zone is dead, replication stops simply by meeting of opposite replication forks, suggesting that the termination zone is not essential. The DNA molecule of *E. coli* is circular. Two replication forks, moving in opposite directions, eventually meet and replication terminates. Two termination regions, known as terD/terA and terB/terC are responsible for the termination of fork 1 and 2, respectively. These regions have 23 bp. There is a specific protein, which recognizes these sites and binds there. After binding it interacts with DNA polymerase and causes the termination of the replication fork.

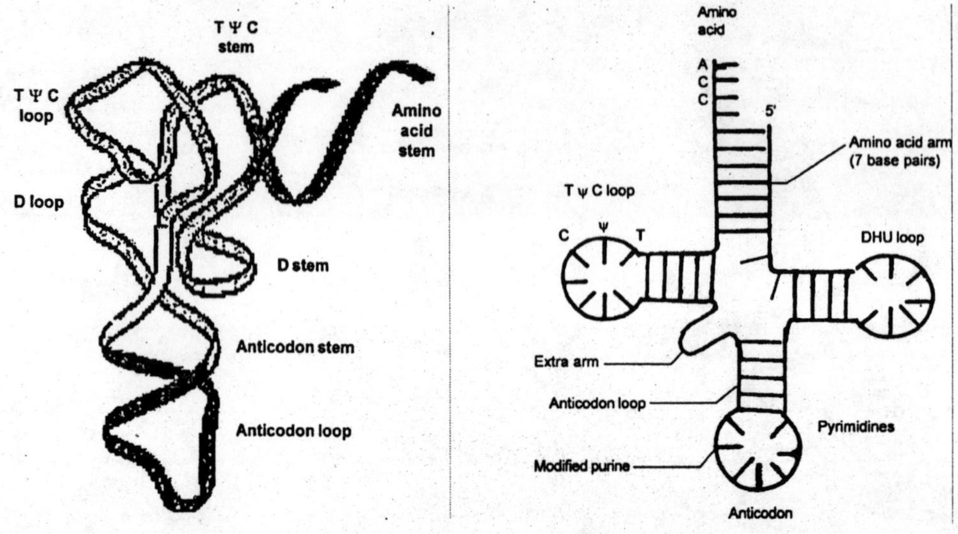

Fig. 16A. **Fig. 16B**. Structrue of tRNA

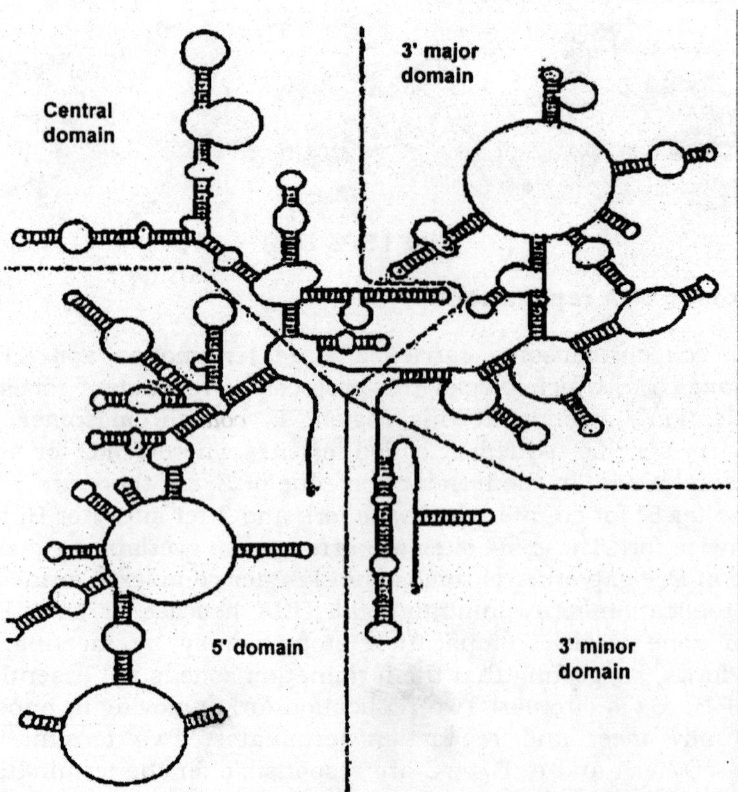

Fig. 16C. Secondary structure of rRNA of *E. coli.*

Z-DNA

Alexander Rich examined Z-DNA also known as d(c-G)3. It is different from the native DNA in following respects. Z-DNA contains about 12 bases/turn. In Z-DNA regions, the base pair residue conformations are alternately, syn- and anti. G-C pairs are distinctly displaced peripherally. Hence, there are deep groves in the Z-DNA forms. Z-DNA is left-handed and the phosphodiester bond has a highly zig-zag orientation. If, Z-DNA is present in plasmid, it modifies super-coiling. Z-DNA is less sensitive to nucleases. One complete helix is 45Å in Z-zone (Fig. 15).

Comparison of helixs of normal B-DNA and Z-DNA

Helixs	B-DNA	Z-DNA
1. Helix sense	right handed	left handed
2. Residues/turn	10	12 (6 dimers)
3. Diameter	-20 Å	-18 Å
4. Rise per residue	3.4 Å	3.75 Å
5. Helix pitch	3.4 Å	45 Å
6. Base pair tilt	6°	7°
7. Rotation per residue	36°	60°
8. Glycoside torsion angle		
deoxyguanosine	anti	syn
deoxycytidine	anti	anti
9. Sugar pucker		
deoxyguanosine	C_2 endo	C_3 endo
deoxycytidine	C_3 endo	C_2 endo
10. Distance of P axis		
dep G	9 Å	8 Å
dep G	9 Å	69 Å

RNA STRUCTURE AND TYPES

RNA is the second type of nucleic acid present in the cell. It is about ten times more than that of DNA. In general, RNA is single stranded with some exceptions of phage RNA that, are double stranded. The basic difference between RNA and DNA is that in RNA has uracil pyrimidine instead of thymine and the pentose sugar is ribose in place of deoxyribose as found in DNA. There are three types of RNA in a cell, ribsomal RNA, transfer RNA and messenger RNA. Molecules of all these RNA resemble single-stranded DNA with reference to the sequence of nitrogenous base. The single stranded regions are interspersed with intramolecular double-stranded regions. Between 1/2 and 2/3 of the bases are paired. The double stranded regions of RNA may contain up to 50 base pairs and a particular molecule has a definite base-pairing pattern.

In *E. coli* (all prokaryotic cells), there are three types of RNA, viz., mRNA, tRNA and rRNA (Fig. 16 A, B, C). While, all eukaryotic cells have four major classes of RNA. Besides the three types as present in prokaryotes, eukayotes have a fourth class known as nuclear RNA (sRNA). It is involved in splicing of mRNA.

MESSENGER RNA

mRNA carries genetic information contained in DNA & makes a small fraction (5%-10%) of total RNA present in the cell. Messenger RNA (mRNA) is very heterogenous, varying less than 5S to more than 30S. It is synthesized by DNA template by the process of transcription. It contains information for amino acid sequence of several polypeptide chains. This is known as polycistronic mRNA. The genes contained in a polycistronic mRNA molecule often encode the different proteins of a metabolic pathway. All base of mRNA, are not translated into polypeptide chain. The polypeptide synthesis initiates from initiation codon. These codons may be present at a distance in 5'-P terminus of the RNA. This section of RNA present before the initiation codon is known as "leader". In certain cases, it contains regulatory sequences that influence the rate of protein synthesis. Polycistronic mRNA molecules contain spacer sequences tens of bases long, which separate the coding sequences. This RNA is rather short lived and therefore has a rapid turnover. The molecular weight of this RNA varies, usually from 500,000 - 2,000,000.

TRANSFER RNA

Transfer RNA (tRNA) are small, single-stranded molecules, having 4s and 73 to 93 nucleotides. Amino acid attaches at 2' or 3'terminus of the tRNA molecule, which is always adenine. Amino acid is linked by as ester bond from its carboxyl group to the ribose sugar. Internal complementary base sequences form short double stranded regions. These double stranded stems separate some folded structures known as loops. In two dimensions a tRNA molecule is drawn as a planar cloverleaf. Its three dimensional structure is more complex. In cloverleaf model, the complementary base paired region is called as stems and the single stranded regions as loops. The base pairing is virtually invariant or conserved. The base pairing is conventional type (A/U and C/G). However, unusual base pair are occasionally found. All tRNA have unusual bases. These bases are formed by modification of the four standard bases. The acceptor arm consists of a base-paired stem. The 3'OH group of unpaired region of this stem are the sites of amino acid linkage. When a tRNA is charge with the amino acid corresponding to its anticodon, it is called aminoacyl~tRNA. The anticodon arm has a loop containing anticodon. It contains base sequence complimentary to codon. There are no anticodons for nonsense codons. The T φ C arm is named for the presence of pseudouridine, a modified base. The D arm has another unusual base, dihydrouridine. In D loop, there is variation of up to four residues. The extra nucleotide relative to the most common structure is denoted as 17 : 1 and 20 : 1 and 20 : 2. It shows that the extra nucleotide is lying between 17 and 18 positions and between 20 and 21 positions of nucleotide sequences. There is an extra arm lying between T φ C and anticodon arm. Depending on the nature of the extra arm, tRNA can be divided into two classes; Class 1 tRNA and class 2 tRNA. Class 1 represents about 75 % of the total tRNA. It consists 3-5 bases. Class 2 has 13-21 bases and a stem of upto 5 base pair stem.

The secondary structure of each tRNA folds into a compact L-shaped tertiary structure in which the 3' end that binds amino acid is distant from the anticodon that binds mRNA. Double stranded stems are present in tRNA. Two double helices are present at right angle to each other. The acceptor stem and T φ C stem form another continuous double helix with a single gap; the D stem and anticocon stem form another continuous double helix, also with the gap. The region between the double helices, where the turn in the L-shape is made, contains the T φ C loop and the D loop

GENETIC CODE

There are four nitrogenous bases in DNA and their different combination code for twenty amino acid taking part in polypeptide synthesis (Table 2). An RNA sequence corresponding to a particular amino acid is called a codon. Combination of three bases (triplets; 4^n = 64), forms a codon. The different combinations of four nitrogenous bases forms total sixty four codons, out of which sixty one codon code for twenty amino acids. Three codons do not code for any amino acid, they are known as nonsense codons. The nonsense codons are UAA, UAG, and UGA. While AUG is the initiation codon and it is responsible for the initiation of polypeptide chain.

Table 2. Genetic codes and aminoacids coded by these codes

	U		C		A		G	
U	UUU	Phe	UCU	Ser	UAU	Tyr	UGU	Cys
	UUC	Phe	UCC	Ser	UAC	Tyr	UGC	Cys
	UUA	Leu	UCA	Ser	UAA	End	UGA	End
	UUC	Leu	UCG	Ser	UAG	End	UGG	Trp
C	CUU	Leu	CCU	Pro	CAU	His	CGU	Arg
	CUC	Leu	CCC	Pro	CAC	His	CGC	Arg
	CUA	Leu	CCA	Pro	CAA	Gln	CGA	Arg
	CUG	Leu	CCG	Pro	CAG	Gln	CGG	Arg
A	AUU	Ile	ACU	Thr	AAU	Asn	AGU	Ser
	AUC	Ile	ACC	Thr	AAC	Asn	AGC	Ser
	AUA	Ile	ACA	Thr	AAA	Lys	AGA	Arg
	AUG	Met	ACG	Thr	AAG	Lys	AGC	Arg
G	GUU	Val	GCU	Ala	GAU	Asp	GGU	Gly
	GUC	Val	GCC	Ala	GAC	Asp	GGC	Gly
	GUA	Val	GCA	Ala	GAA	Glu	GGA	Gly
	GUG	Val	GCG	Ala	GAG	Glu	GGG	Gly

General features of genetic code

The genetic code is nonoverlapping and there is no punctuation. Genetic codes are universal. The same codon is used to code an amino acid in all the organisms. Only few exceptions are mitochondria and a few unusual microorganisms. The frequency of use of each amino acid codon varies considerably between species. The code is highly redundant. One amino acid may be coded by more than one codon. Only tryptophan and methonine are specified by one codon. The genetic code is degenerate i.e., multiple codons must decode the same amino acid. The genetic code is unabmiguous i.e., given a specific codon; only a single amino acid is indicated. To discriminate between degeneracy and ambiguity is important. Each codon has anticodon. For a given codon in the mRNA, only a single species of tRNA molecule possesses the proper anticodon. Each tRNA is charged with only one amino acid; hence only one amino acid will be coded by the codon of mRNA.

Redundancy and the Wobble hypothesis

While, understanding codon-anticodon binding two concepts has developed. The code is redundant. The identity of third nitrogenous base of codon is unimportant. The number of distinct tRNA molecules that have been isolated from a single organism is less than the number of codons. Thus, the anticodons of some tRNA molecules must be able to pair with more than one codon. These features are explained by Wobble hypothesis. It refers to the less stringent requirement for base pairing at the third position of the codon than at the first two positions. The first two bases (between codon and anticodon) should pair in usual type while; the third base pair can be of a different type (like G can pair with U). This concept came into existence while studying anticodon of $tRNA^{Ala}$ of yeast. It contains the nucleoside inosine, I, in the position that pairs with the third base of the codon (the first position of the anticodon).

CHAPTER - 15

GENE EXPRESSION

A gene is a segment of DNA and the part of gene that codes for a polypeptide is known as cistron. Gene regulates biological diversity and specific characters to different organisms even up to cellular level. The central dogma expresses the entire mechanism of these processes. The genetic information is transferred to cytoplasm through mRNA. The latter is being translated in the language of protein. This whole concept is said to be the central dogma of molecular biology.

Polypeptide synthesis is a complex process. It completes in two stages. Transcription : synthesis of RNA (primary transcript) from DNA template and Translation : Translation of the nucleotide language (codons) into the sequence of amino acid (polypeptide chain).

TRANSCRIPTION

Transcription means synthesis of an RNA chain representing one strand of DNA molecule. The identical strand of DNA is called coding strand while the other strand is known as template strand. The enzyme RNA polymerase is implicated in all three stages of transcription (Initiation, elongation and termination). Transcription commences when RNA polymerase binds to a special region, the promoter, at the start of the gene, during elongation the enzyme moves along the template, synthesizing RNA, and termination of RNA takes place when this enzyme reaches a terminator sequence. This action is known as a transcription unit. A transcription unit may include more than one gene. The base pair sequences prior to the startpoint are described as upstream while those after the startpoint are known as downstream. Transcription takes place within a "transcription bubble" (Fig. 1). At this site DNA is transiently separated into its single strands, and one strand is used as a template for synthesis of the RNA strand. The length of the transcription bubble varies with the phase of the elongation reaction from 12-20 bp. The immediate product of transcription is called the primary transcript. It is generally unstable. In prokaryotes, it is rapidly degraded (mRNA) or cleaved to give mature products (rRNA and tRNA). In eukaryotes, it is modified at the ends (mRNA) and /or cleaved to give mature products (all RNA).

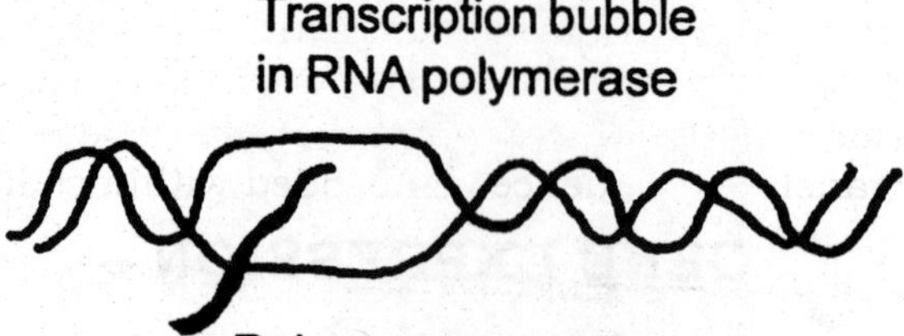

Fig. 1. Transcription bubble

Bacterial RNA polymerase has been purified and its catalytic activities has been studied. About 7000 RNA polymerase molecules are present in an *E. coli* cell. The complete enzyme or holoenzyme in *E. coli* has a molecular weight of ~465 kD. The ionic strength of the incubation medium is a critical parameter of enzyme activity. For *E. coli* RNA polymerase exist as a dimmer but dissociate into monomer when the ionic concentration is raised. Travers and Burgess (1969) found that monomer is active and comprise of four types of polypeptide chain. It consists of five polypeptide chains including two chains of alpha(α) polypeptide and one chain each of beta (β) and sigma (σ) polypeptides. The RNA polymerase molecule thus can be represented as $\alpha_2\beta\beta^1\sigma$. It comprises of 13-14S sedimentation in which (α-2 enzyme, Mol. wt –40,000; β'-1 enzyme, mol. wt.- 1,55,000; β-1 enzyme, mol. Wt. –1,65,000; σ- 1enzyme, mol. Wt. –1,95,000). The holoenzyme can be separated into two components, the core enzyme ($\alpha_2\beta\beta^1$) and the sigma factor (σ). The sigma factor binds at promoter site of DNA. After initiation sigma factor is released when the RNA chain reaches 8 to 9 bases, leaving the core enzyme to undertake elongation. Sigma factor is involved only in initiation. Free sigma factor becomes immediately available for use by another core enzyme. Transcription proceeds from left (upstream) to right (downstream). This corresponds to writing the mRNA in the usual 5'—3' direction. All nucleic acids are synthesized from nucleoside 5' triphosphate precursor. The incoming nucleotide joins at the 3'-OH end of the last nucleotide of the chain. During the condensation reaction the incoming nucleotide loses its terminal two phosphate groups, the group is used in the phosphodiester bond linking it to the chain. Only those nucleotides form diaster bonds that matches with the template base. During elongation transcription bubble moves by a disruption of DNA molecule and the enzyme moves forward 1 bp along DNA for every nucleotide added to the RNA chain. The RNA emerges as a free single strand (Fig. 2). The termination site involves recognition of the point at which no further base should be added to the chain. It is recognized by rho factor. When the last base is added to the RNA chain, the transcription bubble collapses as the RNA-DNA hybrid disrupted. Rho factor was first recognized by Roberts (1969-70). It is a protein factor which enables RNA polymerase to complete transcription. The RNA transcribed, remain associated with its template unless the rho factor is present to cause its release.

The prokaryotes and eukaryotes show some major differences in the relation between the transcript and the mRNA used for polypeptide synthesis. RNA transcribes from DNA is known as primary transcript. In prokaryotes, primary transcript directly takes part in polypeptide synthesis. While in eukaryotes, the primary transcript must be converted to mRNA by a process known as RNA processing. The latter consists of two types of events-modification of the termini and excision of untranslated sequences embedded within coding sequences.

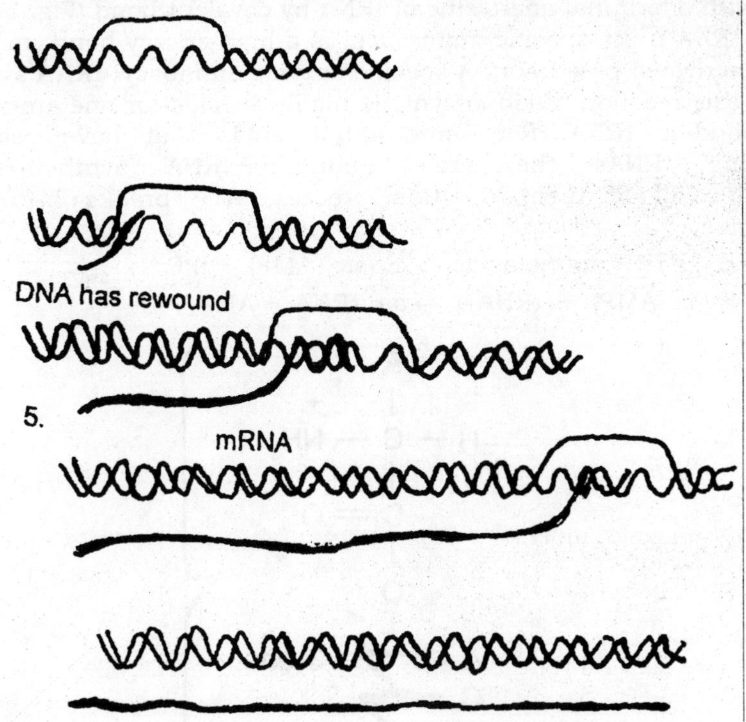

Fig. 2. Steps of Transcription (Initiation, elongation and termination of mRNA)

TRANSLATION

The messages transferred from DNA to cytosol (in the form of mRNA), is translated in the language of polypeptide chain. This is a complex biochemical process and is completed in four stages. The four stages and their requirements are as following:

Stages	Components required.
1. Activation of amino acids	20 amino acid, 20 aminoacyl~tRNA synthetase, tRNA, ATP, Mg+2
2. Initiation of polypeptide chain	mRNA, initiation codon (AUG), N-formylmethionyl~tRNA, 30S ribosomal subunit, 50S ribosomal subunit, Initiation factors (IF-1, IF-2, IF-3), GTP, Mg+2.
3. Elongation of chain	Functional 70S ribosome (initiating complex), Amino-acyl~tRNAs specified by codons, elongation factors (EF -Tu, EF-Ts, EF-g), peptidyl transferase, GTP and Mg+2
4. Termination and release	Termination codon in mRNA, polypeptide release factors (RF1, RF2, RF3), ATP.

ACTIVATION OF AMINO ACID

The twenty different amino acids functioning as building blocks of proteins are esterified to their corresponding tRNA. It takes place in cytosol. The amino acid links with 3'terminal adenosine of tRNA by covalent bond (Fig. 3). The bond linking the tRNA to its specific amino acid is a high-energy bond, making these complexes 'activated precursor'. A specific enzyme aminoacyl-tRNA synthetase is involved in this reaction. Each enzyme is highly specific for one amino acid and its corresponding tRNA. For some amino acids that have two or more corresponding tRNAs the same aminoacyl~tRNA synthetase usually aminoacylates all of them. In this process, ATP breaks into AMP and pyrophosphate.

(1) E + ATP + amino acid → E (aa AMP) + iPP

(2) E (aa AMP) + tRNA → aa tRNA + AMP + E

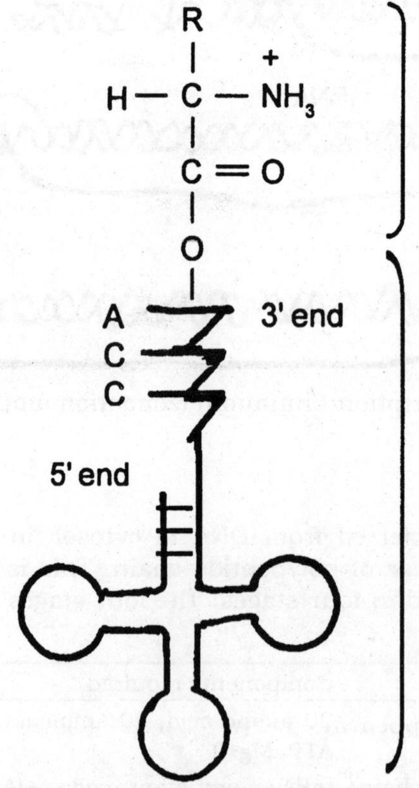

Fig. 3. Aminacyl - tRNA

In the first step ATP reacts with the amino acid with displacement of pyrophosphate to form an enzyme bond intermediate an aminoacyl adenylic acid. In the second step the amino acyl group is transferred from AMP to 3' end to its tRNA followed by release of the products an aminoacyl tRNA and adenylic acid. The enzyme aminoacyl-tRNA synthetase has two different combining sites: one that recognizes the side group of an amino acid and another that recognizes the tRNA specific for that amino acid. Aminoacyl-tRNA synthetase attaches amino acid to the 3'OH, while others attach it to the 2'OH of the terminal ribose sugar of the tRNA. There are usually several types of tRNA for the same amino acid. They are known as isoacceptors. Such isoacceptors tRNA molecules that bind to different codons are activated by the same aminoacyl-tRNA synthetase.

An individual aminoacyl~tRNA synthetase is specific for a single amino acid and certain tRNA as well. The interaction between aminoacyl~tRNA synthetases and tRNAs has been referred to as the "second genetic code". These interactions are due to aminoacid arm and the anticodon arm as well as many other parts of the molecule. The tRNA conformation is also important for such recognition.

1. INITIATION OF POLYPEPTIDE CHAINS

Initiation of polypeptide chain begins with the amino acid methionine coded by initiation codon AUG. In bacteria, the first amino acid is methionine that take part as N-formylmethionyl-tRNA. The formyl group is enzymatically added to methionine after the methionine has become attached to its tRNA. Not all methionine tRNA molecules can be formulated. Instead, there are two types of tRNA methionine, only one of which (known as tRNA formylmethionine) permits the formulation reaction. Some molecules of tRNA~meti, designated as tRNA~meti permits formylation reaction. The other type is tRNA met which incorporates in the elongation of polypeptide chain. Both of these types have same anticodon sequence. There are two differences between the initiating and elongating met-tRNA: the tRNA moieties themselves are different and the second difference is that the amino acids differ in the state of the amino group. An enzyme deformylase that removes the formyl group from the growing chain very soon after synthesis commences. As there is no formylated methionine group in pure protein, it shows that in addition to deformylase another enzyme amino-peptidase subsequently removes the terminal methionine from many proteins. During initiation, first mRNA binds to 30S subunit of ribosome. It is interesting that only specific mRNA remains binded to the ribosome subunit and other mRNA are trimed by ribonuclease. Now the question is that how does the ribosome distinguish between AUG codons at the start of a gene and those coding for internal methionines. When sequences of mRNA roughly 30-nucleotide fragments were analyzed, they were found to contain AUGs roughly near the middle. Following this initiating AUG the other codons code for first three or four amino acids of their respective polypeptide chains. A group of 3 to 9 purine nucleotides appearing in the 5' end of the mRNA are critical for ribosome binding. These nucleotides are upstream from the initiator codon. The polypurine stretch is known as the Shine-Dalgarno sequence. These nucleotides form base pairs with complementary residues found in a pyrimidine-rich sequence at the 3' end of 16S rRNA. Hence it is clear that initiation starts with binding of mRNA-rRNA pairing. During elongation of polypeptide chain the 16S rRNA-mRNA base pair dissociates.

There are atleast three Initation Factors (IF1, IF2, and IF3). These factors are found only on 30S three separate proteins, which are only loosely bound by ribosomes, called subunits,and they are released when the 30S subunit associates with 50S subunits to generate 70S ribosome. IF-1 is required for 30S subunits to bind specifically to initiation sites in mRNA. IF-2 binds a special initiator tRNA and controls its entry into the ribosome. IF-1 binds to 30S subunits only as a part of the complete initiation complex. It may be involved in stabilizing it, rather than in recognizing any specific component.

The process of initiation (Fig. 4) may be described in following steps:

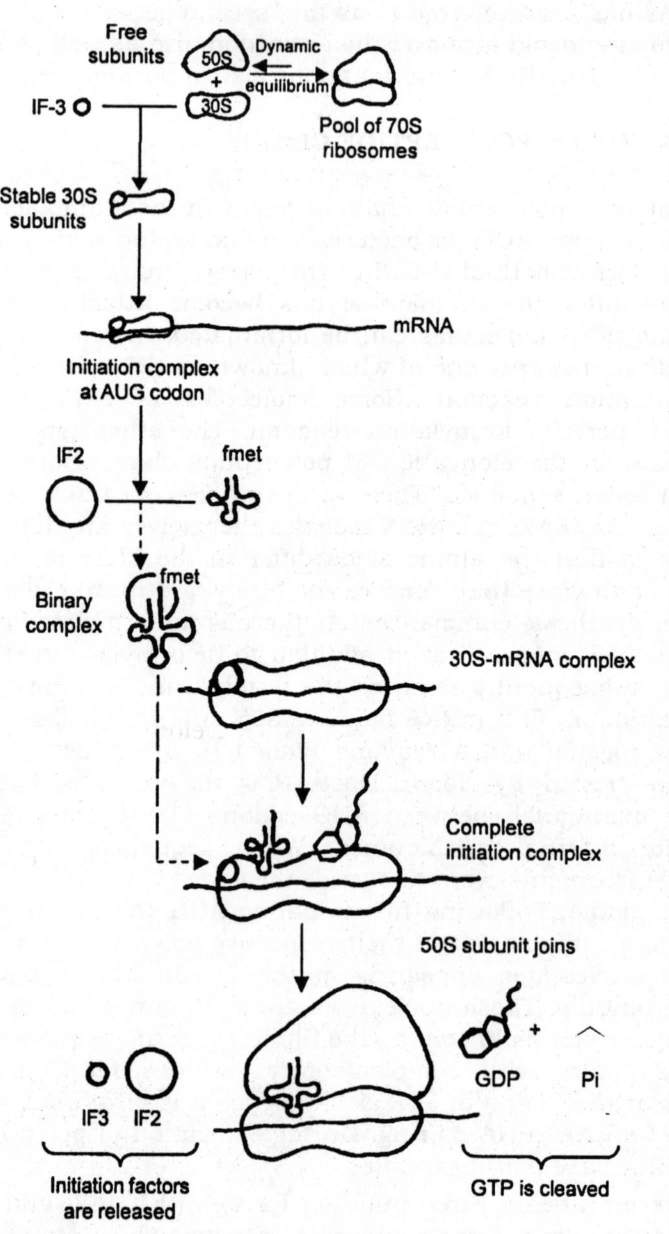

Fig. 4. Steps of initiation of polypeptide chain

At first the 30S subunit forms a complex with IF3. It prevents the 30S and 50S subunits of ribosomes to unite. The mRNA binds to complex I, placing initiation codon in a proper place. To this complex IF1 binds.

In the next step, f-met-tRNA is incorporated to initiation complex. Initiation codon, AUG recognizes this species of tRNA. GUG or occasionally UUG may also take part in initiation. The codons are not recognized equally well. The extent of initiation declines about half when AUG is replaced by GUG and declines by about half yet again when UUG is employed. The initiating tRNA and GTP binds to IF2 and the resulting complex then combines with the 50S subunit to form a complete 70S initiation complex. GTP is hydrolyzed to GDP + iP.

When the initiation process is complete IF-1, IF2 and IF3 dissociates.

Initiation of polypeptide synthesis in eukaryotes is almost the same as bacteria. However, there are some differences in order of events and the number of accessory factors. Eukaryotic cells have 40S ribosomal subunit instead of 30S as in bacteria. In eukaryotes, small subunits first recognize the 5' end of the mRNA, and then move to the initiation site, where they are joined by large subunit. In eukaryotes, the nontranslated 5' leader is relatively short. It is generally less than 100 bases. The nontranslated 3' trailer is often rather long, sometimes a 1000 bases. Initiation is by met-tRNA. It is not formylated. It is known as tRNA~Meti. In yeast, the latter has two unique characteristics. First it the unusual tertiary structure and the second, it is modified by phsophorylation of the 2' ribose position on base 64. There is more initiation factors-at least nine has been well studied.

2. ELONGATION OF POLYPEPTIDE CHAIN

Each 70s ribosome contain two cavities into which tRNA molecule is inserted (Fig. 5). They are P or peptidyl site and A or aminoacyl site. It is not clear that the initiating fmet~tRNA enters directly to the P site or to A site and then subsequently moves to P site. During elongation of polypeptide chain the incoming aminoacyl~tRNA enters to the A site and form the peptide bond with aminoacid of P site. The incoming amino acid binds to a specific protein known as elongation factor (EF-T), GTP acts as energy source (Fig. 6). Two molecules are hydrolyzed for each amino acid subsequently added to the growing polypeptide chain. Elongation factor is made up of two units EF-Tu and EF-Ts. EF-Tu binds with GTP and AA~tRNA to form an AA~tRNA-Tu-GTP complex. This complex is transferred to the ribosomal A site with release of a free EF-Tu-GDP complex and phosphate. The GTPase activity of EF-Tu makes an important contribution to the rate and fidelity of the overall biosynthetic process. EF-Ts is responsible for regeneration of GTP. EF-Ts displaces GDP from EF-Tu by itself forming a complex, EF-Tu.Ts. The latter is known as T factor. The T factor encounters GTP, a GTP-EF-Tu complex is formed, releasing EF-Ts. EF-Tu-GTP can then bind another AA~tRNA complex and thus repeats the cycle (Fig. 7). The incoming aminoacyl~tRNA joins to mRNA by hydrogen bond. The reaction is catalysed by enzyme peptidyl transferase. The bond formation is due to nucleophilic attack of the amino group of aminoacyl~tRNA to carboxyl group of fmet~tRNA. Peptidyl bond formation does not require energy source (GTP or

ATP). In the third step translocation occurs and the ribosome moves towards the 3'end of mRNA due to this the aminoacyl~tRNA of A site is shifted towards the P site. This is a complex process and requires elongation factor G (EF-G) also known as translocase. An EF-G-GTP-ribosome complex first forms translocation then occurs coupled with ejection of the free tRNA from the P site. GTP is hydrolyzed to GDP and inorganic phosphate and EF-G is gained in reusable form. After this process the A site of 70S ribosome is open and is ready to receive another aminacyl~tRNA.

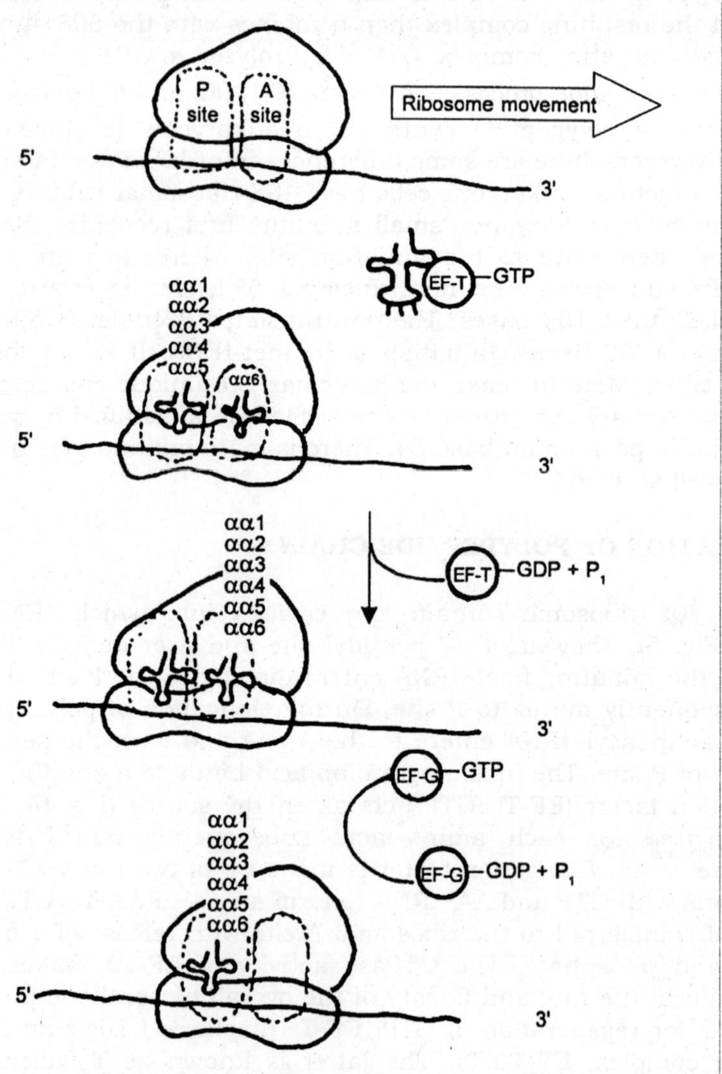

Fig. 5. Steps of elongation in Prokaryotes

In eukaryotic cell, elongation of polypeptide chain takes place in similar mechanism. Elongation factor eEF-1 (formerly known as aminoacyl transferase I) consists of two major subunits. These are eEF-1. It functions as EF-Tu, and binds with GTP. The other subunit of eEF-1 is eEF-1. It functions like EF.Ts i.e.,

GTP-GDP exchange. A third subunit eEF-1 is also reported. However, its function is not established. The functions of EF.G are performed by eEF-2. It was formerly known as aminoacyl transferase II.

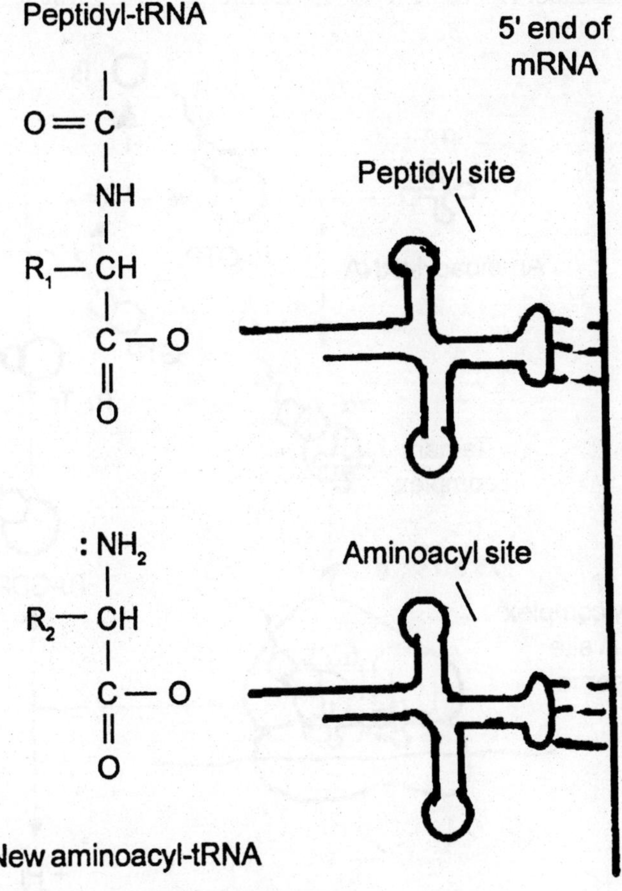

Fig. 6.

3. TERMINATION OF POLYPEPTIDE CHAIN

Two conditions are necessary for the termination of polypeptide chain. The first one is presence of nonsense codon and the second is the release factor (RF). There are three nonsense codons UAA, UAG and UGA, which do not code for any amino acid. These codons are named by three names. These casual names are from the history of their discovery. The UAG is called the amber codon; UAA is ochre codon; and UGA is opal codon. Similarly, there are release factor R1, R2 and R3 or RF1, RF2 and RF3. In *E. coli*, there are two well-characterized release factors. RF1 recognizes the stop codons UAG and UAA and RF2 that acts with UGA and UAA. The third factor (RF3) does bind GTP, and GDP appear to stimulate the reaction. It is however, not clear whether GTP hydrolysis is required for termination in *E. coli*. When the polypeptide chain has reached its full length, its carboxyl end is still bound to its tRNA adaptor. During termination this tRNA splits of from the polypeptide chain. The 30s ribosomal subunit is released from mRNA. Another protein factor may be involved in ribosomal release.

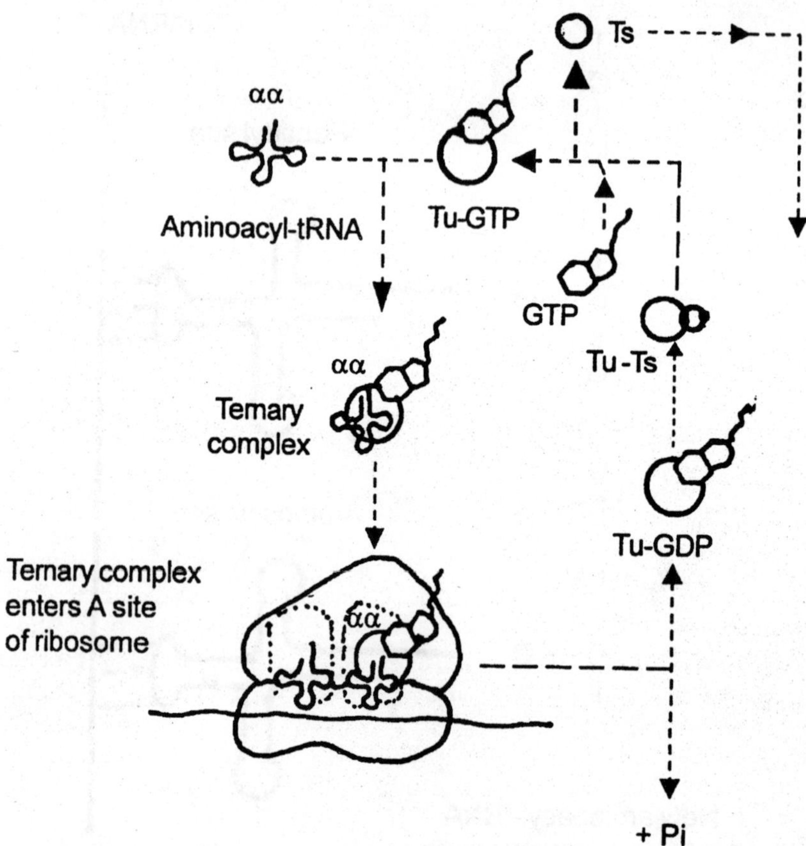

Fig. 7. Elongation factor cycle in Prokaryotes

POLYPEPTIDE SYNTHESIS IN POLYSOME OR POLYRIBOSOMES

When more than one ribosome take part in polypeptide synthesis it is known as polysome or polyribosome. It is rare that a section of mRNA molecule is in contact with only one ribosome. The same mRNA gives template to several polypeptide chains and at any moment, the polypeptide chains being produced along the length of the mRNA are shortest at the 5'end and gradually lengthen toward the 3' end. There is great variation in polysome size, which depends both on the size of the mRNA chain and how closely the ribosomes are packed. As many ribosomes take part in the synthesis of same polypeptide chain it is beneficial to cell and it requires relatively little number of mRNA.

CHAPTER - 16

REGULATION IN GENE EXPRESSION

Enzymes regulate all the activities in a cell. Enzymes are proteins in nature. Synthesis of protein is regulated by gene. A gene is a sequence of DNA that codes for a sequence of amino acid in a protein. Thus, synthesis of protein is regulated by a gene. Many enzymes are needed only on occasion. These are inducible enzymes. Their production is regulated by an on-off type of mechanism. This mechanism enables such products to be present only when demanded by external conditions. More subtly regulated systems can adjust the intracellular concentration of a particular protein in response to needs imposed by the environment.

The synthesis of a particular gene product (enzyme) is controlled by mechanisms known as Gene regulation. Regulation may take place at any stage of transcription and translation. The control generally occurs at the level of initiation or termination in these two processes. The regulatory mechanism of enzyme synthesis in prokaryotes and eukaryotes is somewhat different. In bacteria, RNA transcribed from DNA directly takes part in translation. While, in eukaryotes, processing of the DNA product may be regulated at the stages of modification, splicing, transport, or stability.

SOME DEFINITION AND CONCEPTS

Cis and Trans sequence of DNA

Jacob and Monad (1961) discriminated two types of sequences in DNA. A cis-acting sequences that function exclusively within the DNA. Cis acting sequence is not converted into any other form. A trans-acting sequence of DNA codes for diffusable product. Diffusable product is synthesized by DNA and acts elsewhere. This product may be protein or may be RNA.

Structural and regulator gene

A gene, that codes for a mRNA is known as structural gene. The genes coding for enzymes of the same pathway or proteins whose function is related are found in cluster. While, a gene that codes for a protein involved in regulating the expression of gene is known as regulatory gene. This protein controls transcription.

Constitutive and regulated expression

The genes are transcribed irrespective of the stage of cell cycle or presence or absence of substrate. Gene expression is regulated by binding of specific regulatory factors to the promoter. While the genes expressed in both constitutive and regulated manner is known as dual control. The constitutive nature of these genes shows protein synthesis in all the stages of cell cycle; however, a regulatory factor changes the level of expression.

Positive and negative control

Positive and negative control systems (Fig. 1) are defined by the response of the operon when no regulator protein is present. A positive control means expression of gene is possible only when an active regulator protein is present. The regulator is essential for transcription. The regulator or also known as activator, it interacts with DNA and with RNA polymerase to assist the initiation of RNA synthesis. Genes under negative control are expressed unless they are switched off by a repressor protein. A repressor binds to DNA and inhibits transcription. It may also bind with mRNA and may inhibit translation.

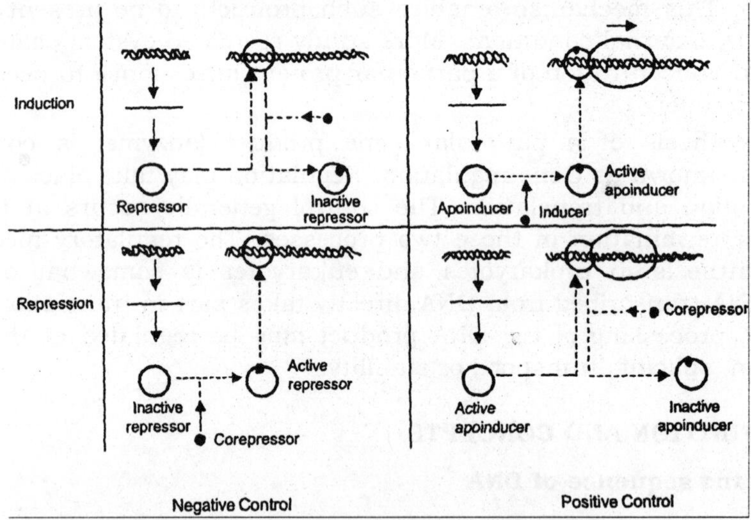

Fig. 1. Positive and Negative Control.

Promoter, Operator and Repressor

Promoter serves to initiate transcription only of the gene or genes physically connected to it. It is associated site of operator and is essential for expression of gene. It also determines the rate of expression of the operon, under conditions of full induction.

Operator is a naturally occurring regulatory element. It modulates the function of a strong promoter. In the original operon hypothesis, it was proposed that the operator has three functions. These functions are (1) It forms the first part of the z gene; (2) it is the site of repressor action and (3) it is the site of initiation of transcription. The second function is confirmed.

A repressor is a protein that prevents transcription of gene. The genetic and chemical evidences have suggested that repressor is a product of i gene.

DNA binding motifs

The trans acting factor is generally a protein, which binds with DNA. Specific structures or domains of the DNA binding proteins are known as DNA binding motiffs. The specific three-dimensional folding of the proteins provides the quality to protein to bind with DNA. DNA motifs are of following types.

Helix-turn-helix (HTH) motif (Fig. 2) Two alpha helices (made up of about 7-9 amino acids) and a short region made up of about 20 amino acids present between these helices form motif. The space between the two alpha helices is β-sheet. The operator sequence of DNA is recognized by one of the alpha helix is known as the recognition helix. The β-sheet turns itself in such a way that the recognition helix interacts with DNA molecule and can bind to it.

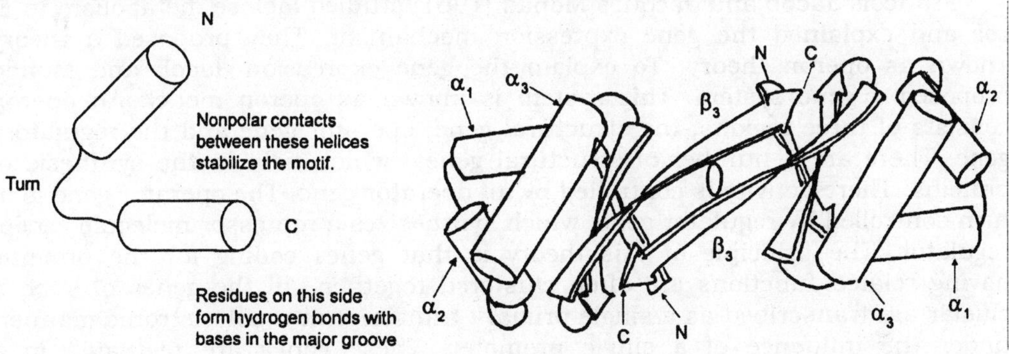

Fig. 2. Helix-turn-Helix motif Digramatic and three dimensional structure of Cro protein

Helix-loop-helix (HLH) motif (Fig. 3). As it is clear by the name, two helices and one connecting loop form forms a motif. Each helix is made up of about 15 amino acids and is separated about 50-60 amino acids. Each helix has a hydrophobic region and a charged region. The mechanism of DNA binding by these motifs is not clear.

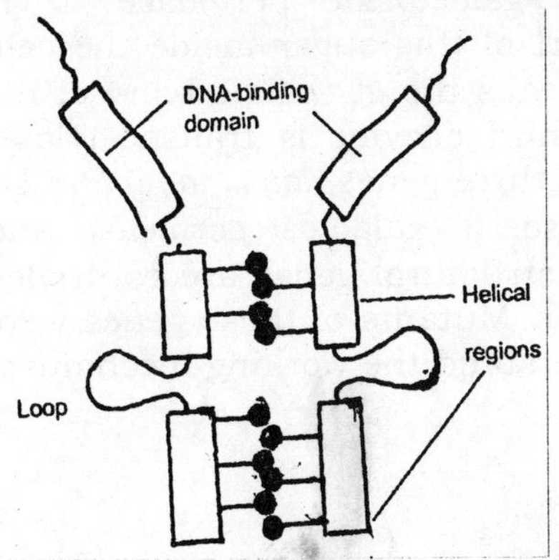

Fig. 3. Helix-loop-Helix motif

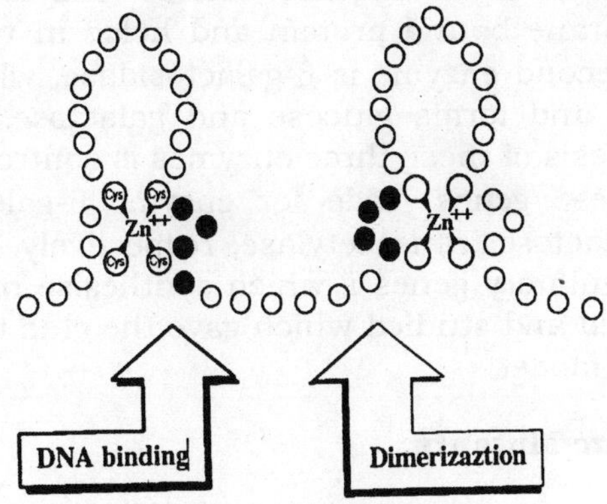

Fig. 4. Zinc finger motif

Zinc finger motif (Fig. 4). The shape of motif is like a finger, made up of 25-30 amino acids. This motif is named as zinc finger because zinc radical is present between two cysteine and two histidine amino acids. It is known as Cys2/His2 finger. Sometimes, only four cysteine molecules are present and such motif are known as Cys2/Cys2 finger. The C-terminal part of motif is α-helix while; the N terminal part is may form a β-sheet. The alpha helix binds to the major groove of DNA.

Leucine Zipper These motifs are formed by 8 helical turns and leucine is present at every 7th position, so that a projection is formed having leucine at side chain. A dimer is formed by two such proteins. Leucine residues interlock with each other just like a zipper. This is the position, which gives the capability to motif to bind with DNA.

THE OPERON THEORY

Francois Jacob and Jacques Monad (1961) studied lactose metabolism in *E. coli* and explained the gene expression mechanism. They proposed a theory known as operon theory. To explain the gene expression Jacob and Monad proposed a gene system. This system is known as operon model. An operon consists of three regions, the structural gene, operator gene and the regulatory gene. There are a number of structural genes, which help in the synthesis of proteins. There activity is controlled by an operator gene. The operator gene is in turn controlled by regulator gene, which synthesizes a repressor molecule except regulator. The principle of this theory is that genes coding for the proteins having related functions are often clustered together. All the genes of such a cluster are transcribed as a single primary transcript in a polycistronic manner, under the influence of a single promoter. These genes are regulated in a coordinated manner. The regulation is mediated by a single operator. An operator is present in a close vicinity of the structural genes. The regulatory gene sends a regulatory protein, which controls the activity of operator.

Lactose metabolism in *E. coli* and Lac operon model

E.coli is capable of utilising lactose as carbon source. Its metabolism requires three enzymes. These enzymes are galactoside permease. It is membrane bound protein and helps in transport of this sugar inside the cell. The second enzyme is β-galactosidase, which breaks the glycosidic bond of the sugar and forms glucose and galactose. The third enzyme is transacetylase. Synthesis of these three enzymes is controlled by three genes, *lac z, lac y and lac a.* These genes, code for enzyme β-galctosidase, β – galactose permease and β – galactose transacetylase, respectively. These structural genes are controlled by regulatory genes *i,* which synthesise repressor. Mutants of these genes were isolated and studied which gave the clue to understand the working mechanism of the model.

The *lac*-Mutants

The knowledge about the existence of enzymes in the lactose-utilizing system is due to a combination of genetic experiments and biochemical analysis.

Some mutants for *lac* gene (*lac*-) were isolated. These mutant genes were transferred from chromosome to plasmid (F'*lac*) so that recombinants may be prepared. The genes were transferred from one cell to another combination of different types of partial diploids were obtained. While, studying these mutants it made easy to conclude the presence of different types of genes associated with lactose metabolism.

Some observations led to conclude that the regulation of enzyme production, essential for lactose metabolism have on-off mechanism. If wild types of cells are grown in medium without lactose (or any other β-galactoside), there is a very small intracellular concentration of enzymes involved in lactose metabolism. However, if this sugar is added in the medium the intracellular concentration immediately goes up. Analysis of the total mRNA present in the cells before and after addition of lactose shows that no lac mRNA is present before lactose is added and that the addition of lactose triggers synthesis of lac mRNA. The observation that lactose system is inducible and lactose (galactoside) is inducer was proved by the use of sulfur-containing analogue (usually isopropylthiogalactoside, IPTG) of lactose. Such substance is said to be a gratuitous inducer. The constitutive mutants were isolated. These mutants produce the enzyme even in the absence of substrate (inducer) and the genes transcribed continually hence, the name constitutive. Complementation tests with partial diploids have two constitutive mutations, one in the chromosome and the other in a plasmid. This showed that the mutants fall into two groups termed lacI and lacO c. The wild type of cells fails to transcribe in absence of inducer but the recessive mutant (lac-) showed just opposite behaviour. Thus, the laci gene is apparently a regulatory gene whose product is an inhibitor that keeps the system turned off. The product of lac I is a protein. It is known as lac repressor. To establish that lacO c is cis dominant, two combinations of genes were examined. They are (i) F'lacO clacz-/lacO$^+$lacZ$^+$ and (ii) F'lacO clacZ$^+$/ lacO$^+$ lacZ$^-$. The former is inducible while the latter is constitutive in nature. There is difference between the two combinations in entries (i) and (ii). In (i) mutation is carried on a DNA molecule that has lacO^c mutation alongwith lacZ$^-$ mutations. Whereas in (ii) the same DNA molecule have both lacO^c and lacZ$^+$ mutations. Thus, a lacO^c mutation causes constitutive synthesis of β-galactosidase only when the lacO^c mutation is said to be cis-dominant. It is because only genes cis to the mutation are expressed in dominant fashion. With all such studies the gene order of the different elements of the lac system (regulatory and structural genes) are as following

lac i lac o lac z lac y lac a

The Model

The Jacob-Monad described the *lac* system as follows. In the absence of any regulation, the expression of the three *lac* structural genes involves two steps. First, the information from these genes is transcribed into a single mRNA molecule. The mRNA is translated into three enzymes. The transcription is initiated at the promoter site, *p*. This site is adjacent to, or part of the *z* gene. The structure of the promoter, or of a site in the same region, determines the rate at which the message is transcribed from these genes under conditions of maximal expression. The control of the expression of the lac structural genes is effected by a repressor molecule, the protein product of the closely linked *i* gene. (Fig. 5). The repressor acts on the DNA to inhibit the transcription of these genes. The combination of the repressor with the operator site, *o*, which is adjacent to *p*, in some way inhibits the initiation process. Compounds that cause induction of the expression of the *lac* genes act by either destroying or altering the repressor-operator complex, thus allowing initiation of mRNA synthesis at the promoter (Fig. 6).

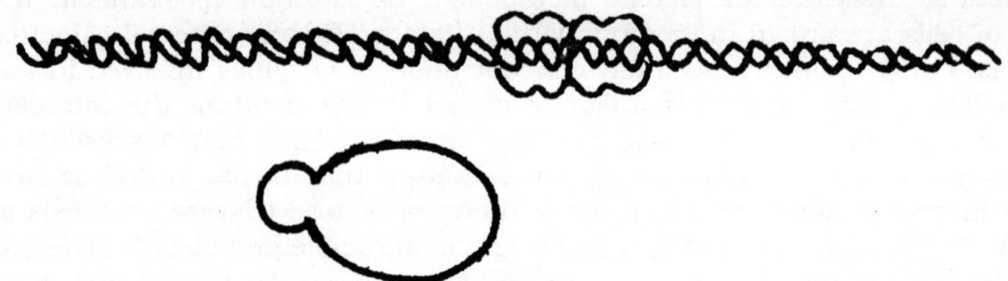

Fig. 5. Repressor inhibits RNA polymerase to act on DNA molecule

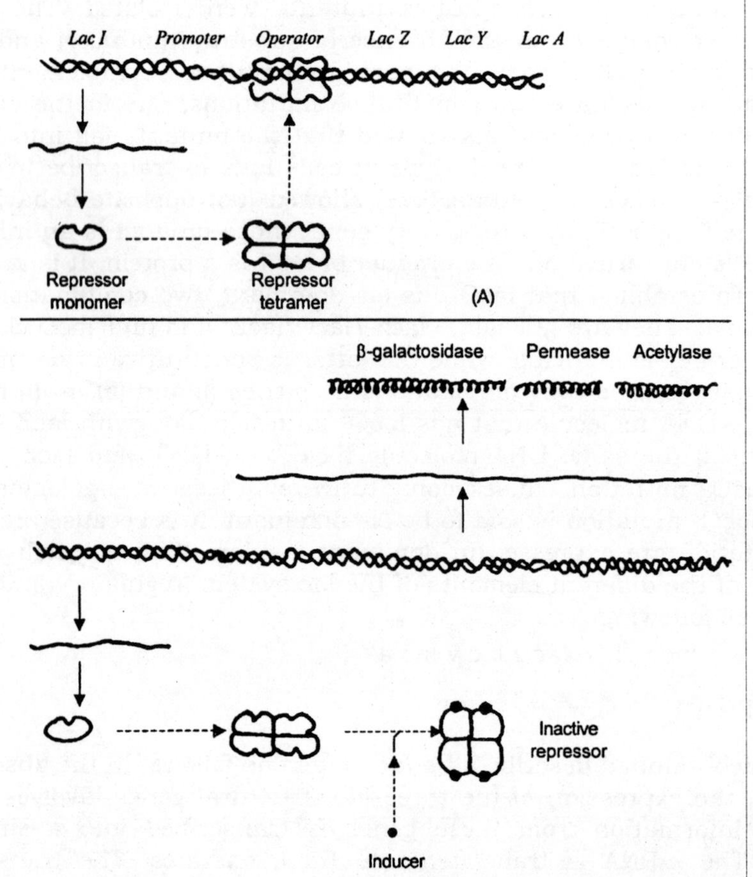

Fig. 6. A. If repressor is attached to operator gene, it prevents RNA polymerase activity **B.** If inducer is attached to repressor, it inactivates latter and RNA polymerase is active.

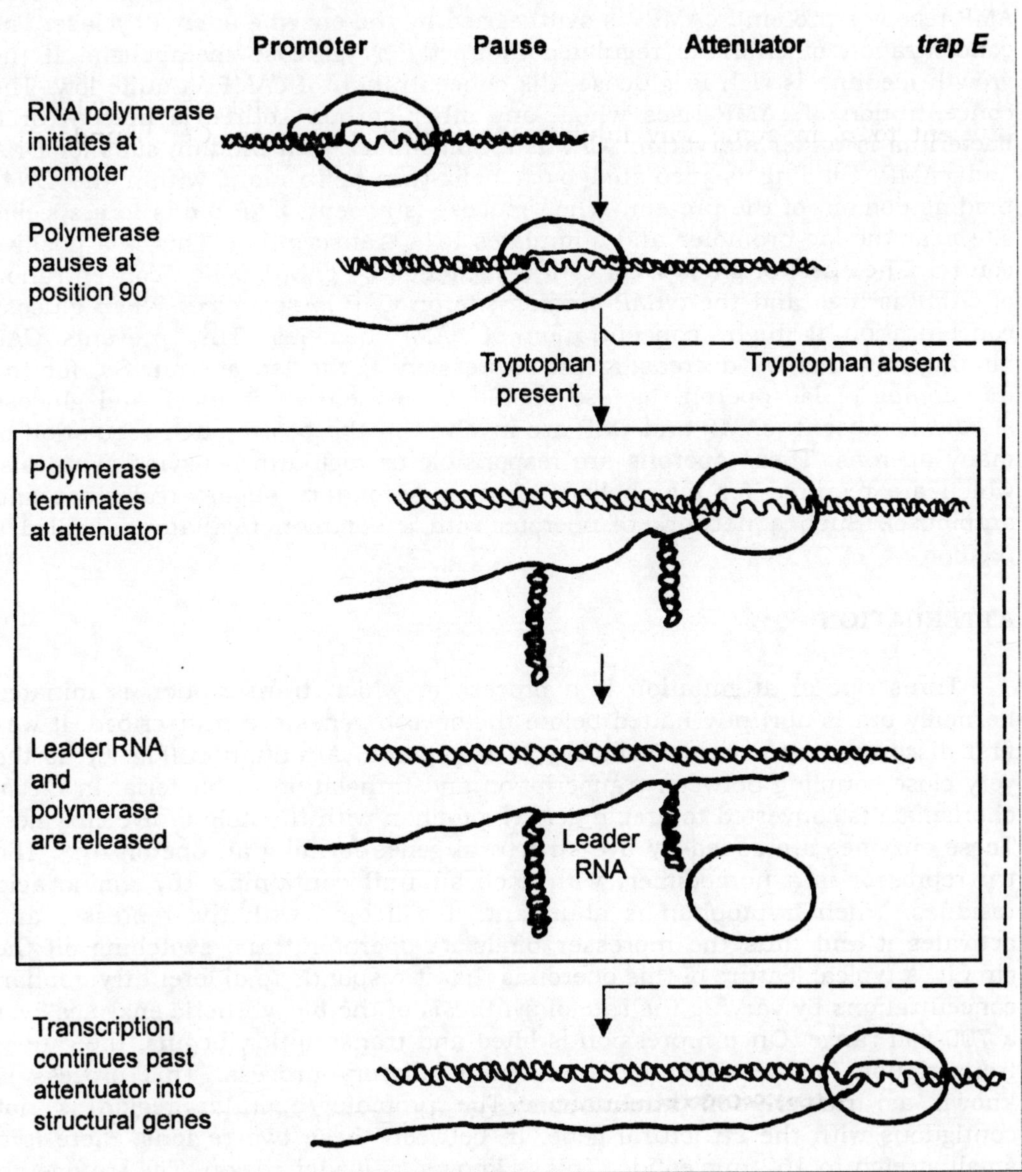

Fig. 6

Positive Regulation of lac operon

E. coli first prefers glucose as sugar or ultimately the carbon source for energy. If glucose is present in the medium, the cells will not utilize other carbon sources like lactose, galactose, and arabinose. If glucose and lactose both are present in the medium, activity of lac operon is suppressed. The catabolic repression takes place due to which other sugars than glucose are not utilized by the cell. This repression is mediated by cyclic AMP (cAMP) and a protein known as catabolite gene activator protein (cga) protein or (CAP) or CRP (cyclic

AMP receptor protein). cAMP is synthesized by the enzyme adenyl cyclase. The concentration of cAMP is regulated indirectly by glucose metabolism. If the growth medium is rich in glucose, the concentration of cAMP is quite low. The concentration of cAMP rises when, any other carbon source is present or if bacterium is under starvation. CAP is a homodimer with binding sites for DNA and cAMP. Binding is mediated by a helix-turn-helix motif within the DNA-binding domain of the protein. When glucose is absent, CAP binds to a specific site near the lac promoter and stimulates RNA transcription. This is a positive control. The effect of glucose on CAP is mediated by cAMP. When concentration of cAMP is high and the cAMP-binding site on CAP is occupied. When glucose concentration is high, concentration of cAMP declines. This prevents CAP binding and thereby decreasing the expression of the lac operon. So, for the functioning of lac operon, lactose should be present as inducer and glucose should be absent. cAMP and CAP are involved in the coordinated regulation of many operons. These operons are responsible or regulating enzyme synthesis, which are required for metabolism of other secondary sugars (galactose and arabinose). Such a network of operons with a common regulator is called a regulon.

ATTENUATION

Transcription attenuation is a process in which transcription is initiated normally but is abruptly halted before the operon genes are transcribed. It was first discovered by Charles Yonofsky in *trp* operon. Attenuation (Fig. 6) is the very close coupling between transcription and translation in bacteria. In *E.coli*, chorismate is converted to amino acid tryptophan with the help of five enzymes. These enzymes are coded by five structural genes forming an operon (*trp*). The trp repressor is a homodimer, with each subunit containing 107 amino acid residues. When tryptophan is abundant, it will bind with the repressor and activates it and thus the repressor binds its operator thus, switching off the circuit. A typical feature of this operon is that it responds to different tryptophan concentrations by varying the rate of synthesis of the biosynthetic enzymes over a 700-fold range. Once repression is lifted and transcription begins, the rate of transcription is fine-tuned by a second regulatory process. This process is known as transcription attenuation. The promoter/operator region is not contiguous with the structural gene. In between these two regions, there is a small stretch to 162 nucleotide. This is known as leader region. The trp operon attenuation mechanism uses signals encoded in four leader region at the 5' end of the mRNA that precedes the initiation codon of the first gene. At the end of the leader is a sequence called the attenuator. It is made up of sequences 3 and 4. An attenuator is described as an intrinsic terminator located at the beginning of a transcription unit. It acts by the formation of a hairpin structure resulting in rho independent termination. Sequences 3 and 4 base-pair to form a G C rich stem and loop structure followed by a series of uridylate residues. This structure resembles a transcription terminator. The sequence analysis of the leader region revealed that there is a putative ribosome binding site followed by an open reading frame which codes for a 14 amino acid long polypeptide, the 'leader peptide'. Other important features are that there are two tryptophan codons within this leader peptide. Four regions with high degree of complementarity in the nucleotide sequence were also seen. These are known as 1,2,3 and 4, respectively. Region 2 is complementary to the regions 1 and 3. Region 3 is complementary to the regions 2 and 4. Thus, due to complementarity, mRNA may have two types of confirmation. In the first type of confirmation, there may be two types of stems. The first complementary double stranded stem lies in region 1 and 2 and the second in region 3 and 4. In second confirmation, region 1 and 2 form a double stranded stem while 3 and 4 remain single stranded. Due to this alternate confirmation, RNA behaves as attenuator. The regulatory sequences 1 or leader peptide is the key element in sensing tryptophan concentrations. When region 2 and 3 has base pairing, transcription continues and if region 3 and 4 are paired (attenuating transcription) transcription is stopped. In the absence of tryptophan in the medium, the trp-tRNA cannot be charged with the amino acid and tryptonyl tRNA is not available in the cell. Thus, the polypeptide synthesis is paused when the chain reaches at the sequence where tryptophan is to be added in the chain. The region 1 of mRNA is rich in trp codon. At this stage, region 1 becomes embedded within the ribosome and is not available for base pairing with region 2. Region 2 is free from ribosome and it is in a position to base pairing with region 3. The base pairing occurs when region 3 is synthesized by growing transcription bubble. While region 2 and 3 have undergone pairing and region 4 is synthesized. Region 4 will remain unpaired due to unavailability of region 3. At this stage, as the confirmation of RNA is entirely different from the structure of a terminator, transcription continues. The structural gene will transcribe mRNA for all the five genes used in tryptophan synthesis. If tryptophan is very well available, there is no shortage of charge trp-tRNAs and the ribosome does not pause region 1 (the trp codons), translation continues. Region 1 and 2 are within ribosome and do not have base pairing. Region 3 will have base pairing with region 4 resulting in alternate confirmation. This confirmation is very similar to a rho-independent terminator. At this stage there is premature termination of transcription and only 140 base long leader mRNA is formed. This do not covers the mRNA of five genes (five enzymes) essential for tryptophan synthesis. Thus, tryptophan synthesis is stopped.

Each amino acid biosynthetic operon uses a similar attenuation strategy. Amino acid leucine (Leu) operon has leader peptide with four contiguous Leu residues, Thrionine (Thr) operon has 9 Thr residues, Phenylalanine (Phe) operon has 7 Phe residues, Histidine (His) operon has 5 His residues. The amino acid synthesis is regulated by the attenuation system.

EUKARYOTIC OPERON SYSTEM

Transcription in eukaryotic cells is divided into three classes. Genes are classified by their promoters and each class is transcribed by a different RNA polymerase which can be distinguished according to their sensitivity to certain inhibitors e.g. α amanitin:

RNA poylmerase I present in nucleoplasm, transcribes rRNA; RNA polymerase II present in nucleoplasm, transcribes mRNA and RNA polymerase III also present in nucleoplasm transcribes tRNA and other small RNAs. The promotors for RNA polymerase I and II are upstream of the startpoint but some promotors for RNA polymerase III lie downstream of the startpoint. There are large numbers of factors, which can act in conjugation with RNA polymerase. These factors may be divided into three groups.

1. The general factors bind with RNA polymerase and form the basal transcription apparatus. These factors join with RNA polymerase to form a complex surrounding the start point and determine the site of initiation.
2. The upstream factors recognizes specific short consensus elements located upstream of the startpoint. These factors are the DNA binding proteins and their activity is not regulated.
3. The inducible factors are same as general factors with only difference that their activity is controlled. The sequences that they bind are called response elements.

A promoter that contains only elements recognized by general and upstream factors should be transcribed in any cell type. Components of the promoter must be located in the general vicinity of the start point and are required for initiation. Another type of site involved in initiation is identified by sequences that stimulate initiation but that are located a considerable distance from the startpoint either upstream or downstream of it and in either orientation. The common mode of regulation of eukaryotic transcription is positive. Only a single gene is transcribed and termination occurs beyond the end of the coding region. Eukaryotic cells often exhibit cell-specific gene expression. The exogenous signals have their own significance. Cell to cell signaling also occurs. It is also specific that a particular cell would express a set of genes and not express others. The cell specific gene expression leads to differentiation and development.

Exons and introns and RNA splicing

Exons are the sequence represented in the mature RNA. While, the introns are the intervening sequences that are removed when the primary transcript is processed to give the mature RNA. RNA transcribed by a gene has much more nucleotide than the structural gene. These, additional sequences represents the introns and are removed from the primary transcript. This process is known as RNA splicing. The mechanism of RNA splicing has been studied *in vitro*. During RNA splicing, a cut is made at the 5' splice site separating the left exon and the right intron exon molecule. The left exon takes the form of a linear molecule. The right intron exon molecule from a lariate in which the 5' terminus generated at the end of intron becomes linked by a 5'-2' bond to a base within the intron. The target base is an A in a sequence that is called the branch site. The cut at the 3' splice site releases the free intron in lariat form. The right exon is ligated to the left exon. The lariat is then debranched to give a linear excised intron, which is rapidly degraded.

Galactose metabolism in *Saccharomyces cerevisiae*

Yeast cells have intracellular organelles, which are membrane bound. These organelles are much more like other nonphotosynthetic cells. The proteins synthesized on ribosomes attached to endoplasmic reticulum pass through golgi apparatus to other other intracellular membranes, to the periplasmic spaces and to the outermost layer of the cell wall. Yeast cell also contains the protein clathrin. This protein enables yeast cells for selectively endocytose those proteins of their external membranes to which extracellular molecules are attached. Operon are absent in yeast. Instead, there are alternate promoters (structural gene) and control elements in a chromosome. Yeast promoters have upstream activating sequences (UAS s, the positive control) and upstream repressing sequences (URS s, the negative control). These sequences may be several hundred base pairs 5' to the start of transcription. In addition to UAS, yeast promoters have two other essential components. (1) T-A-T-A sequences, usually placed some 20 to 40 base pairs 5' to the transcription start site. (2) An initiator element encompassing the start site, which precisely determines where transcription begins. The control elements are specific groups of base pairs that by binding to specific cellular proteins direct the starting and stopping of RNA chains at correct positions along the chromosomes. Gene regulation in yeast (*S. cerevisiae*) shows that cis-acting elements lie hundreds of nucleotides upstream (100 or 200 base pairs) of the region of transcription initiation and activate transcription in response to particular physiological signals. These, upstream activation sites (UAS s) or enhances act in a different manner as compared with prokaryotic cells. There are also present certain regulatory elements known as silencers which act as sites for binding enhancers and can function at great distances from genes they repress. Initiation of transcription has positive control sites. These sites may be single or multiple, and in all cases, are preceded by one or more T-A-T-A boxes (T). The latter are canonical sequences found in many eukaryotic promoters recognized by RNA polymerase II. The GAL10 gene bears a site more than 100 base pairs upstream of the T-A-T-A box, which mediates induction by transcription by galactose.

S. cerevisiae utilizes exogenous supply of galactose with the help of enzymes galactokinase, glactose-1-phosphate uridyltransferase and uridine diphospho-glucose 4-epimerase. These enzymes convert galactose to glucose-1-phosphate, which is then converted to glucose 6-phosphate. The latter enters the glycolytic pathway. These enzymes are coded by GAL7, GAL10 and GAL1 genes, respectively (Fig. 7 A, B). Although these genes (the three GAL genes) are located close together, they do not comprise an operon. They are transcribed into separate mRNA molecule. These enzymes are induced by galactose and repressed by glucose. The action of positive regulatory gene, GAL 4, coordinately mediates the induction of these genes. GAL4 gene produces a protein of 99Kdal. It binds to specific "upstream" sequences within the GAL1-GAL10 and GAL7 promoters. The GAL4 protein is normally found within the yeast nucleus. Each promoter contains two 30-base pair blocks that are protected from nuclease digestion when bound to the GAL4 protein. GAL4 binding is mediated by a stretch of 74 amino acids at its amino-terminal end. The structure of yeast regulatory regions and the molecular mechanism show similarity with higher

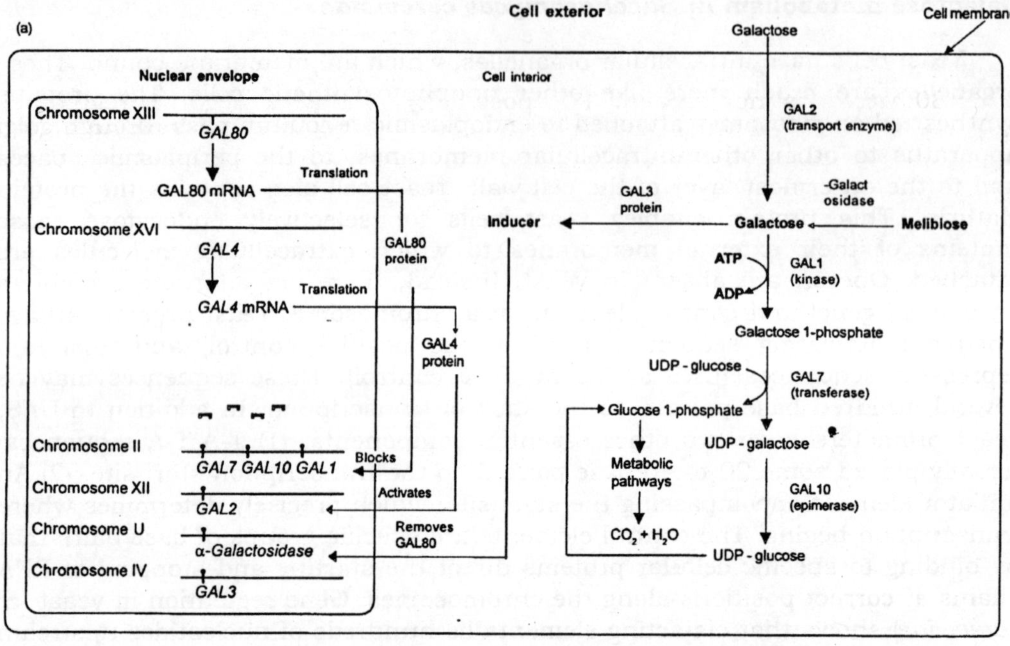

Fig. 7A. Galactose metabolism in yeast. Regulation level of various proteins.

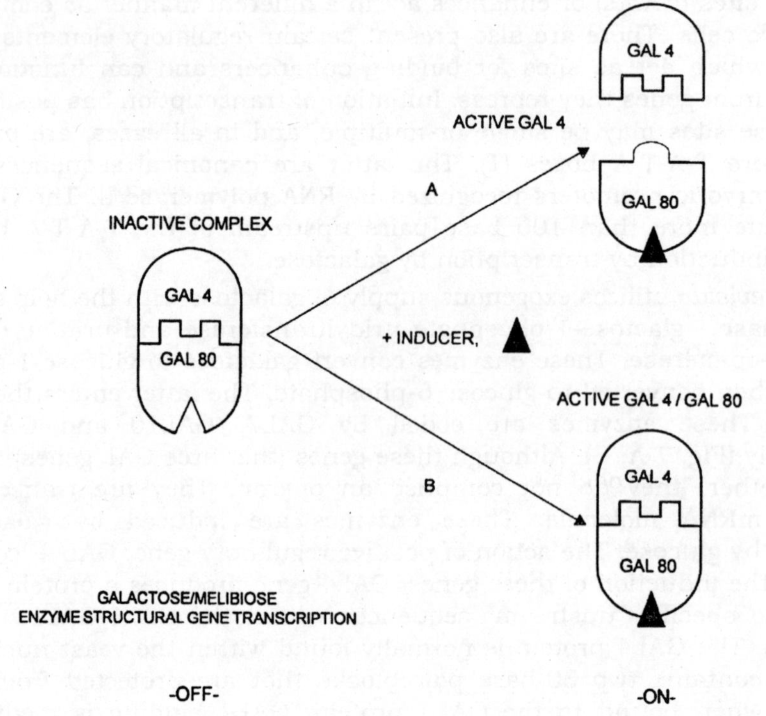

Fig. 7B. Participation of GALBO in galactose metabolism in yeast.

eukaryotes. One key problem under study is whether the regulatory proteins direct RNA polymerase II entry to the DNA template at the UAS s or govern polymerase entry at the T-A-T-A box. How the regulatory proteins encoded by GAL80 works is not known. The possibility is that a negative regulator of transcription of the GAL genes because yeast mutants that lack GAL 80 function express the GAL genes constitutively. However, in absence of galactose, the product of GAL80 prevents GAL4 from functioning. It might code for a repressor like protein that binds to the GAL1-GAL10 promoter region.

Britten and Davidson Model

R.J. Britten and E.H. Davidson (1969) studied the regulation in gene expression in eukaryotes and proposed a model, which can explain the general mode of expression of many inducible genes (Fig. 8). In the cells of eukaryotic organisms, many structural genes respond to different cell signals. They are regulated in an intergrated manner. The regulatory genes are present in a moderatory manner and are known as sensor genes. The latter have specific binding sites. They respond to many exogenous cell signals. The cellular response in relation with any given external signal will depend upon the presence of a corresponding receptor gene, which is associated with any producer gene. These signals may be hormone or other messenger and regulate the transcription of the genes (the integrated genes). The integrated and sensor gene, are present adjacent to each other. Both the receptor genes and the integrating genes are redundant, and then it is possible to devise integrating circuits for the gene expression. On transcription, integrated genes produce RNA known as activator RNA. The activator RNA interacts with receptor genes, which is the last in the series of the regulatory genes. The interaction of activator RNA with regulatory genes is in a sequence specific manner. Instead of RNA if protein is the product of integrator gene, it will behave in the same manner. The interaction of activator gene RNA with receptor gene acts as a trigger for the transcription of the producer gene. The producer genes are the counterparts of structural genes of prokaryotes i.e., they code for proteins, which regulate metabolism. The structural genes are present as constitutive transcription units and these are expressed in all the cells at a basal level. Integral regulatory transcripts are produced in a cell specific manner. The presence of an integral regulatory transcript is essential before a primary transcript of the structural gene can be processed to form respective mRNA.

As cell differentiation is a typical feature of a eukaryotic cell, so to understand this model, let us consider two cell types I and II. The cell type I respond to signal 1 through the sensor 1 and express the integrating regulatory transcription unit 1. The integrating regulatory transcript 1 has repeatitive sequences 'a' and 'b' which are complementary to the regulatory regions a and b present in the structural gene transcript. As result of hybridization of the structural gene transcript and the integrating regulatory transcript, a complex structure is produced. This complex is used in processing of the primary transcript of the structural gene. In cell type II, necessary receptors for signal 1 are absent and receptors of signal 2 are present. It is received by sensor 2 and its integrating regulatory transcription unit 2 is transcribed. This integrating regulatory transcript 2 has complementary sequences for c and d region of the structural gene and a and b are lacking. As result, the primary transcript for structural gene can not hybridize with the integrating regulatory transcript 2 and the secondary structure essential for its processing is not formed. No mRNA is formed and the structural gene of cell type II is not expressed. The cell type II will express other specific genes by the same mechanism.

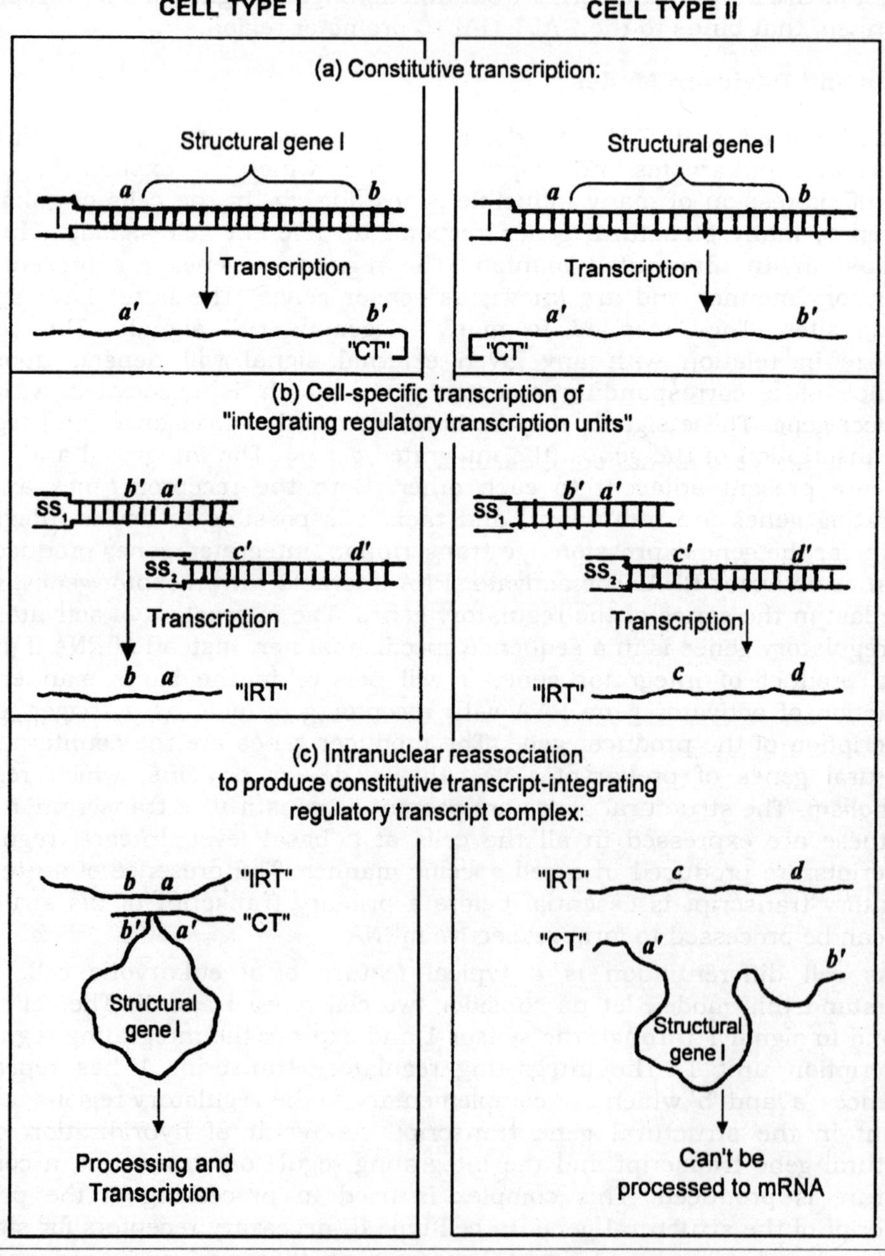

Fig. 8. Davidson and Britten Model

CHAPTER - 17

GENETIC RECOMBINATION

In eukaryotic cells haploid gametes fuse to form a diploid zygote. Zygote has two complete set of genome. Bacteria show the tendency to transfer its genome from one cell (donor) to another cell (recipient). As they are prokaryotic and unable to form gametes and so no complete zygote is formed. During recombination, only a small portion of genome is transferred and the zygote is incomplete. The recipient cell will have its complete genome (endogenote) and a part of donor genome (exogenote). The zygote formed is partially diploid and partially haploid. Hence, it is termed as merozygote. There are three mechanisms of recombination of bacteria.

Transformation

DNA is released from one cell into the surrounding medium. The recipient cell accepts this DNA from the medium and incorporates it into the cell.

Conjugation

Donor and recipient cells come in direct contact and the genetic material is passed from the former to the latter.

Transduction

Genetic material is transferred from the donor to the recipient through phage virions. During replication, bacterial genome attaches with the viral genome resulting in an infected viral genome. When this aberrant phage virion attaches to the recipient cell, it introduces this infected genome to the recipient cell and recombination takes place.

TRANSFORMATION

In 1928, F. Griffith performed an experiment on *Streptococcus pneumoniae* the causal organism of pneumonia (Fig. 1). The bacterium has two types of strains. The capsulated cells (S cells smooth and glistening appearance due to presence of capsule) which are virulent and able to cause disease. The second type of strain is rough, devoid of capsule and unable to cause disease (avirulent). Griffith injected avirulent cells and heat killed cells of virulent strains separately in mice. No symptoms developed in these mice. Avirulent cells and heat killed cells of virulent strains, were together injected in a mice. Disease symptoms appeared in these mice. On autopsy, these mice were found to contain live virulent cells. It was found that surviving cells of bacteria in these mice were recombinant and genetic exchange had occurred between the dead cells and the live cells. In vitro experiments have established that a particular substance is being exchanged amongst cells. This substance was named as transforming principle. O.T. Avery, C.M. MacLeod (1944) purified and examined this transforming principle and found it as DNA. Transforming principle was not digested by enzymes degrading proteins of by RNase. However, DNases enzyme degraded this principle completely. As the transforming substance (DNA) dissolved in the medium and transferred from one cell to another hence, the process is known as transformation. Transformation takes place in species of *Streptococcus, Bacillus, Thermoactinomyces, Neisseria, Acinetobacter, Azotobacter, Haemophilus* and *Pseudomonas*.

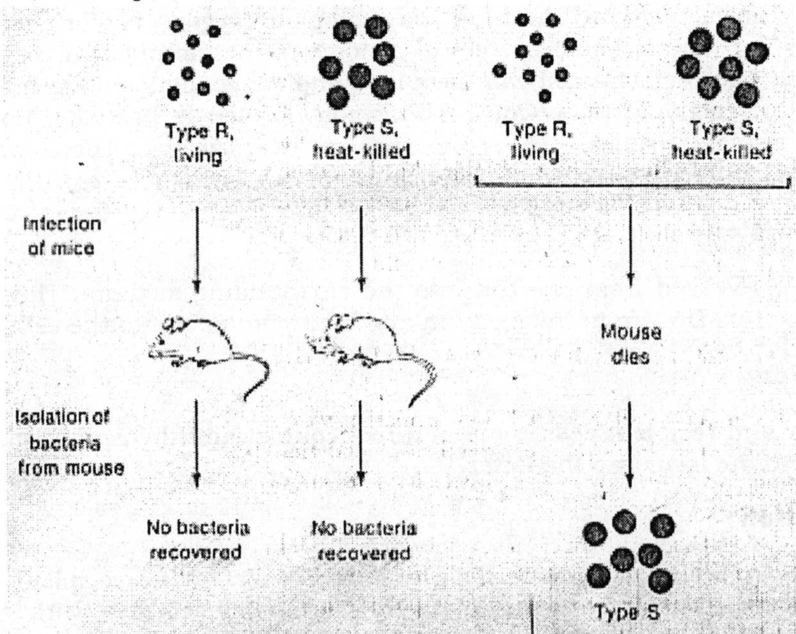

Fig. 1 – Experiment performed by Griffith to demonstrate transformation in *S. pneumoniae*

Steps of transformation

1. *Competence* — The recipient bacteria must be competent if they are to accept DNA and undergo transformation. The competence of bacteria is not a permanent feature but occurs only at certain times in the life cycle. During competence stages growth of the cell wall and chromosome replication occur either at a reduced rate or not at all. Protein synthesis however continues at the normal rate. The number of competence cells doubles every 3-5 minutes. Physical contact of cell is not essential for transfer of competence.

2. *Binding* — DNA comes into contact with the bacterial cell surface as a result of random collision. Donor DNA first binds reversibly and can be removed by exposure to DNase or extensive washing. This reversible period is very short, sometimes lasting just 4-5 seconds. The binding then becomes irreversible or permanent.

3. *Penetration* — Permanently bound nucleic acid penetrates the bacterium dsDNA is converted into DNA upon entering the pneumoccocus cells.

4. *Synopsis* — The recipient chromosome DNA unwinds and initiates pairing with the single stranded donor DNA. This pairing or synopsis is the result of base pairing between donor and recipient DNAs. The donor ssDNA is homologous but not identical with the recipient ssDNA strand with which it pairs.

5. *Integration* — The unsynapsed recipient DNA strand is cut in two places by an endonucleus. An exonucleus then degrades the ssDNA strand. Integration of the donor DNA strand involves a small amount of repair synthesis to close the gaps present, followed by final sealing with a ligase.

Transformation in pneumococcus

Transformation in *Streptococcus pneumoniae* (Fig. 2) is natural, i.e., the cells become competent by certain environmental conditions and artificial treatment is not required. Cells in culture of this bacterium become competent during the exponential phase of growth. A protein called competent factor converts noncompetent cells into competent cells. A set of about 12 proteins is synthesized that mediate the process of transformation. The competent factor is excreted in the medium. With the help of these enzymes, competent cells can absorb double stranded DNA to their outer surface at several sites and cleave it through the action of surface bound enzymes into smaller fragments. One strand is digested by the nuclease. The competence-specific DNA binding protein carries the undigested DNA inside the cell. DNA homologous to endogenote integrates and thereby genetically alters the recipient cell. Integration occurs by strand replacement process. This results in the formation of heteroduplex in a region with one foreign DNA and the other endogenote. At the time of replication, two homoduplex copies are made one, of which is identical to the DNA taken from outside the cell. As consequence transformation brings about the genetic in recipient cell.

Transformation in Haemophilus influenzae

Haemophilus influenzae exhibits natural transformation. In this bacterium, competent cells are due to growth rich media. Transforming DNA enters the recipient in double-stranded form. It maintains its integrity up to the time of its integration into the endogenote. The outer membrane of competent cells has 10 blebs. The latter are the vesicular structures appearing as the localized extensions of the outer membrane. There are present pores at the base of blebs. Certain proteins are present at these regions, which binds specifically with an 11-base pair sequence of DNA. These sequences occur at about 600 sites on the

Haemophilus chromosome. When homologous DNA comes in contact with vesicle, the latter invaginate, trapping the molecule of DNA within it. The DNA containing vesicular membrane structure is termed as transformasomes. DNA inside the transformasomes is protected from DNase, restriction enzymes and modification enzymes. When these transformosomes travels through the cytoplasm and reach upto the endogenote chromosome, a single stranded DNA is released which combines with endogenote.

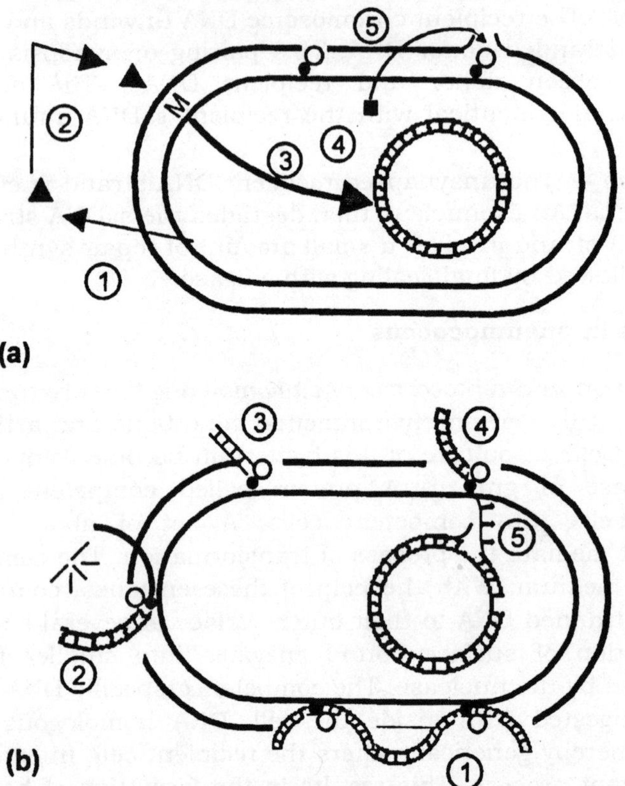

Fig. 2. Schematic representation of transformation in *S. pneumoniae*
(a) 1. Development of competence factor, 2. Site on cell to absorb competence factor, 3. Certain genes to be expressed, 4. Autolysis, 5. DNA binding protein,
(b) 1. Nucleasease, 2. Single Strand, 3. DNA binding protein, 4. Integration to chromosome, 5. Single strand replacement

Molecular mechanism of transformation

To understand the molecular aspect of transformation, experiments were carried out with pneumococcus. There may be two mechanisms to explain the behaviour of donor DNA. (I) The incoming donor allele replaces the allele of the recipient, that is, DNA substitution occurs. (II) The DNA fragment containing the donor allele is added to the genome of the recipient, by insertion adjacent to the recipient allele, forming gene duplication. Reciprocal transformations with X^- - and X^+ strains have confirmed the first opinion.

CONJUGATION

In 1946, J. Ledrberg and E.L. Tatum performed experiments with *E.coli* to understand recombination in this bacterium. During recombination, the DNA passes from F^+ strains to F^- strains by direct contact. F^+ (fertility plus) behaved as donor cell and is considered as male cell, while the F^- (fertility minus) behaved as recipient cell and is considered as female cell. To distinguish it from transformation, the F^+ and F^- cells were grown in a medium containing DNase and still the cells polarized. Later, it was discovered that F^+ strains contain F plasmids (fertility plasmids), which passed from F^+ to F^- strain.

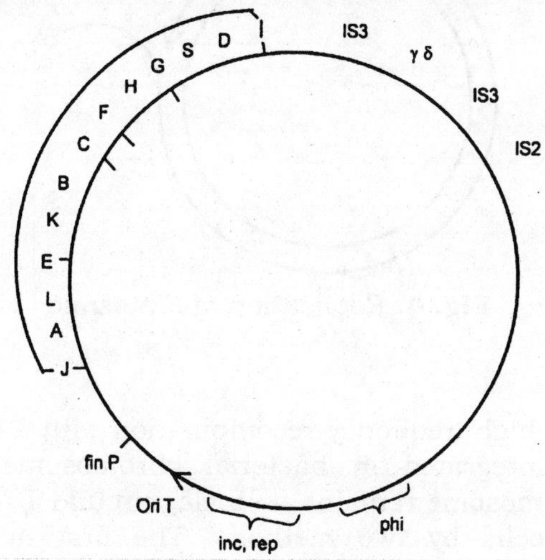

Fig. 3. F plasmid.

F Plasmid

F plasmid (Fig. 3) is termed due to its fertility factor. It passes from F^+ cells to F^- cells during conjugation. After conjugation, the F^- cells are converted into F^+ as they gain F^+ plasmid. The F plasmid carries genes like *rep, inc,* and transfer encoding genes. The latter are set of 13 genes, *tra A* to *tra L* (12 genes) and *tra S,* that forms an operon. F also contains three transposons. These elements do not participate in any of the systems responsible for replication, maintenance, or transfer. They provide a mechanism for integration of F into the chromosome. Gene *rec* is for replication, inc is for incompatibility. The phenomenon of incompatibility inhibits the replication of closely related plasmid present in the cell. Plasmid that codes for self-transfer are termed as conjugative plasmids. The genes coding for transfer have various functions and one of which is formation of sex pilli or F pilli. The sex pilli is present on the surface of F^+ cells. It is very long and is broken at the time of conjugation. It binds with a protein of outer membrane of F^- cells. This protein is termed as OmpA. F pilus forms a conjugative bridge between the F^+ and F^- cells. This bridge holds the two cells at the time of conjugation and forms a passage for transfer of single stranded DNA. Transfer strand originates from *oriT* site of F plasmid (Fig. 4). Site *oriT* stands for origin of transfer. A nick is formed at *oriT* site and DNA replicates by rolling circle mechanism. The 5' end of the nicked single strand is displaced and is transferred to the F^- cell. Within the F^- cell the single stranded molecule of DNA is duplicated by a chromosomally encoded DNA polymerase and recircularized. The ends of the transferred fragment of DNA become attached to a protein located in cytoplasmic membrane. The ends are holded in appropriate juxtaposition and are joined by enzyme DNA ligase. After circularization, the transferred DNA is able to replicate in the recipient cell. The genes of transferred DNA strand are expressed and the F^- cell is converted to F^+ cell.

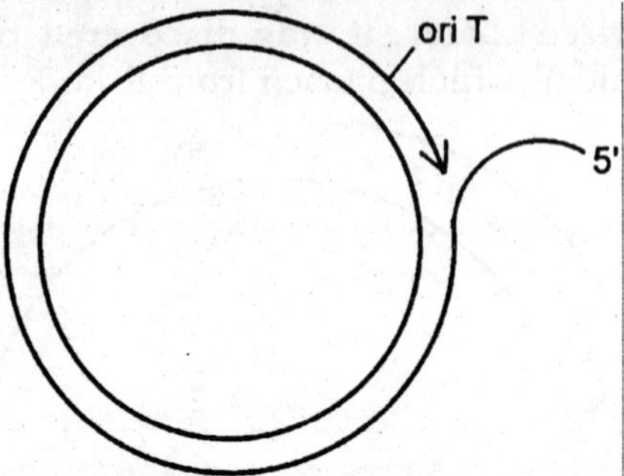

Fig. 4. Replication of F plasmid

Hfr Strains

Hfr strains have high frequency recombination with a special character that the F plasmid is integrated in bacterial chromosome (Fig. 5 A,B). After integration, the chromosome remains a single, circular DNA molecule. Hfr cells may arise from F cells by two methods. The first method is homologous recombination between two identical IS elements, one in the chromosome and the other in F. The second mechanism is replicon fusion. An Hfr cell could arise by formation of a cointegrate mediated by an IS element in F, and by a duplication of a target sequence in the chromosome. Stages of Hfr transfer are pairing of donor and recipient cells, rolling circle replication in the donor conversion of the transferred single-stranded DNA to double stranded DNA by replication in the recipient. The mechanism of these stages is similar to F^+ cells. The major difference is that the transferred DNA does not circularize and is not capable of further replication in the recipient. Replication transfer begins from *oriT* site and continues into the chromosomal region of the large circular molecule. Thus, chromosomal genes as well as F plasmid genes are transferred to the recipient F^- cell. There are some differences between Hfr transfer and F transfer.

1. Hfr needs a considerable time for transfer (about 100 minutes at 37°C). While F transfer requires only 2 minutes.
2. The conjugative bridge is somewhat fragile and hence it frequently breaks even before the complete transfer of DNA. On the average, several hundred genes are transferred before the cells separate.
3. In conjugation between Hfr and F^-, after the F^- cells remains usually F^-.

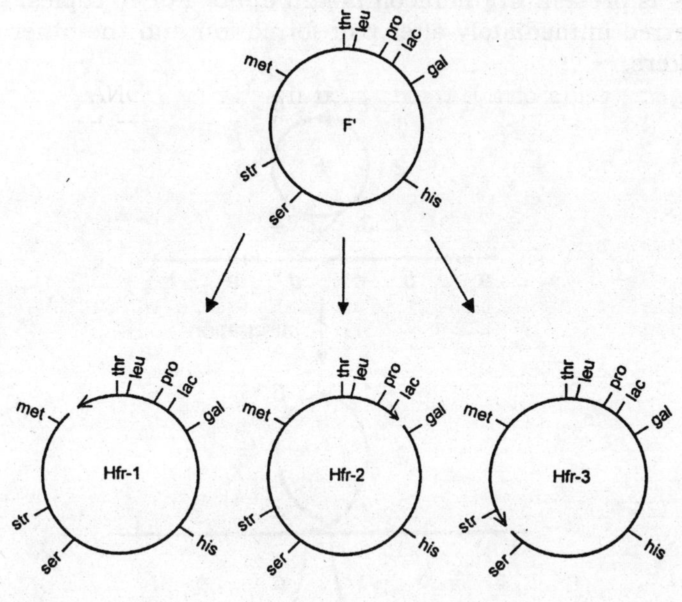

Fig. 5A. Schematic representation of F^+ plasmid forming three different HFr strains

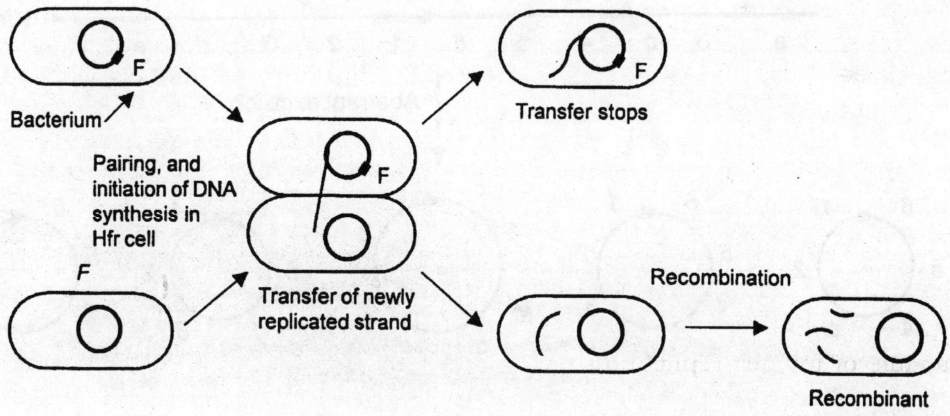

Fig. 5B. Schematic representation of conjugation between HFr male and F-female cells.

Chromosome transfer mediated by F' plasmids

Chromosome transfer by F'-containing strains occurs at a considerably greater frequency than by F^+ strains. This phenomenon is called chromosome mobilization. It is due to reciprocal recombination between the chromosomal segment of the F' and the homologous region of the chromosome itself (Fig. 6). The efficient transfer requires an active *recA* gene in the F' strain. The integrated F' plamid in chromosome differs from a typical Hfr in that the chromosomal genes of the F' is present in the recombinant donor in two copies. Out of the two, one is transferred immediately after pair formation and the other transferred as the final markers.

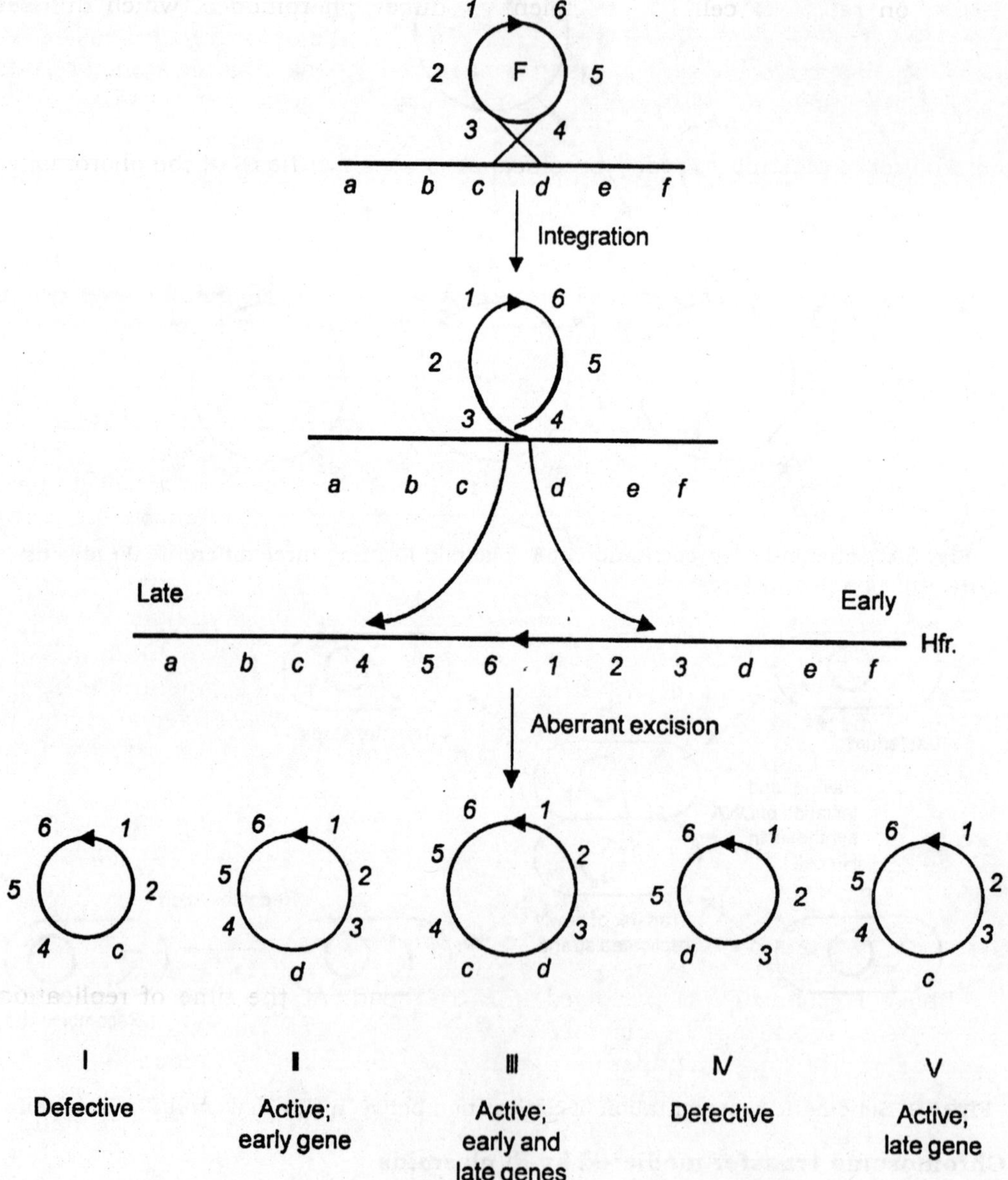

Fig. 6. Formation of F⁺ plasmids by HFr strains.

Conjugation in *Strptococcus faecalis*

Streptococcus faecalis is a Gram-positive bacterium and exhibits some difference in conjugation process from *E. coli.* Pili play no function during pairing. The plasmid containing cells form an association or clump with the cells lacking plasmids. There is aggregating substance in donor cell and binding substance in recipient cells. These substances help in forming clump. Binding substance may be present on the donor cell but aggregating substance is not present on recipient cell. The recipient produces pheromones, which diffuses through the medium and reach the donor cells. Pheromones are chromosomally encoded small molecule and induces a plasmid-encoded gene (in donor cell) to synthesize aggregation substance. Donor cells also have chromosomal gene encoding for pheromones. However, the donor cells fail to produce pheromone as the product of plasmid encoded gene (*Ica*) represses synthesis of the pheromone.

TRANSDUCTION

In 1952, N. Zinder and J. Lederberg discovered transduction. They found a bacteriophage particle carries a segment of bacterial chromosome. In transduction, bacterial DNA is transferred from one bacterial cell to another by a phage particle containing bacterial DNA. During phage replication, occasionally by mistake a defective virion is produced. It is known as transducing particle, it contains DNA; from the bacterial genome replication part or all of the normal complement of phage DNA. There is no change in the protein capsid of these transducing particles. The capsid has the property to identify another sensitive bacterial cell in which the transducing particles can introduce its nucleic acid containing bacterial DNA.

There are two types of transducing particles and two types of transduction.

Generalised or nonspecialized transduction Exchange of any bacterial gene is mediated by phage particle. A random or nearly random segment of bacterial DNA is "wrapped up" during phage maturation. The bacterial genome may be carried with complete phage genome of it may replace a segment of DNA of the phage.

Restricted or specialized transduction A limited number and specific genes are mediated. The transducing particle will always contain both phage and bacterial DNA.

Generalized transduction mediated by Phage P22

Phage P22 has DNA, packaged into the head. At the time of replication, phage DNA synthesizes concatemers by rolling circle mechanism (Fig. 7). Conctamers are long stretches of phage DNA composed of tandemly repeated phage genomes. At the time of assembly, the concatemers are packaged in head by a typical mechanism. Phage encoded endonuclease cleaves the concatemers at pac site (package site). A headful amount of DNA enters inside the assembling head. The length of headful amount of DNA is about 2% more than that of the original phage DNA. The packaged DNA is terminally redundant. Additional DNA is present at one end and the sequence of bases is same as the sequence at the other end of the molecule. Bacterial genome may have site resembling the pac site of phage. This site of bacterial DNA is cleaved by endonuclease. Headful of the bacterial genome, from plasmid or chromosome enters inside the head of the phage. This results in formation of transducing particles.

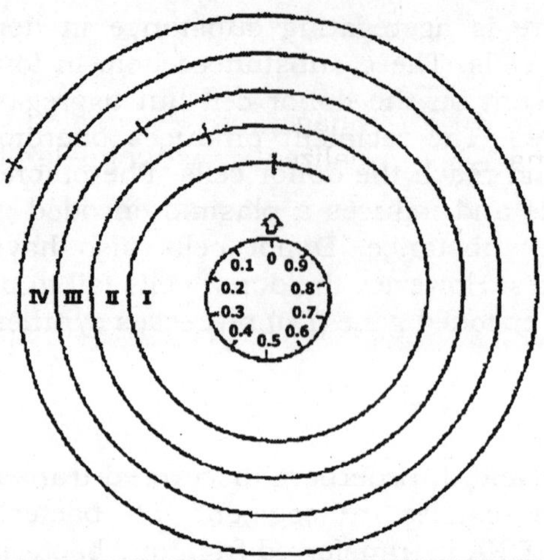

Fig. VII - Generalised transduction mechanism of packaging DNA from a conctamer

Transduction by phage P1

Phage P1 infects *E. coli.* After induction of a P1 lysogen, phage encoded nuclease breaks bacterial DNA into fragments. These fragments remain large and occasionally single strands of bacterial DNA, which is equal to the size of phage DNA is packaged instead to phage DNA. Enzyme nuclease randomly breaks bacterial DNA and thus even in the transducing principle the packaged DNA may contain atleast one bacterial gene. When the transducing particle adsorbs to a bacterium, its DNA is injected inside the bacterium. The homologous regions of bacterial DNA are now available for crossing over. If transducing particle is produced after the infecting wild type of bacterium for gene leu^+ and the packaged head contains bacterial DNA (gene leu^+). Now if this transducing principle infects another bacterial cell mutated for gene leu (leu^-) it will release this gene into the newly infected bacterial cell. The latter will not die because no phage genes are present. As result of infection, the host bacterium will now contain gene leu^+. Hence, this cell will convert into a wild cell.

Specialized transduction

Specialized transduction is mediated only by temperate phage. The chromosomes of this type are capable of autonomous and integrated replication. The replication independent of the replication of the host chromosome is known as autonomous replication. While, the phage replication that takes place as a segment of the host chromosome is known as integrated replication. The genetic element is then known as episomes. Specialized transducing particles differ from generalized particle in following ways.

1. The transducing particle contains both phage DNA and bacterial DNA.
2. One or two regions of host DNA are present in the transducing principle.
3. A single transducing particle can serve as a template, for production of a homogenous population of identical transducing particles.

Specialized transduction by phage lysogen

Specialized transducing principle contains prophage genome and a segment of bacterial DNA (Fig. 8). These phages are designated by the bacterial genes e.g., gal and bio designates a specialized transducing lambda phage that carries gal and bio genes, respectively. It is also called as defective because genes required for growth and maturation under lytic conditions have been replaced by bacterial DNA. The defective phage for gal and bio genes is designated as dg and bio, respectively. The head and tail genes are essential, so gal-transducing particles are unable to form plaques.

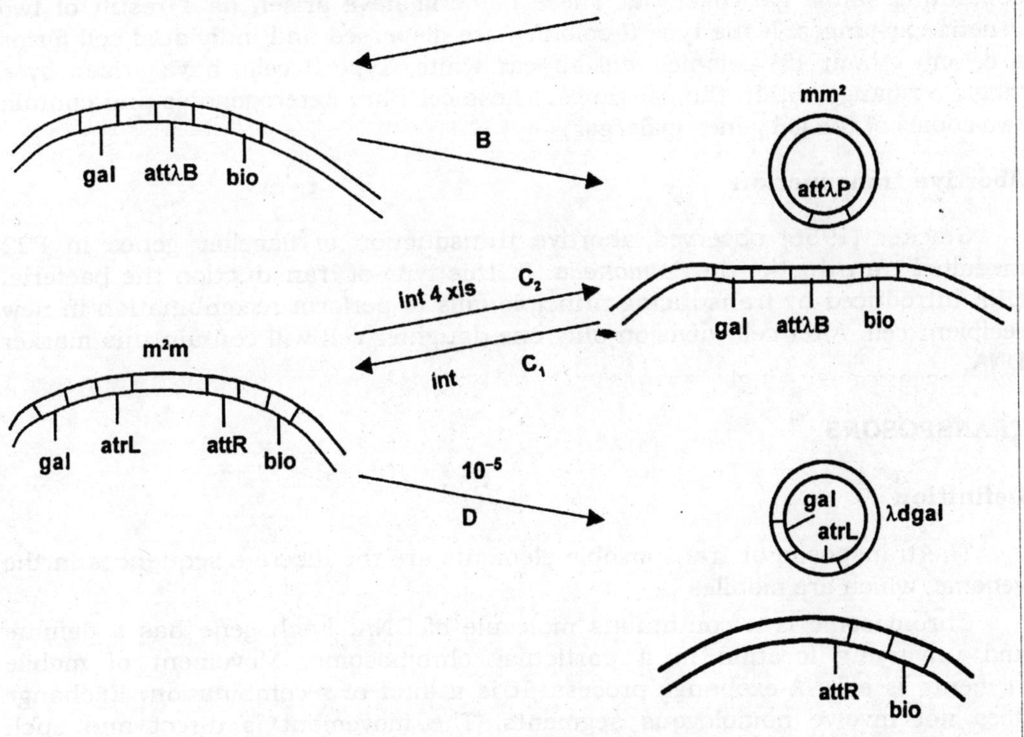

Fig. 8. Specialised transduction by lamda phage

When gal⁺ lysogens are induced by ultraviolet light, rare dg particles are formed. When such particles infect bacterial cell mutated for gal gene (gal⁻ i.e., cells are not able to utilise glactose as carbon source), the dg integrates by a crossover event within the gal region or at the attachment site. Thus the cells are partially diploid for gal⁺/gal⁻. These diploids, are now known as gal⁺/gal⁻ heterogenote and contain a gal⁺ exogenote and a gal⁻ endogenots. The primary heterogenotes are unstable. They segregate gal⁻ cells at a frequency of about one in 1000 cell divisions. Since dg cannot reproduce in the absence of a helper phage, it may simply be lost or diluted out during cell division. Recombination may also occur between the gal⁺ exogenote and gal- endogenote. Such recombination transfer gal⁺ marker to the endogenote and producing stable gal⁺ transductants.

Specialized transduction of a nonlysogen

Specialized transducing particles can transduce a mutant bacterium. Certain conditions of cell are used that will establish immunity response. Thus, the cell is not killed by a lytic response. When these cells are plated on galactose colour-indicator plates purple colonies are formed. The purple colonies are produced by Gal^{+} cells. While, Gal^{-} cells produce white colonies. The purple colonies thus produced are of two types. Type I consist of nonlysogenic cells all of which are Gal^{+}. Type II colonies contain cells with prophage. If the type I colonies are dispersed and individual cell forms a colony. All colonies are purple containing stable gal^{+} bacteria. These bacteria have arisen as a result of two genetic exchanges. If the type II colonies are dispersed and individual cell forms a colony. About 1% colonies will appear white. Type II cells have arisen by a single exchange within the gal genes. These cells are heterogenotes and contain two copies of the gal genes (gal^{+}/gal^{-}).

Abortive transduction

Stocker (1956) observed abortive transduction of flagellar genes in P22 medaited transduction in *Salmonella.* In this type of transduction the bacterial DNA introduced by transducing principle fails to perform recombination in new recipient cell. After cell division only one daughter cell will contain this marker DNA.

TRANSPOSONS

Definition

The transposon or transposable elements are the discrete sequences in the genome, which are mobile.

Chromosome is a continuous molecule of DNA. Each gene has a definite and unvarying location in a particular chromosome. Movement of mobile elements is a DNA-exchange process. It is a kind of recombination. Exchange does not involve homologous segments. The movement is direct and such elements do not take help of extra chromosomal DNA (plasmids etc.). The frequency, with which the elements move in both bacteria and eukaryotes is fairly low. The transposons first discovered did not contain any known host genes. They were called insertion sequences or IS elements and designated as IS1, IS2 and so on. Transposons are divided into following classes.

Insertion sequences

Insertion sequences are short DNA molecules (~ 1000 bp). They have short inverted repeats at their ends. They are autonomous units, which code for the proteins necessary for their own transposition. During insertion, a small sequence at the site of insertion is duplicated which generates short direct repeats in the target DNA. The length of the repeat (repeat of 9 bases) at the site of insertion is always the same for a particular IS element. The termini of each IS element have inverted-repeat sequences consisting of 16-41 base pairs, the number depending on the element. The inverted terminal repeats of an IS element is specific for each IS. These repeats define the boundaries of an IS element. The elements contain at least two apparent coding sequences initiated by an AUG and terminated with an in-phase stop codon. The main body of IS element codes for the protein, transposase. The later takes part in transposition.

Composite transposons

These are large units with complex structure. Composite transposons carry antibiotic resistance genes. They are flanked by two identical or nearly identical copies of an IS elements on both sides (arms). These arms of IS elements may be present either as the direct repeat or as the inverted repeats. Tn5, Tn9 and Tn10, are the frequently studied, composite transposons.

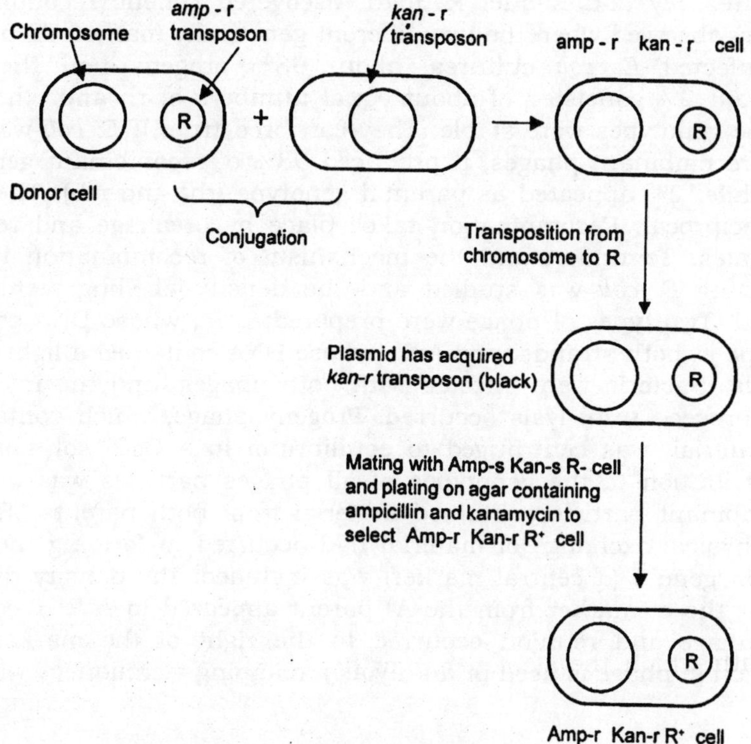

Fig. 9. Experiment to demonstrate transposition.

Transposition

A compsite transposon has a number of different IS elements, which can move independently. Any composite transposon is evolved when two originally independent IS elements got associated with a central region. Therefore, it is possible for a IS element as well as central module to move alone in place of the entire module. It is essential that only the ends (bordered by the IS elements) are required for the mobilization of the gene. The transposase activity creates a target site and recognizes the ends of the transposon. During transposition, an asymmetric nick is created at the target site and the transposon is inserted between the nicks (Fig. 9). The transposition can take place by replicative or non-replicative mechanism. In replicative transposition, the element to be transposed is replicated. The original copy remains attached at the donor site while, the second copy transposes and gets inserted at the recipient site. In non-replicative transposition, the transposing element moves from the donor site to the recipient site as a physical entity. It is cleaved from the donor site and is transported to the recipient site where it gets integrated. The result of the transposition process is the insertion of a transposon between two nucleotide pairs in a recipient DNA molecule.

RECOMBINATION IN PHAGES

Alfred Hershey and Raquel Rotman discovered genetic recombination in phages. They observed when the two different genetically marked T2 phages (rh^+ and r^+h) infected *E. coli* cultures, about 98% progeny had the parental genotypes, but 2% consisted of about equal numbers of rh and r^+h^+ recombinants. These genotypes were stable. They can bred true. If *E. coli* was infected with these recombinant phages, it produced 98% of recombinant genotype (rh and r^+h^+) while, 2% appeared as parental genotype (rh^+ and r^+h). The recombination is reciprocal. Recombination takes place by breakage and rejoining of DNA molecules. To understand the mechanism of recombination in phages, phage infecting *E. coli* was studied and the density-labelling technique was implemented. Two types of phage were prepared: A^+R^-, whose DNA contained a heavy siotope in both strands, and A^-R^+, whose DNA contained a light isotope in both strands. Bacteria were infected with both phages, and the infection was allowed to proceed until lysis occurred. Progeny phage, which contained only parental material, was centrifuged to equilibrium in a CsCl solution and the density distribution of the genotypes of all phages particles was determined. Each recombinant particle contained material from both parents. This shows that that physical exchange of material had occurred in forming the recombinants. When gene c (a central marker) was included, the density distribution showed that the c^+ marker from the A^+ parent appeared in A^+R^+ recombinants only if breakage and reunion occurred to the right of the marker. Genetic recombination in phage is used primarily as a mapping technique.

CHAPTER - 18

MUTATION

Mutation is the change in the base pair sequence of DNA molecule. It is heritable and the cells or organisms, which have undergone mutations, are known as mutants. The chemical or physical agents cause mutations are known as *mutagens.* The process of producing mutants is known as *mutagenesis.*

Spontaneous and Induced mutation

A mutation resulted without a known cause is known as spontaneous mutation. Such mutations are caused due to some mistakes during DNA replication and these mistakes are inherited or by mutagenic agents present in the environment. Induced mutations are caused by some mutagenic agents like ionizing irradiation, ultraviolet light, or chemicals, which bring about changes in DNA molecule.

Replica-plating technique

It is generally observed that a bacterium (e.g., *E. coli*) growing in nature is sensitive to antibiotics (say streptomycin). However, when cultured with this antibiotic the same bacteria develop resistant strains. Now to understand if these resistant cells were present before the treatment and could flourish even when the bacterium is treated with antibiotic. Lederberg developed a technique known as replica-plating technique (Fig. 1). A dilution of bacterial cultures is prepared and inoculated on a semisolid nutrient agar medium. When the bacterial colonies developed properly, each plate is then inverted and pressed onto sterile velvet, which is covering a wood block. Some cells will stick to the velvet. A sterile plate of nutrient agar medium containing streptomycin is then pressed onto the velvet. This process is repeated. Streptomycin-resistant strains will appear on the plates containing streptomycin. When colonies developed on plates without streptomycin are tested. It was found that the colonies of some cells are present, which would have developed resistant cells. This clearly express, that the environmental stress does not direct or cause genetic changes. There are rare preexisting mutants, which are able to flourish in the new environment.

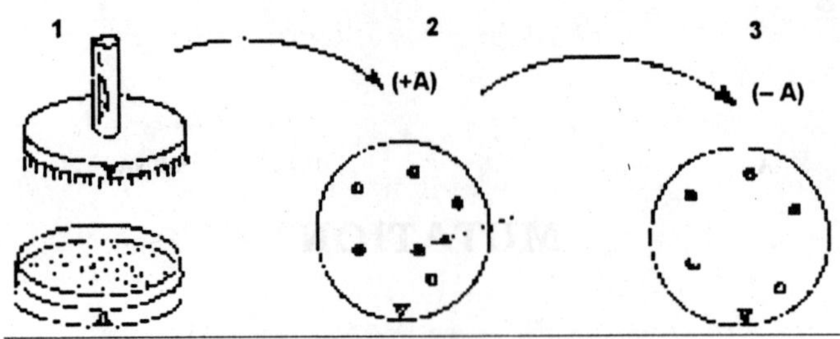

Fig. 1. Replica plating technique

1. A velvet lined block with handle and is pressed on to the petri plate beneath containing discrete colonies. 2. Colonies picked up are replicas stamped on to petri plate containing a metabolite. 3. Cells replica plated on to medium lacking A.

TYPES OF MUTATION

Microlesions

Only one base pair of DNA molecule is changed. There are different types of microlesions. In *frame-shift mutation,* there is a loss or gain of one ni trogenous base pair. This results in reading frame of all codons of the genes. In *base-pair substitution* mutation one base pair is changed by another base pair. This may be explained as A/T base pair is changed by another base pair G/C. If purine base of a pair is changed by another purine, it is known as *transition.* While if purine of a base pair is changed to a pyrimidine base, it is known as *transversion* (Fig. 2). This may be explained in terms of tautomeric shift (Fig. 3). The hydrogen atom of a nitrogenous base can move from one position to another, e.g., from an amino group to a ring nitrogen. The stable keto form of thyamine and guanine may become less stable enol form and similarly, stable amino form of adenine and cytosine may convert into less stable imino forms. If these forms are replicated and inherited into a nascent chain of DNA, it results in mutation.

Macrolesions

More than one or even genes are changed. Macrolesions may be again of different types. *Deletion* One gene of a complete segment of DNA is lost. *Duplication* One gene of a complete segment of DNA is added i.e. there is repetition of a gene. Inversion A gene of a segment of DNA is inverted. Insertion The introduction of a new segment of DNA within an existing sequence of DNA. Translocation The movement of a segment of DNA to another site in the genome.

Another concept of classifying mutation is on the basis of the consequence of the change in terms of the amino acid sequence that is affected. If the mutation results in substitution of an amino acid, it is known as *missense mutation.* If a protein is produced due to mutation, which is active at one temperature (typically 30°C) and inactive at a higher temperature (usually 40-42°C) it is known as *temperature-sensitive* or *Ts mutation.* If the mutation results in the formation of nonsense codon, the protein synthesis will stop, this type of mutation is known as *chain termination or a nonsense mutation.* Temperature sensitive and nonsense mutations are *conditional mutations* as they exhibit the mutant phenotype only under certain conditions.

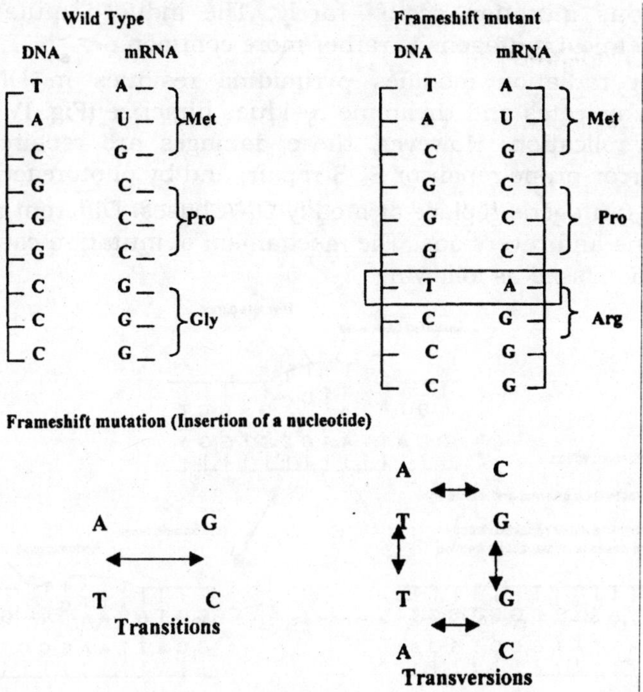

Fig. 2.

Guanin

Thymine (enol form)

Adenine (imino form)

Cytosine

Fig. 3. Mutation due to tautomeric shift.

MECHANISM OF MUTATION

Spontaneous mutation occurs rarely. The induced mutation caused by physical or chemical mutagens is rather more common.

Ultraviolet radiation modifies pyrimidine residues in DNA. Uracil and cytosine form hydrates and thyamine residues dimerize (Fig. IV A,B). The latter blocks DNA replication. However, these damages are repaired by error-free mechanism, error-prone repair or SOS-repair, and by photoreactivation.

Chemical mutagens replace or modify DNA bases. Different mutagens cause point mutations and reversions. The mechanism of mutation caused by different mutagens is discussed as following.

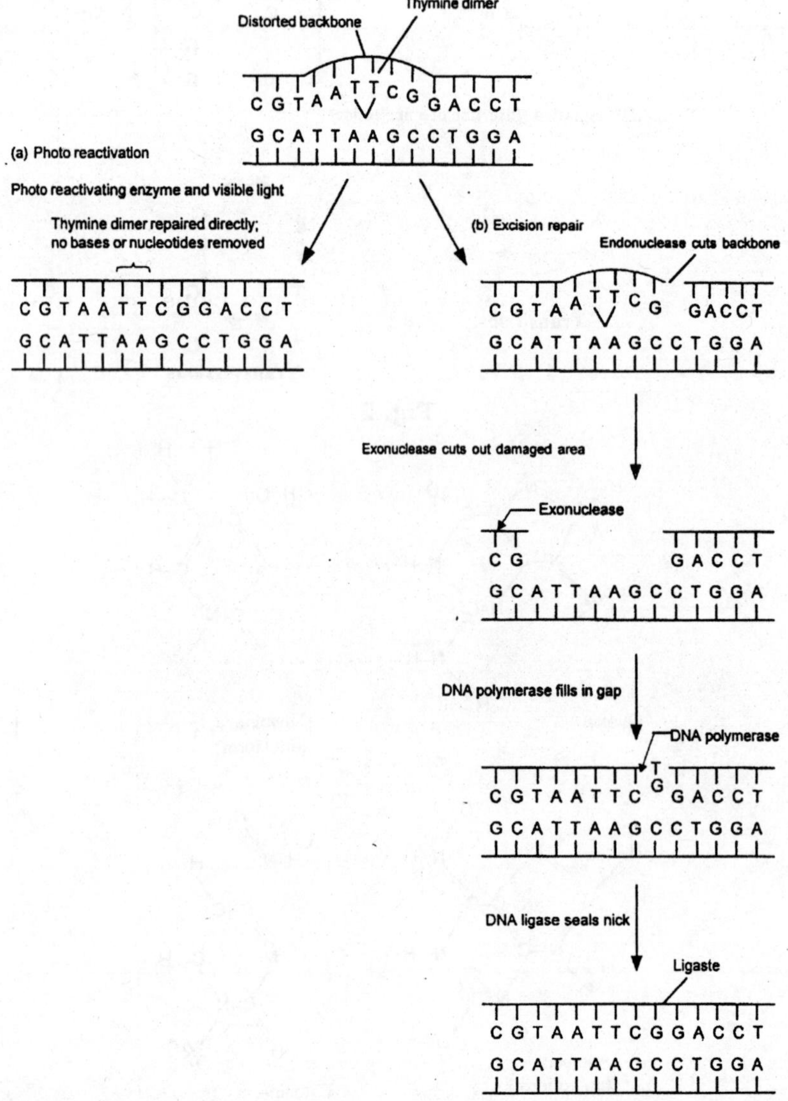

Fig. 4 A. Exposure to ultraviolet light results in the formation of thymine dimers

Fig. 4B. Structure of thiamine dimer

NITROUS ACID AND HYDROXYLAMINE

Nitrous acid (HNO2) causes deamination of base residue in DNA. Hydrosylamine causes one-way transitions, of GC->AT. This results in reversion.

The mechanism of mutation by nitrous acid is well studied. It deaminates adenine to hypoxanthine, guanine to xanthine and cytosine to uracil. Due to such changes, hypoxanthin (formed by deamination of adenine) The The mechanism of mutation by nitrous acid is well studied. It deaminates adenine to hypoxanthine, guanine to xanthine and cytosine to uracil. Due to such changes, hypoxanthin (formed by deamination of adenine) forms base pair with cytosine (as against thyamine in original form). Uracil gives different base pairing properties. All these changes result in altered form of DNA strand, which shows transition or transversion.

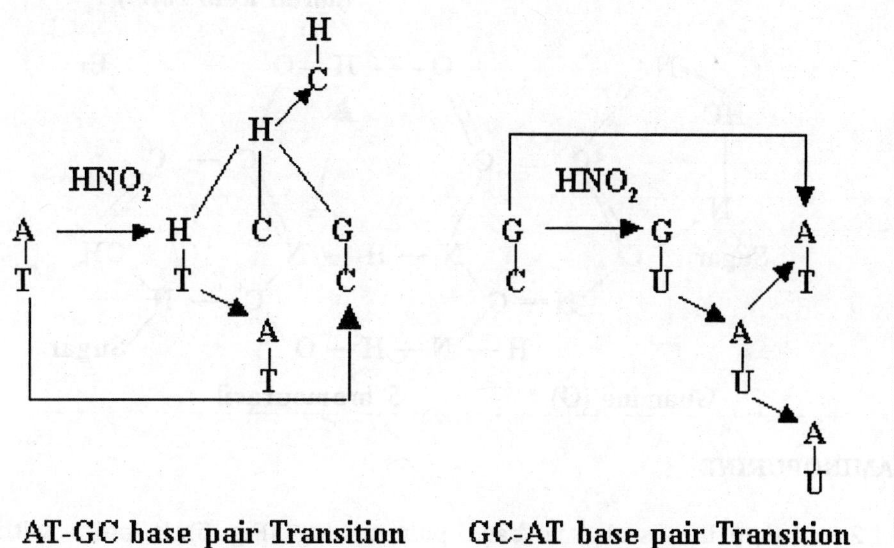

AT-GC base pair Transition **GC-AT base pair Transition**

5-BROMOURACIL

5-bromouracil (BU) causes errors due to mispairing (base pairing analog) during replication. BU is analogous to thymine with respect to the same van der Waals radius, as the methyl group of thymine. Normally, in the keto-form, like thyamine, BU pairs with adenine. Bromine atom results in withdrawal of electron, which causes tautomerism to the enol form. Thus, there is mispairing with guanine. This induces two ways transitions, the template transition and substrate transition. Results from recent experiments have shown that BU causes mutagenesis in another way also. Thymidine triphosphate (TPP) regulates the concentration of the four nucleoside tri phosphates required for DNA synthesis i.e., dNTP. One part of this complex regulatory process is the inhibition of synthesis of deoxycytidine triphosphate (dCTP) by excess TTP. BU nucleoside triphosphate also inhibits production of dCTP. Thus, the ratio of TTP to dCTP then becomes quite high and the frequency of misincorporation of T opposite G increases.

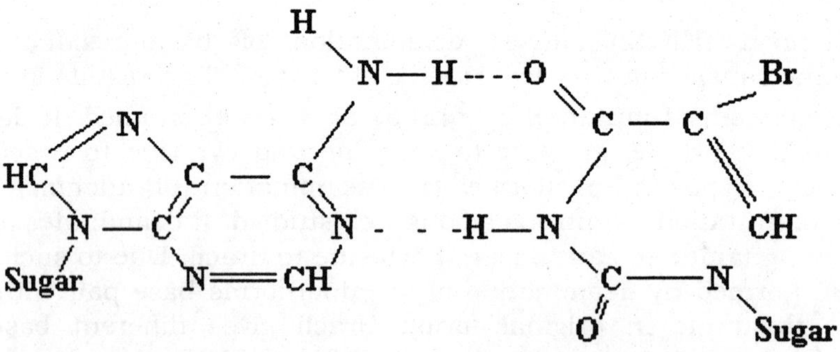

adenine (A) **5 bromouracil (norml keto form)**

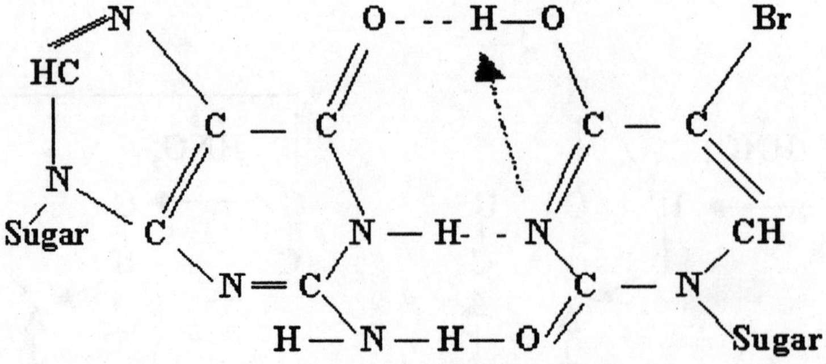

Guamine (G) **5 bromouracil**

2-AMINOPURINE

2-aminopurine, is also a base pair analog (Fig. 5). It pairs with T or C. 2-aminopurine, is also a base pair analog. It pairs with T or C.

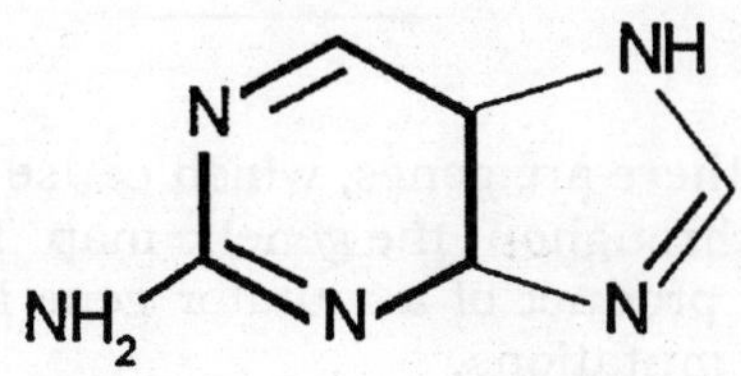

Fig. 5. 2-aminopurine

ALKYLATING AGENTS

Alkylating mutagens causes depurintaion, creating gaps at such sites. These gaps may be filled up by incorrect base resulting in mutation. Methyl methane sulfonate ($CH_3.SO_2.O.CH_3$) and ethyl methane sulfonate ($CH_3.SO_2.O.C_2H_5$) are alkylating agents and depurinate guanosine residues.

INTERCALATING AGENTS

These agents cause frame-shift mutations. Acridine orange, proflavine, and acriflavine are intercalating compounds. These mutagens are substituted acridines, are planar, three-ringed molecules whose dimensions are roughly the same as those of a purine-pyrimidine pair (Fig. 6). It enhances the probability of frameshift mutations by stabilizing the looping out process. Streisinger has explained the mechanism of mutation by such agents. Alkylating agents cause slippage between two strands of DNA during replication and due to this, there is insertion or deletion in ongoing strand synthesis. This phenomenon is known as intercalation.

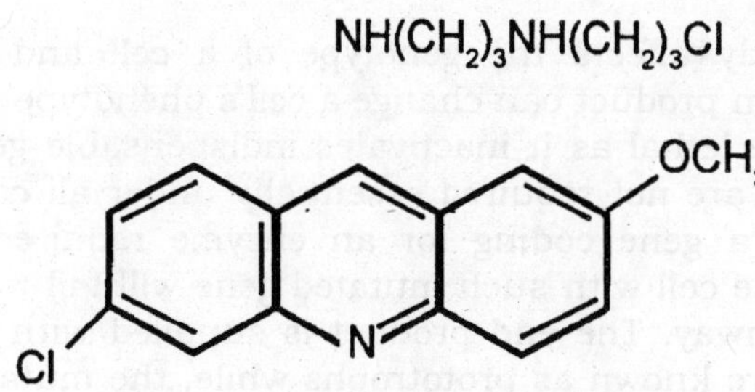

Fig. 6. Intercalating agent ICR-191

MUTAGENESIS

Mutagenesis occurs spontaneously by replication errors or can be stimulated to occur in following means:

(i) removal of an incorrectly inserted base is prevented; (ii) tautomerism in inserted base and allows a substitution to occur in subsequent replication: (iii) a base is chemically altered to a base having different base-pairing specificity; (iv) bases are skipped or inserted during replication.

MUTATOR GENES

In mutants of *E. coli*, there are genes, which cause mutations to appear very frequently in other genes throughout the genetic map. These genes are known as mutator genes. When the product of a mutator gene is itself defective, there is wide spread production of mutations.

MUTATION RATE

The probability of a gene to undergo mutation in a single generation is called mutation rate. It is difficult to estimate mutation rate. Mutation may occur at any stage of growth and the mutated cells may divide with the same rate, as that of wild cells in a culture. The mutation rate may be determined by fluctuation test. A culture may have many mutants, some mutants, or none. If the mutation rate per generation is, the number of cells in a culture is N, the probability of obtaining no mutants is Po and the Poisson distribution, e- N (total number of divisions of individual cells required to produce N cells from 1 cell is N-1, and N is a large number). Thus,

- N = In Po, or

= - (1/N) in Po.

If 10 of the 15 small cultures contained no mutants, and the average number of cells per culture (N) is 5.5 x 10 8. Thus, the mutation rate could be estimated as

= - (5.5 x 10 8) In (10/15). Per cell per round of replication.

EFFECT OF MUTATION ON PHENOTYPIC EXPRESSION

Mutation directly affects the genotype of a cell and thus affecting the proteins. This protein product can change a cell's phenotype in a variety of ways. Many mutations are lethal as it inactivates indispensable gene products. If the inactivated proteins are not required essentially under all conditions of growth, it is non-lethal. If a gene coding for an enzyme required in a biosynthetic pathway mutate. The cell with such mutated gene will fail to synthesize the end product of this pathway. The end product is supplied with culture media. The wild types of cells are known as prototrophs while, the mutated cells are known as auxotrophs. Strains, carrying mutated genes encoding enzymes of catalytic pathway fails to utilize the substrates of the pathway. However, such mutants can be maintained in media provided with other primary substrates. If mutation alters the target protein of an antibiotic or other toxic chemical will result in cell resistant to the antimicrobial agent. Some mutation may result in loss of motility, a tactic response etc.

ISOLATION OF MUTANTS

Mutant cells may be present in small number, in a large number of wild cells populations in a culture. There are two tasks to isolate mutants. It needs detection, or screening and enrichment or increase total number of mutant cells in a culture.

Methods for isolation of different types of mutants have been developed. It is first required to detect these mutants. These mutants are antibiotic resistant, phage resistant or unable to utilize a particular sugar.

For the detection of antibiotic resistant mutant, bacteria are grown on agar plate containing antibiotic. Only a resistant cell can form a colony. Similarly to detect phage resistant mutants, about 10 9 phage are spread on agar plate. After this, bacteria are inoculated. The susceptible cell will lyse while, the resistant cells will develop a colony. Phage mutant are detected by "plaque morphology". The mutants make plaque that are turbid, or clear, or have a halo, or are a combination of these. Sugar-utilization mutants are isolated by means of colour indicator. To prepare colour indicator plate, EMB media is used supplied with dyes (eosin or methylene blue). Dyes are sensitive to pH. Sugar supplied in this medium will be utilized by wild type of strains. This will lower the pH and thus the dye will colour the colonies into intense purple. The mutant cells, unable to utilize sugar will use amino acid of the medium as carbon source. Ammonia is produced by the utilization of amino acid. It will increase the pH. Increased pH will decolourizes the dye and the colonies developed are white in colour. Likewise, if amino acid utilizing mutant is to be detected, there is a separate method. Leucine is an amino acid and the wild type of cells which are able to utilize this amino acid are designated as Leu^+ while, the mutants, unable to utilize this amino acid are designated as Leu^-. To isolate the mutant, replica-plating method is adopted. Leu^+ and Leu^- cells are grown on nutrient agar plates. Many colonies will develop. This plate is taken as master plate. The latter is replicated onto two minimal agar plates, one containing leucine (reference plate) and the other plate is devoid of leucine (test plate). Leu^- mutants will fail to grow on test plate. Thus, a Leu^- mutant is detected by its growth on the reference plate and lack of growth on the test plate. This can be isolated from the reference plate. To isolate temperature-sensitive mutants also, replica-plating method is adopted. The master plate is prepared by growing bacteria at 30°C. The reference plate is developed at 30°C and test plate at 42°C. A colony that forms on the 30°C plate but is absent from the 42°C plate contains a temperature-sensitive mutation.

REPAIR OF DNA DAMAGE

Mutation causes changes in genotype by damaging the DNA molecule. These damages can be reverted and wild types may be regained. This phenomenon is known as reversion. However, reversion is not common in all types of mutations. The mechanism of DNA repair are as following:

Base insertion repair

Deutsh and Linn (1979) isolated an enzyme called insertase. This enzyme is also isolated from *E. coli.* It transfers A from dATP and G from dGTP to the apurinic sites of damaged DNA molecule.

Demethylation of O6-methylguanines

An inducible enzyme demethylase catalyzes such types of repairs. It specifically removes O6-methyl group. The latter damage is caused by alkylating agents like N-methyl-N-nitrosourea and N-methyl-N'-nitro-N-nitrosoguanidine. Enzyme demethylase develops resistant for such alkylating agents. It transfers O6-methyl group to cysteine residue and converts it into S-methylcysteine. This conversion is irreversible and one molecule of enzyme is required for each conversion.

Photoreactivation

The damaged caused by exposure to UV radiation may be repaired by exposure to longer wavelength (300-400 nm). The repair is brought about by monomerization of thymine dimers, by cleavage of the cyclobutane ring. Enzyme photolyase is involved in this reaction. However, the mechanism is still not clear.

UV radiation damages may also be repaired by liquid holding recovery or dark repair. In this process, infected cells are kept simple buffer.

'SOS' Repair

SOS repair allows DNA chain growth across damaged segments at the cost of fidelity of replication. This type of repair is brought about in DNA damages caused by uv irradiation or other mutagens. It specially works in inhibited DNA replication. SOS repair was studied in *rec*A^{-} mutants. Gudas and Pardee explained the mechanism of SOS repair. Enzyme protease is the gene product of rec A. RecA protein has several functions in SOS repair. The operon controlling SOS repair is known as regulon. All the sets of regulon are controlled by a common repressor. This enzyme cleaves the lex A protein, which is a repressor and binds to a common operator sequence adjacent to each gene or operon. This has a central importance in the regulatory circuits of SOS repair. In damaged DNA, some signal molecules activate rec A protein. The activated protease breaks repressor of lex A. As this repressor is cleaved the din genes (damage inducible genes) becomes functional and take part in SOS repair. A set of ten odd genes becomes activated and DNA repair by an error-prone pathway sets in. In error-prone process, intact DNA strands are formed even though, the strands often contain incorrect bases.

RecA protein directly inhibits the editing function of polymerase III by editing tightly in the region of the distortion resulting from a pyrimidine diver. When Pol III encounters a dimer site to which RecA is bound, RecA interacts with the subunit of Pol III, which has editing function. As the editing is inhibited, the replication fork advances.

CHAPTER - 19

GENETIC ENGINEERING

Since the 1970s there have been considerable developments in the methods and techniques used to study biological processes at the molecular level. These have led to many new and powerful ways for the isolation, analysis and manipulation of nucleic acids, which in turn has provided insights and exciting advances in new and expanding areas of biosciences such as biotechnology, drug discovery, molecular medicine and gene therapy. The discovery of restriction endonucleases in early 1970s was a key development which not only led to the possibility of analyses of DNA more effectively, but also provided the capability to cut different DNA molecules so that they could later be joined together to create new recombinanat DNA fragments, a process now termed gene cloning. The newly developed cloning strategies heralded a new and exciting era in the manipulation, analysis and exploitation of nucleic acid molecules.

Although human beings have been altering the genetic makeup of organisations for centuries, only recently has the direct manipulation of DNA been possible. The deliberate modification of an organism's genetic information by directly changing its nucleic acid genome is often called genetic engineering and is accompanied by a collection of methods known as recombinant DNA.

The basic technique is that two DNA molecules are isolated and cut into fragments by one or more specialized enzymes and then the fragments are joined together in a desired combination i.e., specified gene is cloned. The recombinant DNA is restored to a cell where they undergo expression and replication.

Gene cloning is conducted in a stepwise procedure. The steps are —

(1) Identification and isolation of interested DNA (gene).
(2) Insertion of isolated DNA in a vector.
(3) Transformation: introduction of vector in a suitable host (cell).
(4) Selection and multiplication of the transformed host cells.
(5) Expression of introduced gene in the host.

Genetic engineering has wide applications. It is used in —

(1) facilitating the production of useful proteins.
(2) Develop bacteria for the purpose to synthesize economically important products.

(3) Supplying DNA and RNA sequences as a research tool.

(4) Gene therapy: A potential correction of genetic defects in animals.

SOME DEFINITIONS AND CONCEPTS

Genetic engineering is a discipline of Biotechnology. Biotechnology is coming up with revolution in science. There are various disciplines in Biotechnology. *Some important definitions and concepts are discussed below —*

Restriction sites — Loci on DNA molecule at which foreign DNA is integrated.

Restriction endonuclease — Enzymes cleaving DNA polynucleotide at specific sites and thereby permit an insertion of foreign sequences.

DNA Cleavage — Restriction enzymes cut DNA molecules by cleavage. The cleavage takes place in two patterns. *Blunt end style*: DNA polynucleotide is cleaved across both strands at the same position. *Sticky or cohesive-end style:* Two strands of DNA molecule cleave at different position and the protruding termini (ends) are generated.

Linkers — Short double stranded oligonucleotides, ligated at the blunt-ends of DNA molecules to convert it into cohesive end.

Ligation — Direct linkage of nucleic acid fragments. Enzymes, responsible for binding these fragments are known as Ligases.

Chimeric gene — A recombinant DNA molecule, which contains a gene from one organism, joined to regulatory sequences from another organism.

Vectors — Agents used to introduce a foreign DNA molecule into a new host cell.

Some vectors used in genetic engineering are as following.

Plasmids — Extrachromosomal, circular DNA molecules with lower molecular weight. It bears nonessential additional genes, e.g., sex factor (F factor) and resistant factor (R factor).

Phage vectors — Bacteriophage used for cloning in bacteria, e.g., lambda phage.

Cosmid vectors — A type of plasmid vectors containing a fragment of lambda phage including its *cos* site.

Phasmid vectors — It is a type of plasmid vector containing a fragment of phage DNA including its *att* site.

Blotting Technique — The technique of separating samples by electrophoresis. The samples separated are DNA, RNA or protein and accordingly the technique has different names.

Southern blotting — A technique in which a sample of DNA is subjected to electrophoresis using polyacrylamide or agrose gel.

Northern blotting — Technique of separating RNA by electrophoresis using nitrocellulose membrane.

Western blotting — Technique of separating proteins by electrophoresis by using polyacrylamide gel.

Clones — Identical daughter cell derived from a single mother cell by multiplication. Population of cells derived asexually from a single cell also forms a clone.

Cybrids or cytoplasmic hybrids — Cells or plants, have nucleus of one species but cytoplasm from both parental species.

Epigenetic gene control — The control of gene expression in higher eukaryotes is related to methylation of cytosine residues in gene coding sequence.

Polymerase Chain Reaction (PCR) — Technique to produce copies of a piece of DNA from a mixture of total DNA.

Gene Libraries — A collection of DNA fragments which completely spans the entire genome of an organism. A single gene can be studied.

Gene Machine or Automated Polynucleotide Synthesizers — Instrument, which can synthesize a polynucleotide chain. The sequence of base can be arranged as predetermined.

Tissue culture — Technique of growing plant cells, tissues and organs in an artificial nutrient media.

Somatic hybridization — Technique of fusing protoplast of two different cells of different genera or species.

Transgene — Gene introduced in an animal by transfection. Transfection is a technique of introduction of a DNA segment either naked or integrated into a vector, into an animal cell. In plants, the transgenes are from unrelated organism.

Transgenic plants — Plants containing transgenes.

Transgenic animals — Animal containing one or more genes introduced by transfection.

STEPS OF GENETIC ENGINEERING

Genetic engineering is a long and sophisticated process. It is conducted in a step-wise manner. Main steps of this procedure have already been mentioned. These steps are discussed in details.

ISOLATION AND CHARACTERIZATION OF PARTICULAR DNA FRAGMENTS

In recombinant DNA technology, it is important to identify and isolate a specific DNA fragment. DNA fragments are obtained by treatment with a specific class of nucleases. More commonly, the enzymes utilized are restriction endonuclease.

RESTRICTION ENDONUCLEASE

Restriction endonucleases are indispensable for DNA cloning and sequencing. The name restriction enzyme is due to the phenomenon of host restriction of bacterial phages. In 1960s W. Arber postulated the presence of these enzymes while, in 1970, Smith, Nathans and Arber isolated and purified first enzyme of such type. Now hundreds types of these enzymes are isolated, purified and characterized. These enzymes recognize a specific base sequence in DNA molecule and make two cuts, one in each strand, generating 3'-OH and 5'P termini. Most of the sequences are 4-6 base pairs (bp) in length. The sequences recognized by restriction enzymes are palindromes with rational symmetry. It means that the base sequence in the first half of a DNA strand is the mirror image of the sequence in its second half of its complementary strand. In other words, the substrates that are attacked by restriction endonucleases read the same from forward and reverse directions. Longer palindromic target sequences and nonpalindromic sequences, are also known. It is important that the number of cuts made in the DNA from an organism is limited. Because of the sequence specificity, a particular restriction enzyme generates a unique family of fragments from a particular DNA molecule. Another enzyme will generate a different family of fragments from the same DNA molecule. There are two major types of restriction enzymes: Type I enzymes, which recognize a specific sequence but make cuts elsewhere: and type II enzymes, which make cuts only within the recognition site. There are two distinct arrangements of these breaks: (i) both breaks at the center of symmetry (generating flush or blunt ends), or (ii) breaks that are symmetrically placed around the line of symmetry (generating cohesive ends). An important property of staggered ends is that DNA from different sources digested by the same RE will be complementary (or 'sticky) and so will anneal to each other. The annealed strands are held together only by hydrogen bonding between complementary bases on opposite strands. Covalent joining of ends on each of the 2 strands may be brought about by the enzyme DNA ligase. This is widely exploited in molecular biology to enable the construction of recombinant DNA i.e. the joining of DNA fragments from different sources. There are some principles for nomenclature of restriction enzymes. All restriction enzymes are designated by the general symbol R, which is prefixed to their name, e.g., REco. The first letter of the name is from the name of genus and the next two letters are from species. The first letter is written in capital and it is from the genus in which, a given enzyme was discovered e.g., restriction enzyme, discovered from *E.coli* was named as Eco. Strains or type identification is depicted as subscript, e.g., Eco K. In practice it is written in one line. If an organism produces more than one restriction enzyme, they are identified by sequential Roman numerals, e.g., Hind II, Hind III.

Some restriction endonucleases and their cleavage sites

Microorganism	Name of enzyme	Target sequence & cleavage sites
Generates cohesive ends		
E. coli	EcoRI	G A A | T T C C T T | A A G
Bacillus amyloliquefaciens H	BamHI	G G A | T C C C C T | A G G
Haemophilus aegyptius	HaeII	Pu G C | G C Py Py C G | C G Pu

Haemophilus influenzae	Hind III	A A G | C T T T T C | G A A
Generates flush ends		
Brevibacterium albidum	Bal I	T G G | C C A A C C | G G T
Haemohpilus aegyptius	Hae I	(A) G G | C C (T) (T) C C | G G (A)

The isolated DNA fragment is joined in another DNA molecule. This chimeric DNA should be able to replicate. Here, chimeric DNA means that the isolated DNA fragment is introduced in a suitable vector, which is transformed into a new host cell, where this vector is able to multiply.

CLONING VECTOR

For the cloning of any molecule of DNA it is necessary for that DNA to be incorporated into a cloning vector. These DNA elements may be stably maintained and propagated in a host organism for which the vector has replication functions. A typical host organism is a bacterium, such as *E. coli*, which grows and divides rapidly. Cloning vectors are agents, which carry a fragment of foreign DNA. These vectors introduce DNA in the host genomes. A vector should have following properties —

(1) A means of introducing vector DNA into a cell must be available.
(2) Vector is able to replicate autonomously.
(3) Transformants must be selectable by a straightforward assay.
(4) Vector should have suitable marker genes that allow easy detection and/or selection of the transformed host cells.

PLASMIDS

Plasmid is a double stranded small circular, self replicating, extrachromosomal molecule of DNA. They are widely used for gene manipulation. The new genes can be tailored into plasmids and then multiplied repeatedly (Fig 1). Plasmids perform some important functions in a cell. The F and R factors are for fertility and resistance. The former is responsible for fertility while, the latter for resistance to antibiotics and other substances harmful for a cell. They can metabolize sugars and hydrocarbons. They are usually involved in the production of anitbiotics, bacteriocins, enzymes or antigens.

Bacterial plasmid pBR322

Plasmid pBR322 is isolated from *E. coli.* It was discovered by Boliver and Rodriguez. Thus, the name pBR stands: p for plasmid, B and R for Boliver and Rodriguez, respectively. It is widely used plasmid. It is small sized with 4.3 kbp. It has genes resistant for ampicillin (amp-r) and tetracycline (tet-r). It contains only one copy or each of seven different types of restriction-enzymes cleavage sites at which DNA can be inserted. Restriction enzymes PstI or PvuI recognizes gene site Amp-r. By using this enzyme, DNA can be inserted in this region. After insertion, Amp becomes non-functional. Restriction enzyme BamHI and SalI are used to insert DNA at Tet-r site. This plasmid has a nic-bom region. Bom stands for the basis of motility and it is responsible for mobilization (Fig. 2 A, B).

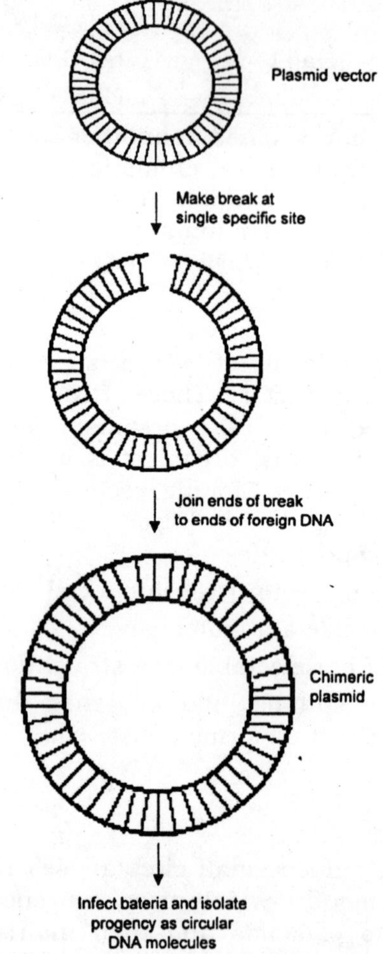

Fig. 1. Plasmid vectors. Schematic representation of insertion of foreign DNA

pUC7

The name pUC stands for plasmid and university of California. It is small sized with 2.7 kb. It has ampicillin resistant gene and Col E1 gene. Isolated DNA fragment is inserted in lacA region of the plasmid.

Ti Plasmids (Ti=Tumour induction)

Plasmids of *Agrobacterium tumefaciens* known as Ti plasmid have been studied. This bacterium causes crown gall of stem. Ti plasmids are large sized measuring 140-235 kilobase pairs. The genes found in Ti plasmids are Agc (agropine catabolism), Agr-s (agrocin sensitivity), Ape (bacteriophage API exclusion), Arc (arenine catabolism), inc (incompatibility), Noc (nopaline catabolism), Nos (nonapline synthetase), Occ (octopine catabolism), Ocs (octopine synthetase), OncH (host range), Ori (origin of replication), Psc (agrocinopine production), Tra (transfer), Tum, [tmr, tms, tml] (tumor morphology), vir (virulence). Ti plasmids are used in the genetic engineering of plants. During the transformation process, the transferred DNA (T-DNA) gets cleavage from the Ti plasmid. It is transferred to a plant cell and inserted into the DNA of the plant cell.

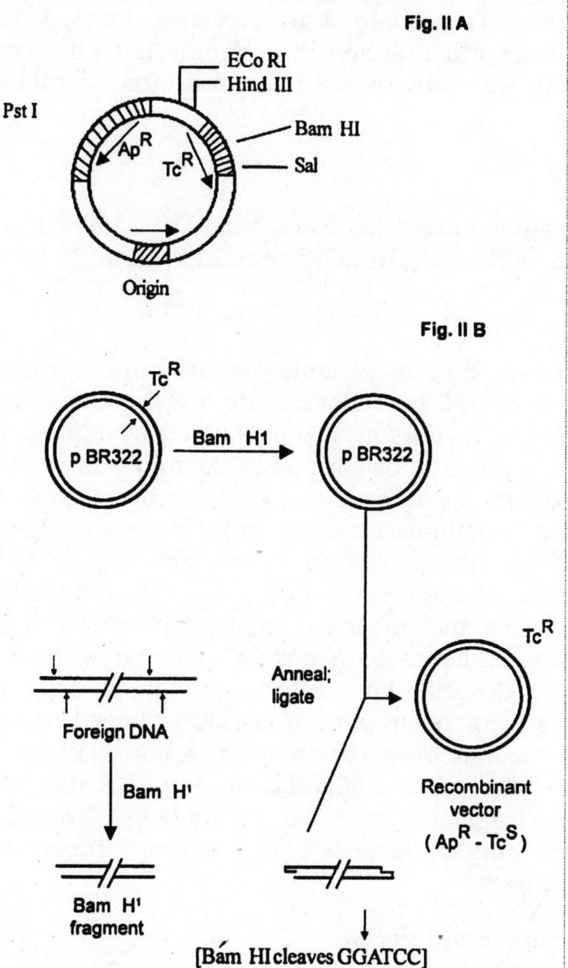

Fig. 2A. Restriction sites of pBR 322, **B.** Schematic representation of chimeric DNA in pBR 22.

BACTERIOPHAGE VECTORS

Lambda phage and M 13 phages are most commonly used cloning vector for *E.coli.* Phage vectors are more efficient in cloning the DNA fragments. To identify recombinant vectors is easy by virtue of plaque counting. A useful feature of any cloning vector is the amount of DNA it may accept or may be inserted before it becomes unviable. Thus bacteriophages (bacterial viruses) have been adapted as vectors in order to propagate larger fragments of DNA in bacterial cells.

Lambda phage vector

The lambda genome is linear and made up of 48,502 bp. The genes present in this genome codes for the structural proteins, enzymes for DNA replication, lysis and lysogeny. Beside these characteristics, the genome also has a site for origin of replication. The chain has cohesive ends with 12 base pairs complementary to each other. When the genome is transferred to the host cell (*E. coli*) the two cohesive ends anneal to form a circular molecule necessary for replication.

Phage M 13 vector

The M 13 genome is linear single stranded DNA molecule. As it enters inside the host cell it forms a double stranded circular replicative intermediate.

Cosmid vector

Cosmid vector is a type of plasmid vectors containing cos site of lambda phage. It is made up of 250 bp. The cos site are the sequence yielding cohesive ends. The cos sites endeavour *in vitro* packaging system to the plasmid vector (Fig. 3). Besides, a replication origin site, a cosmid has unique restriction sites and selectable markers from the plasmid. It is more efficient than a lambda phage vector. It can accommodate DNA fragment up to 32-47 kb. The cosmid vectors are opened by the appropriate restriction enzyme at a unique site, are then mixed with DNA inserts prepared by using the same enzyme and annealed. When injected into a bacterium, the recombinant-DNA of a cosmid circularizes like phage DNA but replicates as a normal plasmid without the expression of any phage functions. Cosmids have selectable markers from the plasmid. Thus, the selection process for recombinant cosmid is based on the type of plasmid used to construct cosmid. Application of cosmids in constructing libraries of DNA fragments of eukaryotes is significant. It is because of their capacity to accommodate large fragments of DNA. Cosmids are especially useful for the analysis of highly complex genomes and are an important part of various genome mapping projects.

Phasmid vector/ phagemid vector

Phasmids are a type of plasmid vector containing shortened lambda phage genome including its *att* site. Phasmids are packaged in lambda particles *in vitro* and used for infection of appropriate host cell. The phasmid may insert into a lambda phage DNA in the same way by which phage DNA inserts into the bacterial chromosomes during lysogenic phase of life cycle. A typical feature is observed in phasmid in relation cI gene of lambda phage. This gene produces repressor for lysis. If this gene is present, the phasmid replicates like a plasmid and if absent, phasmid multiplies like a phage and undergoes plaque formation.

A mutant cI gene produces a temperature sensitive CI protein. If such mutant gene is present in phasmid, it will replicate as plasmid at lower temperature and behave like phage at higher temperature.

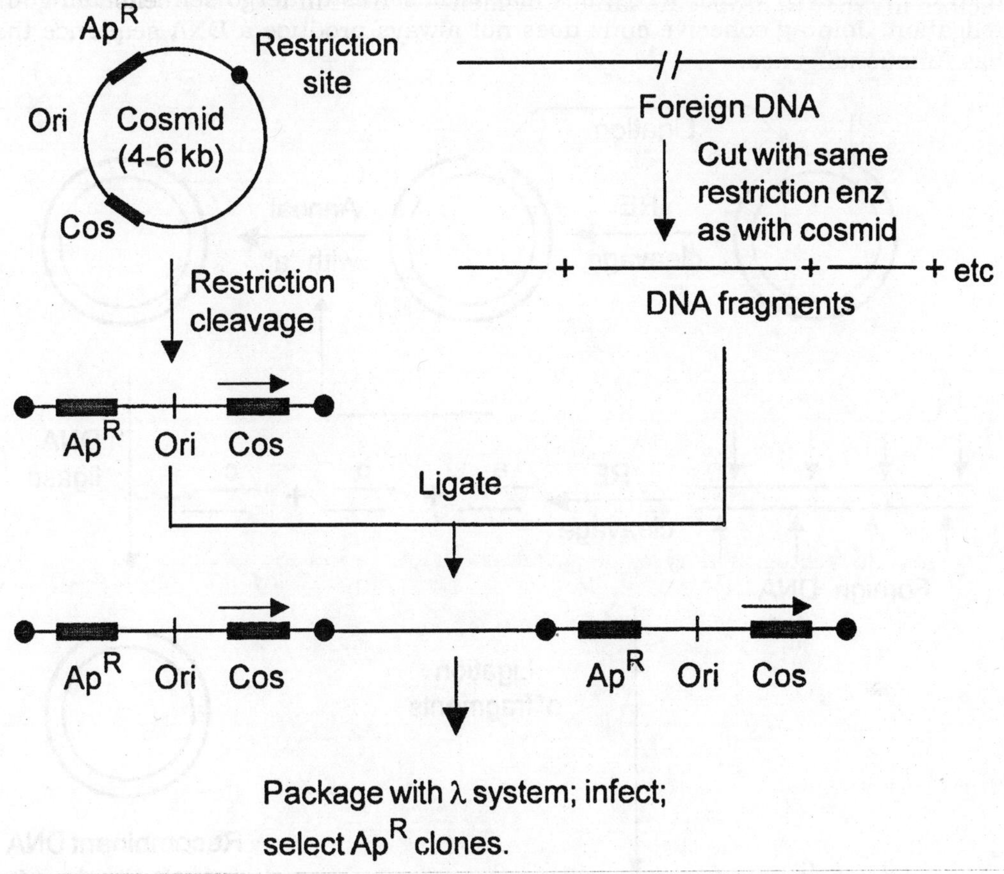

Fig. 3. Cloning with cosmid vectors

LIGATION OF ISOLATED DNA IN VECTOR

Isolated DNA fragment is inserted in the vector, where it links covalently. Different methods used for ligation depends upon the type of DNA inserted has cohesive or blunt termini.

Ligation involving Cohesive Termini

Some restriction enzymes generate sticky or cohesive ended fragments from DNAs (Fig. 4). A particular restriction enzyme produces fragments with the same cohesive ends, without regard for the source of the DNA, i.e., DNA isolated from two different organisms. The sequences at the terminii would be identical if foreign DNA and the vector are cut with the same restriction enzyme. The DNA fragments will soon anneal with the ends of the opened up vector molecule. It can be ligated with DNA ligase. There may be two types of ligation procedure. In the first option, DNA fragments generated by restriction cleavage, by their very nature, may be prone to self annealing and can get ligated since all of them have the same complementary terminal sequences. The alternate possibility is that restriction cleaved vector molecules may themselves undergo self-annealing and religation. Joining cohesive ends does not always produce a DNA sequence that has functional genes.

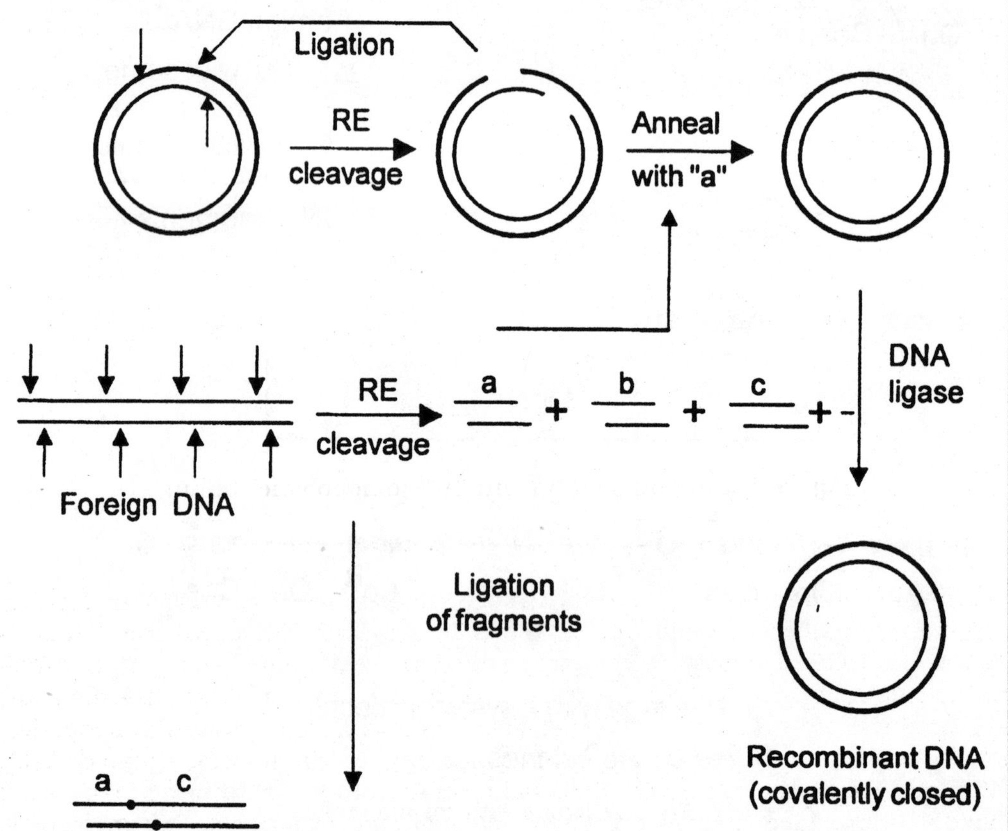

Fig. 4. Ligation through cohesive terminii

Blunt end ligation

Some restriction enzymes generate DNA molecules with (usually isolated from bacteriophage T4) blunt ended termini (Fig. 5). Ligation of such molecules may be effected by T4 DNA ligase or by adding a linker molecule containing within it a site for a restriction enzyme.

T4 phage encodes a DNA ligase, which seals nicks in double stranded DNA having 3'-OH and 5'-P terminii. It has the additional property of joining two DNA molecules having completely base-paired ends.

A linker is a short DNA segment containing a known restriction site. Linkers are coupled to both ends of the blunt-ended molecule. This segment is coupled by T4 DNA ligase to each blunt end of both the fragments i.e., isolated DNA and the vector. Then, both DNA molecules are treated with the restriction enzyme.

The same complementary single-stranded terminii are produced on both the fragment of DNA. Thus, the two DNA are joined.

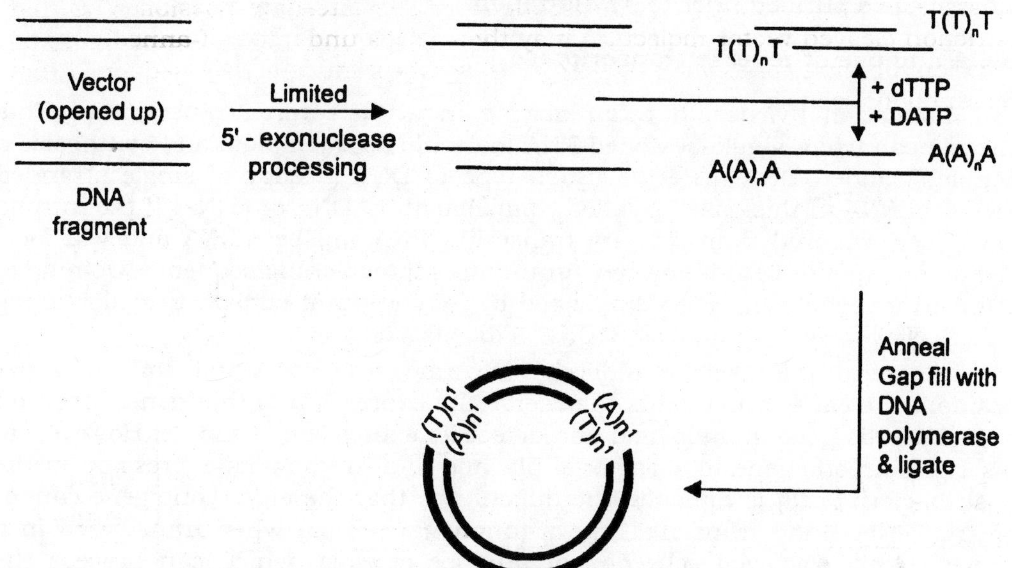

Fig. 5. Blunt end ligation with AT homopolymer tailing

Joining of DNA fragments by addition of Homopolymers

In gene cloning, one important aspect is joining of two DNA molecules. Enzyme terminal nucleotidyl transferase is an unusual DNA polymerase. It adds nucleotides to the 3'-OH group of an extended single-stranded segment of a DNA chain. This reaction does not require DNA template strand. First DNA molecule is treated with 5'-specific exonuclease. This removes a few terminal nucleotides. Now a reaction mixture is prepared consisting of exonuclease-treated DNA, dATP, and the enzyme, poly(dA) tails will form at both 3'-OH termini. Likewise, if poly(dT) is provided instead of poly(A), poly(dT) tail is formed. Two molecules, one with poly dT and another with polydA are easy to join. Joining is accomplished by gap-filling with DNA polymerase I and sealing with ligase. This method is useful when DNA molecules lacking complementary ends are to be joined.

INSERTION OF A PARTICULAR DNA INTO A VECTOR

It is simple to anneal some isolated DNA fragments (collected by digesting a genome with restriction enzyme) into a vector (which is also cleaved and rejoined). However, if a particular DNA segment must be isolated from the set of all vectors possessing foreign DNA, the procedure is typical.

Simple selection technique is adequate for recovery of many genes from a vector containing that gene. If the gene to be cloned were a bacterial Leu gene, one would use a Leu^- host and select a Leu^+ colony on a medium lacking leucine. There are many cases, in which the clone of interest is so rare or so difficult to detect. It would be preferable if, the fragment containing the gene of interest to be purified prior to hybridization.

cDNA and use of reverse transcriptase

Many animal viruses have an enzyme known as reverse transcriptase. This enzyme can use a single-stranded RNA molecule as a template and synthesize a double-stranded DNA molecule. Such type of DNA (a copy of single stranded RNA or mRNA in this case) is called complementary DNA or cDNA. If the introns have been removed from primary transcript, RNA will be mRNA and will form cDNA corresponding full-length of an uninterrupted coding sequence. Joining of cDNA to a vector can be accomplished by joining blunt-ended molecules using linkers. The purpose and use of cDNA is discussed below.

To confirm the insertion of foreign DNA in a vector and that to in a proper position, the easiest method is to observe the expression of that gene. Insertion of prokaryotic gene in a plasmid and detect its expression is simple. However, to insert eukaryotic gene in a bacterial plasmid and to know its expression in the host (bacterial cell) is difficult. The difficulty is that the eukaryotic gene cannot be transcribed and translated into a functional protein when transcribed in a bacterial cell. Bacterial cells do not produce enzyme, which can process the primary transcript from DNA (remove intron and form mRNA from primary transcript). So to test the hybrid clone based on gene expression is difficult. To make this process convenient a processed primary transcript may be for cloning. With the help of enzyme reverse transcriptase, copies of cDNA may be produced.

To explain all this let us take an example of animal cell, reticulocyte. The latter produces only one protein (β-globulin) and some very small proteins. The processed primary transcript will have only one mRNA coding for β-globulin. If purified mRNA is subjected for cloning, it will be an indirect cloning of gene synthesizing β-globulin.

DETECTION OF RECOMBINANT MOLECULE

While breaking a vector and rejoining it inserting the isolated DNA, it becomes necessary to detect the recombinant molecule. The detection of recombinant molecules means to know if the vector contains the inserted DNA after $CaCl_2$ transformation and secondly that the inserted DNA is of our choice. In using the $CaCl_2$-transformation procedure to establish a plasmid in a bacterium, the initial goal is to isolate bacteria that contain the plasmid from a mixture of plasmid-free and plasmid-containing colonies. pBR322 is a useful plasmid. This is a small plasmid consisting of 4362 nucleotide pairs of known sequence. It contains only one copy of each of seven different types of restriction -enzymes. It has two different antibiotic-resistance markers (for tetracycline, *tet-r*, and for ampicillin *amp-r*). The pBR322 plasmid containing cells will form colony if grown in a medium containing any of the two antibiotics. Now, as there are specific known sites for cleavage by restriction enzyme, it is easy to insert a DNA molecule at a proper position. After detection of pBR322 plasmid containing cells, the next important step is to identify plasmid, which contains the inserted DNA. For this purpose, insertional inactivation procedure is employed. In pBR322 the *tet* gene contains sites for cutting by the restriction enzymes BamHI and SalI. Thus, insertion at either of these sites will yield a plasmid that is *amp-r tet-s,* because insertion interrupts and hence inactivates the *tet* gene. The wild cells are sensitive for ampicillin and tetracycline (*amp-s Tet-s*). Such cells are transformed with a DNA sample in which the cleaved pBR322 and restriction fragments have been joined. If the cells are plated on a medium containing ampicillin, all surviving colonies must be *Amp-r* and hence must possess the plasmid. The *Amp-r* colonies may have *Tet-r* and *Tet-s* sites on their plasmids. To confirm that the cells are tetracycline resistant or sensitive, cells are grown on culture medium containing tetracycline. The plasmid, in which DNA fragment is not inserted, will contain Tet-r and if the DNA is inserted, it will inactivate *Tet-r* and convert it into *Tet-s.* Before plating, a proportion of the transformed cells (in which insertion has occurred) are enriched by growing the transformed cells in a liquid medium containing cycloserine and tetracycline. Cycloserine kills growing cells, but *Tet-s* cells are merely inhibited, not killed by tetracycline. Thus, in this medium *Tet-s* cells survive. Plating of cells treated in this way on medium containing ampicillin yields *Amp-r Tet-s* colonies. These colonies possess pBR322 plasmid containing a donor DNA fragment.

SOMATIC HYBRIDIZATION AND PROTOPLAST FUSION

Somatic hybridization is fusion of protoplast of two plants (generally of different species or varieties). There are a few steps in the process of somatic hybridization. In general for somatic hybridization first protoplast from different sources (plants) is isolated and fused. The hybrid cells generated by such fusion is selected and cultured. All these steps are considered as successful when plants are generated from such hybrid clones.

Protoplast isolation

Plants cells have cell wall made up of cellulose, pectic compounds and in some cells hemicellulose. To isolate protoplast first the cell wall is digested by enzymes. The enzymes used are pectinase, cellulase and hemicellulase for pectic compounds, cellulose, and hemicellulose, respectively. The cells are treated for about 16 to 18 hours at 25-30°C and 4.7 to 6.0 pH. To maintain the osmotic concentration sorbitol or mannitol is added in the mixture. Generally, 500-800 m/mol of sorbitol or mannitol is added so that protoplast does not burst after cell wall digestion. In addition, 50-100 m/mol calcium chloride is also added to strengthen the plasma membrane. After the incubation period, the isolated protoplast is washed with a suitable washing medium. Thus, cell wall debris and adhered enzymes are washed. The isolated protoplast is cultured on different media. During culturing isolated protoplast develops cell wall and divides to produce a mass of cells. When colonies of cells develop on this media, they are transferred to normal media.

To isolate protoplast, leaf mesophyll is taken as ideal tissue. Lower epidermis of surface sterilized leaf is pealed by forceps and cut into small pieces. These pieces are incubated in enzyme mixture and washed after a proper time. The debris is removed. Generally, MS and B5 media or their modified forms are used to culture protoplast. Different dilutions of growth hormones, auxins or cytokinins are also added depending upon the source of plants and the type of tissue.

Protoplast fusion

Protoplasts of two different sources to be fused are kept in enzyme mixture and are fused by different techniques. The simplest technique is with high pH and Ca^{++} concentration. The protoplasts are incubated for 30 minutes at 10.5 pH and 50 m/mol solution at 37°C. After this incubation period, fused protoplast is obtained. Another process is by using electroporator. It is a specially designed equipment and protoplasts fuse under electric current. First protoplasts are subjected to low voltage electric, which aligns the protoplasts to be fused. The alignment appears in a form of a chain. Later these protoplasts are pushed between the two electrodes with the help of a micromanipulator. In the second phase, a short pulse of high voltage is released. At this high voltage, there are transient disturbances in the organization of plasma lemma resulting in protoplasts fusion. A third method is Polyethylene glycol (PEG) induced protoplast fusion. The protoplasts to be fused are treated with 28-50% PEG followed by washing with an alkaline solution containing high concentration of Ca^{++}. PEG has negative charge, it binds with Ca^{++}, which in turn, binds to the negatively charged molecules present in plasma lemma. When treated protoplasts are washed, PEG molecules may pull out the plasma lemma components bound to them, disturbing plasma lemma organization resulting in protoplast fusion (Fig. 6).

Selection of Hybrid Cells

The protoplast that has been artificially fused by fusogen (agents which induce fusion) does not undergo complete fusion. After the process is completed, the mixture may have three types of cells i.e., unfused, fusion between the protoplast from the same source (from the same plant) having similar nuclei (homokaryon), and the actually required hybrids having heterokaryons. The fused protoplast may represent only 0.5 to 10% of the total cells of the mixture. Thus, it is important to select out such hybrid cells. Visual markers may help in the selection of hybrid protoplast. Microscopic examination of pigments or other cell organelles is visualized for such selection. After identifying the hybrids, they are mechanically isolated and cultured. Deficiencies of the parental protoplasts (like sensitivity for antimetabolites, temperature etc) may be examined in hybrids. It may be possible that hybrids do not show such deficiencies due to complementation. A simple and general method to identify hybrids is to culture the mixture of protoplasts. All the types of cells will develop callus. Certain characters of callus like morphological characters, chromosome constitution, protein and enzyme bands etc are studied and the hybrids are identified.

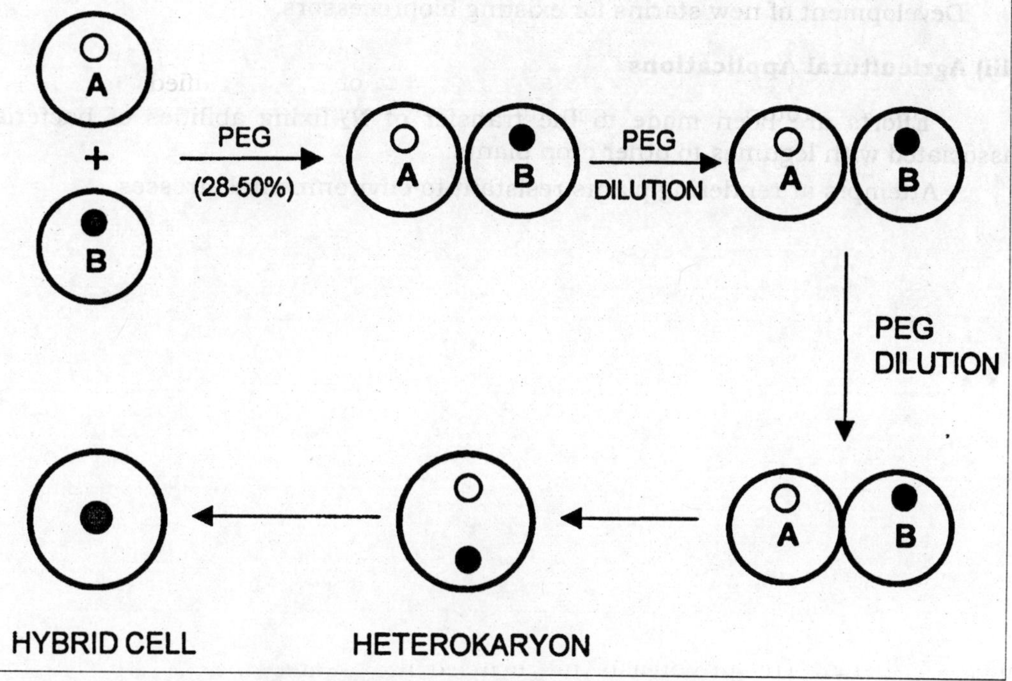

Fig. 6. Steps of somatic hybridization.

Regeneration of Hybrid Plants

The hybrid protoplast, when cultured develops in a mass of cells. This cell mass is known as callus. The calli are induced to regenerate plants from them. In some cases, the fused protoplast may not develop into a calli or if calli is formed, plants do not regenerate from them. This is called somatic incompitibility. Application of somatic hybridization is mainly for crop improvement. Thus, it is important the characters developed in plants regenerated from callus. The hybrids, which retain the almost full somatic complements of the two parental species, are known as symmetric hybrids. Hybrids of sexually incompatible parents are sterile. This is not useful in breeding progamme. Thus, polyploids of such hybrids are prepared which are fertile.

APPLICATIONS OF GENETIC ENGINEERING

(i) Medical Applications

Production of medically useful proteins such as somatostrain, insulin, human growth hormone and interferon.

Probes are used in diagnosis of infections diseases.

(ii) Industrial Applications

Manufacturing protein products by using bacteria. Strain improvement for existing bioprocessors.

Development of new starins for existing bioprocessors.

(iii) Agricultural Applications

Efforts are been made to the transfer of N_2-fixing abilities of bacteria associated with legumes to other crop plants.

Attempts at rendering plants resistant to environmental stresses.

SECTION – IV

CHAPTER - 20

HUMAN DISEASES

A micro-organism, which is able to cause disease is known as *pathogen* and the ability to cause the disease, is called as *pathogenicity.* The degree of pathogenicity is referred as virulence. When a pathogen comes in contact with the body surface it tries to establish, this is known as *colonization* and if the pathogen is able to colonize successfully it proliferates and grows on the body surface, this is known as *infection.* The number of pathogenic cells, which can cause observable infection, is known as *infectious dose.*

At the time of disease the host shows the tendency to resist the disease while the pathogen uses its weapons to cause disease. The outcome of the disease is the dynamic balance between these two efforts.

MECHANISM OF DISEASE

An infectious dose of a pathogen has to establish infection, survive the host's defenses and cause injury in the host tissues. The microbial attributes that increase the severity of a disease are known as virulence factors. The very first virulence factor helps in establishment of the pathogen on the host. Virulent strains of a pathogen show ability to attach to the body surface of the host. Bacteria may attach to the host surface with the help of pilli. Pilli helps in attachment of some Gram-negative bacteria on the mucous membrane of urinary and genital tract. Influenza virus produces an enzyme neuraminidase, which dissolves neuraminic acid of mucous layer of respiratory tract and thus helps in attachment of the virus. When a pathogen is able to colonize on the host surface its second effort is to face defence mechanism of the host. A number of virulence factors protect the micro-organism from host resistance mechanism and may help the pathogen to multiply inside the host tissue.

Toxins : Pathogen causes damage to the host tissues by producing toxins. A pathogen may produce exotoxins or endotoxins or both. *Exotoxins* are soluble proteins. They are secreted by the pathogens and are carried out by bloodstream from the site of infection to the other part of the body. *Clostridium tetani* causes tetanus. The pathogen is localized at the site of infection. The bacteria produce exotoxins, which spreads in the body and attacks motor nerves bringing about muscle contractions. Other pathogens producing exotoxins are *Corynebacterium diphtheriae* causing diphtheria, *Clostridium botulinum* causing botulism,

Bordetella pertussis causing whooping caugh, *Clostridium perfringens* causing gas gangrene, *Shigella dysenteriae* causing dysentery and other food poisoning micro-organisms. Endotoxins are lipopolysaccharides. It is a component of cell wall of Gram-negative bacteria. Endotoxins are confined at the site of infection. It is only liberated from the bacterial cells when the latter is disintegrated. When released from cells, endotoxins triggers white blood cells to discharge chemicals into the blood and surrounding tissues. These released chemicals induce fever and pains; rashes may develop as the result of capillary hemorrhage.

Antiphagocytic factors : The structural contrivance of the bacterial cells protects it from engulfment by host's phagocytic cells. Capsule is one of the antiphagocytic factors. Capsule protects bacteria from phagocytosis either by covering the site of attachment of the phagocytic cells or preventing the binding of a chemical that promotes ingestion of the micro-organisms. Antiphagocytic capsule is present covering the cells of pathogens like *Staphylococcus aureus, Neisseria meningitides, Klebsiella pneumoniae* and *Salmonella typhi.* There are some chemicals produced by the pathogens, which protect them from phagocytosis e.g., Leukocidin, Coagulase or antichemotactic factors, which resists chemotactic movement of the phagocytic cells towards the pathogen.

Spreading factor : Pathogens secrete some enzymes, which destroy physical barriers that inhibit spreading of the former. These enzymes may be lecithinase, collagenase, hyaluronidase, fibrinolysin,hemolysin etc. Lecithinase depolymerizes host cell membrane, collagenase attacks on protein of skin, bone and cartilage, hyaluronidase dissolves hyaluronic acid of connective tissues, fibrinolysin dissolves fibrin clots and haemolysin destroys red blood cells.

BACTERIAL DISEASES

STAPHYLOCOCCUS

Staphylococcus (Gr. *staphyle,* bunch of grapes and kokkos, meaning a berry) was first differentiated by Koch (1878) and Rosenbach (1884), differentiated species of staphylococci based on colonial pigmentation. These bacteria are Gram positive cocci occur in bunches or clusters. The diameter of individual cocci is 0.7 µm-1.2 µm. They are non-motile, non-spore forming and non- capsulated bacteria. On nutrient agar medium their colonies produce different colours e.g., *Staphylococcus aureus* produces golden, *Staphylococcus albus* produces white and *Staphylococcus citreus* produces yellow colour. On NA slants, the confluent growth presents a characteristic "oil paint" appearance. *S. aureus* are spherical cocci arranged in grape-like clusters. Staphylococci contain proteins, antigenic polysaccharides as well as other substances important in cell wall structure. *S. aureus* produces a number of exotoxins. Many toxic effects can be demonstrated using cultural filtrates. On the basis of these biological activities, staphylococcal toxins may be classified into 3 groups:

1. Cytolytic toxins consisting of haemolysins and leukocidins
2. Enterotoxins
3. Exfoliative toxins.

They produce various toxins like haemolysin, leukocidin, fibrinolysin and enterotoxin. Haemolysin has four antigenic forms called *alpha, beta, gamma* and *delta* lysins. Alpha toxin is highly pathogenic. The role of leukocidin in pathogenesis is not very clear. Antibodies to leukocidin may play a role in resistance to recurrent staphylococcal infection. Enterotoxin is responsible for nausea, vomiting and dioarrhoea during food poisoning. Bacterium has enzyme coagulase, which is responsible of clotting human or rabbit plasma. Enzyme coagulase along with coagulase reacting factor (CRF) converts fibrinogen into fibrin. Staphylococci cause the majority of acute lesions in man. Staphylococcal diseases may be classified as cutaneous and deep infections. The common cutaneous lesions produced are furuncles, carbuncle, impetigo contagiosa, while the deep infections are characterised by septic tonsillitis, otitis media, mastidis, pulmonary abscesses, epipyema pyogenic meningitis, acute soteomyelitis and septicaemia. Staphylococcal septicaemia is a rare, but serious disease. Consumption of food contaminated with enterotoxin-producing staphylococci results in food poisoning. In food poisoning the patient suffers from diarrhoea and vomiting within about six hours of taking the contaminated food. Toxic hock syndrome (TSS) is a multisystem illness characteristic by acute onset of high fever, hypotension, vomiting, diarrhoea and a scarlatiniform rash with subsequent desquamation. Sever cases lead to death. In sever cases there is acute renal failure, disseminated intravascular coagulation, peripheral gangrene. Benzyl penicillin is generally used for the treatment of staphylococcal diseases. Some strains are penicillin resistant as they produce penicillinase. Methicillin and cloxacillin are effective against such strains.

STREPTOCOCCUS

Streptococci are Gram-positive cocci occurring in chains. Billroth (1874) coined the name streptococci (Gr. *streptos,* twisted or coiled). Streptococci form a heterogenous group. Some important pathogenic species are *Streptococcus pyogenes, S. viridans* and *S. faecalis.* Medical bacteriologists classify Streptococci into obligate anaerobes and facultative anaerobes. Brown (1919) employed meat infusion peptone agar with 5% horse blood and recognized three types of reactions. (a) *Alpha* haemolytic streptococci produce a greenish discolouration with partial haemolysis around colonies. (b) *Beta* haemolytic streptococci, most of the pathogenic streptococci fall in this group, they produce a sharply defined, clear, colourless zone of haemolysis, and (c) *Gamma* or nonhaemolytic streptococci produce no change in the medium. Streptococci are nonmotile and nonsporing. The cells of *Streptococcus pyogenes* are spherical or oval with 0.5 to 1.0 μm in diameter. *Streptococcus pyogenes* produces several types of exotoxins and enzymes. The important toxin produced by *S.pyogenes* is Haemolysin, Leucocidin, fibrnolysin and erythrogenic toxin. Haemolysin is of two types streptolysin 'O' and 'S'.

Streptolysin is so called because it is oxygen labile. Streptolysin O is antigenic and antistreptolysin. Streptolysin S is oxygen stable and soluble in serum. Haemolytic streptococci possess a group specific polysaccharide C and three type of specific protein antigen M, T, R and nucleoprotein. *Streptococcus pyogenes* is placed in Group A. It causes pyogenic infections. The common lesions caused are cellulites, boils, septic tonsillitis acute pharyngitis sinusitis otitis media, mastoiditis pulmonary abscess, septicaemia and meningitis. *Streptococcus pyogenes* causes two skin infections erysipelas and impetigo. *Streptococcus pyogenes* is also found inhabiting in female genitalia and causes puerperal sepsis. Bacterium is also concerned to two important nonsuppurative sequelae- acute rheumatic fever and acute glomerulonephritis. Penecillin GT and erythomycin or cephalexin is generally used for the treatment of group A streptococcal infections. *Streptococcus agalactiae* is placed under group B, it causes mastitis in cow. Group B streptococcus is the single most common cause of neonatal meningitis. Streptococci of group C are predominantly animal pathogens. *Streptococcus equisimilis* was isolated from horse and cow. Streptococci of group D are also known as enterococci, it includes *Streptococcus faecalis, Streptococcus faecium* and *Streptococcus durans*. All these species are faecal stretopcocci. *Streptococcus bovis* and *Streptococcus equines* are nonfaecal bacteria. These bacteria may or may not be haemolytic.

PNEUMOCOCCUS

In 1880 Pasteur first isolated pneumococci from human saliva. Pneumococcus (*Diplococcus pneumniae*) is Gram positive, lancet-shaped diplococci. Each pair is surrounded by capsule. The pneumococci have been reclassified as *Streptococcus pneumoniae* because of its genetic relatedness to streptococci. Pneumococci possess a number of antigens (a) Nucleoprotein, (b) Species specific polysaccharide hapten and (c) Capsular polysaccharide.The most important antigen of the pneumococcus is the type specific capsular polysaccharide. Pneumococcal capsules are composed of large polysaccharide polymer. The composition of theses polysaccharides has been worked out. However, the structure of only some types has been determined, e.g., Type 3, consists of repeating cellobiuronic acid units (D-glucuronic acid 1-4 linked to D-glucose) joined by 1-3glucosidic bonds. The antigenic property of the capsular polysaccharide varies in different species; pneumococci of different types with capsules of the same size may vary widely in virulence. The toxins and other virulent factors produced are oxygen labile Haemolysin, Leucocidin which kills Leucocytes and large amount of enzyme resembling receptor destroying enzymes of influenza virus. The toxins are weak and are not responsible for virulence. The virulence depends only on the production of the capsular polysaccharide. Acidic and hydrophilic properties of these substances save pneumococci from phagocytosis. The important lesion produced by pneumococci is lobar pneumonia. A few produce bronchopneumonia in man. Bacteraemia is frequent during the early stage of lobar pnemonia. Sometimes, the organism may produce meningitis, purulent arthritis, peritonitis, pulmonary abscesses, empyema, pericarditis, otitis media and conjunctivitis. Penicillin is generally used for treatment. Erythromycin may be used for penicillin sensitive patients.

NEISSERIA

Neisser (1879) described *Neisseria gonorrhoeae*. About 30 species of this genus has been identified but only two species *Neisseria meningitidis* and

Neisseria gonorrhoeae are pathogenic for man. They are commonly called as meningococcus and gonococcus, respectively. These bacteria are Gram negative, aerobic, nonsporulating, nonmotile cocci occurring in pairs or small groups (tetrads or clusters). In 1887 Weichselbaum isolated and described *Neisseria meningitides.* It is the causative agent of acute pyogenic meningitis or cerebrospinal fever, which may occur sporadically or as epidemics. The meningococcus usually inhabits the human nasopharyngeal. Infections may lead to bacteremia, which is ordinarily followed by acute purulent meningitis. The antigenic property lies in capsular polysaccharide. Beside this antigen, meningococci possess carbohydrate and protein somatic antigens. Penicillin is used for treatment of meningococcus. A combined therapy with erythromycin and chloramphenicol is used for patient allergic to penicillin. *Neisseria gonorrhoeae* causes the venereal disease gonorrhoe and its complications. Gonococci are antigenically heterogenous. There are at least 16 distinct serotypes of gonococci, differing in their principal outer membrane proteins. The pilins of pilli is antigenic surface protein. Other surface proteins are Protein I, II and III. There are several other poorly defined proteins, which play important role in pathogenicity. The lipopolysaccharide of gonococcus has endotoxic effects. Sulfonamide was previously used for the treatment of this disease but bacteria developed resistance to this drug. Antibiotic penicillin is administered to such patients. Recently, strains are isolated, which produce penicillinase producing *N. gonnorhoeae* (PPNG). Streptomycin is used for the treatment in such cases.

CORYNEBACTERIUM

Corynebacteria (Gr. *coryne,* club) are Gram positive, nonacid fast, nonsporing and nonmotile bacilli occurring in palisade. The rods show club shaped swelling at the ends and irregular staining. The pathogenic member is *Corynebacterium dephtheriae* causing diptheria. The disease was first recognized by Pierre Bretonneau (1821) who called it diphtheritis (Gr. *diphthera,* skin or membrane). The diphtheria bacterium was first described by Klebs (1883). *Corynebacterium dephtheriae* are bacilli. The rods vary from 2 μm to 6 μm in length and from 0.5 μm to 1 μm in diameter. They are pleomorphic. On the basis of colonial morphology on the tellurite medium and other properties McLeod subclassified this bacteria into three types- *gravis, intermedius* and *mitis.* The diphtheria bacilli produce a very powerful exotoxin. The diphtheria toxin is a protein and consists of two factors, factor A (lethal factor) and factor B (spreading factor). The latter is responsible for binding the toxin to the cells. The disease begins in the upper respiratory tract. The common sites of lesions are pharynx, tonsil, larynx, nose, trachea etc. The virulent diphtheria bacilli invade the throat and multiply in the superficial layers of the mucous membrane. The toxin produces a local lesion characterized by inflammation and necrosis. The inflammatory response results in the accumulation of grayish exudates, which eventually forms the characteristic diphtheritic pseudomembrane. The absorption of toxin causes toxaemia and systemic lesions in the nerve cells, toxic myocarditis, degenerative changes in the muscle and parenchymatous organs. Non pathogenic members are known as diphtheroids. The diphtheroids usually do not show metachromatic granules and do not produce exotoxin. They occur as normal commensals in the throat, skin, conjuctiva and other areas. The common diphtheroids are *C. hofmanni* found in the throat, *C. xerosis* found on skin.

Specific treatment of diphtheria consists of antitoxic and antibiotic therapy. *C diptheriae* is sensitive to penicillin and can be cleared from the throat within a few days by penicillin treatment.

MYCOBACTERIUM

Mycobacteria are Gram positive, nonmotile and nonsporing rods. They may sometimes show branching filamentous forms resembling fungal mycelium. Mycobacteria are acid fast and are defined on the basis of a distinctive staining property. The organism stain with difficulty, but when once stained do not decolorize with acid. *Mycobacterium tuberculosis* (type *hominis* and *bovis*) and *Mycobacterium leprae* are pathogenic species. Hansen (1868) discovered lepra bacilli. In 1882 Koch discovered the tubercle bacilli and proved its causative role in tuberculosis and proved Koch's postulate. *Mycobacterium tuberculosis* is a straight or slightly bent or curved rods, 2 µm 4 µm long and 0.2 µm - 0.5 µm wide. About 20 antigens have been identified in *M. tuberculosis*. Some are species-specific; others are common to many mycobacteria. Group specificity is due to polysaccharide and type specificity due to protein antigens. *M. tuberculosis* causes tubercular lesions (*L. tuberculum*, small lump) in man and produces spreading diseases in guinea pigs. The tubercle is an avascular granuloma composed of a central zone containing giant cells, with or without caseation necrosis, surrounded by epithelioid cells and a peripheral zone of lymphocytes and fibroblasts.

HAEMOPHILUS

Haemophilus are very small (cocco-bacilli), highly pleomorphic, non-motile, non-sporing, Gram-negative bacilli. *H.influnzae* (Influenza bacillus; Pfeiffer's bacillus) is a small encapsulated coco-bacilli (1 µm-1.5 µm X 0.3 µm), but short chains and rods often found. It is a facultative anaerobe and requires two growth factors. These growth factors are present in blood (factor X and V) hence the name, *Haemophilus* (blood loving). It is found on the mucosa of the upper respiratory tract. *H. influnzae* causes bronchopneumonia, pyogenic meningitis, conjunctivitis and subacute bacterial endocarditis. The disease usually begins as a nasopharyngitis. In children *H. influnzae* causes bacterial meningitis. The antigenic principle is found in capsule. Six types of capsular carbohydrates designated from a to f have been described. Out of which type b is most important pathogenically. Sulfonamides and antibiotics are used for treatment. Ampicillin and choramphenicol are effective.

Haemophilus aegypticus (Koch-Weeks bacillus: *H.influenzae* biotype III) causes infectious conjunctivitis. It is common in the tropics and subtropics and may occur in epidemic forms.

Haemophilus ducreyi (Bacillus of Ducreyi) causes soft chancre. Chancroid or soft sore is a venereal disease. It is antigenically homogenous. Sulphonamides and many antibiotics are used for treatment.

VIRAL DISEASES

DNA VIRUSES

POX VIRUS

Pox viruses are the largest viruses that infect vertebrates, large enough to be seen under light microscope. Based on morphological and antigenic differences, family pox viridae is divided into two sub-families: Chordopoxvirinae and Entomopoxvirinae. The viruses of these sub-families infect vertebrates and insects, respectively. Sub-family chordopoxvirinae contain six genera: (1) Orthopoxvirus- virus infects mammals and cause generalized infection with rash-Variola, Vaccinia, Cowpox, Ectromelia, Rabbit pox, Monkey pox. (2) Parapoxvirus- infecting ungulates-Contagious pustular dermatitis, Milker module, Bovine pustular stomatitis. (3) Capripoxvirus-infecting goats and sheep. (4) Leporipox virus- viruses infecting leporids (rabbits, hares, squirrels) (5) Avipoxvirus infecting birds- Fowl pox, Turkey pox. (6) Suipoxvirus: Swinepox. Miscellaneous viruses- molluscum contagiosum, tanapox and the yaba monkey tumor.

Poxvirus is brick shaped structure, measuring 300 x 200 x 100 nm in size. DNA is present as a biconcave structure covered by two coats. The inner core is adhered to nucleoprotein and the outer core is in the form of irregular layer. There are two lateral bodies on either side of nucleoid. All poxviruses produce nucleoprotein antigen. The antigens produced by poxviruses are L (heat labile) S (heat stable) antigen, Agglutinogen, Nucleoprotein, Hemagglutinin etc.

VARIOLA MAJOR (SMALL POX)

Small pox occurs in fatal and mild forms. The virus causing these disease are named as variola major and variola minor, respectively. Virus enters inside the mucosa of upper respiratory tract. It propagates in regional lymph nodes and reaches upto reticuloendothelial cells where it multiplies. It again reaches upto mucosa through blood stream. During disease skin lesions start from macular and pass through papular, vesicular and pustular stages. The characteristics at the initial stage of disease are malaise, fever, vomiting and headache. Rashes on skin appear after 2-3 days, and after 3-5 days, skin eruptions appear. Scab starts forming after 12 days. Vaccination is important to control the disease. Commercial vacccine may be prepared by inoculating vaccinia virus on scarified skin of calves, sheep and buffalo.

MOLLUSCUM CONTAGIOSUM

Generally, children and young adults suffer from this disease. The symptoms of disease are pink or pearly white wart like nodules on the skin. The viral particle embedded in protein's matrix forms molluscum. In the genital areas, molluscum contagiosum become inflamed and ulcerated.

ADENOVIRUS

This virus has spherical (icosahedral symmetry) with 70-90 nm in diameter. The capsid is made up of 252 capsomers (240 are hexons and 12 are pentons) arranged as an icosahedron with 20 triangular facets and 12 vertices. Each penton has a projection consisting of a rod-like portion with a knob attached at the distal end. These projections are known as fiber. Virions are non-enveloped. About 33 serotypes of adenovirus are identified, which infect man. The diseases caused by adenovirus are pharyngitis, tonsilitis, pneumonia, acute respiratory disease, pharyngoconjunctival fever, epidemic keratoconjectivitis, acute folicular conjectivitis and intestinal lesions causing gastroenteritis.

HERPES VIRUS

Based on physical, biological and genetic properties family Herpesviridae is divided into three sub-families: (1) Alphaherpesvirinae, (2) Betaherpesvirinae, and (3) Gammaherpesvirinae. Herpes viruses of medical importance are: Herpes simplex, Varicella, Herpes zoster, Cytomegalo virus, EB (Epstein Barr) virus. It has a double stranded DNA. Herpes simplex is of two types 1 and 2. Type 1 causes infection of mouth, eyes, central nervous system and type 2 strain causes infection of genitals. Type 1 is transmitted by direct contact or droplet spread, while type 2 is transmitted venereally. After entry virus invades in the skin or mucous membranes. It multiplies and reaches upto lymph nodes. They ultimately reach to cutaneous and mucosal region and cause lesions. The lesions are thin walled umbilicated vesicles. These vesicles form pustules and scabs, and heal without scarring. *Varicella* morphology of this virus is similar to Herpes simplex virus. It causes chicken pox in children. In adults prodormal fever and malaise with respiratory symptoms are common. The rash appears usually on the trunk. The rashes of chicken pox have some characteristics, which differentiates it from small pox. These differences are of some tendencies that the rash appears in crops during the first three four days of the disease, so that lesions of varying age can be noticed on the same patients. Live vaccine is administered by subcutaneous injections. *Herpes zoster* is disease of old age. Virus causes skin disease with eruptions over the distribution of sensory nerves. The virus is neurotropic and is localized in nerve cells. The virus is believed to remain latent in the sensory ganglia. It forms a vesicle on skin at the end of that particular nerve. Though the rash disappears within two weeks but the pain and paraesthesia may persist for months together. *Cytomegaloviruses* three serotypes have been recognised by the virus neutralisation test. Infections are inapparent, leading to prolonged latency. Virus causes cell enlargement with acidophilic or basophilic intranuclear inclusion body. It may spread through salivary secretions or venerally. The infection spreads by urine or saliva droplets via respiratory route and through placenta from infected mother to fetus. Subclinical localized infections are found in salivary gland and kidney. The generalized infection is associated with hepatosplenomegaly, jaundice, thrombocytopenic prupura, hemolytic anemia and microcephaly. *Epstein Barr (EP) virus* causes infectious mononucleosis.

Human Herpes Virus (HHV-6) also known as B-lymphotropic virus is a recently discovered virus. It causes exanthem subitum, which is manifested with rash in childhood with lymphadenopathy.

RNA VIRUSES

PICO RNA VIRUS (pico = small)

The virus is naked with 20-30 nm in size. Pico RNA viruses are of two types: Enteroviruses and Rhinoviruses. *ENTEROVIRUSES* include polio, echo and coxsackie virus. *Polio virus* is spherical with icosahedral symmetry measuring 30 nm in diameter. Based on neutralisation test, the poliovirus strains have been classified into three types, 1, 2 and 3. The virus enters inside the body by ingestion or inhalation and multiplies in lymphatic tissue of alimentary canal. It enters regional lymphnodes and through bloodstream, it reaches to the spinal cord and brain. In the central nervous system, the virus multiplies selectively in the neurons and destroys them. The neurons are destroyed with degeneration of Nissel body. Lesions are mostly in anterior horn of spinal cord. The earliest manifestations are associated with the phase of viraemia and consist of fever, headache, sore throat and malaise lasting 1-5 days. This is minor illness followed by major illness. In non-paralytic poliomyelitis, disease is restricted to aseptic meningitis. However, in paralytic stage, flaccid paralysis develops. The disease is controlled by vaccination. Vaccine (salk killed polio) is prepared by inactivating 3 types of polioviruses in formalin. Sabin live polio vaccine is prepared by growing the attenuated strain in monkey kidney cells. Generally, live vaccines are used and are administered orally. *Coxsackie viruses* are classified into A and B groups. By neutralisation tests, group A viruses are classified into 24 types and group B into six types. Group A cause vesicular pharyngitis, Group A and B causes aseptic meningtis and Group B causes Bronholm disease, myocarditis and pericarditis. *Echoviruses* (enteric cytopathogenic human orphan viruses, ECHO) causes fever with rash and aseptic meningitis. *RHINOVIRUSES* are more acid labile and heat stable. They are divided into H and M strains. H strains grow on human being and M strain grows in human beings as well as on monkeys.

ORTHOMYXOVIRUSES

It is spherical 80-120 nm in diameter. They have a character to adsorb on to mucoprotein receptors on erythrocytes, causing haemagglutination. *Influenza viruses* causes disease in respiratory tract. This disease occurs in sporadic, epidemic and pandemic forms. It is spherical virus, however, pleomorphism is also observed. Virus is also found in filamentous forms. It has single stranded RNA with helical symmetry of ribonucleoprotein. There is layer of lipid on the protein capsid. The lipid is derived from host cells. Haemagglutinin spikes and neuraminidase peplomers are present on this lipid layer. There are two types of antigens: internal antigens and surface antigens. The internal antigen is ribonucleoprotein (RNP antigen). While, the soluble antigen is freely occurring in infected tissues and occurs in the supernatant during centrifugation. The haemagglutinin is a glycoprotein composed of two polypeptides-HA1 and HA 2. The neuraminidase is a glycoprotein enzyme, which destroys cell receptors by hydrolytic cleavage. Virus shows antigenic variations and based on this property, it is classified into antigenic drift and antigenic shift types. Antigenic drift types show gradual changes in antigenic structure regularly, resulting in periodical epidemic. Antigenic shift types show abrupt, drastic, discontinuous variation in antigenic structure causing major epidemic. Haemagglutination is an important characteristic. When virus is mixed with a suspension of fowl erythrocytes, the virus is adsorbed on to the mucoprotein receptor on the cell surface. The virus links together adjacent cells producing haemagglutination. After sometime the virus is detached from the cells due to the activity of enzyme neuraminidase. This process is kwon as elution. Vaccine is used as control measure. The change in antigenic make up of the virus is a hurdle in immunoprophylaxis. Amantadine hydrochloride is fruitful in treatment of influenza.

PARAMYXOVIRUSES

Paramyxoviruses are pleomorphic with size ranging from 100-300 nm, and giant forms of upto 800 nm. RNA is single stranded with 18 nm wide nucleocapsid. Antigens are haemagglutinins, neuraminidases and haemolysins. This group includes mumps, parainfluenza, respiratory syncytial and measles viruses. They are pleomorphic *Mumps viruses* cause acute infectious disease with parotitis. Skin over the enlarged parotid glands may be stretched, red and hot. Other organs are involved due to viremia. Orchitis and viral meningoencephalitis are important complications of mumps. It is spherical to pleomorphic with 100-600 nm in size. The outer membrane is covered by projections extending 12-15 nm from the virion surface. It has five major structural proteins. Virus enters through inhalation or through conjunctiva. Vaccine is prepared from killed virus grown in allantoic cavity. Live attenuated vaccine is prepared by Jeryl-Lynn strain. One attack confers lasting immunity so that second attack does not occur. *Parainfluenza viruses* cause febrile respiratory infections. They are classified into four types: Parainfluenza 1, Parainfluenza 2, Parainfluenza 3 and Parainfluenza 4. 'Sendai virus' is the first virus isolated as parainfluenza virus. The virus is spherical with 125-250 nm in diameter. It is enveloped by proteins and lipids. Two glycoproteins protrude from the envelope as 10 nm spikes. No vaccine is available. *Respiratory syncytial virus* (RSV) causes bronchiolitis and pneumonia. This virus was first isolated from chimpanzees and named as 'chimpanzee coryza agent (CCA)'. The name RSV is due to its property to cause cell fusion and the formation of multinucleated syncytia in cell cultures. It is pleomorphic ranging from 150-300 nm in diameter. It lacks either haemagglutinin or neuraminidase. In children RSV causes otitis media, in adults RSV infection may present as a febrile common cold. It can cause pneumonia in the elderly. *Measles* (Rubeola) causing virus have haemagglutinin no neuraminidase. The symptoms of disease are maculopapular rash proceeded by fever, cough, nasal and conjunctival catarrh. Some atypical measles symptoms are associated with measles virus infection. The symptoms are characterised by high fever, pneumonia and an unusual rash without Koplik's spots. The virus is pleomorphic with 120-250 nm in diameter. It is enveloped with 10-12 nm lipoprotein membrane. It has projecting peplomers of haemagglutinin. It is transmitted by nasopharyngeal secretions. Live attenuated vaccine is prepared by Edmonston B strain. This vaccine is given in combination with mumps and rubella vaccines (MMR). Normal human gamma globulin given within six days of exposure can prevent or modify disease. This depends upon the dose.

RUBELLA VIRUS

The virus is pleomorphic, spherical, 50-70 nm in diameter. It causes rash and lymphadenopathy in children; in adults infection develops in joint and purpura. It enters inside the body by inhalation and replicated in lymphnodes. Live attenuated vaccines are Cendehill and HPV77. It is administered subcutaneously. RA 27/3 vaccine is administered intranasally conferring local immunity.

RHABDO VIRUS

Rabies virus is transmitted by bite (saliva) of rabid animals. The disease caused in man is hydrophobia. It is because patient shows fear of water and incapable to drink it. Virus multiplies in muscles, connective tissue or nerves at the site of bite. It penetrates in nerve endings and travels in the axoplasm towards spinal cord and brain. It reaches central nervous system through nerve fibers. The prodormal phase of disease occurs within 2-4 days. In this phase, there are manifestations like malaise, anorexia, nausea, vomiting, headache and fever. The following phases are sensory and excitory phase. In sensory phase, there is painful spasm of muscles and in the last phase, patient gets generalized convulsions and coma. Death is due to respiratory arrest or other complications. Rabies virus bullet shaped with 180 x 75 nm in size. One end of the virus is round or conical and the other is planner or concave. The outer lipoprotein envelope carries spikes. Spikes are absent on the planar or concave end of virus. The surface spikes are composed of glycoprotein. Virus has haemagglutination activity. To control disease, prophylaxis is, washing wound with water and soap. This releases spikes from envelope. After washing, the wound is to be treated with quaternary ammonium compounds, tincture or aqueous solution of iodine, or alcohol. Antitetanus measures and antibiotics may also be used. Vaccination is also recommended. Neural and nonneural vaccines are also available.

OTHER VIRAL DISEASES

HEPATITS VIRUS

Hepatitis viruses are categorized as hepatitis A and hepatitis B, causing infectious hepatitis and serum hepatitis, respectively. *Hepatitis A* virus belongs to family Picornavirus. It is nonenveloped and contains RNA. Fecal virus is most infectious. Infection occurs with intake of sewage contaminated water. Disease is transmitted by the faecal-oral route. Disease symptoms are characterized by two stages: the prodromal or pre-icteric stage and the icteric stage. The general symptoms are fever, malaise, anorexia, nausea, vomiting and liver tenderness. Prophylaxis is to maintain cleanness and faecal discontamination of food and water. Passive prophylaxis with human gamma globulin is effective. *Hepatitis B* is a DNA virus. It has double stranded molecule of DNA. It also contains DNA polymerase. Electron microscopic examination of sera of a patient suffering from Hepatitis B shows three types of particles. These are Spherical particle (22 nm in diameter), tubular particle (22 nm in diameter and of varying length) and Dane particles the double-shelled particles (42 nm in diameter). The first two types of particles produce "Australian antigen". The virion consists of a core of 27 nm diameter, with a 2 nm thick shell and an outer coat or envelope 7 nm thick. The core possesses icosahedral symmetry. The inner core of virus is antigenically distinct. Hepatitis B virus has different antigens. They are HBsAg (Hepatits B surface antigen), HBcAg (Hepatitis B core antigen), and HBeAg (Hepatitis B e antigen). HBsAg has antigenic diversity. Immunisation is important as prophylactic measures. Passive immunisation may be employed by human serum immune globulin. Inactivated human carrier plasma (as a source of HbsAg) is used for active immunisation. *NonA:NonB hepatitis* a types of hepatitis with clinical and epidemiological similarity but is not caused by either of these viruses. There are different types of such viruses. These viruses are transmitted either by transfusion or by faecal-oral route (water borne disease). During disease, acute illness is mild and may last many months with fluctuations in serum transaminase levels.

ONCOGENIC VIRUSES

Viruses that produce tumour in their natural hosts or in experimental animals, or induce malignant malformation of cells on culture, are known as oncogenic viruses. Oncogenic viruses have ability to produce tumours in their hosts. Rous (1911) observed virus associated solid malignant tumour, fowl sarcoma. *RNA containing oncogenic viruses* are placed in family Retroviridae. The retrovirus causes leukaemia and sarcoma. These viruses have an unusual enzyme reverse transcriptase. This enzyme is RNA directed DNA polymerase and it produces a DNA copy of the RNA genome of the virus. This newly formed DNA molecule integrates into the DNA of the infected host cell. It forms proteins of the virus. Retrovirus is of 100 nm size. They are enveloped virus and have single stranded RNA molecule. The envelope develops due to budding through the host cell membrane. Oncogenic retroviruses are classified into five groups:

1. Avian leukosis complex- induces avian leukosis.
2. Murine leukosis viruses.
3. Mammary tumour virus of mice.
4. Leukosis-sarcoma viruses of other animals.
5. Human T cell leukaemia virus (HTLV).

Oncogenic DNA viruses are Papovaviruses, Poxviruses, Adenoviruses and Herpes viruses. The mechanism of malignancy is not yet very clear. RNA has reverse transcriptase. There are specific genes known as oncogenes, responsible for cancer. Viral oncogenes (V-onc) are present in retroviruses. The genes present in host cells, which have similar sequence as that of oncogenes are known as cellular oncogenes (C-onc). These genes are not of viral origin and may be present in normal cells. It is presumed that retroviruses pick up these genes from host cell during infection. If the oncogenic virus is DNA virus, its DNA is integrated with the host cell genome. It influences host cell to be transformed into malignant cells.

HUMAN IMMUNODEFICIENCY VIRUS : AIDS

The disease has a characteristic that the patient loses immune competence. This results in a condition where invasion of even relatively avirulent pathogen becomes infectious or even fatal. All this is expressed by term '*Acquired Immune deficiency Syndrome*' (AIDS). The first case of AIDS was observed in 1981 from New York and California (USA). In 1983, Luc Montagnier and colleagues isolated *Lymphadenopathy Associated Virus* (LAV) i.e., HIV. In 1984 first serology developed to identify for detection of anti-HIV antibodies. In1986, International Committee on Virus Nomenclature decided on the generic name '*Human Immunodeficiency Virus* (HIV)'. HIV belongs to family Retroviridae. This virus is 90-120 nm in diameter. It is enveloped and has nucleoprotein (single stranded RNA and protein). The outermost envelope contains of glycoprotein (gp 41, gp120, gp160) and inner core has two proteins (p18, p24). It has three structural genes (*gag, pol* and *env*) required for replication. The *gag* gene determines the core and shell of the virus. The *env* gene determines the synthesis of envelope glycoprotein. The *pol* gene codes for the polymerase reverse transcriptase. Enzyme reverse transcriptase is characteristic feature of retro-viruses. In host cell, this enzyme first transcribes RNA into single stranded DNA and then to a double stranded DNA. Besides these structural genes there are five genes recognized as non-structural and regulatory gene. These are: (1) *tat* gene (trans-activating factor) enhancing viral protein synthesis. (2) *art* gene (anti-receptor transactivating) promoting viral mRNA translation. (3) Two LTR genes (Long terminal transactivating) contain sequences giving promotor, enhancer and integration signals. (4) 3' *orf* gene (open reading frame) exerting a regulatory effect reducing viral replication. (5) *sor* gene (short open reading frame) encoding a poorly immunogenic protein. The primary clinical manifestation in this disease is due to viral cytopathology. HIV affects T4 lymphocytes and their number decreases in infected persons. When virus binds with CD4 receptor, it looses outermost cover at the site of infection. Infected cells are unable to produce normal amounts of Interleukins-2, gamma interferon and other lymphokines. This results in failure of immune response (It is the secondary clinical manifestation of disease). Symptoms include fatigue, unex-plained weight loss, fever, shortness of breath, chronic dioarrhea, white patches on tongue and lymphoadenopathy. In the last stages of disease, there is non specific unrestrained proliferation of B lymphocytes producing excessive functional incompetent wasteful immunoglobulins (IgG, IgA) and non reactivity to recall antigens. Reconstituting basic T cell dysfunction is helpful in treatment of disease. For this purpose, thymus and bone marrow transplantation, interferon and interleukin therapy is recommended. Use of monoclonal antibodies and other drugs inhibiting reverse transcriptase is also in practice.

FUNGAL DISEASES

Diseases caused by fungi are known as mycoses. Based on the mode of entry, site of infection and the depth of tissue involved, mycoses is classified into: (1) superficial, (2) cutaneous, (3) subcutaneous, (4) systemic and (5) opportunistic mycoses.

SUPERFICIAL MYCOSES

The fungal infection is confined to outer surface of hairs and skin. The infection is collectively known as Piedras (stone from Spanish language, as hard nodules are formed at the base of hair shaft). It is also named as tineas (from Latin, meaning grub, larva, worms). The common types of superficial mycoses are: (1) Black piedra or Tinea nigra, causal organism-*Piedraia hortae.* Fungus forms hard black nodules on the hairs of the scalp. (2) White piedra (tinea albigena), causal organism - *Trichosporon beigelii* a yeast. Symptoms are light coloured nodules on the bread and mustache. (3) Tinea versicolor, causal organism - *Malassezia furfur* another yeast. Symptoms are brownish-red scales on the horny layers of skin (the trunk, neck, face and arms). Mild itching and loss of skin pigment also occurs.

CUTANEOUS MYCOSES

It is also known as dermatophytoses (Gr. Derma-skin + phyton-plant), ringworms or tineas. Cutaneous mycosis are: (1) Tinea barbae, causal organism-*Trichophyton mentagrophytes* or *T.verrucosum.* Disease is confined to beard hair. (2)Tinea captitis, causal organism *Trichophyton* or *Microsporum* speices. The symptoms are loss of hairs, inflammation, and scaling. (3) Tinea cruris, causal organism- *Epidermophyton floccosum, Trichophyton mentagrophytes,* or *T. rubrum.* It is a dermatophytic infection of groin. (4) Tinea pedis (athlete foot) and Tinea manuum (dermatophyte of hands), causal organisms are same as Tinea cruris. The fungi prefer to grow in the keratinized layer of the skin. Symptoms vary from fine scale to a vesiculopustular erruption. Itching and inflammation are frequently present.

SUBCUTANEOUS MYCOSES

Subcutaneous mycoses are caused by soil inhabiting fungi. The entry is through wounds. The subcutaneous mycoses are: (1) Chromomycosis, causal organism- *Phialophora verrucosa* or *Fonsecaea pedrosoi.* The symptoms are formation of dark brown-pigmented nodules. Infection is confined upto legs and feet. The lesions are warty, ulcerating and cauliflower-like growth. (2) Maduromycosis, also known as mycetoma or fungal tumour. Causal organism-*Madurella mycetomatis, M.grisea, Phialophora jeanselmei* (all fungi), *Nocardia asteroides, N.brasiliensis, Streptomyces somaliensis* (all actinomycetes). The disease symptoms appear on foot with destruction of subcutaneous tissue and production of serious deformities. Disease spreads from skin and subcutaneous tissues and reaches upto bone. In advanced stage, mycetoma is chronic suppurative, granulomatous lesion with progressive destruction of contagious tissue and vascular dissemination. (3) Spirotrichosis, causal organism- *Sporotrichum schenckii.* The primary symptom appears as red papule, which begins to ulcerate. In advanced stage, new lesions appear along lymph channels and can remain localized or spread throughout the body. The stage is extracutaneous sporotrichosis.

SYSTEMIC MYCOSES

It is also called as deep mycoses. The fungi causing systemic mycoses are present in two morphological forms i.e; they are dimorphic (exception- *Cryptococcus neoformans*). The two forms are parasitic yeast like phase (Y) and a saprophytic mold of mycelial phase (M). Systemic mycoses are: (1) Coccidiomycosis is also known as valley fever, San Joaquin fever or desert rheumatism. Causal organism- *Coccidioides immitis.* In humans, the fungus grows as a yeast- forming, thick-walled spheule filled with endospores. The fungi enter inside the body by inhalation of spores (arthrospores) present in soil or dust. Three distinct forms are known with this infection. These are acute, chronic pulmonary and disseminated. Mostly infections are asymptomatic or indistinguishable from ordinary upper respiratory infections. The lesions of lesions of lung may heal up by calcification or thin walled cavity may appear. Few percent of patients may develop chronic pulmonary disease. The inflammatory lesions of acute pneumonitis are histologically similar to those of pyogenic bacteria. In chronic pulmonary disease and in disseminated lesions the inflammation is granulomatous characterized by abundant histiocytes, giant cells and caseation necrosis. The fungus can also spread throughout the body, involving almost any organ or site. (2) Histoplasmosis, Causal organism- *Histoplasma capsulatum.* Fungus is transmitted by inhalation of spores. Fungus is present in soil, while microconidia are widely present in air. In culture, fungus forms septate hyphae with short conidiophores. Microconidia are smooth, round or pyriform (1-5 μ in diameter), while macroconidia are thick-walled, round and tuberculate (7-15 μ in size). In human beings, the fungus appears as small budding yeast. Reticuloendothelial system is the site of infection. There may be no symptoms or mild symptoms such as coughing, fever and joint pain are observed. Lesions may appear in the lungs and show calcification and fibrosis. In disseminated cases, which are very less, lesions appear in practically all tissues and organs. (3) Cryptococcosis, causal organism - *Cryptococcus neoformans.* Aged, dried pigeon droppings are an apparent source of infection. After entry, fungus invades respiratory tract causing pulmonary infections. The disease spreads to the skin, bones, viscera and central nervous system. In severe, chronic and disseminated form of disease, meningitis is observed. (4) Blastomycosis, causal organism- *Blastomyces dermatitidis.* The disease is transmitted by inhalation of conidia. Disease occurs in three forms. These are cutaneous, pulmonary and disseminated. Disease initiates from lungs and spreads in bones, skin, prostrate, and other viscera. The lesions are characterized by granulomatous inflammation, microabscesses and extensive tissue destruction.

OPPORTUNISTIC MYCOSES

Some saprophytic fungi, which are ubiquitous in the environment are important in medical mycology and for 2 reasons. Firstly, they are common laboratory containments on culture media – *Aspergillus, Penicillium, Mucor*, and *Rhizopus* species grow on virtually everything. Secondly, they can produce serious and even fatal infections in persons who are otherwise debilitated. Aspergillosis and mucormyosis are important opportunistic systemic mycoses.

Opportunistic mycoses are generally harmless, it becomes pathogenic in a host, which is seriously debilitated and has a lowered resistance to infection. Such hosts are said to be compromised host and non resistance may be due to malnutrition, alcoholism, cancer, diabetes, leukemia prolonged use of antibiotics, trauma from surgery or injury etc. (1) Aspergillosis, causal organism- *Aspergillus fumigatus.* Colonies grey green in colour with septate hyphae. Conidia borne on sterigmata, the latter is arranged on a vesicle on conidiophore. Conidia present is air enter inside the body through respiratory tract, resulting in many types of pulmonary aspergillosis. Allergic aspergillosis is caused by inhalation of spores. Patient suffers typical asthmatic attacks with mucifibrinous sputum containing eosinophils. In bronchopulmonary aspergillosis, the major clinical manifestation of the allergic response is a bronchitis resulting from both type I and type III hypersensitivities. In another type of aspergillosis, *Aspergillus* forms colonies within the lung cavity and develop into "fungus ball" called aspergillomas. From lungs, fungus spreads producing disseminated aspergillosis in a variety of tissues and organs. At this stage, symptoms are granulomatous and in necrotizing forms. (2) Candidiasis, causal organism- *Candida albicans.* It is a common member of microflora of a healthy individual. It is present in gastrointestinal tract, respiratory tract, vaginal area and mouth. Other members of microflora check the growth of this fungus and when this microflora upsets, *Candida* multiplies and causes disease. In oral candidiasis, or thrush, white, creamy patches cover the tongue and mouth. It is common in new borne. If mother's vaginal area is highly invaded by *Candida*, the upper respiratory tract of the new borne becomes colonized during passage through the birth canal. As soon as the new borne develops its microflora, thrush disappears. Paronychia and onychomycosis are associated with *Candida.* These infections occur in fingernails and toenails, respectively. There is inflammation of subcutaneous tissues at the base of nails. *Candida* causes vulvovaginitis. Low pH created by *Lactobacilli*, controls *Candida.* Vulvovaginal candidiasis occurs from diabetes, antibiotic therapy, oral contraceptives, pregnancy or any other factors that compromises the female host. Under these conditions, the cells of bacteria decreases and the fungi dominates causing curd-like, yellow-white discharge from the vaginal area. *Candida* can be transmitted to males during intercourse. In males, it causes balanitis. Infections are found on glan penis. It occurs primarily in uncircumicised males. In early stages, vesicles are formed on penis, which lead to severe itching and burning.

PROTOZOAL DISEASES

AMEBIASIS is also known as amebic dysentery. Causal organism is *Entameba histolytica.* It is trasmitted through water contaminated by mature cyst. When these cysts, reach lower region of small intestine, excystation takes place. The metacyst thus formed divides rapidly to give rise to eight small trophozoites. The latter has finger like pseudopodia, it multiplies and moves to large intestine and feeds on erythrocytes and mciroflora of intestine. The proteolytic enzymes secreted by trophozoites damages epithelial lining of intestine. The pathogen reaches to lungs and other organs through blood stream. The symptom of amebic dysentery is fulminating dysentery, exhaustive diarrhea with frequent passing of stools mixed with mucous and blood. Intestinal lesions are characterized by minute points of entry in the mucosa, and extensive enlargement of the lesion after penetration into the submucosa. Symptoms also appear on other organs like liver, lungs, spleen and brain. Liver suffers in the form of amebic hepatitis and amebic liver abscess. Abscess also occurs in lungs, brain. Amebic vaginits and amebic ulcer of penis are also observed.

GIARDIASIS is a common intestinal disease. Its causal organism is *Giardia lamblia.* It is transmitted through contaminated water and food. Cyst entering the duodenum gets excystation and trophozites are formed. The trophozoites attach to the mucosa of small intestine. Trophozoites are 10-20 μ in length and 6-15 μ in width. The sucking disk is of trophozoites help in attachment. The trophozoites feed on mucosa and interfere with the nutrition absorption. Trophozoites undergo encystment to complete life cycle. Acute giardiasis is characterized by severe diarrhea, epigastric pain, cramps, voluminous flatulence, and anorexia. In chronic giardiasis, there is intermittent diarrhea, with periodic appearance and remission of symptoms.

MALARIA is caused by *Plasmodium.* Four species are reported to cause malaria, viz. *P.falciparum, P.malariae, P.vivax,* and *P.ovale* (this species is not reported in India). It is transmitted by female *Anopheles* mosquito. Life cycle completes in two hosts. The asexual cycle (schizogony) takes place in human beings and other vertebrates, while the sexual cycle (sporogony) takes place in mosquito. When infected female *Anopheles* mosquito bites a healthy human being, it releases with its saliva, a number of sporozoites. The latter reach upto the liver cells through blood stream where they divide asexually by schizogony and forms merozoites. These stages are called preerythrocytic cycle. The merozytes may infect other liver cells or penetrate in the erythrocytes. First, it takes signet ring shape and then forms amoeboid shape. The latter is called trophozoits, it has a single nucleus and feeds on erthrocytes. The trophozoite divides asexually to produce schizont that has 6 to 24 nuclei. The schizont divides and produces mononucleated merozoites. Erythrocytes lyse releasing merozoites into the blood stream, which is capable to attack other erythrocyte cells. These stages are called erythrocytic cycle. The toxin (hemozoin) released by merozoites and the cell debris of lysed erythrocytes results in chills and fever. In the end of erythrocytic cycle, the merozoites differentiate into macrogametocytes and microgametocytes. They are not released from erythrocyte cells. They enter in mosquito when it bites infected person and develop into female and male gametocytes, respectively. In mosquito's gut, the infected erythrocytes lyse and gametocytes fuse to form a diploid zygote called the ookinete. The ookinete penetrates the mosquito's stomach wall, round out and develops into a stationary form known as an oocyst. The oocyst undergoes meiosis and forms sporozoites. This process is known as sporogony. The sporozoites migrate to the salivary glands of the mosquito. Now this mosquito is able to infect another healthy person. The malarial paroxysms are shaking chills, and then burning fever with sweating. Anemia results due to loss of erythrocytes. Spleenomegaly and hepatomegaly are also observed.

LEISHMANIASIS Diseases caused by different species of *Leishmania* are collectively know as Leishmaniasis. *Leishmania* is a flagellate protozoan and its pathogenic species are *L. braziliensis, L. tropica* and *L. donovoni.* The latter is reported in India. The canines and rodents are primary reservoirs and *Phlebotomus* (Sandfly) is the vector of this pathogen. The fly while sucking on human's blood, introduces *promastigotes* (pear shaped to long slender shaped flagellate bodies) into the wounds. Within the skin, the promastigotes are engulfed by macrophages; multiply by binary fission and form *amastigotes* (small aflagellar forms, found in reticuloendothelial system). The host cell is destroyed and amastigotes are engulfed by other macrophages. *L. donovoni* is viscerotropic and invades reticuloendothelial system. It causes Kala-azar (*visceral leishmaniasis*) disease. Spleen, liver, bone marrow, intestinal mucosa are affected. The pathogen may also be found in endothelial cells of kidneys, suprarenal capsules, lungs, meninges, cerebrospinal fluid and also in the macrophages of intestinal wall. The symptoms of disease are irregular fever, loss of weight, skin dries and hairs become brittle, spleen enlarges, emaciation and anemia, epistaxis, lung infection, etc. *L. braziliensis* is found in the forest regions of tropical America. It is dermotropic, causing mucocutaneous leishmainasis. *L. tropica,* and *L.mexicana* cause cutaneous leishmaniasis.

TRYPANOSOMIASIS is a disease of Africa. *Trypanosoma brucei gambiense* causal organism of Gambian trypanosomiasis, and is fond in the rain forests of west and central Africa. It is transmitted by a vector tsetse fly, *Glossina palpalis. Trypanosoma brucei rhodensiense* causal organism of African trypanosomiasis, and is found in upland savannas of east Africa. It is transmitted by vector *G.morsitans,* and *G.pallidipes.* Fly sucks the pathogen from the infected person while feeding on blood. The Trypanosomes pass into the insect's intestine and other regions, where they develop into metacyclic trypanosomes. These are infective units and are able to cause disease when the fly bites a healthy person. The disease is commonly known as sleeping sickness. The central nervous system is involved. There is enlargement of lymph nodes. Initially patient feels fever and headache. The meningoencephalitis manifestation results in sleeping sickness. *Trypanosoma cruzi* causes American trypanosomiasis commonly known as Chagas' disease. Triatomid bug is the vector of this pathogen. The insect while sucking blood from healthy person discharges some feces and it is this, which carries pathogen. The trypanosomes enter the bloodstream and spread in the body. It invades the liver, spleen, lymph nodes and central nervous system. The pathogen multiplies and destroys the infected cells. The symptoms of disease are characterized by cardiac arrhythmia and neurological problems (psychic change and spastic paralysis).

TOXOPLASMOSIS is caused by *Toxoplasma gondii* and is transmitted through cysts contaminated food. It enters body by mouth or nose. Cats, dogs, cattle, sheep, and humans are common hosts. In man, the pathogen completes its exoenteric cycle. In man, pathogen forms merozoites by asexual reproduction. The latter enters blood and lymphatics. Pathogen can also be transmitted through placenta resulting in congenital toxoplasmosis. In new borne, the symptoms are chorioretinitis (occular toxoplamsosis), blindness, epilepsy, and mental retardation. Toxoplasmosis is fatal especially in AIDS patients.

TRICHOMONISIS causal organism is *Trichomonas vaginalis.* It is a sexually transmitted disease. The pathogen is found in genitourinary system. In females, it is found in vagina and in male, it is present in urethia. *T.vaginalis* has a pear shaped structure with 10 to 30 x 5-10 μ. The pathogen causes degeneration and desquamation of the vaginal epithelium. The body accumulates leukocytes at the site of the infection. The vaginal secretion containing leukocytes is greenish or yellow and covers the mucosa down to the urethral orifice, vestibular glands and clitoris. The purulent discharge has disagreeable odor. There is itching at the site of infection. No symptoms appear in males. The infection can be acquired by newborn babies by way of an infected birth canal.

CHAPTER - 21

IMMUNOLOGY

We live in a potentially hostile world filled with a bewildering array of infectious agents of diverse shape, size, composition and subversive character, which would very happily use us as rich sanctuaries for propagating their "selfish genes", had we not also developed a series of defense mechanisms at least their equal in effectiveness and ingenuity (except in the case of many parasitic infections where the situation is best described as an uneasy and often unsatisfactory truce) ? It is these defence mechanisms which can establish a state of immunity against infection (Latin, immunitas, freedom from) and whose operation provides the basis for the delightful subject called "Immunology".

Immunology is science dealing with study of immunity to infection. Immunity (L *Immunis* not serving or exempt) is ability of an organism to resist infection (or injury) caused by pathogens or its products. Immunity may be innate or acquired.

Innate or native immunity is the ability to resist infections, which an individual possesses by virtue of genetic and constitutional make-up. It may be specific or non-specific. Specific immunity is for a particular pathogen. Non-specific immunity indicates a degree of resistance to infections in general. Innate immunity is for species, racial and individual.

Acquired immunity is the resistance that an individual acquired during life. It is of active and passive types. Active immunity is the resistance developed by an individual as result of an antigenic stimulus. It may be natural or artificial. In passive immunity, there is no antigenic stimulus. Preformed antibodies (readymade) are administered. It is also natural or artificial.

Vaccines are preparations of live or killed microorganisms, or their products used for immunisation. Vaccines used for human interest are following:

1. Bacterial vaccines

Live- BCG for tuberculosis. Killed- TAB for enteric fever.

2. Viral vaccines

Live- Oral poliomyelitis. Killed- Salk vaccine for poliomyelitis.

3. Bacterial products

Toxoids for diptheria and tetanus.

Toxoids are inactivated toxin that has lost its toxigenicity but retains its immunologenicity and thus can be used for active immunization.

DEFENCE MECANISM

Human body has the tendency to resist a pathogen and thus controlling the disease to spread. There are three lines of defense (i) Defenses on the body's surface (ii) phagocytic cells and (iii) specific immune mechanism.

Body surface defenses

In the first step of a disease, a pathogen attaches on the body surface of the host. Body surface applies mechanical, chemical and microbial mechanisms to resist the pathogens to establish on the host surface. Skin and its mucous covering on certain organs act as a mechanical barrier for pathogens invasions. However, even though the structure of the skin itself undoubtedly provides a deal of protection, considerably more important are the fatty acids secreted by the sebaceous glands in the skin. This secretion contains both saturated and unsaturated fatty acids, which are lethal for many bacteria and fungi. A stiking example of this type of resistance to infection is seen in the case of the fungus causing ringworm of the scalp. This infection is very difficult to cure in young children, but after puberty it disappears without treatment, presumably as a result of a change in the amount and kinds of fatty acids secreted by the sebaceous glands. A weakened skin is susceptible for disease as is common in malnutrition. The nasal hairs and ciliated mucosa covering of respiratory tract acts as a barrier for air borne respiratory diseases. Coughing and sneezing are other mechanical reflexes, which throw out the pathogens. Lungs are provided with epiglottis, an additional mechanical barrier. Body hairs prevent the pathogens to settle on the body surface. Pathogens of eyes and urinary tract are flushed by flow of tears and urine. Chemical defenses includes enzymes, fatty acids etc. Mucous secretions, sweat and saliva contain an enzyme known as lysozyme. It destroys the cell walls of many Gram-positive bacteria. Fatty acids in sweat and earwax also act as a chemical barrier. High concentrations of hydrochloric acid in stomach destroy food pathogens. There is a resident flora of micro-organisms in the body, which compete with pathogens and thus protects body from disease. Lactobacilli are found in adult human vagina. It produces lactic acid and lowers the pH upto 4.0. This acidic pH inhibits the growth of pathogens.

In addition to these the body apparently produces a large number of substances capable of reacting with, or preventing the growth of invading microorganisms. However, most of these systems are poorly understood and descriptions are frequently based on the killing effect of whole blood or serum on some particular microorganism. One such system, which has antiviral activity, has been extensively studied. This antiviral substance has been named interferon. It is a small protein molecule secreted in exceedingly small amounts by the virus infected cells. It is, by itself not active against viruses and only after it leaves the cell in which it was produced and enters the uninfected cell can it exert its antiviral activity. There are at least 3 types of human interferon produced by different cell types:

1. Fibroblast – epithelial cells
2. Leucocytes
3. Immune T lymphocytes

One of the interesting properties of interferon is that it is not virus- specific; rather it is cell specific.

ANTIGENS

Definition: Any substance which, when introduced parenterally into the body, stimulates the production of an antibody with which it reacts specifically and in an observed manner. The term antigen is use in two senses, the first describe a molecule, which generates an immune response (also called an immunogen) and the second, a molecule, which reacts with antibodies or prime T-cells irrespective of its ability to generate them. The contact area with an antibody is called epitope and the corresponding area on an antibody a paratype. Some antigens may not induce antibodies but may lead to cell mediated immunity or immunological tolerance. There are two attributes of antigenicity : (i) immunogenicity and (ii) immunological reactivity. Immunogenicity refers to induction of an immune response while immunological reactivity means specific reaction with antibodies or sensitized cells. The specificity of antigen-antibody reaction is determined by a portion of antigen molecule called as Antigenic determinants. It is also known as epitope and corresponds to an area of about 6 to 10 amino acids. A complete antigen is able to induce antibody formation and produce a specific and observable reaction with the antibody. Haptens (Gr. *haptein-* to fasten) are substances, which are incapable of inducing antibody formation but can react specifically with antibody. Haptens may be of two types: (i) Complex haptens and (ii) Simple haptens. Complex haptens are relatively higher molecular compounds, which can precipitate with specific antibodies e.g., Wassermann antigen, polysaccharide etc. Simple haptens are low molecular weight compounds, and are non-precipitating with specific antibodies, e.g., picric chloride, para aminobenzoic acid.

PROPERTIES OF ANTIGENS

An antigen has a number of properties. These properties attribute to any substance as antigen.

1. *Size*: Antigenicity bears a relation to molecular size. Usually antigens have a minimum molecular weight of 10,000 or more. Any substance below this molecular weight may not act as antigen.
2. *Chemical nature*: Most of the naturally occurring antigens are proteins and polysaccharides. Lipids and nucleic acids are less antigenic. Not all proteins are antigenic e.g., gelatin.
3. *Foreignness:* Only antigens, which are foreign to an individual, induce an immune response. An individual does not normally give rise to

immune response against his own normal constituent antigen. The antigenicity of a substance is related to the degree of its foreignness.

4. *Susceptibility to tissue enzymes*: Only substances, which are metabolized and are susceptible to the action of tissue enzymes, behave as antigens. Polystrene latex is insusceptible to tissue enzyme and hence it is not antigenic.
5. *Antigenic specificity*: Active sites (antigenic determinants) are present at certain places in antigen molecule. The basis of antigenic specificity is stereochemical. The influence of spatial configuration of the determinant is also important. Antigenic specificity is of following types:
   a. Species specificity: Tissues of all individuals in a species contain species specific antigens. The immunological relationship parallels phylogenetic relationships. It helps in tracing of evolutionary relationship. Species specific antigens also have forensic applications in the identification of species of blood and seminal stains.
   b. Isospecificity: Isoantigens are antigens found in some but not all members of a species. A species may be grouped depending on the presence of different isoantigens in its members, e.g., Human erythrocytes antigen. On the basis of this antigen individuals can be classified into groups (blood group). HL-A (human luecocyte antigen system) is histocompaitibility antigen. It has its application in organ transplantation from one individual to other.
   c. Auto specificity: A number of self antigens, may behave as autoantigen, e.g., lens protein. The latter is confined within its capsule. When these antigens are released into the tissues (due to injury, infection or drug therapy) antibodies are produced against them.
   d. Organ specificity: They are restricted to particular (same) organs or tissues of different species. Some organs, such as the brain, kidney and lens protein of different species, share the same antigen.
   e. Heterogenetic (Heterophile) specificity: The same or closely related antigens may occur in different species, e.g., Forssman antigen found in the tissues of guineapige, cat, horse, sheep. Other heterophile antigens are responsible for some diagnostic serological reactions e.g., Weil Felix reaction in typhus fever, Paul Bunnel reaction in infectious mononucleosis, Cold agglutinin test in primary atypical pneumonia.

ANTIBODY

Definition: They are humoral substance (γ-globulin) produced in response to an antigenic stimulus. Antibodies are found in serum, lymph and other body fluids. Serum proteins are of two types: soluble albumins and insoluble globulins. Most of the antibodies are euglobulins. In 1964 under the auspices of the WHO term 'immunoglobulin' was coined for proteins of animal origin endowed with known antibody activity and for certain other related to them by chemical structure. Immunoglobulins constitute 20-25 percent of the total serum proteins. They are synthesized by plasma cells. Lymphocytes are also capable of synthesizing immunoglubulins. All antibodies are immunoglobulins, but all immunoglubulins and not antibodies. Immunoglobulin refers to structural and chemical concept while antibody is a biological and functional concept. Based on physico-chemical and antigenic differences, immunoglobulins are classified into five classes: IgG, IgA, IgM, IgD and IgE.

STRUCTURE OF IMMUNOGLOBULIN

Immunoglobulins are glycoproteins. There are two pairs of polypeptide chains in each molecule (Fig. 1). These chains are of different sizes and are held together by disulphide (S-S) bonds. Four to twelve of these bonds link the four chains together. The remaining disulphide bonds are distributed within the chains, stabilizing their respective conformation. The smaller chains are called light (L) chains and larger ones are known as heavy (H) chains. The molecular weight of L chains and H chains is approximately 25,000 and 50,000, respectively. The L chains are similar in all classes of immunoglobulin. They are of two types *kappa* (κ) and *lambda* (λ). These chains are named after the investigators Korngold and Lapri. Only one type of this may be present in each molecule i.e., either *kappa* or *lambda* but never both together. In an immunoglobulin molecule carbohydrate forms about 3% -12% of the molecular weight. Oligosaccharide units are attached to one or more positions in the C region of all Ig heavy chains. The antigens combine at the aminoterminus (N terminal) portion of immunoglobulin. It is composed of both L and H chains. The amino acid sequence at this portion shows variability, which determines the specificity. Due to this variability it is also known as "variable region". The other half of immunoglobulin, i.e., carboxyterminus (C terminal) is also known as "constant region". Of 214 amino acid residues that make up L chains, 107 constitute carboxyterminal half of the molecule. In 1959 Portor showed that digestion of rabbit IgG with papain split the molecule into fragments with loss of only about 10% of the protein as small dialyzable peptides. The immunoglobulin molecule when digested by papain in the presence of cysteine splits into two fractions the Fc and Fab fragments. *Fc fragment* (Fc for crystallisable) is an insoluble fraction, which crystallized in the cold.

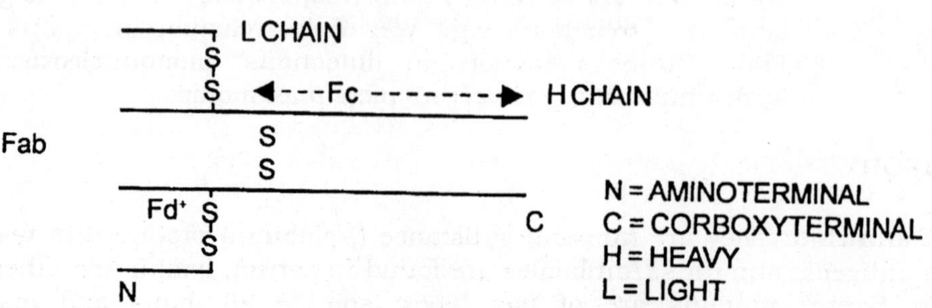

Fig. 1. Structure of immunoglobin

It is composed of carboxyterminal portion of H chain. It determines biological properties of immunoglobulins molecule. *Fab fragment* (Fab for antigen-binding) is a soluble fragment and does not crystallize. It is amino-terminal half of heavy chain and one light chain. A portion of H chain of Fab fragment is called *Fd fragment*. Its function is not known.

Immunoglobulin classes

Human sera contain five classes of immunogloublin. They are named as IgG, IgA, IgM, IgD and IgE.

(1) **IgG:** It contains γ (gamma) heavy chain and constitutes about 80 per cent of serum immunoglobulin. It has molecular weight of 1,50,000 and sedimentation constant is 7s. It has a half life of 23 days. It is distributed equally between intravascular and extravascular compartments. It is the only maternal immunoglobulin that is normally transported across the placenta and provides natural passive immunity in the new born. It produces passive cutaneous anaphylaxis and participates in immunological reactions like precipitation, complement fixation, neutralization of toxin and viruses. In IgG molecule, heavy chains are about twice as long as the L chains (Fig. 2). The first completed H-chain sequence revealed linear repeats of similar segments: each segment had about 110 amino acids with an S-S bond that established a loop of approximately 60 residues. Moreover, similarities in sequence suggested that each of these domanins has an approximately similar shape. Six domains are recognized in IgG molecules: two in L chains and four in H chains. Four classes of IgG have been recognized viz. IgG1, IgG2, IgG3 and IgG4, each possessing a distinct type of gamma chain identifiable with specific antisera.

(2) **IgA:** It contains α (alpha) heavy chain and constitutes about 10-13 percent of serum immunoglobulin. It has molecular weight of 160,000 with sedimentation constant of 7s. It has a half-life of 6-8 days. Cells that produce IgA are particularly abundant around the mucin-secreting glands of the intestinal, respiratory and genitourinary mucosa. It is a major immunoglobulin in colostrums, saliva and tears. IgA found in secretions contain additional structure unit called transport (T) or secretory (S) piece. Two classes of IgA have been recognized viz. IgA1 and IgA2.

(3) **IgM:** It contains μ (mu) heavy chain and constitutes about 5-8 percent of serum immunoglobulin. It has molecular weight of 9,00,000 to 10,00,000 with sedimentation constant of 19s. It has half-life of about five days. It is more efficient in agglutination, cytolytic and cytotoxic reaction. It is largely confined to intravascular space. IgM deficiency is often associated with septicaemia. The human IgM molecule is a pentamer made up of five four-chain subunits, each with a pair of heavy chains and a pair of either kappa or lambda chains. In addition to the ten L and H chains per

molecule, there is one extra chain, known as J chain (Fig. 3). The latter is bonded to heavy chain by disulphide bond.

(4) **IgD:** It contains δ (delta) heavy chain and is present in concentration of 0.03 mg per ml. It has molecular weight of 180,000 and sedimentation constant of 7s. It has half life of 3 days. Mostly it is intravascular. It occurs on the surface of unstimulated B-lymphocytes and serves as recognition receptors for antigens. Two classes of IgD have been recognized IgD1 and IgD2.

(5) **IgE :** It contains ε (epsilon) heavy chain and is present it is present in concentration of 0.00004 mg per ml. It has molecular weight of 1,90,000 and sedimentation constant of 8s. It has half life of 2 days. It is heat labile and becomes inactivated at 56°C in one hour. It mediates Prausnitz-Kustner reaction. It does not pass through placenta or fix complement. It is reaginic antibody responsible for immediate hypersensitive. It has been suggested that it may lay some as yet unidentified role in the defense against intestinal parasitic infection.

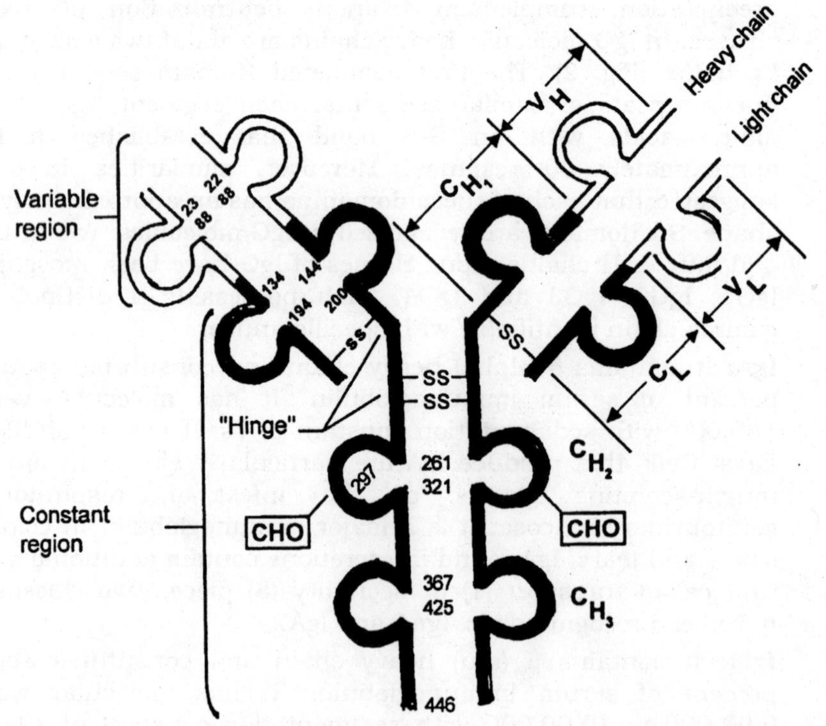

Fig. 2. Structure of IgG showing domains : L - Light chain, H - Heavy chain

ANTIGEN–ANTIBODY REACTION

Antigen and antibody reaction is a fundamental reaction of immunology. It is highly specific and in an observable manner. Antigen-antibody reactions *in vitro* are known as serological reactions. In the simplest reaction a small univalent antigen binds reversibly to a specific site on the antibody. At low concentrations of antigen only a small proportion of the antibody combining sites are occupied by antigen molecules. As the antigen concentration increases the number of occupied antibody sites rises until all are filled. At saturation the number of univalent antigen molecules bound per antibody molecule is the antibody valence. In the primary stage antigen and antibody react without any visible effects. This reaction is reversible and is effected by the weaker intermolecular forces such as VanderWaal's forces, ionic bonds and hydrogen bonding. During secondary reaction, precipitation, agglutination, lysis of cells, killing of live antigens, neutralization of motile organisms and enhancement of phagocytosis takes place. The antibody causing agglutination is known as agglutinin and antigen is known as agglutinogen, while that causing precipitation is known as precipitin and its antigen as precipitinogen.

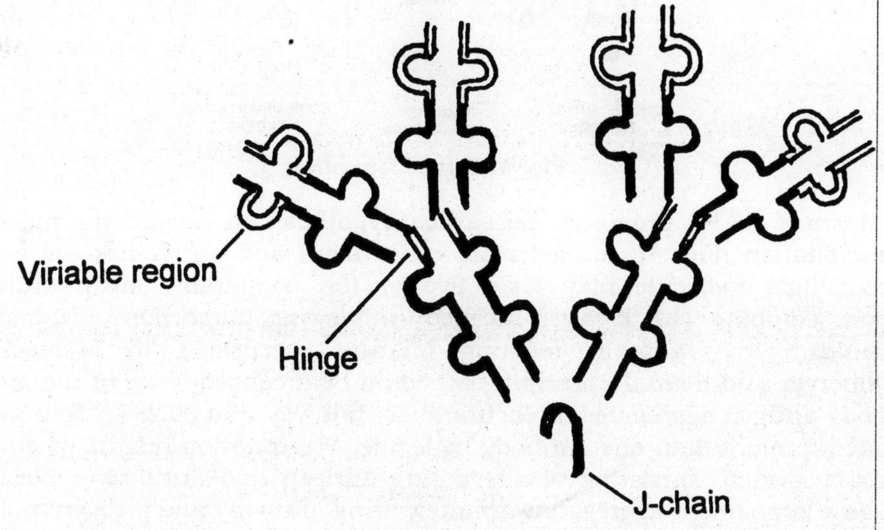

Fig. 3.

Antigen-antibody reactions have the following general features —

The reaction is highly specific. Antibodies react more effectively with the antigen that stimulated their formation i.e., with their homologous antigens. Other antigens with which the antibody may bind are known as heterologous antigens.

Entire molecule reacts, and not its fragment.

There is no denaturation of antigen or antibody during the reactions.

Combination occurs at surface.

The combination is firm but reversible. The firmness of combination is influenced by affinity (intensity of attraction between antigen and antibody molecules) or avidity (strength of the bond after the formation of antigen antibody complex).

Both antigen and antibody participate in the reaction.

Antigen and antibody may combine in varying proportions.

PRECIPITATION REACTION

When a soluble antigen combines with its antibody in presence of it antibody in the presence of electrolytes (NaCl) at a suitable temperature and pH the antigen-antibody complex forms an insoluble precipitate.

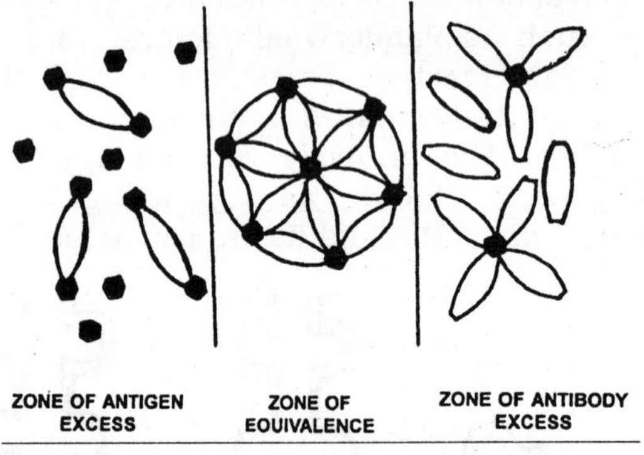

Fig. 4. Lattice formation

Marrack (1934) proposed the Lattice hypothesis to explain the mechanism of precipitation (Fig. 4). Antigens are multivalent and antibodies are generally bivalent (IgM molecule may have five or ten combining sites). Multivalent antigens combine with bivalent antibody in varying proportions, depending on antigen-antibody ratio in reacting mixture. According to Marrack, and Heidelberger and Kendall, precipitation could be a consequence of the growth of antibody-antigen aggregates. It occurs in such a way that each antigen molecule is linked to more than one antibody molecule. Precipitation results when a large lattice is formed consisting of alternating antigen and antibody molecules. It may be a large linear aggregate with alternating antibody and antigen molecules. Other mechanism could arise when some bivalent antibody molecules with a high affinity for antigen preferentially combine with two determinant groups of a single antigen particle forming cyclic complexes. Karush named this type of multivalent binding as monogamous bivalency. Precipitin reaction is possible only in the zone of equivalence. In zone of antigen or antibody excess, the lattice does not enlarge as valencies of the antibody and the antigen.

The precipitation test may be carried out either as a qualitative or as a quantitative test.

AGGLUTINATION REACTION

The agglutination reaction is a very sensitive serologse technique for the detection and approximate quantitation of serum antibodies. It is widely used in clinical laboratories for the tentative diagnosis of infectious diseases, for blood grouping and for non infectious diseases such as rheumatoid arthritis and lupus erythematosus.

When a particulate antigen is mixed with its antibody in the presence of electrolytes at a suitable temperature and pH, the particles the particles are clumped or agglutinated. Bacterial and other cells in suspension are usually clumped (agglutinated) when mixed with their antisera. In agglutination reactions the antigen groups are components of the cell wall or cell surface membrane. When a mixture of readily distinguishable particles is added to a mixture of the respective antisera, each clump consists of cells of one or the other type. Thus each cell (antigen)-antibody system agglutinates independently of the others in the same mixture. Agglutination is more sensitive than precipitation. Agglutination occurs optimally when antigens and antibodies react in equivalent proportions. Incomplete or monovalent antibodies do not cause agglutination. However, they combine with the antigen.

Serology deals with the study of antigen antibody reaction *in vitro*. It includes identification and quantification of antigen or antibody using its known counterpart. The antigen-antibody reactions are termed based on the observable effect of the reaction or the technique used. Following are some technique of serological reactions.

COMPLEMENT FIXATION TEST (CFT)

This test is considerably more difficult to perform but is used frequently because of its great sensitivity; it can detect very small amounts of antibody. The basis of this test is to measure the disappearance of a complex substance collectively called complement, a material present in all normal human and animal sera. The secret to understanding this test is to remember that whether or not the particular reaction requires complement, if present, it will be incorporated and used up during the antigen-antibody reaction.

The term complement means to a system of factors occurring in normal serum that are activated characteristically by antigen-antibody interaction. In the presence of the appropriate antibodies, complement lyses erythrocytes, kills and in some cases lyses bacteria, immobilizes motile organisms, promotes phagocytosis and immune adherence and contribute to tissue damage in certain types of hypersensitivity.

NEUTRALISATION TESTS

Specific antibodies are able to neutralize the biological effects of viruses, toxins and enzymes. Neutralisation of viruses by their antibodies can be demonstrated in various systems. Viruses when mixed with immune serum lose their capacity to infect host, e.g., vaccinia, influenza and poliomyelitis. Bacterial exotoxins are good antigen and induce the formation of neutralizing antibody (antitoxin). These antibodies protect from diseases like diphtheria and tetanus.

OPSONIZATION

It is evident that immunoglobulin antibody can combine with homologous bacterial surface antigens, the association of which in turn make the Ag-Ab complexes more susceptible to phagocytosis. The immunoglobulins involved in this are known as "opsonising antibodies" and the process involved is called "opsonization". The term 'opsonin' was coined by Wright (1903) to a heat labile substance present in fresh normal sera, which facilitated phagocytosis. Opsonization is serological reaction, which sensitizes bacteria for phagocytosis. The opsonic index is defined as ratio of phagocytic activity of patient's blood having bacterium, to the phagocytic activity of blood from normal individual. It is measured by incubating fresh citrated blood with bacterial suspensions at 37 C for 15 minutes and estimating the average number of phagocytosed bacteria per polymorphonuclear leucocyte (phagocytic index) from stained blood films.

IMMUNOFLUORESCENCE

This method was introduced by Coons in 1942. It is widely used to identify and localize antigens in cells and on cell surfaces to identify bacteria. Fluorescence is the property of absorbing light rays of one particular wavelength and emitting rays with a different wavelength. Antibodies are intensely fluorescent and retain their specific reactivity. Fluorescent dyes (like fluorescein isothiocyanate) can be conjugated to antibodies. Such labeled antibodies can be used to locate and identify antigens in tissues. Fluorescein-labelled antibodies can be used in both direct and indirect reaction sequences to detect and localize antigens.

RADIOIMMUNOASSAY (RIA)

Radioimmunoassay was described by Berson and Yallow (1959). In this method radioisotopes are conjugated to antigens and antibodies. The term binder-ligand assay has been used for these reactions. The antigen whose concentration is to be determined is termed as analyte or ligand, while the antibody is called as binder. RIA is a competitive binding assay. A fixed amount of antibody and radiolabelled antigen react in the presence of unlabelled antigen. After the reaction, the antigen is separated into the free and bound fractions and their radioactive counts measured.

ENZYME LINKED IMMUNOSORBENTASSAY(ELISA)

These assay are based on the use of covalently linked enzyme-antibody complexes (E-Ab). Substrates that yield easily detected coloured or fluorescent products measure the enzyme and thereby its linked antibody. This is discussed in details in chapter 9.

SEROLOGICAL TESTS (Application of Agglutination Reactions)

Agglutination reactions are applied for diagnosis of certain diseases, identification of bacteria and certain antigens.

THE WIDAL TEST

This test is performed for the diagnosis of enteric fever or other salmonella infection. The bacterial suspensions are incubated with serial dilutions of patient's serum to find out the titre of agglutinin in the latter. Conventionally H and O suspension of *Salmonella typhii* and paratyphi are used. The titer of an antiserum is a measure of the number of antibody units, per unit volume of original serum. Double dilution method or drop method is used.

BRUCELLA AGGLUTINATION TEST

This test is performed using suspension of *Brucella abortus*. The procedure for this test is exactly similar to Widal test for TO antigen. Demonstration of rising titer of 1 : 100 or more is the diagnostic value of test. A rising titre is always more significant. Prozone phenomenon is very frequently observed in this test. It is due to blocking action of non agglutinating IgG antibodies. In chronic brucellosis the titre may fall while the patient is still suffering from the disease.

WEIL-FELIX REACTION

This test was described by Weil-Felix. It is used for the diagnosis of typhus fever. This test is based on the sharing of an alkali stable carbohydrate antigen by some *rickettsiae* and by certain strain of *Proteus*. *Rickettsial* species causing epidemic, endemic and scrub typhus or spotted fever share a carbohydrate hapten with somatic constituent of non-flagellate strains of *Proteus*. The procedure of the test is similar to Widal test.

STREPTOCOCCUS M.G. AGGLUTINATION

This test is applicable for diagnosis of primary atypical pneumonia caused by *Mycoplasma pneumoniae*. The unheated serum of patients agglutinate *Streptococcus* MG strain. The antigen is obtained from the heat killed saline suspension prepared from 48 hours culture. The test is performed by serial dilutions of patient's serum. 0.5 ml of serum is incubated with 0.5 ml of antigen at 37°C for two hours. After this it is refrigerated overnight. The reading is collected as in Widal test.

PAUL BUNNEL TEST

This test is performed for the diagnosis of infectious mononucleosis. Patient's serum is inactivated at 55°C for 20 minutes and is diluted. To these, dilutions 1 ml of washed sheep's red blood cells (one percent) is added and incubated for 4 hours at 37°C. After incubation, the tubes are examined for any evidence of clumping of red blood cells. Titer of 1:100 confirms infectious mononucleosis.

COLD HEMAGGLUTINATION TEST

This test is performed for the diagnosis of primary atypical pneumonia caused by mycoplasma. The patients of primary atypical pneumonia contain cold agglutinin capable of agglutinating RBC at low temperature. Serial dilutions of patient's serum is prepared and mixed with equal volume of 0.2% suspension of group 'O' RBC (of human). The tubes are incubated at 4°C for overnight. Reading is noted while the tubes are cool. Clumping of RBC is labelled as positive reaction.

IMMUNE RESPONSE

The specific reactivity induced in host by antigenic stimulus is known as immune response. Immune response is characterized by specificity, memory, and the acquired ability to detect foreign substance. It is of two types (1) Humoral and (2) Cellular.

HUMORAL IMMUNE RESPONSE

Humoral immune response is antibody mediated. Antibody production requires, three steps:

(1) The afferent limb-Entry and distribution of antigen in tissue and its contact with immunocompetent cells;

(2) Central limb-Processing of antigen, and control of antibody forming process;

(3) Efferent limb-Secretion of antibody, its distribution in tissue, body fluid and manifestation of its effects.

The antibody production in tissues shows lag phase, log phase, plateau or steady state and the phase of decline. As the antigen is introduced in tissue, first the antibody production shows qualitative followed by quantitative response. The former is known as primary response, while the latter is secondary response.

Primary and Secondary Response

Primary response is slow, sluggish and short lived. It has a long lag phase and low titre of antibody. The duration of lag phase vary with the nature of the antigen. IgM class of immunoglobulin predominantly forms antibody in this phase.

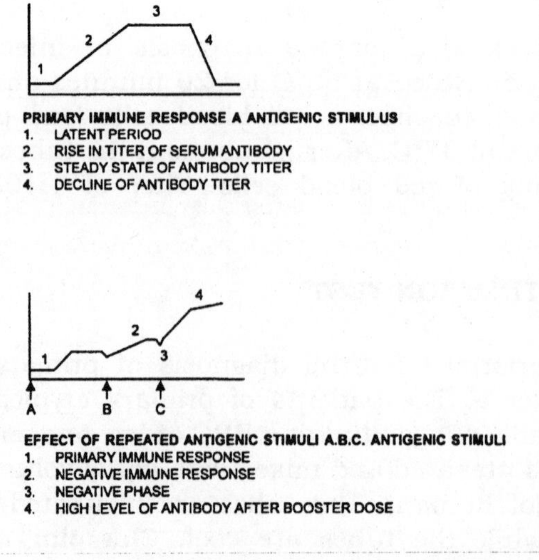

Fig. 5. Primary and Secondary immune response.

Secondary response is prompt, powerful and prolonged. Lag phase is almost negligible. The antibody formed is predominantly IgG (Fig. 5).

The physical and chemical nature of antigen, its dose and route of entry in the body are some factors, which matter in processing the antigen. The antigen with specific antibody if introduced in circulatory system, there is a fall in the level of circulating antibody. This fall is due to combination of antigen and antibody. It is known as negative phase. Removal of antigen much depends whether it is particulate or soluble. The particulate is removed in two phases. In the first phase, the antigen is engulfed by the phagocytic cells. It is broken down and eliminated. It is nonimmume phase. The immune phase starts with the appearance of the specific antibody. The antigen-antibody complex formed is phagocytosed. Thus, antigen is removed from circulation. If the antigen is soluble, it is removed in three phases: (1) equilibration (diffusion of the antigen to the extravascular spaces). (2) Metabolism (the level of antigen falls due to catabolic decay) and (3) immune elimination (there is rapid elimination of the antigen with the formation of antigen-antibody complexes).

Antibody synthesizing cells

Antibody is formed by B lymphocytes of plasma. The immunocompetant B lymphocytes carry surface receptors. These receptors are IgG or IgM in nature. They have the power to detect the foreign molecule, which is antigenic in nature. In antigenic stimulus, the cells proliferate and undergo blast transformation. Another class of lymphocytes, are T lymphocytes. Sensitized T lymphocytes interact with foreign cell to bring about the destruction of those cells. Antibody formation by B cells is subjected to the regulatory effect of T cells. Both B (refers to bursa-dependent lymphocytes) and T (Thymus dependent or thymocytes) lymphocytes originate from bone stem cells. They are differentiated during maturation.

Cellular Immune Response

It is also referred as cell mediated immunity (CMI). It is the specific immune response and does not involve antibodies. Induction of CMI consists of specifically sensitizing T lymphocytes against the antigen. CMI plays role in:

(1) Delayed hypersensitivity.
(2) Immunity in disease caused by intracellular parasites.
(3) In pathogenesis of certain autoimmune disease and
(4) Transplantation immunity and graft-versus-host reaction.

LYMPHOKINES

Lymphokines are biologically active substance (regulatory proteins) released by activated T lymphocytes. Their main function appears to be regulation of the immune response and the growth and functions of cells of the reticulo-endothelial system. Their nomenclature is based on the biological activities they produce. The different lymphokines are as following —

Affecting macrophages

Migration inhibiting factor (MIF): inhibits the migration of normal macrophages.

Macrophage activating factor (MAF) Increases macrophages phagocytic capacity.

Macrophage aggregation factor (MAF) causes aggregation of macrophages.

Macrophage chemotactic factor (MCF) brings about migration of macrophages through millipore membranes.

Macrophage resistance factor (MRF) makes macrophages nonspecific resistant to infection.

Macrophages anti-spreading factor inhibits spreading of macrophages on glass surface.

Crytophilic factor (CF) enables microphages reactivity specific with antigen.

Affecting lymphocytes

Blastogenic or Mitogenic factor (BF or MF) induces blast cell transformation in lymphocytes.

Potentiating factor (PF) may result in transformation in mixed lymphocytes culture or antigen stimulated culture.

Cell co-operation or Helper factor (CC or HF) Increases number of antibody forming cells.

There are also T cell growth factor and B cell growth factor, stimulating the growth of T and B cells, respectively.

Affecting granulocytes

Inhibition factor (IF) inhibits migration of lymphocytes.

Chemotactic factor (CF) responsible for migration of granulocytes along a gradient.

Colony stimulating factor (CSF) stimulates growth of granulocytes and macrophages.

Affecting cultured cells

Lymphotoxin cytooxic for certain cultured cells.

Proliferation inhibiting factor or colony inhibiting factor (PIF or CLIF) inhibits proliferation of cultured cells without lysing them.

Gamma interferon protecting against virus infection.

Producing effects *in vitro*

Skin reactive factor (SRF) is capable of inducing skin lesion of delayed hypersensitivity. It is due to combination of several factors.

Inflammatory factor (IF) causes increased vascular permeability.

Transfer factor (TF) mediates passive transfer of CMI.

INTERLEUKINS

Interleukins are substances produced by leucocytes (leukin) that regulate other cells. The 6th International Conference of Immunology (1986) has assigned numbers to interleukins after the complete knowledge of amino acid sequence and characters. At present six interleukins are studied.

Interleukin-1: It was previously named as a leucocyte activating factor (LAF) and B cell activating factor (BAF). It was numbered in 1979. Its production is induced by antigens, toxins, injury and inflammatory processes. IL-1 activates T-cells and stimulates B cell proliferation and antibody synthesis. Activation of T cell promotes synthesis of lymphokines. Besides, mediating inflammatory activity, it provides nonspecific resistance for general infection.

Interleukin-2: It is a glucoprotein (MW 15,000) and a powerful modulator of immune response. It causes production of T helper cell, T cytotoxic and NK cells. IL-2 is liable for adverse reactions due to increased capillary permeability, leading to pulmonary and systemic oedema.

Interleukin-3: It is growth factor for stem cells of bone marrow.

Interleukin-4: It was previously known as B cell stimulating factor-1 (BSF-1). It activates B cells and act as B cell differentiation factor. It is also a growth factor for T-cells and mast cells.

Interleukin-5: previously known as B cell growth factor II. It may induce maturation of eosinophils and proliferation of activated B cells.

Interleukin-6: previously known as B cell stimulating factor-2. It is produced by stimulated T cells, fibroblasts and macrophages. It may induce immunoglobulin synthesis by active production of autoimmune processes.

THEORIES OF IMMUNE RESPONSE

Theories explaining Immune response are classified into two heads.

Instructive- an immunocompetent cell is capable of synthesizing antibodies of any specificity. The antigen encounters an immounocompetent cell and instructs it to produce the complementary antibody. The template of antigen must be present for the synthesis of antibody.

Selective- immunocompetent cells have only a restricted immunological range. Antibodies are produced in the similar way as other proteins are synthesized.

Side chain theory Ehrlich (1900) opined that cells have surface receptors. These receptors are capable of reacting with substances having complementary 'side chain'.

Direct template theory: proposed by Breinl and Haurowitz (1930), Alexander (1931) and Mudd (1932). The antigen enters antibody forming cells and serves as a template against which antibody molecules are synthesized. Thus, the antibody molecules have combining sites complementary to antigen.

Indirect template theory: proposed by Burnet and Fenner (1949). The antigen enters in the antibody producing cell. It produces a heritable change so that a 'genocopy' of the antigenic determinant was incorporated in its genome and transmitted to the progeny cells.

Natural selection theory: proposed by Jerne (1955). Approximately a million imunoglobulin molecules are present in an embryo. They are considered as "natural antibodies". When antigen is introduced in the body, it selects its antibody. These antibodies are complementary fit to antigen. There is stimulation in synthesis of same kind of antibody.

Clonal selection theory: proposed by Burnet (1957). During immunological development, as result of somatic mutation, cells capable of reacting with different antigens were formed. Clones of cells that had immunological reactivity with self-antigens were eliminated during embryonic life. These are forbidden clones. In the later part of life, these clones by somatic mutation could lead to autoimmune processes. When antigens are introduced into body, each immunocompetent cell may react with one or more antigens. The immunocompetent cell is capable to recognize and combine with its specific antigen. When such cells comes in contact with its specific antigen, there is cellular proliferation and clones are formed.

FACTORS INFLUENCING ANTIBODY PRODUCTION

(1) **Age:** Embryo is immonologically immature. An infant gains ability to produce antibodies with the development and differentiation of lymphoid organs. Till then, child depends on maternal antibody.

(2) **Nutritional status:** Malnutrition (deficiencies of some amino acids, protein and vitamin consumption) leads to suppression of immune response.

(3) **Genetic factor:** The *Ir* gene controls the property for responding (responder) or not responding (non-responder) to a particular antigen.

(4) **Size and number of doses:** An antigen is effective only above a minimum critical dose. The massive antigen stimulus appears to inhibit the response. This is known as immunological paralysis.

(5) **Route of administration:** Antigen administered paernterally are responded immediately than administered by oral or nasal routes.

(6) **Multiple antigen:** When two or more antigens are administered at a time, it shows various effects: (1) antibody production for each antigen as if they are administered separately. (2) Production of one of the antibody is enhanced, or (3) response may be diminished.

(7) **Adjuvants:** Any substance that enhances the immunogenecity of an antigen is known as adjuvant. The adjuvants commonly used in human vaccine are aluminium hydroxide, phosphate, endotoxin and mineral oils.

(8) **Immunorepressive agent:** inhibits the immune response. X-ray irradiation, radiomimetic drugs, corticosteroids, antimetabolites and other cytotoxic chemicals and antilymphocytic serum act as immunorepressive agents.

HYPERSENSITIVITY

Hypersensitivity and allergy are confusing and sometimes used as synonyms. Hypersensitivity is an immunological state of an individual to particular antigens, interactions of which with specific antibodies or sensitized cells lead to harmful or occasionally fatal results. Allergy is altered reactivity on second contact with antigen, now usually refers to a type 1 hypersensitivity reaction (atopy).

Chase classified hypersensitivity into immediate and delayed. This classification is based on time required for sensitized host to develop clinical reaction upon reexposure to the antigen.

Immediate Hypersensitivity: Immediate hypersensitivity is an immunologic sensitivity that manifests by tissue reaction within minutes of antigenic administration. Some characteristics of immediate hypersensitivity are:

(1) It is antibody mediated and induced by antigen or hapten.

(2) Passive transfer with serum.

(3) Desensitization is easy but it is short lived.

(4) Lessions are acute exudation and fat necrosis.

Delayed Hypersensitivity: manifest slowly following antigenic administration in sensitive individuals. The differences with immediate hypersensitivity are:

(1) It is cell mediated. Induced by infection, injection of antigens with Freund"s adjuvants or by skin contact.

(2) Transfers with lymphocytes or transfer factor.

(3) Desensitization is difficult but long lasting.

(4) Mononuclear cell collection around blood vessels.

Coombs and Gell classified hypersensitivity into four types. Later two types were added to it. The classification is as following:

Immediate hypersensitivity

Type 1 or anaphylactic reactions

Type 2 or cytotoxic reactions

Type 3 or immune complex reactions

Delayed hypersensitivity

Type 4 or delayed hypersensitivity

Type 5 or stimulatory response

Type 6 or antibody dependent cellular cytotoxicity.

Type 1 Reaction or Anaphylactic, reagin dependent, e.g., anaphylaxis, atopy etc.

Anaphylaxis, is antigen specific immune reaction mediated primarily by IgE. It results in vasodialation and constriction of smooth muscles. The anaphylactic reaction is the result of sensitization of tissue fixed mast cells by cytophilic antibodies (IgE). This is after following exposure of an antigen and subsequent exposure of same antigen leading to antigen antibody interaction on mast cell surface, causing prompt release of chemical mediators like histamine and consequently their systemic or local effects.

Type 2 Reaction or cytotoxic reaction, e.g., thrombocytopenia, hemolytic anemia.

Reaction involves interaction of antibodies (IgG, IgM) with antigenic determinants present on cell membrane, which leads to cell damage i.e., cell lysis,cell inactivation or phagocytosis of target cells. Complement system may or may not take part in this reaction.

Type 3 Reaction Toxic complex or Immune complex reaction, e.g., arthus reactions, serum sickness, etc.

These reactions are secondary to deposition of immune complexes in tissue, particularly in and around blood vessels. IgG, IgM and complement system take part. With activation of complement system and consequent inflamation which may lead to blockage of the vessels and hence interference to blood supply of the part.

Type 4 Reaction Delayed or Cell mediated hypersensitivity.

This is a slow immunologic response to an antigen. T cells, lymphokines and macrophages take part, e.g., tuberculin type and contact dermititis.

Type 5 Reaction

It is antibody dependent cell mediated and cytotoxic type of reaction. In certain situations, there is interaction of cell bound antigen with antibodies. Cell are not damaged, instead the cell function may be stimulated, e.g., long acting thyroid stimulators (LATS) of thyroditis.

Type 6 Reactions ADCC reaction

In vitro studies have established K or killer cells. The cells are involved in lysing the target cell have receptors for Fc part of antibody. They are not identified as 'B' or 'T' cells.

CHAPTER - 22

FOOD MICROBIOLOGY

Food is the basic need of all individuals. It consists of carbohydrates, protein, fats and other nutrients. The food microflora consists of bacteria, molds, yeasts and yeast like fungi. Bacteria associated with food are *Pseudomonas, Halobacterium, Vibrio, Escherichia, Serratia, Erwinia, Proteus, Shigella, Salmonella, Staphylococcus, Lactobacillus, Streptococcus, Alcaligenes, Bacillus, Clostridium, Streptomyces, Mycobacterium, Acetobacter, etc.* They generally form endospores and their cells are covered with capsules. A number of molds are associated with food. The hyphae form a cottony or fuzzy appearance. The common molds growing on food are *Saprolegnia, Pythium, Mucor, Rhizopus, Zygorrhynchus, Absidia, Thamnidium, Aspergillus, Penicillum, Trichothecium, Geotrichum, Neurospora, Sporotrichum, Botrytis, Cephalosporium, Trichoderma, pullularia, Cladosporium, Helminthosporium* and *Alterneria.* The yeasts have spherical or ovoid cells producing false or true mycelium. Species of *Schizosaccharomyces, Saccharomyces, Zygosaccharomyces, Pichia, Hansenula debaryomyces, Hanseniaspora* are found associated with food.

FOOD SPOILAGE

Food spoilage is defined as any change in a food that renders it undesirable or unsafe for human health. Most of the foods are perishable. They are most susceptible for microbial decomposition. A semiperishable food like potato may have a prolonged microbial activity. The nonperishable foods like sugar and other dry foods are not spoiled under normal conditions. The extrinsic factors like temperature of storage, relative humidity, and oxygen concentration influences the microbial growth in food. Different types of food are deteriorated by different species of microorganisms. It is discussed below.

Breads

Bacteria and yeasts ferment the doughs of various kinds of breads. Bread and most bakery products are spoiled by bread molds, i.e., molds associated with bread and responsible for its deterioration. *Rhizopus nigricans, Penicillium expansum, P.stoloniferum, Aspergillus niger* and *Monilia sitophila* are common bread molds. Many species of *Mucor* also spoils bread. Bacteria *Bacillus subtilis* and *B.licheniformis* causes ropiness of bread. Ropiness is caused due capsulation of bacterial cells and protein gluten. Bacterial cells produce enzyme proteinase, which hydrolyses flour protein, yielding gluten. Similarly, enzyme amylase produced by bacterial cells digests starch yielding sugars, which enhances rope formation. Red or 'Bloody' bread is the result of growth-pigmented bacteria, *Serratia marcescens*. However, red bread is not common. Yeast like fungi *Endomycopsis fibuligera* and *Trichosporon* variable grow on bread giving chalk like spots, hence the name 'Chalky bread'. It is also rare.

Sugar and sugar products

Cane and beet sugar, molasses, syrups, maple sap and sugar, honey and candy are the chief sugar products. Cane sugar extracts and juices are more susceptible for spoilage. Microorganisms produce slime. Yeast and bacteria spoil sugars. Species of bacteria *Leuconostoc, Bacillus, Micrococcus, Flavobacterium, Alcaligenes* and *Enterobacter* and of yeast, *Saccharomyces, Candida* and *Pichia* and a few molds spoil these foods. Spoilage of maple sap and syrup by microorganisms are grouped into five categories: (1) Ropy or stringy sap, usually caused by *Enterobacter aerogenes.* (2) Cloudy: caused by *Pseudomonas fluorescens*, and species of *Alcaligenes* and *Flavobacterium.* (3) Red sap, red pigments produced by bacteria *Micrococcus roseus* or by yeast. (4) Sour sap caused by a variety of bacteria and yeast. (5) Moldy sap caused by molds.

Vegetables and fruits

Raw vegetables and fruits are spoiled by various factors like physical factors, action of their own enzymes, microbial action or combination of the two. Both pathogens and saprophytic organisms deteriorate these foods. The most common type of spoilage of vegetables and fruits is rot. Rotting is caused due to the microbial degradation of pectin. The extra cellular enzymes, pectinases are responsible for such type of degradation. Different types of spoilage caused by microorganisms are classified as (1) Bacterial rot caused by *Erwinia carotovora.* (2) Gray mold rot caused by *Botrytis cinerea.* (3) Rhizopus soft rot caused by *Rhizopus nigricans.* (4) Anthraconse caused by *Colletotrichum lindemuthianum.* (5) Alternaria rot caused by *Alternaria*, Blue mold rot caused by *Penicillium.* (6) Downy mildew caused by *Phytophthora, Bremia.* (7) Pink mold rot caused by *Aspergillus niger.* (8) Sliminess or souring caused by saprophytic bacteria in piled, wet, heating vegetables.

Fruit juices contain sugar and some quantity of pulp. Juices have more or less acidic pH. The most common spoilage by yeast is alcoholic fermentation producing alcohol, which may be further oxidized to acetic acid. The other types of spoilage are

(1) Lactic acid fermentation caused by species of *Lactobacillus* and *Leuconostoc.*

(2) Fermentation of organic acids of fruits caused by lactic acid bacteria.

(3) Slime production by different species of *Leuconostoc* and *Lactobacillus.*

Meat and Meat products

Meat is an ideal culture medium as it is nutritionally rich. It has high moisture content, rich nitrogenous compounds, minerals, other growth factors and an important carbohydrate glycogen. It is contaminated during processing. Microorganisms found associated with processed meat are species of the genera *Cladosporium, Sporotrichum, Geotrichum, Thamnidium, Mucor, Penicillium, Alternaria* and *Monilia*. Species of bacterial genera *Pseudomonas, Alcaligenes, Micrococcus, Streptococcus, Sarcina, Leuconostoc, Lactobacillus, Proteus, Flavobacterium, Bacillus, Clostridium, Escherichia, Samonella* and *Streptomyces* also occur in such foods. Raw meat may be autolysed by its own enzymes resulting in 'souring'. The unsaturated fats may be oxidized in air and may be catalyzed by light and copper. Microbiological spoilage of meat may be grouped in different types: (1) Surface slime caused by *Pseudomonas, Alcaligenes, Streptococcus, Leuconostoc, Bacillus* and *Micrococcus*. (2) Change in colour of meat: Red colour (bloom) of meat may be changed to shades of green, brown or gray. This change is caused by bacterium *Lactobacillus* and *Leuconostoc*. (3) Changes in surface colours: various pigmented bacteria changes surface colours like the red spot may be caused by *Serratia mascescens*, blue colour by *Pseudomonas cycnaea*, yellow colour by *Micrococcus* or *Flavobacterium, Chromobacterium lividum*. (4) Change in fats: Lipolytic bacteria cause lipolysis, also accelerate the oxidation of fats. The enzymatic activity releases free fatty acids and glycerols, bring about undesirable flavours. Rancidity is caused by yeast or species of *Pseudomonas* and *Achromobacter*. (5) Phosphorence is caused by Photobacterium spp. (6) Putrefaction: caused by bacterium *Pseudomonas, Clostridium, Alcaligenes* and *Proteus*. It is the anaerobic decomposition of protein by enzymes proteinases. The subsequent degradation of amino acid produces low-molecular weight sulphur and nitrogen containing compounds like hydrogen sulphide, mercaptans, indole, skatole, ammonia, amines etc. are formed giving foul smell to the deteriorating meat. (7) Taints: off odours and off tastes are caused by bacterium. It is due to formation of volatile acids, e.g., formic acid, acetic acid, butyric acid and propionic acid.

Poultry and Eggs

Like meat, dressed poultry are also deteriorated by their own enzymes. Bacterial spoilage is caused by *Alcaligenes* and *Flavobacterium*. Eviscerated poultry stored at 10°C is spoiled by *Pseudomonas*. Yeast *Torulopsis* and *Rhodotorula* are able to spoil poultry even below 10°C.

Bacteria cause rotting of eggs. There are different types of rotting of eggs. Three types of most commonly occurring rotting: (1) Green rots caused by *Pseudomonas fluorescens*. (2) Colourless rot caused by *Pseudomonas, Alcaligenes*. (3) Black rot caused by *Proteus, Pseudomonas* and *Aeromonas*. Fungal deterioration of egg is caused by species of *Penicillium, Cladosporium, Mucor, Thamnidium, Botrytis* and *Alternaira*. Fungal spoilage initiates with pin-spot molding: a small, compact colony of mold appearing on the shell and just inside it. The final stage of spoilage is fungal rotting.

FOOD PRESERVATION

Food preservation requires the methods for keeping away microorganisms from food. It is conducted by maintaining conditions by physical or chemical means, which may kill or control the growth of microorganisms responsible for spoilage of food. There are two principles of food preservation (1) Prevention or delay of microbial decomposition and (2) Prevention or delay of self-decomposition. The chief methods of food preservation are (1) Asepsis. (2) Remove of microorganism. (3) Maintenance of anaerobic conditions. (4) Use of high temperature. (5) Use of low temperature. (6) Drying. (7) Use of chemical preservative. (8) Irradiation. (9) Mechanical destruction of microorganisms. (10) Combination of above methods.

Asepsis

Asepsis is keeping out microorganisms. The inner tissues of healthy plants and animals are generally free from micoorganisms. In this process, food's products are preserved free from contamination. It is important for a food technologist to consider the type and number of microorganisms present in food material. Asepsis is a general practice in packing of food products. Wrapping or loose carton prevents primarily contamination during handling. The canned foods are hermetically sealed. In the canning industries, special attention is given on heat treatment. It is because to avoid recontamination from heat resistant spore forming thermophiles.

Remove Microorganisms

Filtration, centrifugation, sedimentation, washing or trimming the simple methods to remove the microorganisms from food materials. The clear liquids are filtered through sterilized "bacterial proof" filters. These filters are made up of asbestos pads, sintered glass, diatomaceous earth, unglazed porcelain or membrane pads. Generally, centrifugation is not applied, as it is not helpful to keep away all the microbes. Washing may be beneficial for raw foods to remove soil microbes. However, it may be harmful if water itself acts as a source of contamination of pathogenic microbes. In addition, washing results in retaining moisture and thus creating a favourable condition for the growth of microorganisms. Trimming of infected portion is of some help as the spoiled portion is discarded.

Maintain anaerobic conditions

In the sealed, packaged foods, anaerobic conditions are maintained in the unfilled space. The empty space of food container is evacuated or air is replaced by carbon di oxide or inert gases like nitrogen.

Use of high temperature

High temperature denatures proteins, i.e., inactivates enzymes helpful in metabolism. The heat treatment necessary to kill microbes or their spores varies with the type and the environment during heating. Heat treatment necessary to kill microbes is known as Pasteurization. Food is heated at temperatures below 100°C. This temperature may not kill all microorganisms and spore former survive after pasteurization. The high-temperature-short-time (HTST) method employs a comparatively high temperature for a short time. Whereas, the low-temperature-long-time, or holding (LTH), method uses a lower temperature for a longer time. Pasteurization is applied to kill microorganisms, which are not very heat resistant like yeast. Some food product like cheese is pasteurized, for killing microorganisms, which compete with fermenting microbes. Pasteurization may be followed by refrigeration, or aseptic sealing or maintaining anaerobic conditions in packing of food materials, or by addition of high concentration of sugar or preservative.

Use of low temperature

Low temperature retards metabolic activity and growth of microorganisms. It may even prevent growth of microorganisms, which spoils food. Chilling of food products by ice or refrigeration may be used as a temporary preservation of perishable food. Temperature, relative humidity and ventilation etc are controlled during freeze storage. Most of the food materials are stored at a temperature just higher to its freezing point. The optimal relative humidity of the atmosphere in chilling storage varies with the food stored and with environmental factors like temperature, composition, moisture much below relative humidity results in loss of moisture and hence of weight. Ventilation or control of air velocities maintains the uniformity of the moisture.

Drying

Some foods e.g., grains are very well dried for storage purpose. Drying simply means removal of moisture. There are different methods for drying, i.e., by sunrays, drying by mechanical dryers, freeze-drying etc. Drying in sun light food is exposed to sun rays without any artificially produced heat and without controlled temperatures, relative humidity or air velocities. Some fruits like, raisins, figs, apricots, pears and peaches are sun dried. The mechanical dryers are evaporators or kiln. The heated air with controlled relative humidity is passed over the food. Mild, juices and soups are evaporated with the use of low temperature and vacuum devices. Meats, poultry, seafood etc are sublimated or freeze-dried.

Use of chemical preservatives

In 1965, WHO defined food additives as a substance or mixture of substances, other than the basic food stuff, which is present in food as a result of any aspect of production, processing, storage or packing. The food additives, which prevent the deterioration on decomposition of food, are known as chemical preservatives. The preservatives should not harm the human beings or animals. These compounds act on the cell membrane of the microorganisms or on their enzyme system or on the genetic material.

Applications of food preservative

Chemical agents	Uses
Acetic acid	Bakery goods, beverages, confections, cheese and syrups as acidulant, i.e., maintaining acid level.
Citric acid	Jams, jellies, and soft drinks as acidulant.
Lactic acid	carbonate soft drinks and cottage cheese, as acidulant.
Benzoic acid (sodium benzoate)	carbonated soft drinks, confections, fruit juices, jellies, jams.
Sodium propionate	bakery goods and processed cheese, as mold inhibitor.
(Sodium nitrate]	curing of meat and fish, as bacterial inhibitor.
(Sodium nitrite]	
Sorbic acid (potassium sorbate)	bakery goods, fresh fruit, salads and syrups, fungus inhibitor.
Carbon di oxide	beer, controlled gas storage of fruits, fruit juices, soft drinks and
(carbonic acid)	vegetables.
Sulphur di oxide	dried fruits, fruit juices, syrups and wine, as antibrowning agent.

DAIRY MICROBIOLOGY

Milk contains a variety of nutrients including fat, minerals, protein, lactose, vitamins and water. The dairy products of a particular region or a country are known as "indigenous milk products". In India the indigenous products include dahi, makkhan, khoa, paneer etc. Numerous products are made by the microbial fermentation of milk. The main products are butter milk, yoghurt and many cheeses. The fermentation is primarily carried out by lactic acid bacteria. Hence, it is regarded as lactic acid starter. It also includes bacteria like *Streptococcus cremoris, S. diacetylactis* and *S. lactis*. All these bacteria convert lactose sugar, of milk into lactic acid.

Butter, butter mild and sour cream

Butter, butter milk and sour cream are prepared by inoculating pasterurized cream or milk with a specific lactic starter culture. Butter is normally made by churning cream that has been soured by lactic acid bacteria. *S.cremoris* or *S. lactis* are used to produce lactic acid rapidly and *Leuconostoc cremoris* or *S. lactis diacetilactis* enzymes attack citrate in milk, producing diacetyl. The latter gives a typical aroma to butter. The churned acidified cream is worked to remove excess liquid, washed, salted, and packaged. Buttermilk is the liquid remaining after cream is churned for butter production. Cultured sour cream is prepared by fermenting pasteurized light cream with a lactic starter.

Microbiological Examination of Food

Food Sample
(preparation of homogenate)

- Direct microscopic Examination
  - Stained preparations
  - Wet mounts
- A. Total colony count
- B. Differential enumeration
  - categories of microorganisms
    - Bacteria
    - Yeasts
    - Moulds
  - Physiological Types
    - Aerobic
    - Anaerobic
  - Thermal Types
    - Psychrophilic
    - Merophillic
    - Thermophillic
  - Metabolic Types
    - Proteolytic
    - Lipolytic
    - Sacchanolytic
- Identification of specific groups or species of microorganisms
  - Bacteria
    - Coliforms
    - Fecal streptococci
    - *Staphyloccoccus aureus*
    - *Salmonella spp.*
    - *Shigella spp.*
    - *Vibrio parahae molyticus*
    - *Clostridium bolutenin*
    - *Clostridium perfringes*
    - *Bacillus cereus*
  - Viruses
    - Hepatitis A
    - Entero virus
    - Parvovirus
    - Adenovirus
  - Fungi
    - *Aspergillus*
    - *Penicillium*
    - *Fusarium*
    - *Claviceps*
  - Protozoa
    - *Entamoeba histolylica*

Yoghurt

Yoghurt is excellent source of high quality protein and certain vitamin B-complex and minerals. It is produced by fermenting milk with a mixture of *L. bulgaricus* and *S. thermophilus.* Commercially, yoghurt is made from a mix of whole, partially defatted milk, condensed skim milk, cream, and nonfat dry milk. Equal number of the two bacteria is required for flavour and terture production. The characteristic flavour of yoghurt is due to the accumulation of lactic acid and acetaldehyde produced by *L. bulgaricus.* The lactobacilli grow first, liberating the amino acids glycine and histidine and stimulating the growth of streptococci. In addition to milk fat, casein, sodium and calcium caseinates, whey and whey protein concentrates are also present in yoghurt. The acetaldehyde gives a tart taste, which is controlled by adding fruit in this food.

Cheese

Varieties of cheese are produced by microbial fermentation. Based on their texture, cheese are divided into three types (1) soft with 50-80% water content (2) hard with less than 40% water content and (3) semi-soft with about 45% water content. Based on the ripened state cheese are classified into unripened (produced by a single step fermentation) and ripened (additional microbial growth is required during maturation of the cheese to achieve desired taste and aroma). Cheeses consist of milk curds that have been separated from the whey. For cheese production, first milk is curdled (skimmed or whole). Curdling of milk is accomplished by using the enzyme renin (casein coagulase or chymosin) and lactic acid bacterial starter. A natural cheese is directly made from milk or in some cases from whey. Pasterurized process cheese and cheese spreads are made by blending and heating one or more varieties of natural cheese. After curdling, moisture is removed by applying heat, pressure, or cutting and compression of the curd. Sometimes salt is added during cheese production. Besides, removing moisture and providing a particular flavour, it prevents the growth of unwanted microbes. If cheese is to be ripened one or two methods are adopted. Ripening of cheeses involves additional enzymatic transformations after the formation of the cheese curd, using enzymes produced by lactic acid bacteria or enzymes from other sources. To obtain hard cheese, microorganisms are grown into the interior of the cheese. While for soft cheese, microbes are grown only on the surface. The shape and size also matters on the type of cheese to be produced.

Grades of milk

There are certain standards by which milk is judged. The grading of milk has become an important factor in maintaining the health of the population. Grades of milk are based upon regulations pertaining to production, processing and distribution e.g. sanitation, pasteurization and holding conditions as well as bacterial standards. The number of organisms permissible in different grades of milk varies somewhat depending upon standards set up by local public health authorities. However the recommendations contained in the U.S. public health service publication milk ordinance and code are widely used.

Chemical, bacteriological, temperature standards for Grade A milk and milk products

Product	Temperature	Bacterial limits	Chemical & others
Grade A raw milk for pasteurization	Cooled to 50°F or less and maintained there at until processed.	Individual producers milk not to exceed 1000.000 /ml prior to comm-ingling with other producer milk.	Antibiotics less than 0.05 unit/ml by *Bacillus subtilis* method or equivalent.
Grade A pasteurized milk products (except cultived products)	Cooled to 45°F or less and mainta-ined there at.	Milk and milk products 20.000/ ml coliform limit, not exceeding 10/ml.	Phosphotase, less than 1μg by Scharer rapid method (or equivalent by other methods).
Grade A pasteurized cultived products	Same as above	Exempt	Same as above

Quality Control of Milk

The chemical, physical, microbiological, aesthetic and economic characteristics of milk are referred as quality of milk.

Pasteurization of milk — There are methods of pasteurization in general use. One is low temperature holding (LTH) methods. Other is the high temperature short time method (HTST). In the holding method milk is heated to 145°F (62.8°C) for 30 min. This accomplished by the use of commercial pasteurizes large closed vats in which the milk is heated by steam coils, hot water jackets or the continuous spraying of hot water on the sides of the tank containing the milk. In the high temperature short time method, milk is heated to a temperature of 161°F hot water and requires a heat exchange system, which preheats raw cold milk. In both methods of pasteurization it is essential that the equipment must be designed and operated so that every particle of milk is heated to the specified temperature and held for the specific time. It is also necessary to cool the milk immediately and to prevent recontamination following pasteurization until it is delivered to the consumer.

The original time-temperature relationship for pasteurization was worked out with *Mycobacterium tuberculosis* since this was regarded as the most heat resistant pathogen likely to be found in milk. This organism is destroyed when exposed to a temperature of 140°F for 10 min. Thus the "milk ordinance and code" was set at 143°F for 30 min. Later it was discovered that *Coxiella burnetti*, the causative agent of a fever is able to survive in milk heated to 143°F for 30 min. As a result of this finding it became necessary to raise the temperature by 2°F to 145°F to ensure complete destruction of *Coxiella burnetti.* Economic factors dictated the upper limits placed on the heat treatment of milk.

Pasteurization does not sterilize milk. It only reduces the microbial count. The surviving organisms must be kept from multiplying by constant refrigeration. Pasteurized milk usually keeps better when refrigerated than raw milk of a comparable microbial content. This is because pasteurization kills those organisms that grow most readily at low temperature. The surviving organisms are relatively heat resistant thermodemics and require higher growth temperatures.

Microbiological quality control for milk

Milk may be contaminated by microbes at the time it is drawn from the cow, or during handling or processing. The different aspects of microbiological criteria for quality control of milk are as below.

1. Microorganisms and/or their toxins.
2. The sampling plan and methods for detection and quantification.
3. The microbiological limits.

Based on the above aspects, the microbiological criteria in vogue are: (1) Microbiological standards, the microbiological limits incorporated in a law or regulation controlling the milk, processed or stored in, or imported into the area of jurisdiction. (2) Microbiological specifications, is the type of microorganisms and their number in a processed milk or its products. (3) Microbiological guideline, designate the requirements such as sampling and analysis to achieve the prescribed standard.

There are a number of national and international agencies on microbiological quality control of dairy products. In India the agencies for controlling quality of milk are: (1) Prevention of Food Adulteration Act (PFA). (2) Bureau of Indian Standards (BIS)-formerly Indian standards Institution (ISI) and (3) AGMARCK (Agricultural produce). The Indian Standards on milk and milk products have been laid down and the limits for some of the microorganisms, which provide a suitable indicator for possible contamination with unhygenic conditions have been specified.

MICROBIOLOGY, SPOILAGE AND PRESERVATION OF DAIRY PRODUCTS

Dairy products are obtained from milk. In market, the dairy products available are milk, cream, frozen desserts, cheese, condensed milk and dried milk powder. The optimum temperature for growth and their heat resistant quality is important criteria for classifying the microflora of milk. Based on these criteria, the microflora of milk may be classified into four categories: (1) Psychrophilic the microbes grow at temperature just above freezing and at refrigeration temperature, e.g., species of *Pseudomonas, Vibrio* and *Achromobacter*. (2) Mesophilic, e.g., Streptococci, Lactobacilli. (3) Thermophilic, e.g., *Bacillus* and *Clostridium* and (4) Thermoduric the microbes which survive after pasteurization, e.g., *Micrococcus* and *Streptococcus*.

Spoilage of dairy products by microorganisms

In raw milk, *Streptococcus lactis* is found at temperature between 10 to 37°C. It causes souring. Coliform bacteria, enterococci, Lactobacilli and Micrococci grow between 37 to 50°C. Acid production in raw milk is due to growth of *S.thermophilus, S.faecalis* and *Lactobacillus bulgaricus. L.thermophilus* survives above 50°C. The heat resistant lactics cause lactic acid fermentation. The coliform bacteria produce some lactic acid with a large amount of acetic acid, formic acid, alcohol and gases. A small quantity of lactic acid is also produced by species of *Micrococcus, Mycrobacterium* and *Bacillus.* Clostridium spp. produce butyric acid. Gases produced are hydrogen and carbon di oxide. Coliform bacteria, *Clostridium* spp. and *Bacillus* produces both the gases, while yeast, propionics and heterofermentative lactics produces only carbon di oxide. Microorganisms are responsible for hydrolysis of milk proteins. Such proteolysis releases some peptides, which give bitter flavour. *Pseudomonas flourescens* produce enzyme proteinase. This enzyme retains its activity even after pasteruization when the bacterial cells are killed. Other proteolytic bacteria are *Micrococcus, Pseudomonas, Proteus, Flavobacterium* and *Serratia. Streptococcus faecalis* var. *liquefaciensis,* a lactic Pseudomonas acid organism, is also proteolytic. The capsular substances (mucin and gums) of bacterial cells causes ropiness, There are two types of bacterial ropiness. In one type of ropiness, the milk is ropy only at the top, and in the second type, milk becomes entirely ropy. *Alcaligenes viscolactis* causes surface ropiness. Enzyme lipase decomposes milk fats. *Pseudomonas, Proteus, Alcaligenes, Bacillus, Micrococcus, Clostridium* etc. produce lipase. Change in colour of milk may be due to physical or chemical composition. Few bacteria give different colours to milk. Blue milk is caused by *Pseudomonas syncyanea, Streptococcus lactis,* yellow milk is caused by *Pseudomonas synxantha, Flavobacterium,* red milk is due to *Serratia marcescens* and brown milk is due to *Pseudomonas putrifaciens, P.fluorescens.*

Microbiological spoilage of sweetened condensed milk are: (1) Gas formation by *Torula lactis-condensi* and *T. globula* and by coliform bacteria. (2) Thickening is caused by micrococci and (3) buttons, which appear due to growth of the colonies of molds like *Aspergillus repens* or *Penicillium.*

Butter may be subjected to many types of spoilage. Cream is churned to prepare butter. If cream is stored for a long time, it is susceptible for the growth of lactic acid bacteria, gas formers and molds. The undesirable flavour in cream may be due to the feed of the cow or the growth of microorganisms. The bad flavours may be: Cheesiness, caused by Lactobacilli. Rancidity is caused by lipolytic bacteria and molds. Barmy flavour is caused by *Enterobacter,* yeasty flavour caused by yeast, malty flavour caused by *Streptococcus lactis* var. *maltigenes,* musty flavours caused by molds and actinomycetes, flat flavour caused by some *Pseudomonas* due to breakdown of diacetyl. Unclean flavour is caused by coliform bacteria. When the cream has excessive acidity it give highly acid flavour. The dissolved metals in highly acid cream give metallic flavours.

Cheese is spoiled by *Penicillum, Lactobacilli, Streptococci* etc. Frozen desserts are generally not susceptible for microbial spoilage. However, if stored for a long time it may be deteriorated.

Preservation of dairy products

Milk and milk products, may be preserved by general methods of food preservation. However, due to physical and chemical characteristics of milk the methods adopted are quite delicate. The chief types of methods used for preservation are asepsis, removal of microorganisms, temperature treatment, drying and use of preservatives.

FOOD BORNE DISEASES AND TOXINS

The pathogens and their secondary metabolites, if taken with food cause gastrointestinal disorders and food poisoning. F.L. Bryan (1976) classified food borne diseases as "poisoning" or "infection". Poisoning is due to intake of toxins produced by pathogens. While, infection is illness caused by the entrance of the pathogen into the body, through ingestion of contaminated food. *Salmonella*, enteropathogenic *E. coli*, *Bacillus, Clostridium* grow on food and cause disease. Food may not favour the growth of some pathogen but acts as a carrier. Thus, food acts as a carrier of pathogens causing disease like tuberculosis, typhoid fever, cholera etc. These pathogens are discussed in chapter 20.

FOOD INTOXICATION

Food poisoning may be due to intake of poisonous chemicals or by intoxication. The intoxication may be caused by bacteria as well as fungi. There are tow major food intoxications caused by bacteria. These are: (1) Botulism caused by the toxins produced by *Clostridium botulinum* and (2) Staphylococcal intoxication caused by the toxin produced by *Staphylococcus aureus.*

Botulism

It is caused by a neurotoxin produced by *Clostridium botulinum.* Based on serological specificity these toxins are divided into seven types. Symptoms appear within 12-36 hours after the consumption of contaminated food. The symptoms are digestive disturbances followed by nausea, vomiting, diarrhoea, together with drowsiness and headache. In the advanced stage, involuntary muscles become paralysed. Paralysis spreads to respiratory system and to the heart resulting in death. To avoid botulism properly cooked and heated food should be taken. Antitoxins are also available to control the symptoms.

Staphylococcal poisoning

Staphylococcus aureus produces enterotoxins. Five types of toxins are identified. The toxin is protein in nature and resistant for proteolytic enzymes (trypsin and renin). Symptoms appear within 2-6 hours after consumption of contaminated food. The symptoms are onset of nausea, vomiting, abdominal pain and diarrhoea. In advanced and severe cases, dehydration and collapse may occur.

Mycotoxins

Many fungi invade food releasing their metabolites (toxins) known as mycotoxin. The symptoms appear due to ingestion of mycotoxins are known as mycotoxicosis. The first mycotoxicosis was observed with toxins produced by *Cleviceps purpurea* causing Ergot of rye. Aflotoxins and related compounds are produced by *Aspergillus flavus* and *A. parasiticus*. It was first recognized as "Turkey X disease". The term aflotoxins refers to group of bisfuranocoumarin. Fungi invade during harvesting and flourishes in post harvest stages and storage. Aflotoxin is common in all types of cereals. Other mycotoxins are Patulin, Ochratoxin, Sterigmatocystin, Penicillic acid, etc.

CHAPTER - 23

WATER MICROBIOLOGY

Study of microorganisms and their activities in aquatic environment is known as water microbiology. Aquatic environments include fresh water, estuarine and marine water. Natural water includes atmospheric, surface and ground water. The atmospheric water is moisture contained in clouds and precipitated as snow, sleet, hail rain constituted atmospheric water. Air and the dust particles of air are the main source of microbial flora of this water. Lakes, streams, rivers and ocean are the surface water system. Besides having the atmospheric microflora the surface water is contaminated by surface runoff and the waste dumped in it. The subterranean water is ground water. The microorganisms are removed by filtration. If filtration is proper, the microbial population is negligible.

The atmospheric environment is determined by physical and chemical conditions like temperature, hydrostatic pressure, light, turbidity, Hydrogen ion concentration, salinity, organic and inorganic constituents. The temperature varies from 30° to 40°C in equatorial region. In the polar region, temperature is near 0°C and favourable for psychrophilic microorganism. In hot spring, the temperature varies from 75° to 80°C. *Thermus aquaticus* grows at the temperature of 70° to 72°C. *Bacillus stearothermophilus* grows in hot spring at temperature between 55-70°C. The hydrostatic pressure varies from surface water to the depths of ocean. Hydrostatic pressure affects chemical equilibrium. There is sufficient light at surface water, which favours the growth of photosynthetic organisms. In oceans the light penetrates on local conditions and thus the photosynthetic organisms are confined to the upper region of 50 to 125 M. The suspended particles result in turbidity of the aquatic environment. The pH of water may vary from 6.5 to 8.5. The pH of sea varies from 7.5 to 8.5. The fresh water is free from salinity. However, salt lakes occur in arid regions where evaporation exceeds fresh water inflow. Salt lake also occurs when it is fed by a salt spring. Few halophilic and salt-tolerant bacterial species are found in such habitat, e.g., *Halobacterium.* Halophilic microorganisms have high concentrations of potassium chloride concentrations in their cells. The salinity is at saturation in sea. The microbial flora depends upon the organic and inorganic constituents of a water system. Growth of algae is favoured by salts.

THE AQUATIC MICROFLORA

The microorganisms are distributed in aquatic environment under all types of conditions. They are found in top layers of the surface water, up to the bottom of the oceans. The fresh water forms like lakes and ponds may be divided into zones and strata. There is usually a large litoral zone along the shore. Light can richly penetrate in this zone. This zone has rooted vegetation. The open area is limnetic zone and is determined by the light compensation level. The litoral and limnetic zone has a large number of producers and other physiological types. As we go deep into this zone, the photosynthetic activity decrease as light is not available. This is called as profound zone. The bottom zone is benthic zone and is rich in mud. The profound and benthic zones are occupied by heterotrophic organisms. In the benthic zone, there is accumulation of organic material and the anaerobic decomposers are highly populated. Lakes and ponds of the temperate region exhibit interesting seasonal changes in their microbial populations due to stratification of the water, as result of temperature differences. Lakes and ponds with enriched nitrogen and phosphorus nutrients favour the growth of algae, which results in the formation of algal blooms. The streams obtain organic and inorganic nutrients through inflowing water in it. The microflora varies according to the nearby terrestrial environment. As estuary is semienclosed coastal body of water, which, has, free connection with the open sea. It receives inputs from a variety of sources.

Besides a wide group of algae, aquatic systems contain bacteria, viruses and fungi. In areas receiving water effluents, bacteria like coliforms, fecal, streptococci and species of *Bacillus, Proteus, Clostridium, Sphaerotilus, Beggiatoa, Thiothrix, Thiobacillus* and many others. Viruses of the enteric groups are also present. Soil bacteria like, *Azotobacter, Nitrobacter* and *Nitrosomonas* may also be present. Fungi belonging to the group Phycomycetes, Ascomycetes and Fungi Imperfecti are found in aquatic environments.

Nearly three fourths of the earth surface is covered by the sea. The study of microflora and its activities is known as marine microbiology. Microorganisms are found at all depths and at all latitudes in sea. Microorganisms inhabiting surface are known as plankton and those found near the bottom are called as benthos. The phytoplanktonic zone consists of photosynthetic microorganisms. Species of diatoms, cyanobacteia, dinoflagellate, coccolithophores, silicoflagellate, chrysomonads, Crytomonads and chlamydomonads are found in phytoplankton population. Bacterial populations differ widely with prevailing conditions. Species of *Vibrio, Acinetobacter, Pseudomonas, Flavobacterium, Alteromonas* and *Staphylococcus* are commonly found in this region. Certain luminous bacteria, which can produce light in the presence of oxygen, are generally found in symbiotic association with certain species of marine animals. Species of Deuteromycetes, Phycomycetes and Myxomycetes have been isolated from marine environments. The protozoa population includes members of group flagellata and ciliata and species of *Foraminifera* and *Radiolaria.* The benthic populations are inhabited by bacteria and protozoa.

MICROBIOLOGY OF DOMESTIC WATER AND WASTE WATER

The surface water from lakes, ponds and rivers is the main source of drinking water. It is generally polluted by domestic and industrial wastes. It carries pathogenic species and is dangerous for health and life. Water that is free of pathogens and chemical substances, deleterious to health is called potable water. The contaminated and polluted water is known as nonpotable water. The nonpotable water is to be properly purified before it is used for drinking purposes. The water of wells and springs undergoes filtration, which removes suspended particles including microorganisms. However, it may also be contaminated.

There are sanitary survey to inspect water producing system and a proper distribution of water to the public. The sanitary survey includes inspection of the source and condition influencing raw water. It also inspects the operation of the water purification plant or the construction of the well. The drinking water polluted by sewage containing human excreta is most dangerous for health. The diseases transmitted through such water are typhoid fever, paratyphoid fever, amebic dysentery, bacillary dysentery, cholera, tularemia, poliomyelitis and infectious hepatitis. The coliform bacteria (*E. coli, Enterobacter aerogenes* and related organisms), fecal streptococci (*Streptococcus faecalis*) and *Clostridium perfringens* are normal inhabitants of the large intestine. The other enteric bacteria are *Salmonella, Schigella, Klebsiella, Proteus, Serratia* and other Gram negative, non sporulating bacilli. The extracellular excretion products and capsular structure may elaborate gummy or mucilageous materials in water bodies. Another group of bacteria, known as iron bacteria. It also produce slime discolour water and cause undesirable odours and tastes. These bacteria transform soluble compounds of iron to insoluble compounds of iron (ferric hydroxide). The latter may be deposited in a sheath around the cells as in *Sphaerotilus* or secreted to form stalks or ribbons attached to the cell as in *Gallionella.* The sulphur bacteria like *Thiobacillus* oxidizes elemental sulphur to sulfuric acid. *Desulfovibrio desulfuricans* reduce sulphates and other sulphur compounds to hydrogen sulfide. The enteroviruses found in sewage include the polio, coxsackie and echo viruses. The virus that causes infectious hepatitis has been isolated from polluted water. Rotaviruses are also of major importance.

BACTERIOLOGICAL EXAMINATION OF WATER

The water is examined by various methods and they are based on (1) the examination of presence or absence of the enteric microorganisms and (2) an approximate determination of total number of bacteria present in the water sample.

Test for coliform

Test for coliform is completed in three steps (1) the presumptive test, (2) the confirmed test and (3) complete test.

In presumptive test, a series of fermentation tubes containing lactose broth or lauryl broths are inoculated with definite amount of water to be tested. The tubes are incubated at 37°C for 24 to 48 hours. If gas accumulates as result of fermentation confirms the test.

In confirmed test, two procedures are employed. In the first method, a drop of culture from a positive lactose broth tube is transferred to brilliant green lactose-bile fermentation broth. It is incubated at 35°C for 24 to 48 hours. Formation of gas confirms presence of coliform bacteria. As this media contains brilliant green dye, it inhibits the growth of Gram positive organisms. These organisms may produce fermentation gas in presumptive test, which is to be confirmed by this method. In the second method, a drop of culture from the positive lactose broth is streaked on a petri plate containing Endo eosine-methylene blue agar. After 24 hours, if nucleated colonies appear on the medium it confirms the presence of coliform bacteria.

In complete test, the colonies appeared on the Endo'eosine-methylene medium are transferred into lactose fermentation broth and streaked on to an agar slant. Formation of gas and growth of Gram-negative non-spore forming bacilli on slants, confirms the presence of coliform bacteria.

IMVic test

Water may contain fecal or nonfecal microorganisms. The coliform bacteria, *E. coli* and *E. aerogenes* are considered as fecal and nonfecal contamination, respectively. It needs biochemical test to differentiate them. Parr described IMVic test. The name of the test was coined from the first letters of the four tests, I from Indole, M from Methyl red, Vic from Voges-Proskauer and citrate.

Test	*E. coli*	*E. aerogenes*
Indole	Indole is produced from trptophan (+)	Not produced (-)
Methyl	Methyl red is turned red, which means	Methyl red remains yellow, which
Red	a pH of below 4.5 is produced	means less acid is produced.
Voges-Proskauer	Acetylated methyl carbinol is not produced in glucose peptone medium	Acetyl methonyl produced
Citrate	Citrate as the sole source of carbon does Support growth (-).	Supports the growth (+)

The membrane filter method

In this method, a filter disk is used, which is made of cellulose derivative. This disk is fitted in a filtering apparatus constructed of a glass or stainless steel funnel and a suction flask. A definite volume of water is passed through this disk. The bacteria are retained on this disk, which is later transferred to a petri plate under aseptic conditions. This disk contains an absorbent pad saturated with an appropriate medium. The bacteria present on the filter disk receive medium and develop colonies. These colonies may be easily counted.

The membrane filter technique has several desirable features.

1. A large volume water sample can be examined Theoretically almost any volume of non turbid water could be filtered through the disk, the organisms from any given volume being deposited on the disk.
2. The membrane can be transferred from one medium to another for purpose of selection and differentiation of organisms.
3. Results can be obtained more rapidly than by the conventional MPN standard methods.
4. Qualitative estimation of certain bacterial types e.g. coliforms can be accomplished when appropriate media are used.

WASTE WATER TREATMENT

There are different types of processes for treating waste water. The procedure depends upon the type of water structure, i.e., the structure is a single dwelling or unit structure or it is a community or municipality system. The main methods for water purification are sedimentation, filtration and disinfection.

Sedimentation

In a stagnant water system, suspended particles settle down and thus carry a number of microorganisms with it. It is important that what kind and in how much amount the suspended particles are present in the system. Beside, this the physical, chemical and biological conditions also regulate the rate of sedimentation. Addition of alum, iron, salts, colloidal silicates etc increases the rate of sedimentation. These chemicals produce flocculent precipitates. Sedimentation may help in settling of microorganisms but it is not helpful in making the water free from microbes. However, this procedure may be used as the first stage of purification.

Filtration

During filtration the suspended matter is removed along with the microorganisms. Two types of sand filters are generally used (1) slow sand filtration plants: This process requires a large area and the rate of filtration is slow. The plant is constructed of drainage tiles covered with coarse gravel, fine gravel, coarse sand and the lower most layer of 2 to 1 feet of sand. The water steeps through these layers slowly and is collected by tile drain pipes at the bottom. It is further pumped into a reservoir. The floccules along with microorganisms, accumulates and a thick gelatinous film is formed on the surface layer. This is removed. (2) Rapid sand filters: Such plants are also constructed like slow sand filters consisting of different layers. Water is pretreated with coagulants such as alum or ferrous sulphate and passed through the layers of gravel and sand. The precipitates settle out and the remainder is pumped on the filter. Rapid sand filters are back washed, i.e., it is cleansed by forcing cleaned water backwards. Along with the bubbling air passes in backward direction through the gravel and sand. This water rises through the filter and carries the accumulated material to the sewer.

Disinfection

Disinfection means treatment of filtered water to make it free from microorganisms. For this purpose, solutions of calcium or sodium hypochlorite are used. Now a day's chlorination of the public water supply is a general practice. Chlorine reacts with water to produce hypochlorous acid. It is immediately decomposed and oxygen is released. The amount of chlorine taken up is called as chlorine demand. The point at which the available chlorine becomes proportional to the added chlorine is called the break point. Chlorine destroys microorganisms but does not kill spores. Filters with germicidal ultra violets rays are now available which is commonly used to purify drinking water in houses.

SEWAGE DISPOSAL

In general terms, sewage is used water supply of a community. It may consist of domestic, agricultural and industrial wastes. Sewage has a large part of water (99.5 to 99.9%) and 0.1 to 0.5% of inorganic and organic matter either suspended or soluble form. The suspended solids may be cellulose, lignocellulose, proteins, fats and other inorganic chemicals present in their colloidal state. The soluble materials are sugar, fatty acids, alcohols, amino acids and inorganic ions. Domestic sewage contains human excrements. Detergents, antiseptics, and pesticides are also important constituents. Many types of saprophytes and pathogens are found in sewage. The common types of bacteria found in sewage are coliform, Streptococci, Clostridia, Micrococci, Proteus and Lactobacilli. The bacterial diseases transmitted by sewage are bacterial dysentery, cholera and typhoid fever. Viruses and protozoa are also present in sewage. Besides, different types of enteric viruses, coxsackie, hepatitis, polio and adenoviruses appear in sewage. Amebic dysentery is a common protozoal disease transmitted by sewage. The microorganisms found in sewage may be aerobes, facultative anaerobes and strict anaerobes. Initially aerobic and facultative anaerobes are present which are replaced by strict anaerobes during the process of decomposition. Organic matter is oxidized by microorganisms under aerobic conditions. Thus, oxygen is consumed resulting in anaerobic conditions. The waste water has a high biochemical oxygen demand (BOD). The latter is defined as the quantity of oxygen required for decomposition of organic and inorganic matter by aerobic biological action. Sewage with high BOD expresses the strength of sewage and the efficiency of sewage treatment processes.

Sewage treatment requires physical, chemical and biological treatments. In physical processes, the suspended and floating solids of sewage are removed by various mechanical devices. The screenings are collected and disposed by incineration or burial. Following this, the sewage is passed through grit chambers to allow grit and gravel to settle. It is further processed to allow the solid matter to settle out as much as possible by passing it through sedimentation tanks. The effluent is collected and is treated with chlorine. The treated liquid is discharged into a large body of water or into porous soil. Biological treatment is used for primary treatment of sewage.

Septic tanks

Septic tanks are constructed for individual family (Fig. 1). It is generally a concrete vat (sometimes a metal vat) placed below the ground level. All the domestic waste water flows into this vat and is retained for a long time to permit adequate decomposition. Sedimentation and biological degradation of sludge takes place in a septic tank. The organic material is hydrolyzed and fermented by anaerobic bacteria. The products are sugars, alcohol, organic acids, amino acids, amines, glycerol, fatty acids and other chemicals. The gases produced are H_2S, CH_4, CO_2 and H_2. The gases are passed out through the outlet pipes and the effluents are distributed under the soil surface through perforated pipes. The material which, is not decomposed by microbes, is called as sludge. It can be thrown into the soil, where it contributes to the humus. The pathogenic microbes are not destroyed in septic tanks.

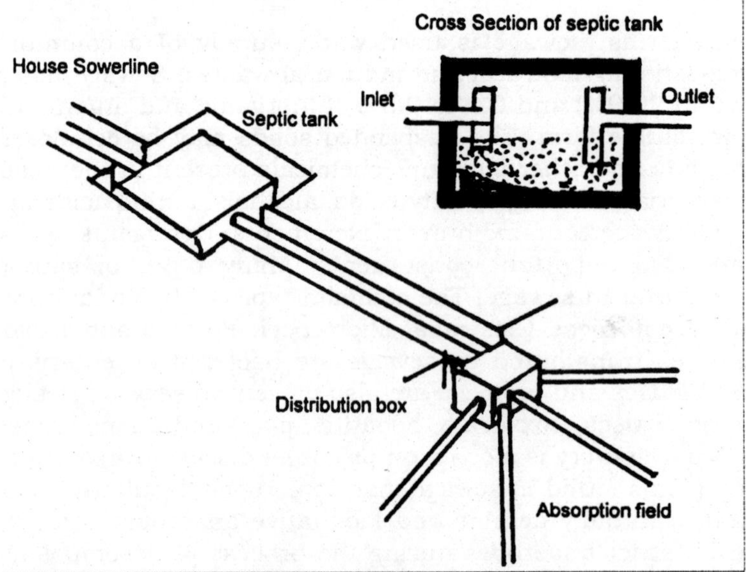

Fig. 1. The septic tank

Imhoff tank

Imhoff tank is a large scale sewage treatment plant (Fig. 2). It is used for treating the sewage or a community or small city. It consists of two chambers an upper and a lower chamber. As the sewage passes through the upper chamber the solid matter settles out and falls through the bottom opening and collects in the lower chamber. The sediments are decomposed by anaerobic micro-organisms. Gases produced during decomposition are collected and burned for heat and power. The undecomposed sludge is collected and is further treated. Organic matter in liquid stage is subjected to aerobic processes in oxidation pond, aerating (trickling) filter and activated sludge. On oxidation these materials yield inorganic materials like carbon di oxide, sulphate, nitrate, phosphate and water.

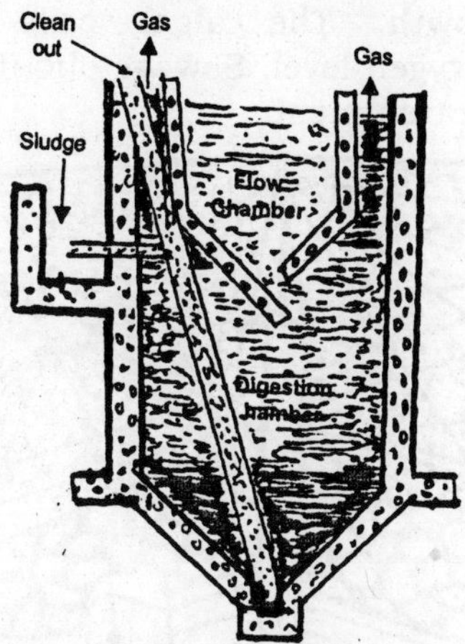

Fig. 2. The Imhoff tank

MUNICIPAL TREATMENT PROCESS

Municipal waste water is treated in a five step process viz., primary treatment, secondary (biological) treatment, advanced treatment, final treatment and solids processing.

Primary treatment

In this step, the coarse solids are removed. It is carried out by mechanical processes like screening, grinding and grit chamber. Sedimentation units/tanks or mechanical devices bring about collection of sludge.

Secondary (biological) treatment

The organic matter is absorbed and oxidized by microorganisms. There is reduction of BOD. The methods employed are (1) filtration, (2) aeration and (3) oxidation. Filtration is carried out in intermittent sand filters, contact filters and trickling filters. *The trickling filters* (Fig. 3) are piles of rocks over which sewage or organic wastes slowly trickle. It consists of a bed of crushed stone, gravel, slag or synthetic material with drains at the bottom of the tank. The sewage is sprayed intermittently over the surface of the bed. Rotating arm device or nozzles are used for this purpose. Air also passes with sewage so that aerobic conditions are maintained. This enables the growth of aerobic microorganisms on the sewage medium. These microbes oxidize and hydrolyze the organic matter during filtration process and comparatively clear effluent is obtained. *Activated sludge process*: sludge is prepared by aerating sewage vigorously resulting in the formation of floccules, which is allowed to settle. The floccules are aggregates of suspended and colloidal particles along with microorganisms. The process is repeated and sedimentation floc (activated sludge) is collected. Different species of bacteria, yeast, moulds and protozoa are present in the sludge. Some species of bacteria form zooglea (Gr. Zoos=living; Gloea=glue, living glue). Zooglea is formed as a sticky and gummy mass of polysaccharide covering the bacterial cells. Oxidation ponds (lagoons) or stabilization ponds are shallow ponds favourable for algal growth. The algal cells produce oxygen during photosynthesis increases oxygen level. Sewage should be subjected to oxidation pond preceding the primary treatment.

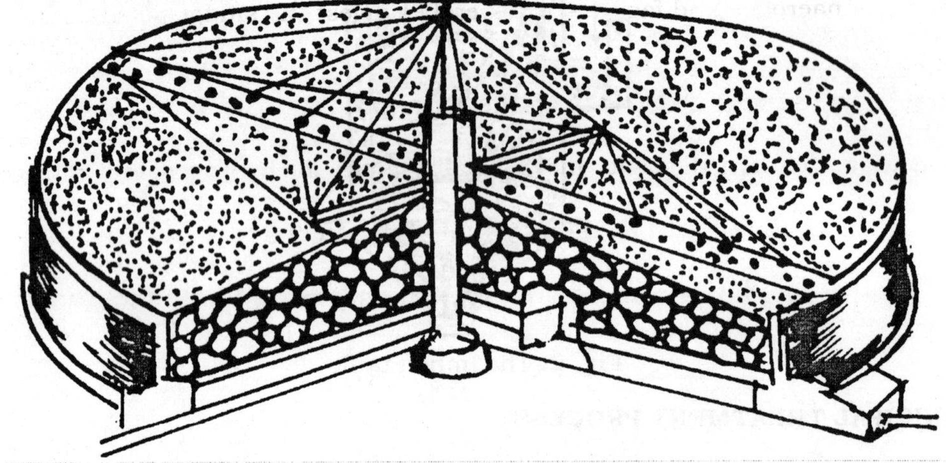

Fig. 3. The Trickling filter

Advanced treatment

The additional organic and inorganic wastes and nutrients (like phosphorus and nitrogen) are removed to further reduce BOD. This may be conducted biologically. However, physical-chemical methods predominate. Unit processes have been developed for biological nitrification-denitrification, filtration, reverse osmosis, carbon absorption, chemical addition and ion exchange. This type of treatment is a costly affair.

Final treatment

The treated liquid is disinfected and disposed into a body of water. The disinfected water is harmless for public health. Treated water may be used for down-stream water supply, recreation, irrigation or shellfish harvesting. Before disposal water can be post aerated by dissolving oxygen to it.

Solid processing

It is thickening, stablization, dewatering and disposal disposal of sludge. Solids are removed from liquid processes (primary, secondary and advanced treatment). It is followed by thickening, i.e., further concentrating the solids or sludge. Thickening may be accomplished by gravity thickening or by dissolved air flotation. The next step is stabilization. Aerobic and anaerobic digestion, composting, chemical addition and heat treatment processes are employed for stabilization. Dewatering is accomplished by physical method with the help of vacuum filters, belt filter presses, plate and frame presses and centrifuges. Addition of polymers or other chemical coagulant aids enhances dewatering process.

ANAEROBIC SLUDGE DIGESTION

The sludge is subjected into a separate tank (sludge digestor) for its digestion. Anaerobes and facultative anaerobes digest the sludge generally at 50° to 60°C and pH 7.0. This temperature is required for thermophilic bacteria. However, 30°C temperature is maintained, as it is not practical to operate thermophilic bacteria. Digestion is completed within 2 to 3 weeks. The sludge is digested into soluble substances and gases are produced. The gasses produced are methane, carbon di oxide and a small amount of hydrogen and nitrogen. These gases can be used as a fuel for heating purposes and for operating power. Fresh sludge is mixed with a small amount of digested sludge (ripe sludge). The latter acts as an inoculum. In a modern waste water treatment plant the sludge remaining in the digestor is removed to a mechanical drying unit. The dried sludge (cakes) is carried by conveyor belt to incinerators. The cakes are turned into ashes in the incinerators.

Composting

Dewatered sludge is mixed with wood chips. The latter enhances air circulation. This mixture is placed in aerated piles and allowed for decomposition for 3 weeks. After this, the wooden chips or the bulking agent is separated from the sludge. It needs several weeks for biological decomposition of such sludge. Upon final curing, the sludge has been transformed to a humus-type material and is suitable for use as a soil conditioner.

EUTROPHICATION

Sewage contains phosphates, nitrates and other matters which when added in water bodies makes it productive or eutrophic. This phenomenon is known as eutrophication. Stagnant water bodies in their early stages of formation do not contain substances, which favour the growth of biological forms. This unproductive state is known as oligotrophic. The algal forms contribute the maximum in the flora of eutrophic water bodies. Later, cyanobacteria occupy the major area of these water bodies. They form algal blooms, which is discussed later. Due to enhanced algal growth, water bodies appear as big mounds producing unpleasent odor, clogging pipes and interfering with fishing and navigation. Following strategies will enable to control eutrophication.

1. Control of addition of substances in water bodies favouring eutrophication. For this purpose, water must be properly treated before discharge.
2. To remove dissolved nutrients in water by physical or chemical methods.
3. To check recycle of nutrients into the water through harvest and removal of algal blooms upon their death.
4. Encourage growth of bacteria, which may reduce the amount of nutrients solubilised in water.

ALGAL BLOOMS

Profuse growth of some species of cyanobacteria in water bodies is known as algal blooms. The blooms appear as floating scums or blankets of algae of water bodies. The bloom forming algae compete for light with other photosynthetic plants. They cause suffocation to the animals living in the water body. They release some toxic chemicals, which kill fish, birds and other animals. The blooms are not utilized by zooplanktons.

WATER POLLUTION

Introduction of any foreign substance, which changes the physical and chemical characteristics of water making it useless for legitimate purposes, is known as water pollution. Polluted water is turbid with false odour and is unpotable. Polluted water carry many pathogens, which cause diseases known as water borne disease. In earth, water sources do not contain pure water. Some minerals, gasses and suspended particles are always present in it. Such, naturally found components, which may not pollute water but make an environment for the growth of microorganisms.

Pollutants: Sources and Effects

Water is polluted by different source carrying different types of pollutants. The main sources of pollution are sewage, industrial effluents and agricultural wastes. The domestic waste water is called sewage. It has been already discussed.

Different industries have different effluents varying in their compositions. The effluents are inorganic or organic pollutants. This includes solvents, acids, salts, oils, metallic wastes, suspended solids, phenols, plastics, plasticizers, DDT etc. Industrial effluents may also contain some radioactive substances. These pollutants are toxicants. The heavy metals like lead, mercury, zinc, copper, cadmium, chromium, sodium etc are harmful for plants, animals and for human health. Generally, gastrointestinal and neuromuscular systems are most effected systems. The enzymes inhibited by toxic metals are adenosine triphosphatase, amylase, carboxypeptidase, carbonic anhydrase, glutamic-oxaloacetic transaminase (inhibited by Cadmium), alkaline phosphatase, glucose-6-phosphatase, lactate dehydrogenase (inhibited by mercury) and pyruvate dehydrogenase (inhibited by Arsenic). Flourine is also a major industrial effluent of different industries. It causes dental fluorisis and stifness of joints. The insecticides and pesticides are not decomposed readily. They become more harmful by virtue of biological magnification or food chain concentration. In simple terms, this phenomenon means that there is accumulation of such compounds at higher levels of food chain. It becomes more harmful because, the concentration of these compounds keeps on increasing, as it takes a long time for degradation. The toxic chemicals are subjected to microbial action, known as detoxication. In modifying the chemical, the detoxifying microorganisms destroy its actual or potential harmful influence on one or more susceptible animals or plants. Some harmful intermediates may be produced during detoxication.

The agricultural wastes are the chemicals used in fertilizers and pesticides (insecticides, herbicides and fungicides). Pesticides are compounds to control pests. The pests and pesticides are not defined properly. In general, all the organisms causing damage or destroy crops, incite diseases on crop, weeds are said to be pests and the chemicals used to kill them are pesticides. The pesticides are grouped as organochlorine pesticides, organophsophorous pesticides and carbamates. Pesticide include one or more elements such as arsenic, boron, bromine, hydrogen, iron, lead, magnesium, manganese, nitrogen, oxygen, phosphorus, sodium, sulphur, tin and zinc. The benzene ring is quite common component of pesticide. There are various types of pesticides, which are introduced in water as effluent. Species of *Pseudomonas, Aerobacterm Agrobacterium, Achromobacter, Flavobacterium* and *Cornebacterium* helps in degrading pesticides.

WATER BORNE DISEASE

Water is a carrier of many pathogens. The disease causing microbes present in water are bacteria, virus and protozoa. Among bacteria, members of family enterobacteriaceae are found in abundance. *Salmonella, Shigella* and *E.coli* are pathogenic forms of this family present in water. These bacteria colonize in the human gut, hence the name enteric bacteria. Pathogenic forms of virus carried by water are poliovirus and hepatitis viruses. Pathogenic protozoans are *Entameba histolytica, Giardia lamblia* and *Balantidium coli.* All these microbes are described in chapter 20.

CHAPTER - 24

MICROBIAL ECOLOGY AND INTERACTIONS

Physical and chemical properties of soil

Physical properties of a soil type depend on the size of particles in it. As per nature, soil is differentiated into gravel (particles above 2.0 mm), coarse or fine sand (between 0.05-2.0 mm), silt (0.002-0.05 mm) and clay (less than 0.002 mm). Soil is made up of particles and the space between the particles is called pore space. Air and water is present in pore space. The bulk density of soil is defined as the mass of a unit volume of dry soil, which also reflects the total pore space of a soil. The pore size and the bulk density determine the structure of soil. A soil profile consists of the surface litter and horizons. The horizons are classified as A-, B-, and C- horizons. A- horizon is rich in mixtures of minerals and organic matters. B- horizon consists of materials leached from upper horizon. The horizon with parental rocks is regarded as C- horizon; there are no biological activities in these horizons.

Soil is rich in chemicals, which mainly provide nutrients for plants and other organisms. The nutrients are carbon, hydrogen, oxygen, nitrogen, phosphorus, sulphur, potassium, calcium, magnesium, iron, manganese, zinc, copper, molybdenum, boron, and chlorine. Of these sixteen elements, iron, copper, molybdenum, boron and chlorine are considered as micronutrients. The water holding capacity of a soil is governed by the porosity or soil structure. The occurrence of water in soil can be of three types-gravitational, capillary and hygroscopic. Following rains or irrigation excess water is drained in soil as gravitational water. After the drainage, water retained at field capacity in soil pores is known as capillary water. Dry soil particles absorb water from atmospheric humidity. This water is called as hygroscopic water.

Soil microflora

The microflora of soil depends upon the type of soil, depth, temperature, moisture and water regime, pH, type and availability of organic matter, seasonal variations and other environmental factors. The different groups of microorganism present in soil are bacteria, actinomycetes, fungi, algae and protozoa.

The most common soil bacteria are *Pseudomonas, Xanthomonas, Streptococcus, Staphylococcus, Arthrobacter, Clostridium, Achromobacter, Acinetobacter, Agrobacterium, Alcaligenes, Bacillus, Brevibacterium, Caulo-bacter, Cellulomonas, Micrococcus, Flavobacterium, Corynebacterium, Sarcina* and *Mycobacterium.* The myxobacteria common in soil are *Myxococcus, Chondrococcus, Cytophaga* and *Sporocytophaga. Streptomyces, Nocardia* and *Micromonospora* are common actinomycetes in soil. *Anabena, Cylindrospermum, Lyngbya, Oscllatoria, Phormidium, Plectonema, Scytonema* are common cyanobacteria. Cyanobacteria form surface crusts on soils bare of plant growth and contribute to soil stabilization.

Yeasts are common in most soils. Species of *Candida, Rhodotolrula*, and *Cryptococcus* are abundant indigenous yeasts in soil. A number of fungi representing all the groups are present in soil. Fungi are able to metabolize all types of carbohydrates including polysaccharides. They perform various functions. Their role in decomposition of organic matter in soil is most valuable for soil fertility. Numerous fungi occur in soil in the form of dormant structures like sporangiospores, conidia, oospores, ascospores, basidiospores, chlamydospores and sclerotia. *Allomyces, Saprolegnia, Pythium, Abisidia, Cunninghamella, Mucor, Rhizopus, Zygorynchus, Pythium, Aspergillus, Botrytis, Penicillium, Trichoderma, Fusarium, Alternaria, Helminthosporium, Verticillium, Cladosporium* and *Chaetomium* are the most common fungi present in soil. The pathogenic fungi are also abundant in soil.

Algae are also present in soil. However, their number is not like that of bacteria and fungi. Some of the common algae present in moist soil are *Chorella, Chlorococcum, Protosiphon, Oedogonium* etc.

The flagellate protozoa are more common in soil. The common genera are *Allantion, Cercobodo Cercomonas, Monas, Spiromonas* and *Tetramitus.*

Microbial examination of Soil

Microbiological studies of soil need collection of soil sample. Samples (10-25 gm soil) are collected at a depth of 15 cm. Different technique for the isolation of microorganisms is discussed in chapter 3 (Section V). Other methods of microbiological studies are discussed below.

Direct microscopic examination of soil: One gram of soil is mixed with 9 ml of 0.015% agar solution. It is shaken and 1/10 of a ml of the suspension transferred to the center of a glass slide by means of an pipette. The suspension is evenly distributed over a 4 sq cm area of the slide with needle. 0.1 N HCl is used as a fixative. It is applied for a minute and after that slide is washed in water bath. Staining is carried out with rose bengal or erythrosin. The prepared slide is observed under compound microscope. Counting the number of each microbe gives a quantitative estimation. Haemocytometer is also used to count the numbers accurately.

Fluorescent Staining: The cells are stained with fluorescent stains and observed under uv light. It gives different colours. The stains used are acridine orange and fluorescein isothiocyanate.

Chemical methods to determine microbial biomass: Soil sample is treated with chloroform vapours. Due to this treatment, cytoplasm comes out of the cell. Biomass is estimated by carbon content of cytoplasm or by the estimation of ATP. Carbon estimation may be conducted by conventional method or by tracer technique using C14. ATP is estimated by using substrate luciferin. This substrate, in the presence of enzyme luciferase and Mg combines with ATP to give enzyme-luciferin-AMP. The latter breaks down and light is generated. Optical measurements of the light generated gives a picture of ATP content.

Nitrogen, oxygen and carbon di oxide, are the major gasses of soil.

MICROBIAL ECOLOGY

Ernst Haekel (1869) coined the term "Ecology". Ecology is the study of the relationships among different forms of life, with their physical environment. The earth and atmosphere in which organisms live is known as biosphere. The biosphere is made up of ecological units called ecosystems. Groups of individuals having similar properties and living in the same area are called populations. A community is formed by interactions of various populations. In an ecosystem, the biological component occurs in three forms:

1. producers are the photosynthesizing organisms in microbial ecology producers are photosynthetic bacteria, cyanobacteria and algae.
2. Consumers are simply heterotrophs. They cannot convert inorganic compounds into organic compounds. In microbial world, mainly protozoa are considered as consumers, and
3. Decomposers which, breakdown complex organic matter into simple inorganic compounds which can be consumed by producers, bacteria and fungi are the decomposers. The specific physical space occupied by an organism is known as niche. It is also expressed as the specific role or function of an organism in a community. Microbial ecology is studied under autecology and synecology. Autecology is the study of life history and behaviour of individual populations as a means of adaptations to their environment. Synecology is the study of interactions between the various populations within the ecosystem.

TERRESTRIAL HABITATS OF MICROORGANISMS

Soils have many different microhabitats with differing physicochemical properties. The part of the soil microbial community, which is capable of utilizing the refractory humic substance, is designated as "autochthonous" by Winogradsky. This population is rich in actinomycetes and Gram negative bacteria. In contrast, microbial population not able to utilize humic compounds but exhibits high levels of activity and rapid growth on easily utilizable substrates (plant litter, animal droppings). These are designated as "zymogenous" or "opportunistic soil organism". These microbes have characteristic intermittent activity with inactive resting stages. The population is rich in bacteria and fungi like *Pseudomonas, Bacillus, Penicillium, Aspergillus* and *Mucor.* Presently term "indigenous" is used covering both types of populations.

At a particular location there may be several different microenvironmental situations favouring the growth of different indigenous microbial populations. Higher number of microorganisms occurs in the organically rich surface layers than in underlying mineral soils. Soils have both aerobic and anoxic regions. Bacteria found in soils may be obligate aerobes, facultative anaerobes, microaerophilic or obligate anaerobes. Gram positive genera are much found in soil as compared to aquatic environment. However, the Gram negative bacteria predominate in soil. Fungi are most abundant microorganisms of soil. Some microorganisms are capable of surviving in extreme conditions of temperature and pH. Soil receiving less than 25 cm of rainfall per year is called as desert soil. Microorganisms living in dry desert soils must be able to tolerate long periods of desiccation. Many bacteria and fungi living in desert form spores that allows them to survive even for decades between growth periods. When there is adequate moisture, the spores germinate and for a brief period, the organisms can actively grow and reproduce. Lichens are such organisms, which can survive in desiccated soil and at low temperature.

INTERACTIONS

F.E. Clements observed the phenomenon of coactions in ecosystem. Coactions are different types of associations and interactions amongst the micro-organisms themselves or with macro-organisms. He recognised four types of coactional relationships: syntrophism, competition, predation and symbiosis. Syntrophism Organisms are not intimately associated with each other but are benefited by each other. Competition Interaction among organisms, to gain nutrients and energy. Predation The predator is a species, which ingests other species the prey. Symbiosis Two different species coexist in an intimate relationship. Symbiosis shows mutualism, when both symbionts are benefited; Commensalism, when one symbiont is benefited while the other is neither benefited nor harmed; and parasitism, when one symbiont is benefited at the cost of other symbiont.

Interaction amongst the microbial population is an important aspect in an environment. These interactions may be positive or negative. The positive interaction between different populations enables each other to grow in a particular environment. It is a type of cooperation. The populations make resources available to each other. It is typical that population of some species may fail to grow individually but can flourish very well with the vicinity of other species. The negative interaction between the populations of different species hurdles and sometimes even may not permit a population to grow in an environment. It is a type of competition. Based on these interactions, a population may be divided as below.

Neutralism there is lack of interaction among the population of species. It is a rare phenomenon. Resting stage of a species behaves neutral with vegetative cells of other species. There is no competition for the available resources.

Commensalism is a type of unidirectional relationship. The term is derived from Latin word “mensa”, which means table. It describes a relationship in which one organism lives off the “table scraps’ of another one. Population of one species benefits while the other is not. A species by its biochemical processes makes an environment favourable for other species. Commensalism is based on physical or chemical modifications of the habitat. A facultative anaerobe will consume all oxygen making an environment totally lacking oxygen and thus obligate anaerobes may grow easily. A microbial population by virtue of its biochemical reactions produces growth factors (like vitamins), which are taken up by other populations. The other basis of commensal relationship is cometabolism. Cometablism is defined as the failure to assimilate second substrate by the primary organism. A species growing on a particular substrate gratuitously oxidizes it and the product is metabolized by other species. In some cases of commensal relationship, there is removal or neutralization of toxic materials. The ability to destroy toxic factors is widespread in microbial communities. A particular microbial population oxidizes hydrogen sulphide (detoxication) and benefits another population. Detoxification may also be due to immobilization. *Leptothrix* reduces manganese concentrations in some habitats, thus permitting the growth of other microbial populations. In certain types of commensalism, there is transformation of insoluble compounds to soluble compounds and the conversion of soluble compounds to gaseous compounds. *Desulfovibrio* can supply *Methanobacterium* with acetate and hydrogen from anaeraobic respiration and fermentation. The *Desulfovibrio* can utilize sulphate and lactate to generate these products. The *Methanobacterium* can use the products from the *Desulfovibrio* to reduce carbon di oxide. Commensalism may also help in enhancing dispersal.

Synergism (protocooperation) population between two species is for benefit of both the species. However, the relationship is not obligatory and both populations are able to survive independently. In some cases, two populations provide nutrition to each other; this phenomenon is known as cross feeding. Two populations relying on cross feeding cooperate with each other for completion of metabolic pathways. This type of synergism showing cross feeding, is called syntrophism. *E.coli* and *Streptococcus faecalis* show syntrophism. *S.faecalis* converts arginine to ornithine, which is used by *E.coli* to convert it to produce putrescine. *E.coli* alone convert arginine into agmatine. It fails to convert arginine into putrescine. Synergestic relationship has significance in degradation pathways of agricultural pesticides, e.g., *Arthrobacter* and a *Streptomyces* strain isolated from soil are capable of the complete degradation of the organophosphate insecticide. Plants and microbes establish synergistic interaction in rhizosphere. Plant roots provides some important chemicals (including some amino acids) to microbes by exudations. Microbes provide nutrients by solubilization of mineral; help plants for water holding capacity; removing toxic compounds like hydrogen sulphide; synthesizing compounds essential for growth and antagonizing potential plant pathogens through competition and developing amensal relationships. Microbes may produce allelopathic substances in rhizosphere. The allelopathic substances are produced by one organism, inhibits growth of another organism. By the production of such substances at rhizosphere, one plant will inhibit the growth of another plant. Synergestic relationships are also loose in the sense that one member population is often easily replaced by another one.

Mutualism is also named as symbiosis. In this relationship, two populations grow associated with each other and benefit both the populations. The association is obligatory. Symbiosis is discussed in details later in this chapter.

Competition is a negative relationship between populations. When two populations are striving for the same resources, they undergo competition. When the nutrients are low, one microbial population may have a higher growth rate and competitively displace a second population. Competition results in separation of closely related populations, is known as competitive exclusion principle. The establishment of a population after competition may be related to growth rate. Intrinsic growth rates of competing populations will vary under different environmental conditions. This results in coexistence of population though there is competition between them for same resources. Coexistence is observed among competing psychrophilic and psychrotropic bacteria in marine environments. Production of autoinhibitory substances also results in coexistence of multiple populations.

Amensalism (antagonism) is a relationship, when one microbial population produces a substance that is inhibitory to other populations. It is typical that one population (producing inhibitory substance) is not only unaffected by these substances but it may also gain a competitive edge and thus benefited. Production of ammonium, carbon id oxide and production and consumption of oxygen inhibit growth of some microorganisms. Yeast produces alcohol (ethanol) is inhibitory to many microbial populations especially bacteria. However, *Acetobacter* is capable of converting ethanol into lactic acid. The production of inhibitory substances by microorganisms may also act as a preservative for organic compounds in natural habitats. Some microorganisms produce antibiotics.

Parasitism Parasite takes nourishment from host and thus harms it. Species of all types of microorganisms are parasitic. The parasitic association is generally established between microorganisms with higher plants, animals and human beings. However, some microbes are parasite on other microbes, e.g., viruses are parasitic on bacteria, algae and even on fungi. A bacterium *Bdellovibrio* is parasitic on other bacteria. Similarly, protozoa are host for many bacteria, fungi or even other protozoa.

Predation simply means that one organism eats (engulfs) other organism and digests it for the sake of nutrition. The organism, which is engulfed, is called prey.

SYMBIOSIS

Symbiosis is a type of adaptation between two different species; living with each other is close association. Symbiosis is divided into two categories: ectosymbioses and endosymbioses. As the name suggests in ectosymbioses micro-organisms live outside the body of the host while in endosymbioses, microbes live inside the host cells. Depending upon the nature and benefits for the species, symbiosis may be mutualistic or parasitic types. In the former type, both the symbionts are benefited while in the latter type, one is benefited and the other is a looser. Sometimes the host symbiont may suffer damages due to such association.

FUNCTIONS OF SYMBIOSIS

The symbionts perform different functions for their benefits. Some of the important functions are as following

Nutrition In major types of symbiosis, the partners provide nutrition to each other. Mycorrhiza and nitrogen fixing bacteria are the common examples, which provide nutrition to the higher plants. A number of grazing animals, termites and wood boring insects take cellulose in their food but they are unable to digest it. Certain bacteria and protozoa present in the digestive tract of these animals help in the breakdown of cellulose.

Protection Endosymbionts and some ectosymbionts living inside the body cavity of animals feel protected from external environment. Even the micro-organisms provide protection to the host from some toxic materials. In the excretory organs of some insects, bacteria are present as symbionts. These bacteria breakdown urea and uric acid into ammonia, which is assimilate in their cells.

Provision of favourable position it is important in some symbiotic association, one partner by locomotion places the other partner so that the latter is able to get proper nutrition. In marine environment, some ciliated protozoa are present on the body surface of some crustacea. Present on these surfaces protozoans come in contact with their food due to respiratory or feeding current of the crustaceans. Some tridacnid clams carry algal symbiont in such a special organs, which acts as a lens, and provides light to these algal components.

ESTABLISHMENT OF SYMBIOSIS

Symbiosis is established either by direct transmission or by reinfection. In direct transmission, the symbionts are transmitted from one generation to another living associated with each other while in reinfection the symbionts from an association in each generation.

Direct transmission : In this type of establishment of symbiosis, both the symbionts are transmitted from one generation to another simultaneously. A species of alga is found intracellularly in a protozoan. The rate of cell division is almost the same for both the species. Hence, each cell division has the same number of protozoan and algal cells. In case of sexually transmitted symbionts, the endosymbiont goes up to zygote through egg cytoplasm. In certain insects endosymbiont are present in specialized cells, mycetocytes, which form an organ, mycetomes. The symbiont is transmitted from mycetomes to the egg directly. The mechanism of transmission varies from species to species.

Reinfection : In this type of establishment of symbiosis, symbionts infect its host in each generation. Some of the common examples of this type are observed in insects. In some insects, there is a direct anatomical connection between the intestine and vagina. The symbionts form a smear on the surface of egg. When larvae hatch it eats its egg's shell and thus the symbiont reaches in the intestine of the newly hatched larvae. In another example, the hatchling suck up drops of bacterial suspension. This suspension comes out from mother's anus as an exudation during the period of brood care. Nodulation process of symbiotic nitrogen fixation by *Rhizobium* is also an example of reinfection.

MYCORRHIZAS

Frank (1885) coined the term mycorrhiza (Gr. *Mykes* = mushroom + *rhizas* = root) to the composite fungus-root organs of the Cupuliferae. It is a mutulasitic association of fungi with roots/rhizomes of a plant. It involves integration of the plant roots and fungal mycelia, forming integrated morphological units.

Based on morphological and anatomical features, mycorrhizae are of two types (a) ectomycorrhiza and (b) endomycorrhiza.

Ectomycorrhiza

Ectomycorrhiza (Fig. 1) are present in temperate and boreal forests. Ectomycorrhizal fungi generally have optimal growth temperatures between 15 and 30°C, and are acidophilic with optimal growth at pH 4-6. Some ectomycorrhizal fungi grow best below pH 3.0. The fungal partner of ectomycorrhiza belongs to Ascomycetes or basidiomycetes. In ectomycorrhiza, the fungal partners are species of *Boletus, Cortinarius, Russula, Tricholoma, Entoloma, Hebeloma, Amantia, Rhizopogon* and *Scleroderma.* These fungi infect plants belonging to the family *Pinaceae, Salicaceae, Betulaceae, fagaceae, Junglandaceae, Caesalpinaceae* and *Tiliaceae.* In the ectomycorrhiza pseudoparenchymatous fungal tissue forms an envelope in a mantle or sheath on the lateral rootlets. Ectomycorrhizal roots are devoid of root hairs. The apical meristem is not infected. The hyphae penetrate the cortex and forms a network called the Hartig-net (named after the German botanist Hartig), in the middle lamella of the cell walls and the intercellular spaces of the external cortex. There are ectomycorrhiza in which, the hyphae external to the absorbing organs are not aggregated to any extent. There is characteristic intracellular as well as intercellular penetration by hyphae. The ectomycorrhiza are basically coralloid and ball types. Based on the external forms ectomycorrhiza are classified into three types. (1) Short dichotomously branched roots invested with mantles of various colours. The colours come up due to the fungal component. (2) knots or tuber-like growths developed by the fusion of mantles produced by merged clusters of forked roots, and (3) long, thin unbranched structures which occur upon the roots.

Ectomycorrhizal fungi grow well on simple carbohydrates such as glucose. However, some of them are able to utilize complex carbohydrate (like disacchrides and sugar alcohols). Ectomycorrhizal fungi generally utilize complex organic sources of nitrogen, amino acids, and ammonium salts. Still, many require vitamins such as thiamine and biotin for growth.

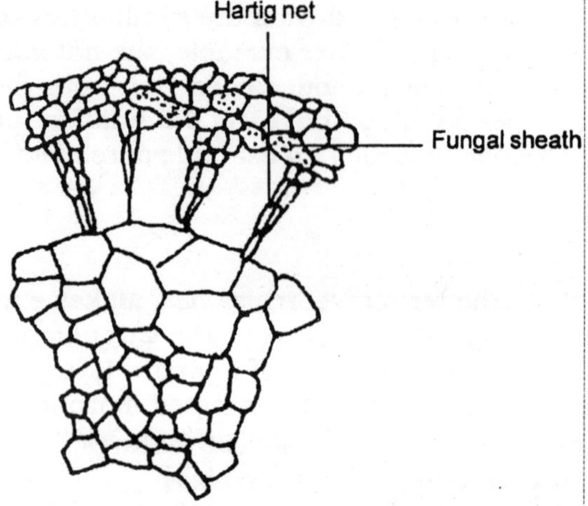

Fig. 1. Ectomicorrhiza

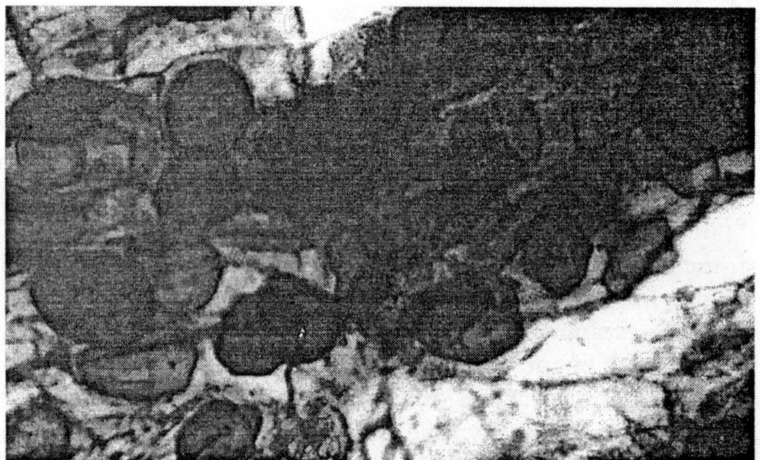

Vesicles of VAM in Root Tissue

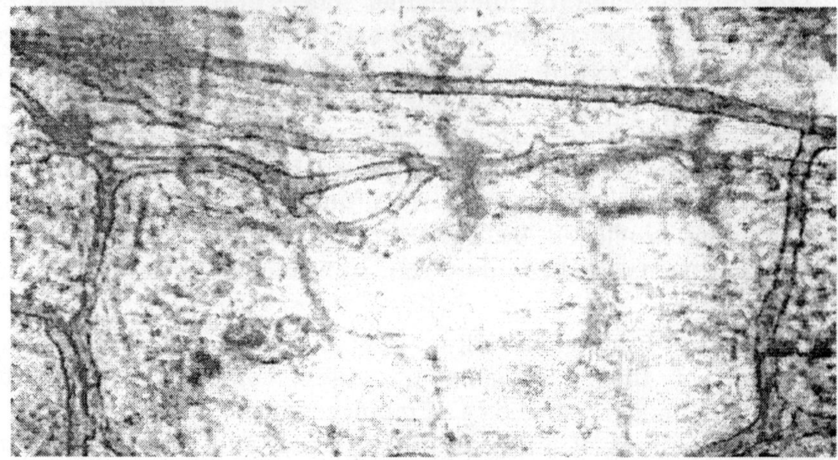

Branched hyphae of VAM fungi

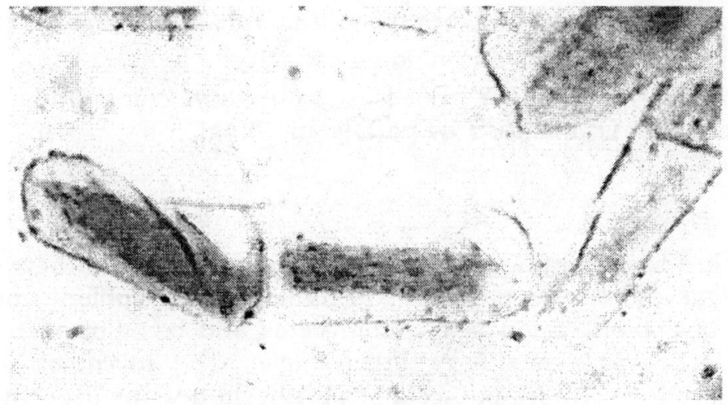

Arbuscules of VAM fungi

Endomycorrhiza

Endomycorrhiza is common in tropical, temperate and arctic regions. As the name suggests, the endomycorrhiza are present in the cortex of roots/rhizomes. The fungi involved in endotrophic association belong to the Phycomycetes or the Basidiomycetes. The fungal partner of endomycorrhiza is even intracellular and may be septate or nonseptate. Hence, they are divided into two types. The mycorrhizae caused by aseptate hyphae are of the vesicular-arbuscular mycorrhiza (VAM) type. It is common in most of the angiospermic plants. It is also found associated with various groups of bryophytes, pteridophytes and gymnosperms. Most major agricultural crops form this type of mycorrhizal association. Mycorrhizal association is established after a few weeks of seed germination. In VAM, inter- and intracellular hyphae are present in the cortex, and the infection inside the root is directly linked to external mycelium forms a loose network in the soil around the VA mycorrhizal roots. The mycorrhizae have vesicles, which are swollen, thick walled structures. The arbuscules are dichotomously branched haustoria. The VA mycorrhizal association results in increased phosphate uptake by the plants. It also improves the uptake of other ions, like zinc, sulphur and ammonium from soil.

Endomycorrhizal associations, which are not of the VA type, occur in Ericales and Orchidaceae groups of higher plants. *Armillaria mellea* and *Rhizoctonia solani* form endomycorrhizal associations with orchids. The orchid roots are internally infected and attacked by the hyphae, which pass through the surface cells into the cortical cells. The fungi from coils within the cells of the outer cortex and later the hyphae lose their integrity and much of their contents passes into the host cell. Endomycorrhizal fungi of orchids help in seed germination. It exhibits nutritional characteristics that are different from those of other mycorrhizal fungi. Endomycorrhizal fungi utilize simple sugars.

Significance of Mycorrhiza

1. Mycorrhizal infection results in profuse branching of rootlets, thus increasing absorption area.

2. Mycorrhiza increases the uptake of minerals and inorganic substances.
3. Mycorrhiza helps in water holding capacity.
4. Mycorrhizal association releases inhibitory compounds to establish barriers against infection by pathogenic fungi.

LICHENS

Lichens are structurally organized symbiotic association between alga and fungi. The algal and fungal partners are named as phycobiont and mycobiont, respectively. Mycobiont belongs to ascomycetes and basidiomycetes. While, the phycobionts are cyanobacteria or green algae. The mycobiont belonging to ascomycetes produce either apothecia or perithecia bearing asci with ascospores and paraphysis in mature lichen thalli. The phycobionts are *Nostoc, Gloeocapsa, Rivularia, Pleurococcus (Protococcus), Cystococcus, Trentepohlia* and *Cladophora.* In a lichen the phycobiont utilizes light energy and produces organic compounds, which are utilized by the mycobiont. The mycobiont provides a form of protection for the phycobiont. The mycobiont may also synthesize growth factors to the phycobiont.

Types of Lichens

Lichens have different forms and colours. Lichen thallus has closely interwoven hyphae of the fungus in association with the algal component. Based on the morphology, lichen is grouped in three forms: Crustose, Foliose, and Fruticose. Crustose lichen (Fig. 2) form a crust closely adpressed to the substrate may be partly or wholly embedded in it, e.g., *Graphis, Verrucaria.* Foliose lichen (Fig. 3) has leaf-like thallus with lobed or irregular margins, e.g., *Physcia, Gryophora.* Fruticose lichens (Fig. 4) are more or less bushy-branched to upright in habit, e.g., *Cladonia, Usnea.* Anatomically lichen thallus shows different zones. These zones are properly differentiated in foliose lichens. Crustose and fruticose lichens show some or no differentiation. The uppermost zone or upper cortex consists of more or less vertical hyphae, with or without intercellular spaces. Next to upper cortex is algal or gonidial layer. It is made up of loosely interwoven fungal hyphae intermingled with algal filaments. Beneath the algal layer is medulla and lower cortex. It is made up of loosely interwoven hyphae. From lower surface of lower cortex, arise some hyphae known as rhizines. These structures help in attachment to substratum.

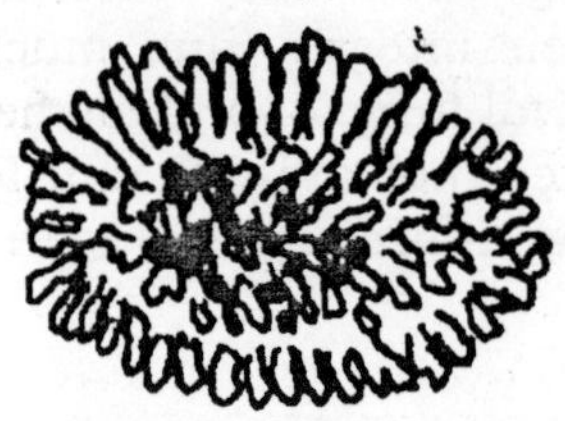

Fig. 2. Crustose lichen

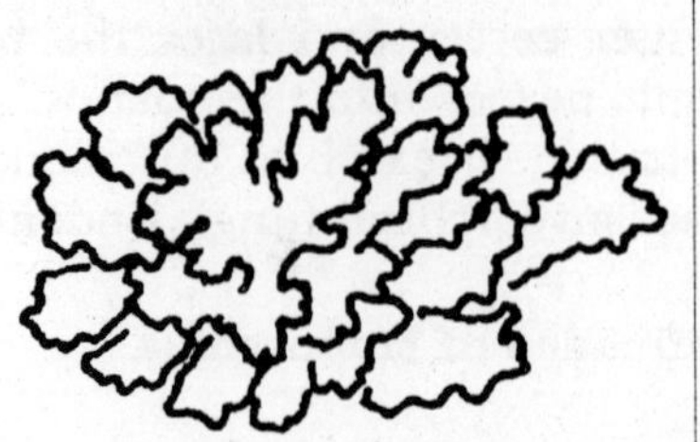

Fig. 3. Foliose lichen

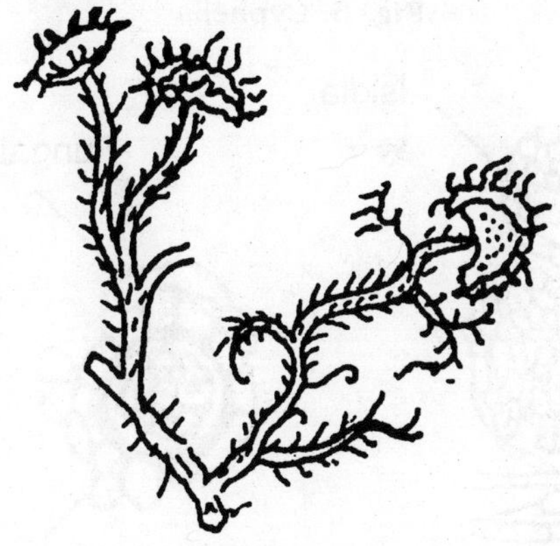

Fig. 4. Fritcose lichen A. External morphology

There are some specialized structures, which perform different functions. Breathing pores (found on upper cortex), cyphellae and pseudocyphellae (Fig. 5) perform gaseous exchange. *Cyphaellae* are circular, concave depression found in lower cortex of some foliose lichen. In these, cup like structures are present hyphae with thin walled rounded cells. The hyphae are originated from medulla and the end cells give white powdery appearance to the cup. *Pseudocephellae* resembles cephellae. The only difference is that no margin is formed. White filaments appear by bursting of cortex. *Isidia* are rod shaped, cigar-shaped, or coral-like outgrowth meant for increasing the photosynthetic surface. Isidia (Fig. 6) are formed on external cortical layer and are covered by this layer. It is made up of outer fungal hyphae with internal algal layer. The fungal and algal components are the same as present in lichen thallus. It also helps in propagation. *Soredia* (Gr. Soros = heap) (Fig. 7) are minute, rounded, structures and take part in reproduction. It is made up of algal cells clasped and surrounded by fungal hyphae (Fig. 7B). They are formed in patches on upper cortex and the latter becomes perforated. Soredia are disseminated by wind or rain. When fall on suitable substrates, they germinate into a new thallus. Pycnidiospores are common asexual spores formed in flask shaped structures, the pycnidia. The spores contain only fungal component. Following germination, pycnidiospores in contact with proper algal component, and forms lichen thallus. Fragmentation of thallus is the most common method of reproduction.

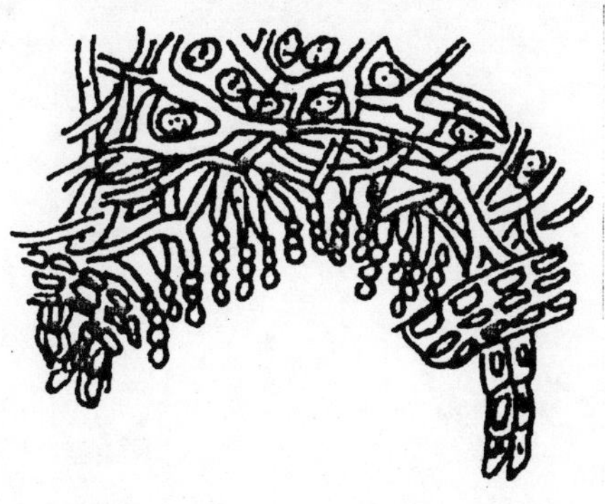

Fig. 5. Cyphella

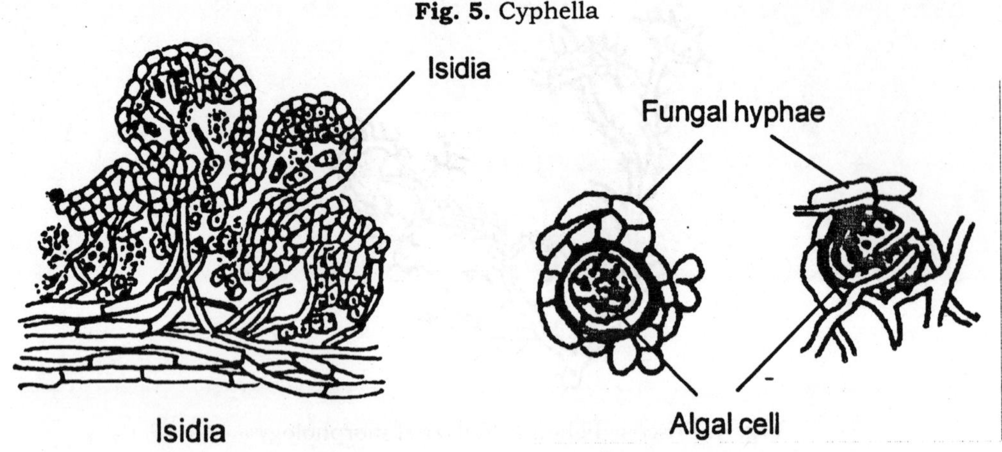

Fig. 6. L.S. Isidium

Fig. 7A. L.S. Lichen with soredia
B. Soridium

Lichens have multiple economic importances. Lichen grows slowly and on habitats in which other microbes fail to grow. They are important ecologically by virtue of forming soil and act as pioneers of rock vegetation. Some lichen, are able to fix atmospheric nitrogen. *Lobaria pulmonaria, Evernia prunastri* and species of *Ramalina* contain a carbohydrate lichenin (similar to starch). These lichens are used as fodder. *Cetraria islandica, Evernia prunsatri, Endocarpon miniatum* are used as food in different countries. In India species of *Parmelia* is used by some natives. Species of *Peltigera, Parmelia, Perusaria, Cladonia* have medicinal values. Lichens are also used for preparation of dyes. They are useful in tanning, brewing and distilling industries.

CHAPTER - 25

BIOGEOCHEMICAL CYCLE

Soil is a complex mixture of inorganic and organic components. There are two main groups of soils. (1) Mineral soils that originate from initial weathering of rocks, and (2) Organic soils laid down in bogs and marshes. Weathering of rocks is brought about by physical and biological processes. The biological activities of lichens, algae, fungi and bacteria bring about weathering of soil by production of organic acids.

NITROGEN CYCLE

Nitrogen is an important structural element of all types of cells. In a cell, nitrogen is present in oxidized state like $R\text{-}NH_2$, NH_3, NH_4+, N_2, N_2O, NO, NO_2- and NO_3-. It is present in gaseous form. However, ammonia and some organic nitrogenous compounds are also utilized by plants to a lesser extent. There is a huge amount of organic nitrogen in soil. The conversion of organic nitrogen to the more mobile, inorganic state is known as nitrogen mineralization. It is conducted by many micro-organisms like bacteria, fungi and actinomycetes.

Ammonification : The transformation of organic nitrogen to inorganic nitrogen in which there is accumulation of ammonia. In other words, it may be defined as release of ammonia from organic nitrogenous compounds.

Nitrification : Oxidation of ammonium salts into nitrates through an intermediate nitrous acid is known as nitrification.

Denitrification : It is a process in which nitrates are reduced.

AMMONIFICATION

A number of ammonifying micro-organisms are present in all types of soil and water conditions, which oxidize proteins. *Bacillus mycoides, B.subtilis, B. putidis* are aerobic bacteria taking part in ammonification while *Clostridium putrifaciens, C. sporogenes* are anaerobic bacteria, which decompose proteins. The anaerobic decomposition of protein is known as *putrifaction.* The end products are generally amines and related compounds and hydrogen sulphides.

Protein catalyzing enzymes are known as *proteases* or *proteolytic enzymes.* Protein molecules first break down into peptides, which is digested into their monomer amino acids. Amino acids are then reduced to certain derivatives, depending on the organisms and the condition of decomposition. Ammonia is ultimate end product. The enzymes catalyzing these reactions are known as *deaminase* and *amidase.*

The biochemical reactions of ammonification are described as follows in atmosphere. Other forms of nitrogen present in biosphere are NH_3, $R\text{-}NH_2$ and NO^-_3. Gaseous from of nitrogen is not utilized by plants or animals. Higher plants generally uptake nitrogen in the form of nitrates

Hydrolytic deamination : The amino acids formed as a result of breakdown of proteins, is hydrolysed to form lower fatty acid and ammonia; or an alcohol, CO_2 and ammonia or an aldehyde, lower acids and ammonia.

$$R.CH.NH_2.COOH + H_2O \rightarrow R.CHOH.COOH + NH_3$$

$$R.CH.NH_2.COOH + H_2O \rightarrow R.CH_2OH + CO_2 + NH_3$$

$$R.CH.NH_2.COOH + H_2O \rightarrow R.CHO + H.COOH + NH_3$$

Decarboxylation and amine formation :

$$R.CH.NH_2.COOH \rightarrow R.CH_2NH_2 + CO_2$$

$$R.CH_2.NH_2 + H2O \rightarrow R.CH_2.OH + NH_3$$

Reductive deamination:-

$$R.CH.NH_2.COOH + H_2 \rightarrow R.CH_2COOH + NH_3$$

$$R.CH.NH_2.COOH + H_2 \rightarrow R.CH_3 + NH_3 + CO_2$$

Aerobic bacteria and specially fungi oxidize amino acids and form ammonia without reduction. The reactions involved are oxidative deamination, decarboxylation and reduction. Organic acids are produced with ammonia. If oxygen is present organic acids are oxidized into CO_2 and H_2O.

$$R.CH_2.CH.NH_2.COOH \rightarrow R.CH.OH.COOH + NH_3$$

$$R.CH.NH_2.COOH + O_2 \rightarrow R.COOH + NH_3 + CO_2$$

Besides, monoamino acids, proteins contain di-amino acids. Di-aminoacids are decomposed into protamines or diamines. In this process, decaroboxylation of diaminoacids takes place without deamination.

Lysine $\rightarrow$ Pentamethylenediamine

Sulphur if present in amino acids is released as hydrogen sulphide.

Aromatic amino acids produce putrification of some characteristic compounds. On putrifaction, phenylalanine produces benzoic acid; tryosine produces phenol and tryptophan produces indole and skatol.

Pasteur discovered bacteria, which takes part in ammonification of urea. Bacterium is found in bacillus and coccoid forms. It was named as *Urobacter.* They produce an enzyme urease, which catalyzes the decomposition of urea. During, reaction ammonium carbonate is produced which is unstable and is soon converted into ammonia and carbon di oxide.

NITRIFICATION

Ammonium salts are oxidized into nitric acid (nitrates). Plants are able to uptake nitrogen in this form. In 1890 S. Winogradsky observed for the first time that nitrification is associated with the metabolism of certain chemoautotrophic bacteria. The bacteria, which take part in nitrification, are known as nitrifying bacteria. In Bergey's manual nitrifying bacteria are placed in family Nitrobacteriaceae of order Pseudomonadales. These bacteria have two types of nature one, which derives energy by oxidation of ammonium and the second derive energy by the oxidation of nitrite.

I. Family Nitrobacteriaceae

Bacteria receive energy by oxidation of ammonium into nitrite.

(a) Zoogloea not formed.

Cells ellipsoidal – *Nitrosomonas*

Cells spherical – *Nitrosococcus*

Cells spiral – *Nitrospira*

(b) Zoogloea formed.

Zoogloea surrounded by common membrane forming a cyst –*Nitrosocystis*

Cells occur in masses in slime layer without a common membrane – *Nitrosogloea*

Bacteria receive energy by oxidation of nitrate.

Zoogloea not formed, *Nitrobacter.*

Zoogloea are formed, *Nitrocystis.*

Nitrification is a strictly aerobic process and does not occur at redox potential lower than +200 mV. Alkaline pH (7-8) favours nitrification. The biochemical reactions of nitrification can be studied in two phases. In the first phase ammonium is oxidized into nitrous acid through the removal of electrons.

$$NH_3 \rightarrow NH_2OH \quad (HNO) \rightarrow \quad NO_2^-$$

In the second phase nitrous acid is oxidized into nitric acid and it takes place in the following manner.

Along with this process carbon di oxide is reduced due to chemosynthesis.

DENITRIFICATION

In general terms, denitrification is a process of reduction of nitrate. In this process nitrogen is lost as gas. Micro-organisms reduce nitrate and nitrite and molecular nitrogen and in some cases nitrous oxide is liberated.

NITROGEN FIXATION

Nitrogen fixation simply means that atmospheric nitrogen is converted in a form, which is easily utilized by plants. This is conducted by micro-organisms.

There are two main groups of nitrogen fixing organisms. These types are based on their mode of nitrogen fixation. They are (1) Symbiotic nitrogen fixers and (2) Non-symbiotic nitrogen fixers.

SYMBIOTIC NITROGEN FIXATION

Boussingault (1837), a French chemist observed that nitrogen present in the leguminous plants exceed the nitrogen present in the seeds and soil. Hellriegel and Wilfarth established the fact that nitrogen is fixed by certain bacteria living in root nodules of the leguminous plants. Beijerinck (1888) isolated pure culture of root nodule forming bacteria and named it as *Bacillus radicicola.* This bacterium was later named as *Rhizobium.* It is classified with *Agrobacterium* and *Chormobacterium* in the family Rhizobiaceae. *Rhizobium* is a Gram-negative, aerobic, non-spore forming rod shaped bacterium. The cells are 4 to 5 μ long and 1 μ in thickness. Some cells may be slightly curved. These curved cells may be T or Y in form and are called bacteriods. Bacteriods ultimately form motile forms. Beijerinck has described all these forms in relation to its developmental stages in three steps. In soil, *Rhizobium* is present in rod shaped cells. These cells can penetrate in the root hairs of leguminous plants. In the second stage bacteria changes into motile form and in the last stage, which is inside the plant tissues, the cells are transformed into bacteroids. This bacteria forms root nodules in leguminous plants. This process is known as nodulation process. Prazmonski (1889) studied the mechanism of root infection by the legume organism. According to him the nodule forming bacteria may be three forms: (1) those, which develop normally by the path of infection threads. This type is found in most of the leguminous plants. (2) Those, which develop through the infection, taking place mainly by the intercellular zoogloea and (3) those, developing from the bacteria with long infection.

The nodulation process

Rhizobia enter the roots of legumes either through root hairs or directly at the point of emergence of lateral roots. The root cap cells secrete large amounts of polysaccharide and globules of material may be formed at the root hair tips. This material forms the 'mucigel' that attracts the rhizobia (Fig. 1). This is the initial stage of nodulation process, by which bacteria establishes at the ends of the root hair and grows rapidly. Enzyme polygalacturonase plays an important role in root hair infection. The plant roots excrete tyrptophan in the soil. The bacterial cells convert tryptophan into indoleacetic acid (a growth hormone). This growth hormone or auxin softens the cell wall of the legume root hair. These bacteria also produce an extracellular polysaccharide capsular material that induces the plant root to excrete the enzyme polygalacturonase. Following infection bacteria soon enters the walls of the root hairs. Inside the root hair, bacterium forms a filamentous structure, which extends up to the parenchymatous tissues of the root. This filamentous form is known as infection threads. This thread branches into the central portions of the developing nodules and the bacterium is released in the cytoplasm of the root cells. After this stage, bacteria cause a rapid proliferation of the surrounding tissue in the innermost cells of the root cortex, which results in the formation of a young nodule. These nodules pushes out the overlying parenchyma and epidermis and produce a swelling on the side of the root. A nodule consists of a mass of thin walled parenchyma cells. These cells are filled by bacterial cells known as bacteriods. Bacteroids have a peculiar morphology and cellular forms. They are swollen and irregular, frequently appearing in star, clubbed or branched in shapes.

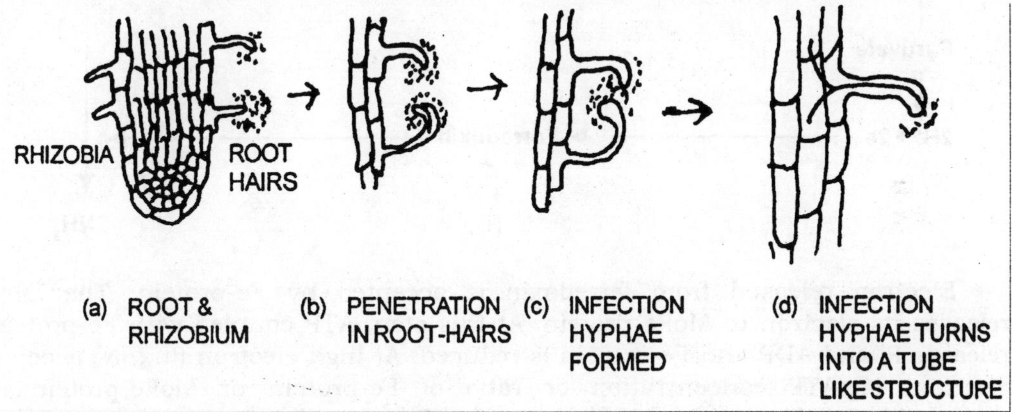

Fig. 1. Rhizobium - Nodulation process

The enzyme nitrogenase

Nitrogen fixation is an anaerobic process. The oxygen supply to bacteroid is excluded due to presence of leghaemoglobin around it. Enzyme nitrogenase catalyzes nitrogen fixation. This is a complex enzyme having two components viz., nitrogen reductase (Fe-protein, molecular weight about 65000) and nitrogenase (MoFe protein, molecular weight 200,000). Both of these enzymes catalyze reduction of N_2 to NH3, acetylene to ethylene, H+ to H_2 etc. The iron of nitrogenase is believed to be involved in the binding of nitrogenase while molybdenum is responsible for the decrease in the strength of nitrogen bonds to an optimum extent to facilitate reduction. Nitrogenase couples ATP hydrolysis to ADP with electron transfer from a reduced electron donor, probably a ferredoxin or a flavodoxin, to reduce N_2 to NH3. The redox activity is brought about by coenzyme ferredoxin (non-haem Fe compound) or flavodoxin (FMN containing protein). It is the first reducer, which in turn reduces nitrogen bound to the enzyme nitrogenase in presence of ATP. The mechanism of electron pathway to reducible substrate through the nitrogenase complex is as following.

Fd (ox) + nH $\rightarrow$ Fd (red)

Fd (red) + nitrogenase + nitrogen $\rightarrow$ Fd (ox) + N2-nitrogenase (red)

N2-nitrogenase(red) + nATP $\rightarrow$ NH_3 + nitrogenase + nADP +nPi

This mechanism may be expressed in details as following :-

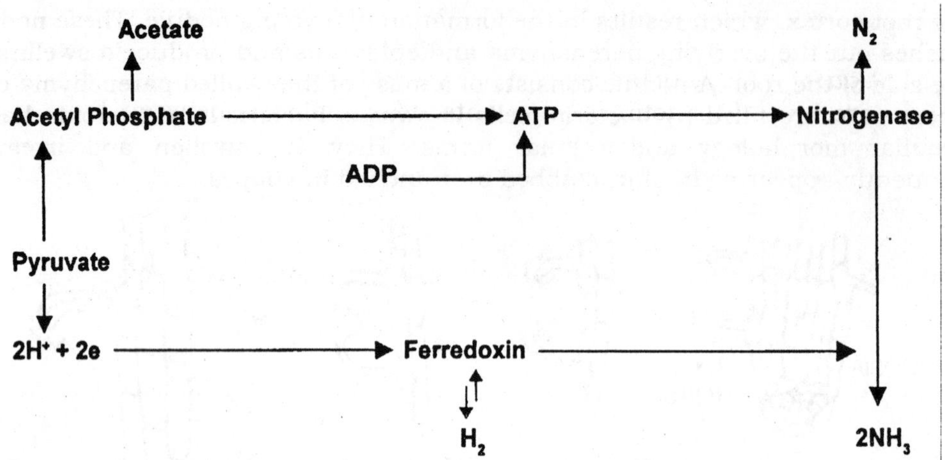

Electron released from ferredoxin is accepted by Fe-protein. The later releases its electron to MoFe protein. At this step, ATP coupled with Fe-protein releases Pi and ADP and Fe-protein is reduced. At high electron flux, N_2 accepts electron. Mg.ATP concentration or ratio of Fe-protein or MoFe-protein is responsible for electron flux. Hydrolysis of ATP is associated with the transfer of electrons from nitrogenase reductase to nitrogenase.

Reduction of molecular nitrogen takes place as following mechanism

$$N_2 + 6e^- + 12\,ATP + 8\,H^+ \xrightarrow{Mg^{2+}} 2NH_4^+ + 12\,ADP + 12Pi$$

Hydrogen production is a constant feature of nitrogenase activity. One molecule of hydrogen is evolved when one molecule is reduced. The mechanism can be expressed as following

$$2H^+ + 2e^- + 4ATP \xrightarrow{Mg^{2+}} H_2 + 4ADP + Pi$$

Sucrose formed by photosynthesis reaches up to the roots from its site of synthesis i.e., leaf. This is the main source of energy for this reaction. The respiring root nodules oxidize sucrose. This metabolic activity yields ATP and reduced ferredoxin. Low oxygen tensions of the developing nodule brings about transformation of vegetative cells of bacteria to bacteriods. Ammonia formed by the bacteroids is transferred to host cells where it is converted into organic form, glutamin.

The genetics

nif genes were isolated and studied from *Klebsiella pneumoniae.* These genes relate to the nitrogenase enzyme, the electron transfer system and the regulatory functions. It is a cluster of 21 genes. In bacterial chromosome this cluster is present after the genes regulating histidine biosynthesis. Five genes (nif B, nifN, nifE, nifV, and nifQ) are involved in the formation of functional Mo-Fe protein. The nifM and nifS are responsible in the processing of the Fe protein. Role of the remaining six genes is still not clear. Ammonia totally represses nif gene products biosynthesis.

Frankia induced nodulation

Frankia is the actinomycete. The plants in which, nodules formed by *Frankia* are known as actinorhizal plants. The symbiosis between *Frankia* and plants is known as actinorhizal symbiosis. About 24 angiospermic plants are reported to have actinorhizal association. It is most common on *Alnus* and *Casuarina* and hence the structure of nodule is named as Alnus type and Casuarina type. The only difference in the structure of these two nodules is that in Alnus type, there are present lenticells for ventilation. It is absent in Cauarina type. Nitrogenase activity is observed in the vesicles of these nodules. Ammonia assimilation is observed by involvement of GS and GOGAT enzyme system. The vesicles produce glutamine. It is transferred to vegetative hyphae, where it is converted glutamate by GOGAT enzyme. The glutamate goes back to vesicles and acts as ammonia acceptor. The second theory suggests that ammonia present in the surrounding of the vesicles is assimilated by the GS-GOGAT system.

Nitrogen fixation by Cyanobacteria

Gunnera is a herbaceous plant and species of *Nostoc* are present in mucilage-filled cavities near the bases of petioles. These cavities are known as glands. Species of *Nostoc* and *Anabaena* are also present in the coralloid roots of Cycads. *Anabaena azollae* is found associated with *Azolla.* These, cyanobacteria fix atmospheric nitrogen in free and in association stage. In heterocysts, ammonia assimilation takes place by GS and GOGAT enzymes system.

NON SYMBIOTIC NITROGEN FIXATION

Azotobacter and *Clostridium pasteruianum* fix atmospheric nitrogen non-symbiotically.

Nitrogen fixation by Azotobacter : In 1901, Beijerinck isolated Azotobacter. This bacterium is found in rhizospheres of many plants, it utilizes nitrogenous compounds like molecular nitrogen, ammonia, nitrate, nitrite, urea and sometimes organic nitrogen. Five species of this bacterium have been reported, viz. *A. chroococcum, A. beijerinckii, A. vinelandii, A. macrocytogenes* and *A. agilis.* These species differ in cell shape, pigmentation and motility. *A. chroococcum* is most widely found in soil of temperate regions. This bacterium has large cells, spherical and slightly oval in shape. Capsule is found covering the cells. Azotobacter is polymorphic, Gram negative and strictly aerobic bacterium. Old culture of this bacterium produces brown pigments.

Previously it was believed that the atmospheric nitrogen is fixed through the hydrolysis of hydroxylamine or to ammonium nitrate produced from nitrogen and water in the following manner

$$N_2 + 2H_2O \rightarrow NH_4NO_2$$

However, as matter of fact, hydrolysis of nitrogen takes place at a very high pressure, which is not possible in bacterial cultures. Bacterial culture produces some amount of hydroxylamine. This is possible with help of hemine enzyme. The reaction takes place in following manner:

Even this hypothesis was rejected due to the fact that respiration and assimilation of atmospheric nitrogen by this bacterium is connected with different catalytic surfaces that the direct combination of it is not possible.

Yet, another hypothesis believes that reduction of nitrogen takes place by the active hydrogen. The mechanism was expressed as following

N_2 + [Activating hydrogen of] first NH3 Growth [Nitrogen of organic]
[organic compounds] products [substance of protoplasm]
enzyme deaminazation

Carboxyl and amine groups of catalyst do not take part in the process of fixation of atmosphere nitrogen from nitrogen. Phenyl hydrogen and hydroxylamine are present in the composition of the catalysts of nitrogen fixation, which are capable of taking part in the process of fixation of nitrogen from the atmosphere. Molecular nitrogen reacts with catalyst (of nitrogen fixation) through the oxygen of carboxyl group and gives the first products of nitrogen fixation of the type of hydrazine

The energy released by burning of aldehydes or ketones is utilized in production of hydrazine. Later is further reduced by the activating hydrogen and is transformed into amino acids entering into the proteins of the protoplasm of bacterial cells. Nitrogen fixation and growth of bacterial cells shows direct relationship.

Nitrogen fixation by Clostridium

In 1895, Winogradsky isolated *Clostridium pasteurianum.* It is a bacillus and an obligate anaerobe having butyric acid fermentation. This bacterium is capable of fixing nitrogen. The mechanism of nitrogen fixation is by direct reduction of elementary nitrogen to ammonia by nascent hydrogen produced during the butyric acid fermentation. It is yet to be proved.

CARBON CYCLE

Plants convert CO_2 into organic from by photosynthesis. It has been estimated that in atmosphere CO_2 content is approximately 2600 X 10(9) t. Plants fix 129 X 10 (9) t. of CO_2 by photosynthesis. A part of this organic carbon goes back to atmosphere by respiration and microbial mineralization. Carbon cycle is completed by biological and geochemical processes.

Organic matters are introduced in the soil via number of agencies. Dead and decaying plant matters are a major source of organic composition of soil. These remains are degraded by micro-organisms and the products are taken up by them for their biological processes. Animal tissues and excretory products are also subjected for microbial decomposition. Plants contain six types of organic compounds. The major category is of carbohydrates. The cell wall components are cellulose (forming about 15 to 60% dry weight of plants), hemicellulose (10 to 30% of dry weight), lignin (5 to 30% of dry weight). Other organic matters of plants are water soluble sugars, amino acids and aliphatic acids (forming about

5 to 30% of dry weight), organic soluble organic compounds like fats, oils, waxes, resins and pigments. Protein also forms a large part of plant cell. All these compounds are decomposed easily. Some compounds like lignin, humins, polyphenols, carotenoids and waxes are resistant for microbial decomposition.

There are three simultaneous processes occurring at decomposition of organic compounds. These are:

1. Microbial enzymes digest plants and animal tissues.
2. energy released and monomers formed as the result of decomposition is utilized in the synthesis of cellular components of micro-organisms and
3. The unutilized end products of decomposed plant and animal tissues remain in the surrounding and thus they are accumulated and are further metabolized.

Decomposition of cellulose

Cellulose is a polysaccharide made up of linear chain of glucose units. The monomers form polysaccharide molecule by α 1 $\rightarrow$ 4 glycosidic linkage. Xylans, mannans and polyuronides are associated with cellulose of plant cell wall. Small quantities of arabans and galactans are also found. The polysaccharides that are structurally linked with the cellulose of the cell wall have been termed cellulans.

Those micro-organisms, which decompose cellulose are known as cellulolytic. Cellulolytic fungi are generally found in humid soils while bacteria decomposing cellulose are more common in semi-arid soils. Common cellulolytic fungi are *Aspergillus, Chaetomium, Curvularia, Fusarium, Phoma* and *Trichoderma. Cytophaga* and *Sporocytophaga Cellulomonas, Cellulovibrio* and *Cellfalcicula* are cellulolytic bacterium. *Acromobacter, Pseudomonas, Vibrio* and *Bacillus* show cellulose degradation tendencies. *Clostridium thermocellum* decompose cellulose at higher temperature (55 to 65°C). This is known as thermophilic decomposition of cellulose. Some members of actinomycetes, such as *Streptomyces, Micromonospora, Streptosporangium* and *Nocardia* produce extra-cellular enzyme, which decompose cellulose.

Bacteria decompose cellulose into CO_2 and cell substance under aerobic conditions. Similarly, fungi and actinomycetes decompose cellulose into CO_2 and cell-carbon. There are a few micro-organisms, which produce small amount of organic acid after cellulose decomposition. The anaerobic micro-organism produce different types of organic compounds after decomposition of cellulose. These organic compounds include CO_2, H_2, ethanol and acids (acetic acid, formic acid, succinic acid, butyric acid and lactic acid. Even the mesophilic and thermophilic anaerobic bacterial produce different types of products.

Enzyme that mainly catalyzes cellulose decomposition is known as cellulase. The enzyme converts insoluble cellulose to water soluble form. However, even this form of cellulose is not permeable. It is degraded by some extracellular enzymes secreted by microbes. The activity of these enzymes produces sugars from cellulose, which enters inside the cell and acts as the carbon source for it. Cellulase enzymes break -1,4-glucoside linkage. This enzyme acts on cellulose molecule and it causes random cleavage or it splits cellobiose (a molecule with two glucose monomers) molecules at the terminal ends of the polysaccharide molecule. Mandels and Reese (1957) have described that cellulose hydrolysis to cellobiose level requires two steps. In the first reaction enzyme C1 breaks native cellulose molecule into a long and linear chain of glucose. The second reaction is catalyzed by enzyme Cx, which breaks the long chain of glucose into cellobiose. Cellobiose is assimilated and is metabolized intracellularly.

Decomposition of Hemicellulose

Hemicellulose is a component of plant cell wall. It is made up of hexoses, pentoses and uronic acids. Sugar constituents of hemicellulose are glucose, galactose, mannose, glucuronic acid, galacturonic acid, xylose and arabinose.

Fungi, bacteria and actinomycetes take part in hemicellulose decomposition. *Trichoderma, Polyporus, Penicillium, Cunninghamella,* and *Rhizopus* are the popular fungi, which degrade hemicellulose. The bacteria capable of decomposing hemicellulose are, *Bacillus, Achromobacter, Cytophaga, Sporocytophaga,* and the actinomycetes, Streptomyces breaks hemicellulose.

Hemicellulose breaking enzymes are collectively known as hemicellulase.

Decomposition of Lignin

Lignin is another important cell wall component. The basic unit of lignin is phenyl-propane structure containing one methosyl-carbon.

Lignin is not decomposed easily by enzymes. Lignin decomposition is affected by temperature. Thermophilic bacteria finds easy to decompose lignin. Lignin decomposing higher fungi is *Agaricus, Ganoderma, Tarmetes, Polyporus.*

Decomposition of Pectin

Pectic substances are components of middle lamella of plant cells. It is also present in the primary and secondary cell wall of plant cell. However, they do not form a major portion of dry matter. Species of *Bacillus, Clostridium* and *Pseudomonas* hydrolyze these polysaccharides. Basically, three enzymes catalyze the decomposition of pectic compounds. (I) protopectinase decomposes protopectin with the formation of soluble pectin; (II) pectin methyl-esterase (PME) hydrolyzes the methyl ester linkage of pectin to yield pectic acid and methanol and (III) Polygalcturonase (PG) breaks the linkages between galacturonic acid units of either pectin or pectic acid with release of smaller chains and ultimately free galacturonic acid.

Decomposition of Chitin

Chitin is made up of amino sugar monomers (N-acetylglucosamine) in a linear arrangement. Fungi, bacteria and actinomycetes are chitin decomposing micro-organisms. Fungal genera decomposing chitin are *Fusarium, Mucor, Mortierella, Trichoderma, Aspergillus, Penicillium, Thamnidium* and *Absidia. Achromobacter, Bacillus, Chromobacterium, Cytophaga, Flavobacterium, Micrococcus* and *Pseudomonas* are common chitin decomposing bacteria. *Streptomyces, Nocardia* and *Micromonospora* are actinomycetes, which are able to produce enzymes catalyzing decomposition of chitin. Enzymes chitinase breaks chitin into disaccharide N-N-diacetylchitobiose, which is converted into glucosamine and acetic acid by enzyme chitobiase. The overall reaction can be expressed as following

Other polysaccharides like starch, insulin; gums are also decomposed by micro-organisms.

PHOSPHORUS CYCLE

Phosphorus is an important element. It is second to nitrogen in respect of importance in relation to mineral nutrition. In the biosphere, phsophorus is found as phosphates in the rock deposits. Plant residues contain 0.05 – 0.5 % phosphorus. In a cell phosphorus is important constituent of protoplasm and present in several compounds or group of substances e.g., in phytin, phsopholipids, phosphosugars, nucleic acids and nucleoprotein and in coenzymes and related compounds. Its major physiological role is in certain essential steps in the accumulation and release of energy during cellular metabolism. In soil 25 to 85% of the total phosphorus is organic found as orthophosphates. This form of phosphorus is not useful and it has to be converted into inorganic form. In inorganic form, phosphorus is present in the form of phosphates of calcium, iron or aluminium. Calcium phosphate is present in abundance in neutral or alkaline soils. In nature, organic phosphorus is mineralized and after this, it is transformed into salts of phosphoric acid. Phosphors cycle completes by two processes. These processes are mineralization of the organic phosphorus and transformation of salts as more soluble forms. Cycling of phosphorus occurs between inorganic and organic forms. Some bacteria are capable of reducing phosphate (valency +5) to phosphite (valency +3) and hypophosphate (valency +1). Such reactions are quantitatively small as compared to phosphate cycling. Phosphorus present in organic form is hydrolytically decomposed into phosphoric acid. Following scheme is to explain this in general terms.

(1) Nucleoprotein → nuclein → nucleic acid → H3PO4

(2) Lecithin → glycerophosphoric esters → H3PO4

Nuceloproteins are molecules of protein and nucleic acid. Lecithin is a phospholipid. Lecithin is made up of glycerol, fatty acids, phosphate and choline.

Bacteria and fungi take part in mineralization and mobilisation of phosphorus. Some of the common bacteria and fungi are *Pseudomonas, Mycobacterium, Micrococcus, Flavobacterium* and fungi like *Penicillium, Sclerotium* and *Aspergillus* etc. VAM fungi are found as symbionts with higher plants. These fungi mobilize phosphorus and provide it to the roots of the host plants.

Enzymes catalyzing mineralization of phosphorus from organic compounds are known as phosphatase. The cleavage of monoesters like ethyl or phenyl phosphates or glycerophosphates require one enzyme and that of diesters require different enzymes. Phytic acid or phytin is decomposed by enzyme phytase. The decomposition of the organic phosphorus compounds produces phosphoric acids, which combine with the bases transforming mainly into less soluble and less available forms of salts of calcium, magnesium and iron. The solubility of phosphorus is called the mobilization of phosphoric acid. Solubility of salts of phosphorus is important as it directly reflects on the fertility of soil.

The amount of phosphates brought into solution by heterotrophs varies with the carbohydrate oxidized. Transformation proceeds only if the carbonaceous substrate is converted into organic acids the carbon di oxide transforms the tri-calcium phosphate, which is easily soluble. The organic and inorganic acids transform the tricalcium phosphate first into monophosphate and then into phsophites.

Nitric and sulphuric acids are produced by the oxidation of nitrogenous materials or inorganic compounds of sulphur. These acids react with rock phosphate and increase the solubility of phosphate.

SULPHUR CYCLE

Sulphur is one of the important elements of a cell. 25 to 40 X 10 (8) t of sulphur is present in living biomass. Sulphur forms about 1% of the dry weight. It is a component of S-containing amino acids (cystine, cysteine, homocysteine and methonine). It is also a building element of vitamin B (thiamine, biotin and lipoic acid). Sulphur is present as sulphate and sulphide in lithosphere. It enters the soil in the form of plant residues, animal wastes, chemical fertillizers and rain water. As result of decomposition of organic sulphur a major part produced are hydrogen sulphide and a small amount of marcaptane. Hydrogen sulphide may also be produced by the reduction of salts of sulphur, sulphurous acid and sulphuric acid. Plants utilize sulphur generally as sulfate ion and sometimes as amino acid.

Sulphur bacteria are placed in six families

I. Family Beggiatoaceae Cell occurs within motile trichomes. Sulphur granules deposited in the trichome. *Beggiatoa, Thiothrix, Thioplaca*

Order Beggiatoales

II. Family Leucotrichaceae Sulphur may be deposited outside of cell. Not autotrophic. *Leucothrix*

III Family Achromatiaceae Globules of sulphur found within cells if sulfide present. Cells occur singly, not in trichome. *Achromatin.*

Order Pseudomonadales

IV. Family Thiobacteriaceae Non-filamentous, colourless bacteria oxidizing inorganic sulfur compounds, chemoautotrophy not established. Elemental sulfur deposited either inside or outside of cell. *Thiobacterium*

V. Family Thiorhodaceae Photoautotrophic, purple sulfur bacteria. Sulfur globules deposited in the presence of sulfides. *Thiobacterium thiodiction, Chromatium okenii*

VI. Family Chlorobacteriaceae Photoautotrophic, green sulfur bacteria. Sulfur globules usually deposited outside of cell.

The oxidation of hydrogen sulfide takes place in two steps. In the first step hydrogen sulfide is oxidized into sulphur and in the second step sulphur is oxidized into sulphuric acid.

$$2H_2S + O_2 \rightarrow 2H_2O + O + S_2 + 126 \text{ Cal}$$

Sulphur is formed due to oxidation of hydrogen sulfide. It deposits in protoplasm of the cell of sulphur bacteria, in the form of semisolid drops. Accumulation of sulphur in the cells of bacteria is like storage product and its oxidation helps in respiration of these bacteria. When the threads of sulphur bacteria fall into water not containing hydrogen sulfide, the sulphur, and the sulphur inside their bodies start oxidizing into sulfuric acid. This acid is released from the cells.

$$S_2 + 3O_2 + 2H_2O \rightarrow 2H_2SO_4 + 294 \text{ Cal}$$

Spirillum desulfuricans reduces sulphate. The process can be expressed by following equation

$$H_2S \rightarrow H_2SO_3 \rightarrow H_2SO_4$$

Thiobacillus thioparus oxidize thiosulphates into sulphates and sulphur according to the following equation

$$2Na_2S_2O_3 + O_2 \rightarrow Na_2SO_4 + S_2 + 375 \text{ Cal}$$

In soil, sulphur is fixed in the form of sulphuric acid. It provides the available form of sulphur to plants. It also converts many insoluble forms into soluble forms thus improving the fertility of soil.

Desulfovobrio, Desulfotomaculum, Desulfomonas etc conduct desulfurication. These bacteria metabolize organic acids, ethanol, higher fatty acids and hydrogen.

CYCLES OF MATTER IN ANAEROBIC ENVIRONMENT

Mesophilic and thermophilic microorganisms carry out decomposition of organic matter under anaerobic conditions. The carbohydrate and protein is degraded into CO_2, alcohol and different organic acids. Primarily, *Clostridium* takes responsibility of degradation and organic acids are formed. In the later stage, strict anaerobic bacteria like methane bacteria degrade these acids by fermentation. Carbon dioxide and methane is produced during fermentation. If volatile fatty acids are produced, it is oxidized by *Syntrophomonas wolfeii* and *Syntrophobacter wolinii.*

Production of biogass by anaerobic decomposition of cattle-dung is a widely used by mankind. Methane bacteria carry out decomposition. A plant is designed for this purpose. Biogass consists of 55-60% methane, 5-10% hydrogen, and 30-35% carbon di oxide. Water is added to cattle dung to prepare slurry. It takes about four weeks for the production of biogass. The undigested organic matter is used as manure.

CHAPTER - 26

GEO – MICROBIOLOGY

Geo–microbiology is the study of role of microorganisms in geological processes, i.e., the formation of coal, petroleum and utilization of raw materials (hydrocarbons) for transformation into valuable chemicals. Organic matters produced by microorganisms, accumulate in mud deposits of the ocean floor. In due course of time this matter is buried and by sedimentary action, after a long period, these matters are converted into oils and gases. The first observation in this field is due to Ehrenberg (1838). He discovered an association of *Gallionella ferruginea* with ochreous deposits of bog iron. Colmer, Templer and Hinkle (1950) observed the role of acidophilic, iron-oxidizing bacteria in acid coal-mine drainage. Tyler and Barghoorn (1954) detected Precambrian prokaryotic fossils in sedimentary rocks. Beerstecher defined geomicrobiology " the study of the relationship between the history of the earth and the microbial life upon it". Kuznetsov defined this branch of microbiology as 'the study of microbial processes currently taking place in the modern sediments of various bodies of water, in ground waters circulating through sedimentary and igneous rocks, and in weathered earth crust [and also] the physiology of specific microorganisms taking part in presently occurring geochemical processes'.

RECOVERY OF PETROLEUM

Petroleum and natural gas are clean, convenient and most commonly used energy source for domestic and industrial purposes. Petroleum is a mixture of aromatic and aliphatic hydrocarbons and various other heterocyclics, including oxygen, nitrogen and sulphur containing compounds. Petroleum is abundantly found in reservoir rocks. These rocks are folded, porous, sedimentary rock strata, or in other fractured rock such as fissured shale or igeneous rock. The origin of petroleum is due to some abiological factors (heat and pressure), which would have converted planktonic debris into petroleum and associated natural gas. Some believe that hydrocarbons are produced by aerobic or anaerobic reduction of fatty acids, proteins, amino acids, carbohydrates, carotenoids, sterols, glycerol, chlorophylls and lignin-humus complexes. The reduction took place along with decarboxylation and deamination. There are some of the bacteria in the initial stages of petroleum formation. However, this is still not clear. Three types of bacteria are observed in the brines associated with petroleum reservoirs. The groups of bacteria are (1) *Desulfovibrio desulfuricans* and *Desulfotomaculum nigrificans*, which reduce sulphates. (2) *Methanobacterium mazei, Sarcina methanica* and *Methanobacterium omelianskii*, these are methanogens. (3) *Rhodopseudomonas palustris,* which is photohe-terotroph. In petroleum recovery, traditional methods like pumping, gas pressurizing, water flooding and thermal methods are popularly implicated. However, such spontaneous flow recovers about one third of the total petroleum deposits. Gas pressurizing and water flooding, are the primary oil recovery technique. The secondary is the thermal method. In this process, steam injection or controlled underground combustion is designed to increase the flow of heavy, viscous petroleum. The so-called tertiary technique is the use of solvents, surfactants or polymers designed to dislodge oil from geological formations.

The role of microorganisms in recovery of petroleum is important economically. *Xanthomonas campestris* produces surfactants and polymers known as xantham gum. The fungi like *Sclerotium, Stromantinia* and *Helotium* produce glucan polymer. This gum has substantial viscosity with characteristic flow rate. *Desulfovibrio* increases petroleum recovery by *in situ* dislodging action. The bacteria grow on rock pore surface and oil by direct or indirect lodging. The direct dislodging is due to physical displacement and the indirect dislodging is by synthesis of surface active metabolites and gases (H_2, CO_2, H_2S and CH_4). Field tests in situ have not yielded promising results due to some unfavourable conditions. As an alternative approach, appropriate organisms may be introduced into the oil well to produce a surface-active agent in situ. Molasses and subsequently *Clostridium acetobutylicum* (gas former) is injected in oil well. This bacterium ferments molasses into acetone, butanol, and ethanol and large amount of gasses (CO_2 and H_2). The organic solvents thus produced decrease viscosity of oil and the gases formed pressurize oil to move.

The alternate sources of oil wells are Tarsands and oil shales. These are the fossil hydrocarbon reserves. Tarsands contain a highly viscous hydrocarbon mixture called bitumen. Bitumens are solid hydrocarbons and are soluble in benzene and other organic solvents and fusible below ~150°C. Oil shales are impermeable fine-grained sedimentary rock formations. It contains kerogen along with bitumen. Kerogen is highly cross-linked and condensed hydrocarbon complex and insoluble in benzene and other organic solvents. These sources may contain 50-60% hydrocarbon material. Out of this about 20-30% is economically useful. Oil shales contain substantial amounts of carbonates and some pyrite. Carbonate is dissolved with acids and results in increased porosity and reducing the bulk of the shale prior to retorting. Pyrite contains sulphur. Other mineral nutrients are derived from the oil shale formation and the leaching water. *Desulfovibrio* takes the responsibility of regenerating sulphur from sulphate in the lechate. *Thiobacillus* species produce sulfuric acid in an aerobic zone and *Desulfovibrio* species reduce the sulphate, regenerating elemental sulphur in an anaerobic zone. The bitumen and kerogen can be converted to a petroleum like substance by heat treatment like retorting.

MICROBIAL DEGRADATION OF PETROLEUM

Petroleum is aerobically degraded. Species of *Pseudomonas (P.oleovorans, P.fluorescens, P.aeruginosa) Nocardia salmonicolor, Brevibacterium erythrogenes, Mycobacterium,* Yeasts *(Trichosporon, Arthrobacter)* and fungi *(Cladosporium* and *Graphium)* utilize hydrocarbons and oxidizes or desaturates it. Enzyme oxygenase is responsible of oxidation of hydrocarbons. During oxidation, terminal carbon of alkanes is attacked and alcohol, ketone or hydroperoxide are formed. During desaturation, alkanes are terminally or subterminally desaturated and alkenes are formed. Alkenes may be attacked and epoxides are formed. The latter are further metabolized to form diols. Bacteria are very specific for chain length and branching of alkanes, e.g., bacteria, which are able to attack on C8-20, may not be able to attack on C1-6 chains. The genes that code for the enzymes, which attack hydrocarbons, are not found on the main bacterial chromosome but on plasmids.

Two types of strategies are adopted since each strain only consumes limited range of hydrocarbons.

1. Using mixture of strains
2. Genetically engineered superbug

Ananda Chakraborty (1974) created a strain of *Pseudomonas* that consume all types of hydrocarbon. Fungi can grow on hydrocarbons up to C34 length.

PRODUCTION OF FUEL

Microorganisms play a wide role in converting organic wastes into liquid or gaseous fuels. Another aspect of fuel formation is that the solar energy is converted into biomass by microbes, which is in turn fermented and yield fuel.

Production of ethanol

Ethanol is mixed with gasoline in the ratio of 1:9 and the mixture is known as "gasohol". It is quite efficient fuel. It may be directly used in internal combustion automobile engines. It reduces the release of atmospheric hydrocarbon. Carbohydrate is the basic substrate, which after fermentation by yeast yields ethanol. However, yeast alone is not able to breakdown polysaccharides. Hence, the polysaccharides are first digested into simple carbohydrates by enzymatic reactions, e.g., if the substrate is rich in cellulose, it is treated by enzyme cellulase produced by microbes (like fungi *Trichoderma viride*). Agricultural wastes may be used as substrate for ethanol production. Forest biomass represents a largely unused resource that could be trapped for energy production. Production of biomass in fresh water is limited by the availability of nitrogen and phosphorus.

Production of methane

Methane may be used as a source of energy by various means:

1. in the generation of mechanical, electrical and heat energy.

2. as fuel source for homes and industry by transmission through natural gas pipelines.
3. as chemical form methanol, used as a fuel of internal combustion engines.

Many of the methane accumulations on earth are of biogenic origin. Bacteria produce methane by decomposing organic wastes and the process is known as methanogenesis. Methane formation is a form of anaerobic respiration, in which CO_2 is often an electron acceptor. Production of methane from the biomass of primary producers (sea weeds) is a promising subject. Methanogenic bacteria (anaerobic photoorganotrophs) are able to couple solar energy with the production of gaseous hydrocarbon. Methanogenic bacteria are *Methanobrevibacterium, Methanococcus, Methanospirillum, Methanosarcina, Methanothermus,* etc. They are thermophilic or mesophilic in nature. Thermophilic microbes are more efficient in methane generation as compared to mesophilic microbes. Anaerobic digesters of sewage treatment plant can be used for the production of methane. Municipal and agricultural wastes are the main substrate for such digesters. In such digesters, CO_2 is also generated along with methane. The ratios of methane to carbon dioxide depend on the chemical composition of the substrate and the environmental conditions of the digesters. Therefore, methane must be trapped and separated from other gases.

Methanotrophs are bacteria, which utilize methane as primary energy source. Some archaebacteria and yeast also utilize methane. *Methylococcus, Methylobacter, Methylosinus* are obligate methanotrophs. These bacteria can use methane, methanol, and methylamines as energy sources by oxidizing them to CO_2, H_2O and NH_3, respectively.

Production of other hydrocarbons

Microorganisms produce various alkanes, including branched alkanes, alkenes, aromatic compounds and alicyclic hydrocarbons. Various fungi, bacteria and algae are able to produce hydrocarbons C13-C20. Genetic transformation (genetic manipulation and selection of strain) in *Micrococcus luteus* has given positive results in hydrocarbon production.

MICROBIOLOGY OF COAL

Coal is a compact, stratified mass of mummified plants, which have been modified chemically in varying degrees, interspersed with smaller amounts of inorganic matter. Coal is of four types:

1. Lignite coal least developed coal with 59.5-72.3% dry weight carbon and high moisture content.
2. Subituminous coal, with 72.3-80.4% dry weight carbon.
3. Bituminous coal, with 80.4-90.9% dry weight carbon.
4. Anthracite coals, with 92.9-94.7% dry weight carbon and least moisture.

Coal formation has occurred by large amounts of plant debris accumulated in swams or shallow lakes. With time, this accumulated debris became buried and covered by clay and sand under water. On this a new layer of plant debris accumulated, followed by subsidence. Role of microbes (bacteria and fungi) at the initial stage of coalification was to degrade organic substances. The substances like sugars, amino acids and volatile acids were degraded earlier, while polymers like cellulose, hemicellulose, lignins, waxes and resins took some time for degradation. It is obvious that first aerobic microbes attacked the debris followed by anaerobic microbes. Hyphal remains, sclerotia and spores found in some coal belts have confirmed the occurrence of fungi in the debris.

Basidiomycete fungi, *Polyporus versicolor* and *Poria monticola*, grow on lignite coal. These fungi secrete extra cellular enzyme lignase (a type of peroxidase), which degrades lignin and its derivatives. Now it is presumed that some lignin derivatives are present in lignite coal. Bacteria are not able to grow on coal. There are some reasons for this. Bacteria do not produce lignin like enzymes. Coal is a water insoluble polymer due to which bacteria cannot adhere on the surface of coal. An early view suggests that coal have an antibiotic in its waxy and resinous part. This part is supposed to be extracted by methanol. However, it is observed in an experiment performed on the coal slurries, that *Escherichia freundii* and *Pseudomonas rathonis* is able to grow on this slurry. It was further observed that the growth was enhanced if the slurry was prepared after washing the coal with H_2O_2.

Bituminous coal contains organic or inorganic form of sulphur. The inorganic form of sulphur is pyrite, marcasite, elemental sulphur and sulphate. Pyrite sulphur in coal is harmful as SO_2 gas is released when such coal is burned. Thus, such coal has less commercial value. Bacteria like *Thiobacillus ferroxidans* and *Sulfolobus* spp. are helpful in oxidizing pyrite. Sulphur present in coal in other inorganic and organic forms is not susceptible for microbial attack. Some of such forms cause air pollution by combusting such coal. All such forms, are not attacked by a single species. Experiments with genetic engineering are being conducted to find out solution of such problems.

CHAPTER - 27

AIR MICROBIOLOGY

Air is not a natural environment for the growth and reproduction of microorganisms. It does not contain the necessary amount of moisture and utilizable form of nutrients. Yet organisms are found in air though they have a transient survival. The organisms come from soil, from organic wastes of men and animals and microbes in air from the lungs through coughs and sneezes. Obviously, most of the organisms come from the soil as dust. It is generally realized that dust bears germs; but it must be noted that many of the organisms especially mould pores are free floating and not attached to foreign particles. Air is present in strata of varying light intensities and temperature. The moisture content also varies in different strata. Seasonal changes have its importance in changing the values of these factors. Organic matter is only present in the lower layers of atmosphere. The population and type of microorgansims varies in different layers of atmosphere.

MICROORGANISMS OF AIR

Air has no indigenous microflora, but still air is rich in some groups of microorganisms. The dispersal of latter takes place by air. The aquatic and terrestrial environments are the main source of contamination of microorganisms in air. The numb of organisms in the air depends upon the activity in the environment and upon the amount of dust stirred up. Microorganisms survive in the air for varying periods of time. This is controlled by a complex set of circumstances, including atmospheric conditions, e.g. speed of the air current, humidity, sunlight,temperature, the size of the particles on which they are attached and the nature of the organism i.e. its ability to survive in a new physical environment. The groups of microorganisms present in air are as following

FUNGI

Fungal spores and mycelia form a major part of air microflora. The density of fungal spores in still air is about 1.0 million. Pathogenic and saprophytic forms of fungi are common in air. Diseased organisms or decomposing organic matters, are the main sources of contamination. Required percentage of humidity and temperature are the main factors, which keep these microbes viable. The ability of fungal spores or mycelia to resist physical environment is the factors of fungi, which keep them alive. Most of the fungal spores cause allergy in human beings. The size of fungal spores varies from 3-200 μm. Members of form class Deuteromycetes (order Moniliales) have highest population in air. Spores of *Cladosporium, Alternaria, Curvularis, Helminthosporium* and *Aspergillus* are most common. Spores of fungi belonging to other groups are also common. *Albugo, Cunninghamella, Mucor, Rhizopus, Sclerospora, Blakeslea, Circinella, Mortierella* belonging to class Mastigomycotina and Zygomycotina are also common. Ascospores are found in abundance in the months of July to September. Even diurnal periodicity has been observed. Spores of *Amphisphaeria, Claviceps, Hypoxylon, Chaetomium, Leptosphaeria, Pleospora* etc are found in air. Spores of the members of order Agaricales, Aphyllophorales, Uredinales and Ustillaginales of class Basidiomycetes are found in air during rainy seasons.

Frutification of members of Agaricales and Aphyllophorales can be well observed in rainy seasons. The spores are released in a very high amount from these frutifications. Uredospores and *Teleutospores* of rust fungi and chlamydospores of smut fungi are common pathogenic spores found in air. Members of class Basidiomycetes *Agaricus, Coprinus, Ganoderma, Melampsora, Puccinic, Ravenalia, Uromyces* and *Ustilago* are quite common.

ALGAE

Algae have no adaptation for air. However, wind current makes unicellular algae and fragments of filaments to float into air. The sources of algae may be aquatic as well as terrestrial environments. Algae from aquatic systems like ponds, lakes reach the air due to eutrophication. While algae from soil, is disseminated by wind currents and thermal uplifts. Heterocyst and other resting spores of alga may be observed in air. Planktonic algae are widely distributed in air. Members belonging to chlorococcales are most common in air. *Oscillatoria, Anabena* are common is air and may be contaminated by house dust. *Chlorella, Chlorococcum, Prototheca,* belonging to class chlorophyceae are also common in air and cause aeroallergy.

Bacteria ranging from 0-3 to 10-15 μm are abundant in air. The source of bacteria in air may be bacteria-laden minute droplets. They are carried on even adhered to the clothes.

Viruses have no mechanism of dispersal but still they are found in air. Bacteria and viruses causing respiratory diseases are common in air. Bacteria and viruses are on the rafts, which is responsible for their transmission. Protozoan in the cyst form may be transported by air.

AIR SAMPLING

Air samplers collect particles from air. Their operations are totally based on physical and mechanical processes, like sedimentation, filtration and centrifugation. These processes have impact by wind, suction and by rotation, respectively. Air borne particles may be collected due to gravitational force, or by horizontal sampling. The latter provides a measure of deposition per unit area. Based on these phenomena, some samplers are developed.

In gravitational samplers microslides coated with glycerine jelly or petroleum jelly are kept in horizontal surface. After sometime, these slides are observed under microscope. The slides are generally kept for 24 hours so that a complete picture of dirunal occurrence of spores etc may be observed. *Durham's gravitational sampler* is a simple and most widely used. This sampler is made up of two discs separated by three struts. Two microscopic slides smeared with glycerine jelly or vaseline are kept for 24 hours an inch above the centre of the lower disc. The upper disc provides protection from rainfall. Two sides can be kept for observation on a sturdy stand, while circular disc being protected by a hood. This sampler can be used for qualitative analysis of various kinds of spores. An idea for seasonal trend of spore occurrence may also be estimated. Tauber (1984) developed a sampler known as Tauber Trap. This sampler is made up of a cylindrical container covered with aerodynamically shaped collar with a central circular orifice through which particles enter. The bottom of the container is covered with glycerol, which traps the air borne particles. The trap may be protected with a roof to avoid precipitation.

Impaction and suction sampler are sophisticated samplers. A definite amount of air is sucked in and air borne particles are impacted on objects of any shape. A circular shaped object smeared with an adhesive is used as impactor. Perkins (1957) developed a rotorod sampler used for indoor and outdoor sampling.

Rotorod Sampler, is a cheap, portable high sampler with greater sensitivity. In this sampler, there is an intermittent rotating impactor. The sampler has collecting arms, made up of brass having 0.159 cm cross sectional area. The vertical arms are 6 cm long and 4 cm from the axis. The arms hold two cubical rods coated with silicon grease. Unexposed arms remain folded. In rotord sampler, the surface is rotated so that it strikes the spores. The volume of air swept can be calculated from the frontal area of the rod, the diameter through which it is turned and the number of revolutions for which it is run.

The sampling rate is the volume of air swept over by the collecting surface per unit time. The volume of air can be calculated on the basis of the dimensions.

= 2 (arms) x 0.159 cm x 8 x 2300 x 10-3

= 48.0 x 10-3 x 2300 liters/min.

= approximately 100 liters/min.

Tilak Air Sampler This sampler was developed by Tilak and Kulkarni (1970). Sampling may be carried out for 8 days. Air is sucked through the orfice of the projection tube. 1.5 cm breadth cellotape are stuck on the rotating drum. The tape is coated with glycerine mixed with vaseline or petroleum jelly. Air is sucked at the rate of 5 liters per minute. As the sampling is carried out for eight days hence, scanning is done by dividing the tape into 8 sections, which are mounted on 8 separate slides.

Calculated conversion factor for Tilak air sampler is = 14

The volume of air sampled per minute = 5 liters/min.

The number of spores scanned is multiplied by conversion factor. This is the number of spores in m^3 of air.

Andersen six stage volumetric sampler. This sampler is more useful for indoor sampling. In this sampler, six petri plates containing suitable medium are kept under sieves of different pore size. Each sieve has 400 pores. The sieves are arranged in gradually decreasing order of pore size. Air is sucked at the rate of 28.3 l/min from the orifice. Spores of different dimension, impact on different petri plate according to the pore dimensions.

AIR MONITORING

Air monitoring programmes have environmental significance. It helps in prevention of environmental biopollution, disease forecasting, and search for microorganisms for reducing pollution level.

Besides the organic and foreign particles causing pollution, spores, pollen grains and other living particles cause hazards for human beings. This is said to be biopollution. This may be controlled by checking the entries of exotic species with potential pests. Control of desertification of landscapes, contamination from water is also to be regulated. Emphasis is also given to the different radiation, which affect biological particles.

Air borne diseases

There are 2 main sources of microorganisms in air. These are saprophytic soil organisms raised as dust and organisms from body tissues introduced into the air during coughing, sneezing, talking and singing. Most dust particles laden with microorganisms are relatively large and tend to settle rapidly. Some droplets are of such size that complete evaporation occurs in a warm, dry climate and before they reach the floor quickly become droplet nuclei. These are small (10 μm) and light and may float about for a relatively long period.

CHAPTER - 28

PLANT DISEASES

Disease is a malfunctioning process that is caused by continuous irritation, which results in some suffering producing symptoms. A pathogen is associated with a disease. In terms of plant pathology, a pathogen is defined as any agent or factor that incites 'pathos' (disease). A plant pathogen may be abiotic or biotic. The abiotic factors include deficiencies or excesses of nutrition, light, moisture, aeration, abnormalities in the soil, and atmospheric impurities. The biotic causes of disease are animate or cellular organisms. The plant disease may be endemic or epidemic. The endemic disease is prevalent in, and confined to, a particular country or place. The epidemic diseases appear very virulent among a large section of the population. If the disease occurs at very irregular intervals and locations and in relatively few instances, it is called as sporadic disease.

The symptoms of disease are the signs appeared on a plant at the time of disease. The symptoms appear due to character and appearance of the visible pathogen or due to some effect upon or change in the host plant. The symptoms appear due to the character or structure or visible appearance of pathogens, are mildews (the pathogen is seen as a growth on the surface of the host). Rust (small pustules with rusty appearance). Smut (pustules with sooty or charcoal-like powder). Blisters (white blister like pustules common in crucifers). Scab (roughened or crust-like lesion or to a freckled appearance of the diseased organ). Sclerotium (compact, hard mass of dormant fungus mycelium). Blotch (superficial growth giving the fruit a blotched appearance as in sooty blotch). Fruiting bodies (large spore bearing structures). Exudations (in bacterial diseases masses of bacteria ooze out to surface or the affected organ) or tar spots (raised, black coated fungus bodies with the appearance of a flattened out drop of tar on leaves).

In some symptoms of disease, there may be marked change in the form, size, colour, texture or some of its organs. If there is yellowing in colour of leaves, it is known as chlorosis. While, the white colour is known as albinism. In some diseases, there is overgrowth of some organs of plant. This may be due to hypertrophy or hyperplasia. Hypertrophy is the increase in the size of cells. While, hyperplasia is abnormal increase in number of cells. There are different names of overgrowth and deformity in plant diseases. These are galls, curl, pocket or bladder, witches broom, etc. Necrosis is death of plant organs or tissues due to the activity of pathogen. A canker is a dead area in the bark or cortex of the stem, the surface may be smooth or rough and is usually somewhat sunken. Blight is a burnt appearance due to sudden death of plant (especially leaves). Rot is the death and decomposition of affected tissues (cell wall and middle lamella is degraded due to enzymes secreted by pathogen). Wilting is loss of turgidity of leaves or succulent parts of plants. Wilting may be due to injury to the root system or partial plugging of water conducting vessels or due to toxic substances secreted by the pathogen and carried to delicate cells with water.

DISEASE OF IMPORTANT CROP PLANTS

The important crops of India are wheat, rice, pulses and sugarcane. Potato is a common ingredient of food in many parts of India. The some important diseases of crop plants are discussed herewith.

DISEASE OF WHEAT

Black or stem rust disease of wheat is caused by *Puccinia graminis tritici.* Disease symptoms appear as an eruption of elongated, brown pustules on the leaf sheath and leaves. The pustules bursts and brown powder (uredospores) comes out. It is surrounded by prominent white epidermal fringes. The pustules turn into black due to emergence of teliospores. In severe form of disease, plant is unable to form ear, grains are shriveled, and lighter in weight. The fungus in biotrophic (obligate parasite) polymorphic (with different types of spores) and heteroecious (life cycle completes in more than one host). *Berberis vulgaris* is the alternate host. There are five types of spores and hence the five stages in life cycle. Pycnia are flask-shaped structures. It consists of pycniosporophores and receptive hyphae with a pycnial chamber in the centre. Aecia are cup-shaped structures and are enclosed by a peridium. The aeciospores are echinulate spores. It infects wheat. On wheat, first uredospores are produced. Uredospores are brown in color oval shape. It has a single spiny thick wall with four germ pores. Teleutospores are two-celled spore. It has a thick and smooth wall. Young teleutospores are dikaryotic and later become diploid due to karyogamy. It germinates forming four celled promycelium. Each cell has a sterigmata which bears sporidia (basidiospores). The later infects Berberis and completes life cycle. The chemical treatment of this disease is by sulphur dusting. Four to five applications of nabam and zinc suplhate reduce disease.

Loose smut of wheat is caused by *Ustilago tritici* (*Ustilago nuda tritici*). The fungus damages the ear of plants. A black mass of spores covers the entire ear with total failure in the formation of grains. Spores produce promycelium and later gives rise to infection threads. The spores develop in the young spikelets and bursts before the ears are completely formed. The fungal spores are olivaceous brown, spherical or sometimes oval. Spores gall on the feathery styles of the stigmas and germinates into a promycelium. It reaches ovary. The diplodization between the cells of the germ tubes take place. The mycelium becomes ramified in ovary and embryo. At the time of ear formation, the fungus enters the reproductive phase. The fungus lies dormant within the mature grain. When infected grains are sown the hyphae becomes active. This disease is seed borne. Seed treatment by hot water is recommended. Seeds dressed with fungicides, such as Carboxin and Benomyl have proved to be beneficial. Cultivation of immune or resistant varieties is also done.

DISEASES OF RICE

Bacterial leaf blight is caused by *Xanthomonas oryzae.* The symptoms appear as linear yellow to straw coloured stripes with wavy margin on both the edges of leaf. The leaf tip dries and undergoes twisting. The blighting extends up to the leaf sheath and culms, killing the tiller or the whole clump. In dry weather, opaque and turbid drops of bacteria ooze out. On drying yellowish beads of bacterial ooze may be seen on the leaf surface. The bacteria is Gram-negative. Cells are rod shaped with a single polar flagellum. Several chemicals including antibiotics are tested for controlling the disease. Satisfactory results have not been observed. Seed treatment with Agrimycin and wettable Ceresan followed by hot water treatment have proved to eradicate disease up to 95 %.

Leaf spot disease of rice is caused by *Helminthosporium oryzae.* The symptoms appear on coleptile, leaves and leaf sheath. Symptoms may be observed on the glumes also. The spots are brown, small and circular to oval in outline. The centre may be pale yellow, dirty white, brown or grey. The fungus belongs to fungi imperfecti. Its perfect stage is reported as *Cochliobolus miyabeanus.* The mycelium has prostrate hyphae with erect conidiophores. It forms greyish-brown to dark brown mat. Conidiophores have characteristic bent and posses knee-joints, the points where conidia were attached. Conidia are olivaceous brown, slightly curved, tapering towards the rounded ends with 5 to 10 septa. The perfect stage shows globose perithecia with ostiolar beak. Asci are long, fusiform, slightly curved and usually contain 4-6 ascospores. The ascospores are filiform. This disease is seed and soil borne. Seed treatment with fungicides like Ceresan or Agrosan GN enables to control disease in seedlings. Spraying of copper fungicides or Dithane Z-78 on crops shows effective control on disease.

DISEASES OF ARHAR

Wilt of pigeonpea Cajanas cajan L. (arhar) is caused by *Fusarium oxysporium f. sp. udum.* Symptoms wilting of seedlings and adult plant. The main roots and the base of the stem tissues are blackened. In partial wilting, stem on one side is blackened and few or all branches on that side wilt and withers away. The fungus is confined to vascular tissues and is both inter- and intracellular. The hyphae plug the vessel and thus choking the water conduction. There are two types of conidia. The microconidia are small, elliptical or curved, unicellular or sometimes with more than one cell. Macroconidia are produced in small cushions of stromatic mycelium (sporodochia) on the surface of the host. The macroconidia are long, fusoid (curved) pointed at tip. Three to four septa are present on the conidia. Chlamydospores are also formed. The chlamydospores are oval or spherical thick walled occurring singly or in chains.

This disease is soil borne. The incidence of disease is always high in soils where the disease has appeared previously and no crop rotation has been followed. Thus, crop rotation is the best way of eliminating soil-borne infection. Infection of seedlings may be avoided by presoaking seeds in zinc or manganese salt solution. Treatment with carbendazim may provide protection in the early stage of plant growth.

DISEASES OF GRAM

Cicer arietinium L. *Rust of gram* is caused by *Uromyces ciceris-arietini.* The symptoms appear on both surfaces of leaves. The pustules are small, round or oval, light brown to dark brown in colour. In advanced stage pustules coalesce and from bigger pustules. Uredia are hypophyllous, scattered. Uredospores are globose, loosely echinulate, yellowish brown in colour. Teleutospores are rounded, ovate or angular in shape with dark brown warty wall. Pycnial and aecial stages are unknown.

DISEASES OF SUGARCANE

Red rot of sugarcane is caused by *Colletotrichum falcatum.* The fungus belongs to fungi imperfecti : *Glomerella tucumanensis* is reported as perfect state. The symptoms are observed after rainy season, when sucrose synthesis is high. In the early stage of disease, there is loss of colour in leaves. Leaves start dropping. In the advanced stage, the canes become shriveled, the rind sinks and becomes longitudinally wrinkled. When canes are split open longitudinally, the pith is found reddened. In very advanced stage, dirty brown colour replaces red colour of pith. The enzymatic activity of pathogen converts sugar into glucose and alcohol. The mycelium is septate. A soluble gummy dark material ooze out of the protoplasm of fungi. It degrades the cell wall of host giving red rot appearance. Conidiophores are found in acervulli. Acervuli are rounded or elongated. In between conidiophores are present brown coloured setae. Conidia are sickle shaped, hyaline and aseptate. A large oil globule is present in the centre. In perfect state, globose perithecia are produced. Asci are hyaline and eight spored. Ascospores are aseptate, hyaline, straight or slightly fusoid, ellipsoid or ovoid. Numerous septate and unbranched paraphyses, extends along the ostiole. Crop rotation and field sanitation is recommended for reducing incidence of disease. Disease resistant host varieties have been developed.

DISEASE OF POTATO

Wart of potato is a fungal disease, caused by *Synchytrium endobioticum.* Warts appear due to hypertrophy and hyperplasia on buds on stems, stolons and tubers. The warts vary in size from small protuberance to large intricately branched systems. In advanced stage, the warts are dark in colour. Thick walled resting sporangia are present in the infected tissues. The zoospores are produced from these sporangia. The zoospores are capable of invading healthy tuber. The zoospores enter inside the host tissues through eye of tubers. In dry conditions or towards the end of the crop season the zoospores function as planogametes. They fuse to form a diploid zygote. The zoospore entry in the host results in hypertrophy while, zygote causes hyperplasia.

Late blight of potato is caused by *Phytophthora infestans*. The initial symptoms appear on leaves in the month of January. The main symptoms are brown dead spots or extended dead areas on foliage. The dead areas appear at the tip or margin of the leaves and spread downward or inward. In early stages, the pustules are faded green but turn to browinsh-black in advanced stage. The leaves are killed. The fungus is present intercellularly in host. The mycelium is hyaline, well branched and coenocytic. Hausotria are club shaped. Sporangiophores are slender, hyaline, branched and indeterminate. The sporangia are oval or pear shaped with a definite papilla at the apex. It is multinucleate, thin walled and hyaline. Depending upon the temperature, sporangia germinate either by germ tube or by zoospores. Low temperature favours zoospore formation. Zoospores are biflagellate, one flagella is tinsel type and the other is whiplash type. Flagella are laterally placed. At higher temperature, sporangium germinates by germ tube. A secondary sporangium develops at the tip of germ tube. Sexual reproduction is not very common. It is oogamous. Various treatments are suggested for control of this disease. The tubers should be treated with mercuric acid before stroage. Fungicides, like Bordeaux mixture should be sprayed on crop. Copper oxychloride, Dithiocarbamates, Diathane etc. are now widely used fungicides.

Early blight of potato is caused by *Alternaria solani*. The symptoms appear as scattered pale brown spots on the leaflets, which become covered with a deep greenish-blue growth of fungus. In the necrotic tissues, concentric ridges develop to produce a target-board effect. Stems may also be affected, showing brown to black necrotic lesions. It leads in collapse of aerial parts of plant. The causal organism belongs to fungi Imperfecti and thus, there is no sexual reproduction. The mycelium consists of septate, branched, light brown hyphae. Conidia are present on conidiosphores, which emerge out of stomata. Conidia are muriform, beaked, dark coloured. Conidia have five to ten septa and a few longitudinal septa. Conidia germinate by germ tube. This disease is soil borne. Crop rotation and sanitation are essential for an effective control of disease. Spraying of fungicides (Diathane Z-78, Dithane M-45, Blitox, Difolatan, Daconil, Brestan, Antracol and Captan), are effective in disease control.

CONTROL OF PLANT DISEASE

Plant disease may be controlled by avoidance, exclusion, eradication, protection and immunization. However, there are three principal methods for controlling plant disease: (1) cultural methods, (2) chemical methods and (3) developing disease resistant varieties.

The cultural methods include agricultural cropping, harvesting and storage, crop rotation. The distance among the plants, irrigation, type and quantity of fertilizers also matter in disease control.

The chemical methods to control disease include use of fungicides and antibiotics, chemical sprays, dusts or seed treatment.

Plant breeding has enabled to produce resistant varieties of plants. Use of resistant varieties is a painless method and minimizes the chemical treatment to control disease. The disease resistance may be governed by one gene (mongenic), or by few genes (oligogenic) or by many genes (polygenic).

CHEMICAL METHODS OF DISEASE RESISTANCE

In 1885, Pierre Alexis Millardet observed that the downy mildew of grapes could be controlled by mixtures of copper sulphate and lime. Later many organic fungicides developed which are used for controlling plant diseases. Few important fungicides (their functional groups are discussed herewith.

Bordeaux mixture

It is a copper containing fungicide. Bordeaux mixture was originally prepared by mixing 15% copper sulphate and 8% lime. This formula is slightly modified with copper sulphate : lime : water (5 :5 : 50). It is prepared by mixing 5 lb (2.268 kg) each of copper sulphate and lime dissolved in 50 gallons of water (Sharvelle, 1961). The copper sulphate and lime react to form cuperous hydroxide and calcium sulphate. The blue gel contains tribasic copper sulphate [Cu_4 SO_4 $(OH)_6$] with absorbed lime and free lime and calcium sulphate. Various substances are formed in absorbed lime. The cupric hydroxide (the active principle) is soluble and is toxic to fungal spores and sporelings. It is sometimes toxic to plants also.

Inorganic mercury compounds

Inorganic mercuric compounds are used for chemical control are mercuric chloride and mercurous chloride. It is highly toxic to fungi and bacteria. Dilution of 1 : 1000 is applied for treatment of seed, rhizome, corms, and diseased flowers. Inorganic mercury compounds are also toxic to animals and human beings E.g. Calomel etc. Now, organomercurials are used.

Sulphur sprays

In 1802, Forsyth observed the effectiveness of sulphur as fungicide. Sulphur is used in mixture form A mixture of lime and sulphur is used as fungicide as well as insecticide. It is a mixture of calcium thiosulphate and calcium polysulphide. When this mixture is sprayed, free sulphur (active principle) is released. The polysulphide are phytotoxic. The wettable or floatation sulphur is finely divided elemental sulphur. It is prepared from hydrogen sulphide gas from the coke industry. It is less injurious to plants.

Organic compounds

Tisdale and Flanner were the first to describe the fungicidal properties of carbamates. It was the pioneering work in the field of use of organic compounds, as fungicides. Carbamic acid is derived from carbonic acid by the replacement of a hydroxyl group by an amino group. Dithiocarbamic acid has two sulphur atoms in place of oxygen atoms. From dithiocarbamic acid, some other derivatives are prepared like, Therum (Tetra methyl thiuram disulphides), Ferbam (Ferric dimethyl dithiocarbamate), Ziram (sinc dimethyl dithiocarbamate), Nabam (Disodium ethylene bisdithiocarbamate), Zineb (Zinc ehtylene bisdithiocarbamate), Maneb (Mangnese ethylene bisdithiocarbamate), Xgrosan, TCNB.

Quinone and Phenolic Fungicides

Derivatives of quinone and phenolic compounds are widely used as fungicides. Chloranil (Tetrachloro-p-benzoquinone), Dichlone (2,3 dichloro-1,4 naphthoquinone) are used as fungicides for seed treatment and other diseases. Phenols are used as fungicide in textiles and woods.

Heterocyclic nitrogen compounds

The fungicides of this group are used as foliage protectants and eradicants of fruits and vegetables. Captan (N-trichloromethyl, thio-4-cyclohexane-1,2 dicarboximide) and Glyodin (2-hepta decyl-2 imidazoline acetone) are important fungicides of this group. Glyodin interferes in the biosynthesis of purines, which are the components of nucleic acid. Captam competes with co-caroxylase (thiamine pyrophosphate) and thereby interferes with the process of decarboxylation.

Benzene compounds

Corden (1969) has discussed the benzene compounds and their ability to control pathogens. Some the benzene compounds are Chloroneb (1,4-dichloro-2,5-dimethoxy benzene), Quintozene (Pentachloronitrobenzene, PCNB), Dexon [p-(dimethylamino)-benzene-diozosodium sulfonate], Dianocap [Karathane, 2-(1-methyl-heptyl),4-6 dinitrophenyl crotonate]

SYSTEMIC FUNGICIDES

The systemic fungicides (compounds) penetrate in a plant and can move from their site of application, to other parts of the plant. Most of the systemic fungicide are absorbed by the roots and transported acropetally to the xylem. The mechanisms of action by systemic fungicides are: Inactivation of toxins secreted by pathogen. It may accumulate in fungal cell wall (as fugal cell wall has more permeability), damage of membranes of fungal cell, inhibition in the germ tube formation, appressoria and haustoria formation is also inhibited, inhibition and destruction of fungal enzymes. Some of the systemic fungicides commonly used are following.

Benlate (Benomyl) was discovered by Delp and Kloppping (1968). It is methyl-10 (butyl carbamoyl)-2 benzimidazole carbamate It is hydrolyzed to methyl benzimidazole carbamate (MCB). This active principle is effective against a wide range of pathogen. One of the derivatives of benzimidazole is Bavistin. It has both prophylactic and curative actions.

Terrazole (5-ethoxy-3-trichloromethyl-1,2,4-thiadiazole) is toxic to phycomycetous fungi.

Oxathiins [carboxin (2,3-dihydro-5 carboxyanilido-6 methyl-1,4,oxathiin)] is effective against basidiomycetes.

Thiophanates (methyl and ethyl thiophanates) are broad spectrum fungicides.

'pp'675 (5-Butyl-2, dimethylamino-4 hydroxy-6 methylpyrimidine) may be used for soil treatment for the control of powdery mildew.

ANTIBIOTICS

Antibiotics are produced by microbes and it acts against some other microorganisms. Some antifungal antibiotics and agricultural antibiotics are developed, which have proved effective in controlling many diseases.

Streptomycin is produced by *Streptomyces griseus* (actinomycetes). Streptomycin sulphate is commercially produced. It is effective against both Gram positive and Gram negative bacteria like Xanthomonas, Pseudomonas. Some fungal diseases phytophthora (caused by phycomycetes) are also suppressed by it.

Cycloheximide is produced by *Streptomyces griseus* and *S.nouresi.* It is effective against many fungi but it is ineffective against bacteria. E.g. Ustilago, Rhizopus etc.

Tetracyclines Tetracyclines (Terramycin, aureomycin and actinomycin) are widely used to control plant disease. It is also effective against disease caused by mycoplasma.

Griseofaluvin is produced by *Penicillium griseofulvum.* It is used against many fungi (powdery mildew and some rusts). Alternaria, Sclerotia.

Nysatin is produced by Streptomyces noursei. Commercially it is named as mycostatin or fungicidin. It is used in some fungal diseases. It is also reported to be effective in post harvest dip.

Selection of fungicide, manner and mode of application are important for successful control of disease. Fungicides may be diluted with a solid, and applied as dust or with a liquid and sprayed on the plant target.

DISEASE RESISTANCE VARIETIES

Most higher plants are resistant to the majority of the pathogenic organisms.

Techniques are used in breeding for disease resistance

(a) Mass selection

(b) Pedigree selection

(c) Back crossing

(d) F1 hybrids

(e) Interchange between species and genera

(f) Mutations and polyploids.

CHAPTER - 29

INDUSTRIAL MICROBIOLOGY

FERMENTORS

Fermentation is derived from the Latin language fervere (= to boil). It describes the appearance of the action of yeast on extracts of fruit or malted grain during the production of alcoholic beverages. In terms of industrial microbiology, fermentation process is applied for the production of a product by the mass culture of microorganism.

The fermentation technique applied for industrial production of various compounds. The fermentation processes are grouped in four categories.

1. The processes produce microbial cells (biomass) as the product.
2. The processes produce microbial metabolites, which are used by man.
3. The processes produce microbial enzymes.
4. The processes in which the compounds added to the fermentation are modified. These processes are known as the transformation processes.

Micro-organisms are grown under artificial conditions and in a specially designed vessel filled with proper nutritive media. This vessel is known as fermentor. Before the fermentation is started the medium must be formulated and sterilized. A starter culture is inoculated. Downstream of the fermentor the product has to be purified and further processed. The products of the metabolic activity are economically useful. Proper application and use of fermentor depends upon the selection of the best type of reactor for the particular reaction and the operating conditions (temperature, pH, aeration etc). Selection of strain of microbe is highly important. While selecting the strain stress is given to the optimum conditions and growth phase in which there is maximum synthesis of product. For the selection of a proper fermentor, dimensions of the vessel, fabrication (so as the fermentor could bear the pressure generated due to biochemical reaction) and the conditions operating is being noted.

The industrial usage of microbes often requires that they be grown in large vessels containing considerable quantities of nutritive media. These vessels are commonly called "fermentors" and they can be quite complicated in design. Since frequently they must provide for the control and observation of many microbial growth and biosynthesis.

Factors involved in fermentor design-

According to Leveni 1971 there are two problems to design reactor-

1. The selection of best type of reaction for particular reaction.
2. The determination of best operating system.

Reaction in Fermentor

Fermentor carry on fermentation process and which are actually biochemical conservation of substrate with the help of microbes this biochemical process differ from chemical process and based-

1. The complexity of reaction, which is highly compressed in biochemical process and less complex in chemical process.
2. The mild condition of pH and temperature
3. The difficulty to maintain biochemical condition
4. The low concentration of both substrate and products.
5. The increase of mass of microbes related with the biochemical process
6. The ability of microbes

General features of ideal fermentors-

1. The vessel constituting enough of agreous medium but at the same time the material come which they are fabricated must be rusted.
2. Fermented product not produces any toxic ion to the growth media.
3. Fermentor must take some provision for control or prevention of the growth of contaminating microbes.
4. Aeration devices provided for aerobic microbes so that there is rapid sterile air and CO_2 resulting from metabolism large affected the media.
5. Facility for addition of antifoam agents
6. Controlling devices for controlling of pH and temperature.
7. Stirring device provided nutrients and microbes.
8. Draining passage for removed of completed fermentation broth.

TYPES OF FERMENTORS: In a fermentor, microorganisms may be grown in batch, fed-batch, or continuous culture. Following types of fermentors are used for industrial purposes.

(a) Batch fermentor
(b) Continuous stirred tank fermentor
(c) Tubular fermentor
(d) Fludised fermentor

Batch fermentor

Fermentation is carried out on a batch basis. The capacity varies from 12 liters to thousand gallons. To provide proper temperature a water jacket is constructed around the vessel. Fermentor is connected with an autotiter, which controls the addition of acid or alkali thus maintaining pH of the medium.

Autotiter is connected with a pH probe, which governs the addition of acid or alkali. An agitating device is used for rotating the medium. If required, fermentor is aerated with a device. Air is passed with a pressure. A ring sparger with orifices is used for this purpose. A simple air injection is aerating the fermentor with an impeller by means of a hollow drive shaft. Sparger air injection is widely used in fermentors. In this technique, fermentor is the stirred, baffled, upright cylinder with air sparger. In other techniques, air and nutrients are fed in the fermentor by two-phase injection. This is conducted in the form of foam or suspension. Still, another method of aeration is designing air-lift fermentor. Air is used to circulate the contents of the fermentor, wither through tubes external to the vessel or internally by using a draft tube. In batch culture, productivity is expressed by following equation

$$R\ batch = \frac{X_{max} - X_o}{ti + tii}$$

Where, R batch is the output of the culture in terms of biomass concentration per hour,

Xo is the initial cell concentration at inoculation,

Xmax is maximum cell concentration obtained at stationary phase,

ti and tii is time during which growing and not growing at maximum, respectively. Here, time includes the lag phase, the deceleration phase, and the periods of batching, sterilizing and harvesting.

Continuous Stirred-tank fermentor (CSTF)

The basic principle of this type of fermentor is that contents, i.e., medium and microbial concentration is kept constant. This is known as steady state. To maintain this state, a detailed knowledge of all the relationship, i.e., the rate of reaction and the operating variables is to be studied. Steady state is achieved by chemostatic or turbistatic principles. Chemostatic methods govern the flow rate. This enables the entry of medium and microbes and outlet of product so that a natural level is maintained. In tubristatic methods, turbidity of medium is measured periodically and thus flow rate is controlled. Production and consumption of gases, heat output, production of acid or alkali is maintained constant.

In continuous culture, productivity is expressed by following equation

$$Rcont = Dx\ (1\text{-}tiii/T).$$

Where, Rcont is the output of the culture in terms of cell concentration per hour,

x is the steady-state cell concentration,

tiii is the time period prior to the establishment of a steady state. It includes time for vessel preparation, sterilization, and other operations prior to continuous culture,

T is time period during which steady-state conditions prevail.

There are certain practical difficulties in continuous fermentor. The biggest disadvantage is the contamination by foreign inoculum.

The tubular fermentor

In a tubular fermentor microbes occur as floccules in different environment. Certain flocculating agents like aluminium and calcium chloride enhances flocculation. Microbes also adhere to an inert surface and form a film. Processes with micro-organism adhering to a support surface are chiefly used in the waste water treatment industry (trickling filter).

STRAIN IMPROVEMENT

The basic need in industrial microbiology is the microorganism and its strain to be used. An ideal strain of microbe is that can be easily cultivated having stable biochemical characteristic, it should have high yielding properties with less production of undesirable byproducts. The species are selected from natural sources and are screened out. Screening is conducted in primary and secondary stages. Primary screening is the test and search out of appropriate species. Secondary screening is the systematic screening in detecting the commercially and economically useful microorganism in the fermentation process. Once the species and strain is screened out, it is important to manipulate it so that it develops into a high yielding and economic type. For improving a strain and make it for potential productive yielding strain, genetic manipulation is conducted. This is carried out by two processes: mutation and recombination. Recombinant DNA technology has also given positive results.

Mutation

There are chances of some genetic changes after each cell division resulting in a degree of mutation. Mutation in a cell may be easy by using mutagens. The mutants prepared may be major mutation or minor mutation. Wild strain of *Streptomyces griseus* has low ability of producing streptomycin and high production of mannosidostreptomycin. The latter is not significant as for as antibiotic activity is concerned. Major mutation in this organism shows reversed results. There are minor mutations, in which some lead to small minor change in the production. A stepwise improvement in these genes results in 10-15% increase in the productivity. There are few genetic mechanisms for controlling the biochemical pathway in wild strains and thus the production of the metabolite. A good commercial mutant should lack the control systems so that there is overproduction of the desirable product.

The mutants developed are of two categories: auxotrophic mutants and analogues resistant mutants.

Auxotrophic mutant has a defect in one of its biosynthetic pathways so that it requires a specific biochemical for normal growth and development. Lysine synthesis in *Corynebacterium glutamicum* is subjected for studies. The first enzyme of the pathway "aspartokinase" is inhibited only when both lysine and threonine are synthesized above the required amount. A mutant of this bacterium is unable to catalyse the conversion of aspartic semi-aldehyde into homoserine. It can grow normally on homoserine supplemented medium. Thus, the mutant is homoserine auxotroph. If the concentration of homoserine is very low in the medium, the endogenous level of threonine would not reach the inhibitory level fro aspartokinase control. Under such condition, aspartate would be converted into lysine, which would accumulate in the medium. The isolation of bacterial auxotrophs may be achieved using the penicillin enrichment technique.

Analogue resistant mutant: An analogue is a compound, which is similar in structure to another compound (analogues of primary metabolites). It is frequently inhibitory to microbial cells. Lysine production in *Brevibacterium flavum* is studied. The biosynthetic pathway is same as in *C.glutamicum.* Lysine analogue S-(2-aminoethyl)cysteine (AEC) only inhibited growth completely in the presence of threonine. Thus, AEC might have combined with threonine in the concerted inhibition of aspartokinase and bacteria would have been deprived of lysine and methionine. After the treatment, the survivors remained, were plated on agar plates containing both AEC and threonine. A relatively high proportion of the resulting colonies over produced lysine.

Yet another category is revertants from nonproducing mutants of a strain are high producers. *Streptomyces viridifaciens* showed over 6-fold increase in chlortetracycline production over the original strain from which the nonproducing mutant was obtained.

Recombination

In simple terms, recombination is the process by which, a new combination of parental genes may be generated. By recombination, desirable alleles present in two or more strains are combined resulting in strain with high potential productivity. Recombination may help in improving strain capable to generate new products.

Parasexual cycle is present in some fungi, which only reproduce asexually. Most of the fungi important in industrial productions reproduce asexually. However, these fungi show tendency to form heterokaryons and the heterokaryons fuse to form diploid cells. In parasexual cycle, there is no definite stage of formation of heterokaryons, diploid nucleus and meiosis. As the diploid cells are formed so there may be recombination. Strain improvement by virtue of parasexual cycle is carried out in *Penicillium chrysogenum.*

Protoplast fusion is a new technique to improve strains, which do not have compatible nuclei. Protoplast of cells is isolated by enzymatic lysis of cell wall in isotonic solution. The fusion is induced by polyethylene glycol treatment. Recombinant DNA technology is widely applied for strain improvement.

Examples

Some high yielding strain of *Penicillium crysogenum-*

Designation of mould	Use
P. crysogenum NRRL 1951-B-25	Submerged culture production. In industry during in the latter half of 1944

P. crysogenum Standford B 25099	The first half of superior sub merged culture production.
P. crysogenum x-1612	Superior submerged culture production
P. crysogenum Q-176 (wisconson)	Superior submerged culture production

YEAST

Yeast is highly significant in fermentation products. The common useful yeasts are *Saccharomyces cervisiae* (in bakery, distillery, alcoholic beverages industry), *S.ellipsoideus* (in wine production), *S.diastaticus* (in distillery industry) and *S.carlsbergensis* (in production of alcoholic beverage from lager and cider). Yeast produces some important enzymes significant in fermentation processes. The enzymes are Invertase, β-Galctosidase and it is also a good source of vitamins. Mass production of yeast for the purpose of inoculation in other industries is important in itself. Cultivation of yeast depends upon the type it is used. Yeast is commercially used in the form of Brewers' yeast, Bakers' yeast, and food and fodder yeast. These are produced aerobically by batch processes.

Production of Brewers' yeast

Brewers' yeast is cultivated in pasteurized hopped malt wort medium. The plant is sterilized by live steam. The plant consists of two propagating closed vessels. The media is prepared with molasses, biotin and ammonium salt. Intermittent aeration is required for the growth of yeast. However, agitation is avoided. The 200 grams of maltose and 1 gm. of ammonium salt yield 10 gms of yeast, 97.5 gms of ethanol and 93.6 gms of CO_2. The temperature required is 20°C and as 100 kCal of heat energy is produced per 10 gms of yeast, hence the temperature is maintained by exterior cooling jackets.

Production of Bakers' yeast

Bakers' yeast is cultivated on beet molasses mixed with black-strap molasses and diluted with water. It is supplemented with ammonia and ammonium sulphate, diammonium hydrogen phosphate and magnesium sulphate, biotin and thyamin. The blended molasses is pasteurized. pH is maintained between 4.0-4.5. The fed batch process is adopted. Aeration is necessary. Vigorously aerated medium and exponentially growing yeast produces a considerable amount of foam. Hence, propylene glycols or silicone antifoams are used. Temperature of 30°C is maintained by circulating coolant through internal coils. 200 gms of sucrose (in molasses), 10.4 gms of ammonia and 102.5 gms of oxygen yields 100 gms of yeast, 77.2 gms of water and 145.6 gms of CO_2. 387 Kcal of heat energy is generated during the production of 100 gms of yeast. Sucrose favours high production of yeast cells. For the maturation of cells, pH is allowed to rise and no nutrients are further added. This treatment enables to yeast cells to detach from the media and stops budding. The yeast cells are separated from the liquid. It gives yeast cream, which is washed and centrifuged. It is further dried on rotary vacuum filters. Before filtration, salt is added so that the cells shrink. Filtered cells are subjected to water spraying so that cell gets swollen. The drier product consists of 30% dry matter content. The product is scrapped and water and plasticizers are added to it and yeast blocks are prepared. The blocks are stored at low temperature.

Active dry yeast

Active dry yeast is like bakers' and much more the production technology is same. It is also cultivated in medium with beet molasses under the same conditions. Here, specially-bred yeasts, which offer stability against drying, are used. By the end of the process, sugar and nitrogen sources are restricted. About 35 % dry matter is gained after filtration. Pellets are prepared by chopping the filtered mass. Pellets are further dried in a tunnel or rotary driers for 6-24 hours.

Production of food and fodder yeast

Yeast is used as dietary supplement by human beings. It is also used in fodder. *Saccharomyces cerevisiae* and *S.carlsergensis* are used for this purpose. The media used for cultivation of yeast for this purpose is same as that of bakers' yeast. The quality is maintained in the production of food yeast. It contains more of protein and vitamins. Vitamins may also be added during the production of food yeast. The yeast cream is centrifuged and dried in rotary vacuum filter. Later on it is desiccated, grounded and packed.

SINGLE CELL PROTEIN

In the modern age of microbiology, microbiologists are in all efforts to explore microbes by all mean (direct or indirect) for the benefit of human kind. Microbial biomass (high grade protein) has yielded promising results as the supplement for diet and nutrition. The concept of single cell protein (SCP) was developed by Wilson (1966) and by Scrimshaw (1968). It is exploration of microorganisms (bacteria, yeast, algae and fungi) for the production of biomass (i.e., their proteins). Microbes may be grown in simple culture media. Microorganisms need less time for their multiplication and growth to produce a huge biomass. The microbial cells have high protein contents and need less time in protein synthesis as compared to higher microorganisms.

Bacterial genera worked out for the study and gain of SCP, are Pseudomonas (*P. facilis, P. saccharophila, P. ruhlandii, P. pseudomallei, P. orvilla, P. methanica); Alacligenes (A. eutrophus, A. paradoxus), Methanomonas (M. methanooxidans, M. capsulatus), Methylococcus capsulatus, Cellulomonas galba, Corynebacterium, Micrococcus,* etc. These bacteria are hydrogen utilizing or may grow on hydrocarbons. The photosynthetic bacteria are also important. Hydrogen utilizing bacteria require hydrogen and carbon di oxide and some salts (ammonium salts, nitrates or urea for nitrogen) and some trace elements. These bacteria contain about 100 grams protein, 15 grams nucleic acid and 20 grams of cell wall materials. Hydrocarbon utilizing bacteria grow on one carbon organic compounds (methane) or complex compounds like n-paraffins. The bacteria grown on methane or methanol have to be handled with care as there are chances of explosions, high heat generation etc. The protein content of these bacteria is 35-70 grams/100 gram of cell weight. n-paraffin utilizing bacteria were discovered in petroleum belts. They yield about 73% protein in their dry weight.

The Spirulina

S. maxima, is a cyanobacteria, grown in an open and shallow (1m deep) pond. It needs carbon di oxide, ammonia and trace elements for their growth. The biomass production is about 5 tons a day. The microbe is harvested by filtration, followed by vacuum filtration and is dried by using drums

Fungal proteins

Fungi decompose organic waste materials. The cellulose decomposing fungi are of great interest. *Trichoderma viridae*, is the best studied cellulose degrading fungi. Other fungi of interest are amylolytic fungi. *Aspergillus oryzae, Rhizopus arrhizus*, are amylolytic fungi grown on barley mash supplemented with urea or ammonium sulphate as nitrogen source. The sugar is converted into glucose and later protein is synthesized. The conversion rate varies from 23 to 40 percent. The fungal proteins are advantageous but mostly it is deficient of methionine.

Biomass from yeast

Yeast produces biomass from starch. *Endomycopsis fibuliger* produces enzyme amylase and converts starch into sugar, *Candida utilis* uses this sugar and 85-90% yeast biomass is produced. Yeast (*Candida lipolytica*) produced SCP utilising hydrocarbons (10-23 carbons). The culture media consists of n-alkanes, minerals and ammonia. About 60% crude protein is produced from this yeast.

BIOMASS RECOVERY

Biomass is called SCP, as it is rich in protein. It is produced by microorganisms is used by man or by animals in the form of food and feed. Biomass is recovered under hygeineic conditions. The microbes are collected by filtration, or by centrifugation, or by flocculation, or both by flocculation and filtration. Followed this, it is dried. Drying is carried out under sun light and finally with heat treatment.

One of the major disadvantages of biomass is the high nucleic acid contents and the digestion-resistant cell wall. The metabolism of nucleic acids yields excessive amounts of uric acid that can cause kidney stones and gout.

HYBRIDOMA TECHNOLOGY

Hybridoma technology was developed by G. Kohler and C. Milstein (1975). It is technique to develop monoclonal antibodies by the somatic hybridization of B-lymphocytes and myeloma cells. The myeloma cell is a tumor cell of B-lymphocyte origin, which has the ability to grow continuously in tissue culture. The hybrid cells are capable of growing fast and producing antibodies. The hybridoma carries some characteristics of parent cells. It synthesizes and secretes antibody identical to that of the plasma cell parent. It can grow in tissue culture conditions. These cell, are cultured *in vitro* or passaged through mouse peritoneal cavity. The B-lymphocyte cells are preprogrammed so that they produce antibodies for only a single antigen. For this purpose, a single antigenic determinant is allowed to react to the cell surface receptor of B-lymphocytes. The latter proliferates to produce a clone of these specific cells producing antibody of the same specificity. This is known as clonal selection. B-lymphocyte cells are isolated from spleen of an animal (generally mouse) immunized by a particular antigen. Immunization regimes may differ according to the nature of the antigen and the type of antibody required. IgM antibodies are generated early in the immune response, whereas, IgG antibodies predominate later. The specific antigen along with adjuvant is injected in the animal subcutaneously or in peritoneal cavity. Generally, one or two infections are made using an emulsion formed of equal volumes of the antigen and Freunds complete adjuvants. These isolated cells are mixed with myeloma cells. The myeloma cells are capable of dividing very fast. The cells selected should have two properties. They must contain a genetic marker and they should not produce antibodies by themselves. If cells of genetic marker HGPRT (hypoxanthine-guanine phospho-ribosyl-transferase) are fused with B-lymphocytes, the resulting population will have three types of cells: (1) B-lymphocytes, (2) myeloma cells and (3) hybrid cells (hybridoma). Fusion among the cells was originally carried out by inactivated Sendai virus. However, now polyethylene glycol (PEG) is used instead of this virus. Myeloma cells in the log phase of growth or washed in serum-free medium or buffer. They are pooled with a ten-fold excess of dissociated spleen cells from the donor animal, and centrifuged to form a pellet of mixed cells. The supernatant is withdrawn and then 1-2 ml of a solution of PEG (50%) in serum-buffer is added to the pelleted cells. After 1-2 minutes of gentle stirring, the PEG solution is slowly diluted by the addition of serum-free medium. The fused cells are gently centrifuged and resuspended in growth medium containing serum and the selective agents. For this HAT (hypoxanthine, aminopterin and thymidine) medium is used. This medium enables only hybrid cells to grow and after a period the unfused cells die. This medium contains aminopterin drug, which blocks the cellular biosynthesis of purines and pyrimidines from simple sugars and amino acids. Hypoxanthine is converted into guanine by the enzyme hypoxanthine-phosphoribosyl-transferase, while; thymadine is phosphorylated by thymidine kinase (TK). These enzymes are of salvage pathway of nucleotide synthesis. Cells having these enzymes (HGPRT+ and TK+) can grow on this medium. It is important to identify and isolate the hybrid cells. Between ten days and two weeks after fusion, it is usual to carry out some form of screening of the hybrid cells to detect those which are producing specific antibody. The hybridoma cells grow and secrete antibodies into the medium. The concentration of specific antibody in medium conditioned by the growth of the cells is of the order of 10 g/ml. Much higher concentrations of antibody may be achieved by growing the cells as an ascitic tumor in mice. The antibody concentration in the ascitic fluid is usually between 2 and 10 mg/ml. The supernatant from each micro-well is sampled and assayed for the presence of antibodies specific to the antigen to be worked out. Agglutination, precipitation and ELISA tests are applied for assay. The desired hybridoma clone is multiplied *in vivo* or *in vitro*. Once cloned, the hybridoma will grow continuously in tissue culture medium. Hybridoma appears to be genetically quite stable.

INDUSTRIAL PRODUCTION

Cultural techniques have been developed to produce a variety of compounds as secondary metabolites. Microbes produce primary products during growth phase (trophophase), which are related to the growth and synthesis of their cells. The secondary metabolites are produced after the active growth phase (idiophase). These compounds have no direct relationship in synthesis of cell materials. Various compounds produced by microbes are important for human beings. These compounds are produced commercially with using specific microorganism.

PRODUCTION OF ALCOHOLIC BEVERAGES

Alcohol is an important product of fermentation.

WINE PRODUCTION

Grape sugar is used as raw material for wine production. It contains carbohydrates like glucose, fructose, pentose, pectin, protein, acids (tartaric and malic) minerals like potassium, calcium and magnesium salts of phosphoric, sulfuric, hydrochloric and salicilic acids, tannins, pigments, vitamins, enzymes and aromatic compounds. The relative percentage of glucose, fructose and mannose present in must, depends upon the maturity of grapes. Different species of *Saccharomyces cerevisiae* is glycophilic and thus fructose is left after fermentation. Acetic acid, lactic acid bacteria are generally, found in must. The contamination is controlled by SO_2. SO_2 also helps in binding of acetaldehyde formed during fermentation. Presence of acetaldehyde results in undesirable organoleptic properties. Different types of yeast are utilized in the production of wine (fermentation). The common species of *Saccharomyces* used in fermentation are *S. cerevisiae, S. bayanus, S. uvarum, S. rosei, S. rouxii, S. ludwijii,* etc. Other types of yeast used in fermentation are *Candida pulcherrima, Torulopsis bacillaris, Kloeckera apiculata, Hanseniospora valbyensis,* etc. Yeast cells if remained in wine after fermentation may harm the quality by further fermentation of residual sugars. Presence of acetic acid and lactic acid bacteria in wine is also harmful. The bacterial defects are fermentation of fructose producing mannitol. Other types of sugars if present, are fermented into CO_2, volatile acids etc. Malic acid, tartaric acid and citric acid present in wine is metabolized by bacteria resulting in undesirable changes in aroma. Acetic acid bacteria oxidize ethanol via acetaldehyde to acetic acid. The acetic acid is esterified to ethyl acetate giving vinegary wines. There are different types of wines produced industrially. White wine production depends upon the types of grapes as well as on its maturity and its health. Grapes are plucked from stem. Special care is taken so that no stem pieces remain attached to the grapes. Grapes are immediately crushed, and the must is treated with SO_2. SO_2 inhibits growth of undesirable microbes and much more the oxidative enzymes are inactivated. Must is filtered or centrifuged or sedimented so that all the particle components may be removed. Must, may be treated with temperature for short time so that the oxidative enzymes are inactivated and the existing microbes which, may interfere during fermentation are killed. Must is now ready to be inoculated with yeast. After fermentation wine is separated, clarified and stabilized. It is stored in bottle. Ethanol content is usually between 8 and 15% by volume. Red wine colour of red wine is due to the skin of grapes (tanins); hence the methods of production of red wine (extraction of colour) are different. Crushed grapes are fermented and the colour is extracted by the alcohol formed during fermentation. Fermentation is carried out in a closed vessel. Continuous recirculation process is also developed. Champagne type wines are sparkling wines. High concentration of CO_2 gives foam. Brandy is produced by distillation of wine. Generally, it is produced by white grapes.

The aroma of wines is due to volatile components like alcohol, acids, esters, aldehyde, ketones, terpenes, lactones and others. Terpens alcohol and terpene oxides are found in higher concentration. Their concentration in different species of grapes varies. All these aroma are primary aroma. While, secondary aroma is due to fermentation and during storage of wine. Metabolites produced by yeast have their own role in the secondary aroma of wine.

PRODUCTION OF DISTILLED BEVARAGES ALCOHOLS

Distilled beverage alcohols are produced by fermentation of grain mash. Fermentation is carried out in a cistern tank. *Whisky* is produced from corn mash. Grains are selected, graded and dried milled. Precautions are taken so that grain is free from bacterial contamination or toxins like aflotoxin. During milling effort is being made so that more starch is produced instead of flour. For this purpose, roller milling and pressure cooking is preferred so that there is gain of more starch. Yeast is inoculated in mash. Distillation is carried out in leettle-column batch to continuous multicolumn vacuum stills. *Rum* is prepared by fermentation of sugarcane juice (syrup), sugar cane molasses or other sugar cane by-products. The molasses is diluted and inoculated with yeast and temperature is maintained between 32-33°C. It takes about 36 hours in the fermentation process to produce this alcohol from sugar molasses. *Gin* is produced from grain neutral spirit. Other plant products are also used for the production of gin. Gin is named after the plant products like juniper berries, dried orgnae peel, cardamom and corriander seed gin. *Vodka* is prepared by charcoal treatment of spirit. *Tequila* is prepared by fermenting agave or mezcal.

PRODUCTION OF ORGANIC ACIDS

A number of organic acids are produced by bacteria and fungi.

CITRIC ACID PRODUCTION

Scheele (1784) isolated and purified citric acid from lemon juice. It is found in many other fruits. Wehmer (1893) found citric acid as product of fermentation in two molds, viz. *Citromyces pfefferianus* and *C.glaber*. This genera, was later classified as Penicillia. The media used for fermentation contained sucrose solution and calcium carbonate. Currie (1917) found that *Aspergillus niger* produced citric acid. Citric acid is used industrial purposes as following.

1. For production of carbonated beverages, and as an acidulant in food and pharmaceutical industries.
2. For chelation and sequester in tanning of animal skin.
3. As plasticizers (citrate and citrate esters).

Surface culture method for production of citric acid

Fungi are used for the production of citric acid by surface culture method. Different species of *Aspergillus* (*A. niger, A. clavatus*) and *Penicillium* (*P. luteum, P. citrinum*), *Paecilomyces divaricatum, Mucor piriformis, Ustulina vulgaris* etc. are used for the industrial production of citric acid by surface culture method. However, *A. niger* is commonly used.

Media

Sugar is best carbon source. Maximum yield has been secured from sucrose and fructose. In commercial scale, beet molases is extensively used as a carbon substrate in the fungal production of citric acid. Nitrogen, potassium, phosphorus, sulphur, and magnesium sources are also needed. Currie (1917) and Doegler and Prescott (1934) have proposed different media for production of citric acid. The composition of media is as following

Currie		Doelger and Prescott	
Sucrose	125 –150 Gms/litre	Sucrose	140G./L
NH_4NO_3	2.0 – 2.5	NH_4NO_3	2.23
K_2HPO_4	0.75 – 1.0	K_2HPO_4	1.0
$MgSO_4.7H_2O$	0.20 – 0.25	$MgSO_4.7H_2O$	0.23
pH 3.4 3.5 adjusted with N/5HCl		pH 2.20-1.60 adjusted with N/1HCl	

Salt and sugars are dissolved and made up to 1 litre with distilled water. The pH of the medium is adjusted and sterilized at 8 to 10 lb. steam pressure per square inch for 30 minutes. If the quantity of metallic salt is increased, it effects adversely (oxalic acid is produced more and citric acid production is inhibited). If the quantity of $MgSO_4.7H_2O$ is more than 0.30 gm/litre, sporulation takes place. Aluminium, chromium, iron and magnesium ions stimulated, while molybdenum, copper, zinc, and calcium ions inhibited citric acid production. Calcium carbonate favours contamination. In the absence of $CaCO_3$, more citric acid is produced. 26 to 28°C, temperature favours citric acid production. Citric acid fermentation is catalyzed by intracellular enzymes. Hence, more and more of mycelial mat are to be produced. For this purpose, a shallow pan culture vessel proves to be more useful as compared to deep vessel. The lower the ratio of the volume to surface area gives beneficial results. Aeration is also essential for fermentation. The rate and time of aeration governs optimum yield. Approximately, 60 to 80 gm of citirc acid is produced per 100 gm of sugar supplied. About 7 to 10 days is needed to complete fermentation. After the fermentation, citric acid is to be recovered. The fermentation solution is drained off. The mat is pressed to remove any acid contained in it. The solution is treated with milk of lime and calcium citrate is precipitated. The precipitate is further treated with sulfuric acid liberating citric acid and calcium sulphate. The precipitate is again filtered and washed. The unconverted sugar may be fermented by yeasts and the citric acid crystallized directly, in an altered method.

Submerged-culture method

In this process also fungus, *A. niger* is used. It is grown dispersed through a liquid production medium. In this process, agitators and spargers are used. Schreyer (1928) discovered that aeration and continuous mechanical agitation and continuous mechanical agitation increased the total production of acid by *A.fumaricus*. High aeration rates producing much foam are necessary. Beet molases supplied with ferrocyanide, gives optimum mycelium. Sugar and acid contents is checked from the sample collected at regular intervals from the sample.

Citric acid production by yeast

Yeast (*Candida guilliermondii*) is used for the production of citric acid by submerged culture method. Glucose or black-strap molasses with an equivalent amount of sugar is used in medium. It takes less time as compared to production of citric acid by fungi by submerged culture process. Hydrocarbons (n-paraffins) are used in medium with yeast, *Candida lipolytica*. Strain ATCC 20114 yields 145 per cent of citric acid. Monofluroacetate is used in culture medium. Strains accumulate 112 gram of citric acid per litre after 3 days incubation at 30°C.

LACTIC ACID PRODUCTION

Blondeau (1847) first discovered lactic acid as a fermentation product. Avery (1881) discovered the production lactic acid in commercial scale. Different species of *Lactobacillus* (*L. bulgaricus, L. delbrueckii, L. leichmannii, L. casei*) *Streptococcus lactic* and *Rhizopus oryzae* are used for the production of lactic acid by fermentation. Crude substrates like corn sugar, molasses, whey and starch (obtained from potato, rice) are used for this purpose. Glucose, maltose, sucrose or lactose is also used as carbon sources. The choice of carbohydrates used will depend upon its availability, its fermentability, with or without preliminary treatment and its cost. The starchy substrates are pretreated with sulfuric acid or enzyme. Thus, starch is hydrolyzed into maltose and sugar. The sugar in mashes is normally adjusted to a concentration of 5 to 20 per cent. Ammonium salts (e.g., ammonium hydrogen phosphate 0.25%) is used as nitrogen source. Some accessory nutrients (like malt sprouts, steep water) and growth factors (vitamin B complex) are supplied to the media for better yield of lactic acid. The pH of the medium is adjusted to 5.5–6.5. This range is satisfactory for lactic acid production. To maintain this pH, neutralization of the accumulated acid is carried out by continuously adding calcium hydroxide in the form of slurry. Ammonia, ammonium carbonate, or mixtures of sodium, potassium and ammonium carbonate may also be used as neutralizing agent. The fermenting solution is gently agitated for a homogenous mixing of neutralizing agents. For most of the lactic acid fermenting bacteria temperature between 45-50°C is required. However, *L. pentosus, L. casei* and *S. lactis* need 30°C for fermentation. As the lactic acid fermenting microbes are facultative anaerobes, hence complete removal of oxygen or supply of oxygen is not essential. The fermentation process completes within 5-10 days, yielding 93-95% lactic acid of the weight of glucose supplied. There are four principal grades of lactic acid (Technical grade-22, 44 and 80% lactic acid strength; edible grade of 50 and 80% lactic acid strength; plastic grade and U.S.P grade). The recovery of lactic acid depends upon these grades. The first step of recovery is by filtration. It is carried out by using industrial precoat filters. For the production of technical grade, calcium present is removed as calcium sulfate dihydrate by precipitation. It is followed by filtration to separate the precipitate of calcium sulphate dihydrate. Finally, the filtrate is subjected to concentration to 35-40% lactic acid by evaporation. For the production of food grade lactic acid, higher-grade sugar source and a mineral protein content is used in medium. The fermentation solution is filtered and filtrate is treated with sulfuric acid. Lactic acid is regenerated; calcium is precipitated as calcium sulfate. This is washed and the wash is combined with the filtrate and is then treated with activated vegetables carbon to remove organic impurities. By evaporating this mixture, 25% of solids of lactic acid are obtained. The process of refining and evaporation are repeated to obtain 50-65% of acid. Sodium ferrocyanide is used to remove heavy metals (like copper) and the acid is discoloured. Ion-exchange resins are applied for any other impurities. To obtain plastic-grade lactic acid, the filtrate is esterified with methanol after concentrating it.

Lactic acid production by whey

Lactobacillus bulgaricus is used for the production of lactic acid from pasteurized whey. Fermentor is made up of wood. Intermittent treatment with lime is essential for neutralization. 43°C, temperature is maintained during fermentation. At the end of the fermentation, the lactalbumin is coagulated by heat. The remaining supernatant, solution of calcium lactate is decanted off, filtered, treated with decolourizing carbon. Further treatment is by filter aids, filtered and washed.

Lactic acid production by corn sugar

Lactobacillus delbrueckii is used in this process. Medium is prepared with 15% corn sugar, 0.375% malt sprouts, 0.25% diammonium phosphate, 74.375% water. pH is maintained between 5.8 to 6.0. 10% calcium carbonate is added for this purpose. Examination of reducing sugar and pH during fermentation is essential. Calcium lactate is recovered and purified.

PRODUCTION OF ACETIC ACID or VINEGAR

Acetic acid is popularly and commercially known as Vinegar (Fr. Vinaigre = sour wine). In 1837, Kutzing reported that some micro-organisms are able to convert ethanol into acetic acid. In 1868, Pasteur further worked on this fermentation. Hansen (1878, 1897) studied and classified bacteria involved in this fermentation. He isolated the bacteria and named it as *Bacterium aceti.* This bacterium was previously named as *M. aceti,* now it is named as *Acetobacter aceti. Acetobacter* has a number of species. These species are grouped into two categories: (1) those, which oxidize acetic acid to carbon dioxide and water, and (2) those, which do not oxidize acetic acid.

There are two stages of acetic acid production. In the first stage, alcoholic fermentation of sugar takes place, which is later converted into acetic acid. Acetic acid is commercially produced by following procedures: (1) Orleans method, (2) Quick vinegar process, (Frings generator) and (3) Deep fermentation process.

Orlean process

It is the "French method", the oldest method. This process takes place in a barrel of 200-liter capacity. Initially one third of barrel is filled with good grade vinegar and about 10 to 15 liter of alcohol is added to it. This much of wine is added for four weeks. After fifth week, 10 to 15 liter of vinegar is withdrawn and again wine is added. This is a continuous process. A thin film of bacteria is formed at the top. Later, this film becomes thick and gelatinous. It is called as 'mother of vinegar'. This film should be supported so that is does not break and sink in the bottom. Some techniques are developed which have even modified this old process.

The quick vinegar process

Boerhave discovered this process and later Schutzenbach modified it. This process is carried out in generator of various size and shape. Frings generator is widely used. This generator is an airtight tank of 14ft in diameter and 15 ft in height. There is wooden grating near the bottom. Above it, is placed packing material, the beech wood shavings. Other packing materials like coke, pumice, rattan, grape and other twigs may also be used. The tank is superimposed on concrete beams; in such a manner that air may circulate beneath the apparatus. There are certain accessories equipped in the tank. At the top, there is a 4 inches ventilator with a damper connected to a vapour-liquid separator. It serves to control the air supply. The sprayer of distributing arm is fitted near the top of the tank. Below the wooden grating at the bottom of the tank is the collection chamber of the generator. At the very bottom of this chamber is a cock or faucet. It serves to withdraw the finished product. The bottom collection chamber has tubular cooling coils. The coils are connected by a centrifugal pump to a feed line supplying the sprayer. The vinegar mix, passes down through the shavings, is circulated through an inner pipe made of stainless steel, or other acid-resistant substance. It is cooled by the passage of water from the cooling coils to the top of the tank, through a rubber line. Thermometers are placed at different levels of generator. The vinegar stock or mix is circulated through the shavings. *Acetobacter* is inoculated in the mix. The composition of mix is as following

Glucose	0.9%
Ammonium phosphate	0.4%
Magnesium sulfate	0.1%
Potassium citrate	0.1%
Pantothenic acid	0.00005%

The mix is permitted to circulate repeatedly through the beechwood shavings. Thus, alcoholic solution undergoes more and more oxidation. During this process heat is generated. The temperature is controlled by passing the nutrient acid-alcohol mix through the cooling coils. The cooled mix is returned to the sparger and sprayed over the surface of the shavings. This process is repeated until the acetic acid of desired strength is obtained. It takes about 8 to 10 days for the conversion of the ethanol in a 2,500 gal. Batch to 105 grains of acetic acid is produced. Slime formation takes place due to mother of vinegar, which may plug the generator. To overcome from this problem, proper balance of nutrients should be maintained. This process is simple, low cost, and easy to control.

Mackin (1947) designed a generator for producing high-grade vinegar. In this generator, wood shaving or packing material is not used. Nutrient solution of dilute alcohol along with acetic acid bacteria is sprayed through the jet nozzles. The mixture falls as a fine mist through air being trubulently circulated.

Deep fermentation process

This process is introduced by Hromatka and Ebner; and Haeseler. Hormatka and Ebner (1950, 1951) have studied the relationship of oxygen with acetic acid production by bacteria. The role of oxygen concentration in beech wood shavings and in submerged culture was also studied. Hormatka, Kastner and Ebner (1953) have studied the influence of temperature and the total concentration of the substrate on the submerged vinegar fermentation. In deep fermentation process, *Acetobacter acetigenum, A.pasteurianum* proved to yield more acetic acid. Richardson proposed media for this process containing Corn sugar, Diammonium hydrogen phsophate, potassium chloride, citric acid powdered whey and autolysed yeast. Temperature and pH were maintained at 76-82 F and 3.9-5.0, respectively. The air flow was regulated at 20 litres/hour of air circulated in the fermentor. The fermenting wash contained 0.25-0.30% of alcohol.

PRODUCTON OF ENZYMES

Enzymes are biocatalyst and take part in all types of conversions. Though, enzymes are produced from living cells but it can exert their specific action independent of living cells. More than 1000 enzymes are known. The important microbial enzymes used in fermentation technology are amylase, proteases, pectinases, invertases, cellulases, glucose oxidase and streptikinase.

AMYLASE PRODUCTION

Amylase enzyme hydrolyzes, 1,4 glucosidic linkage in polysaccharide. It is produced by bacteria and fungi. It is used in starch degradation, designing of textile fabrics, preparation of adhesives sizes and coating for paper industry, manufacture of glucose and glucose syrups and in flour treatment. Amylases are of two type, α-amylase and β-amylase or α-1-4, gluconases and α-1-6 gluconases. Amylase is an endoenzyme and cleaves c-1 linkage. It is produced by bacteria and fungi.

BACTERIAL AMYLASE

Bacterial amylase is used in continuous process for designing of textile fabrics, modification of starches suitable for preparation of adhesives, sizes and coating for the paper industry, manufacture of glucose and glucose syrups and brewing processes. Few bacterial genera produce amylase. *Bacillus licheniformis, B.amylo-liquefaciens, Pseudomonas saccharophila* and *Clostridium* species are used for the production of this enzyme. Fermentation is carried out by submerged culture method. The composition of media is following

Ground soya bean meal	1.85%
Amber BYF	1.50%
Distillers dried soluble	0.76%
N-Z amine	0.65%
Lactose	4.75%
$MgSO_4.7H_2O$	0.04%
Hodag KG-1 antifoam	0.05%
Water	90.40%

FUNGAL AMYLASE

Basic use of fungal amylase is in flour treatment for supplementing the diastatic activity of flour. Species of *Aspergillus* (*A. niger* and *A. oryzae*) is used for amylase production. Solid substrate culture method is applied for fermentation. The composition of media is following

Corn starch	24 grams/liter
Corn-steep liquor	36 gm/lt
KCl	0.2 gm/lt
Na_2HPO_4	4.7 gm/lt
$CaCl_2$	1 gm/lt
$MgCl_2.6H_2O$	0.2 gm/lt.

Growth of mycelium makes the medium viscous thus, creates obstacle in agitation and aeration of medium.

30-40°C, temperature and 7.0 pH are most suitable for enzyme production. During fermentation (both for bacterial and fungal processes) pH shifts towards alkalinity and to overcome from this, salts of organic acids (citrate, gluconte, acetate) are added as pH regulator.

Amylase produced by bacteria does not require much of purification. If highly purified amylase is to be produced, adsorbtion techniques (fractional precipitation and selective inactivation) are applied.

PROTEASE PRODUCTION

This is an extra cellular enzyme. It is produced by bacteria (*Bacillus subtilis* and *B.licheniformis*) and by fungi (*Aspergillus niger* and *A.oryzae*). It is widely used in food industry, like chill proofing of beer, in the production of protein hydrolysates, in the production of condensed fish solubles and as feed supplement. Bacterial proteases are not used in food processing. They are used on a large scale in enzyme containing washing powder. Fungal proteases are also used in modification of wheat protein in bread doughs, in meat tendering. Proteases of microbial origin are classified into four types: serine protease, thiol protease, metalloproteinases and acid proteinases. In general, microbes produce a mixture of these enzymes. The type of protease formed depends upon the composition of the medium. *Bacillus licheniformis* is used in the production of serine protease by submerged culture method. The composition of the medium is

Starch hydrolyzate	50 gm/lt
Soya bean meal	20 gm/lt
Casein	20 gm/lt
Na_2HPO_4	3.3 gm/lt

Temperature is adjusted between 30-40°C and pH 7.0.

Bacillus subtilis, Pseudomonas aeruginosa and *Streptomyces* are used in the production of enzyme metalloprotease. *Aspergillus oryzae, A. niger, Mucor pusillus,* and *M. miehei* are used in the production of acid protease. Semi solid or submerged culture method is applied for fermentation. Acid proteases are pepsin like or renin like acid. Pepsin like acid is produced by *Aspergillus niger* and other species of *Aspergillus* and *Penicillium.* Presence of inorganic nitrogen source inhibits enzyme production. Renin like acid is produced by *Mucor miches, M. pusillus.*

INVERTASE PRODUCTION

This enzyme is β -fructofuranosidase, and catalyzes the hydrolysis of the terminal non-reducing β -fructofuranoside residue. It breaks sucrose (dextro-rotatory) into glucose and fructose (laevorotatory). Thus, the name invertase is due to its activity to change the optical properties of substrate and product. The enzyme accumulates in the cells of yeast, so it is essential to break these cells to gain the enzyme. Yeast is suspended in different amount of ammonium hydrogen phosphate, potassium phosphate, magnesium sulphate, potassium nitrate, etc. Temperature is maintained at 28-30°C, and 4.5 pH. Aeration of medium for 3-8 hours is necessary. After the cells are cultivated, they are mechanically crushed. The crude enzyme is subjected to ultra-filtration, precipitation and adsorbtion for the purification of enzyme.

LIPASE PRODUCTION

Enzyme lipase hydrolyzes, triacylglycerol into glycerol and free acids. It is more useful as digestive aid. It is not much significant for industrial uses. Microbes used for the production of this enzyme are *Candida cylindracea, Rhizopus arrhizus* and *Aspergillus niger.* Thermophilic fungi *Humicola lanuginosa*, produces thermostable enzyme.

CELLULASE PRODUCTION

Enzyme cellulase catalyzes cleavage of -1,4 glucosidic linkages in cellulose. Based on substrate, the enzyme acts cellulases are grouped into C1 and Cx types. Enzyme is used in improving texture and palatibility of poor quality vegetables. Cellulosic materials are digested by these enzymes and glucose is produced, which can be used in culturing microbes for the production of single cell protein. Cellulase is obtained from *Aspergillus niger, Trichoderma viridae, Neurospora* and some other fungi.

PECTINASE PRODUCTION

Pectic compounds are subjected for enzymatic action. The enzymes acting on pectic compounds are divided into two groups: depolymerizing and pectine-sterase. Industrial applications of pectolytic enzymes are in processing of food juices (to increase juice yield on pressing), as aid in clarifying juices and for depectininzing in order to obtain high density for fruit juice concentrates. Another important use of pectolytic enzyme is treatment of coffee beans to liquify mucilagenous coating on the beans. Pectolytic enzymes are produced by a number of bacteria and fungi. *Aspergillus niger, Sclerotinia libertiana* and *Coniothyrium diplodiella* are widely used for the production of pectolytic enzymes. Both solid surface and submerged culture methods are used for the production of pectolytic enzymes. Wheat bran, defatted rice bran fruit pulp are generally used as substrate for enzyme production.

PRODUCTION OF AMINO ACIDS

In 1956, Kinoshita and coworkers laid the milestone in production of amino acid by fermentation. Production of lysine, glutamic acid and tryptophan by microorganism has been well explored. Amino acid production is typically carried out by regulatory mutants.

PRODUCTION OF L-GLUTAMIC ACID

Glutamic acid is used in food industry (mono-sodium glutamate as flavouring agent), as nutritional supplements in bread products. It is produced by *Cornybacterium glutamicum* previously identified as *Micrococcus glutamicum.* Several other amino acids are also produced by this bacterium (e.g., lysine, leucine, homoserine, valine, ornithine, histidine, tyrosine, phenylalanine and tryptophan). L-Glutamic acid may be produced by different ways: by hydrolysis of wheat gluten, soybean cake, or other protein-rich food materials; by cleavage of the pyrrolidone carboxylic acid found in Steffen's molasses, or by a one-stage fermentation process using only one microbe; by two stage process (in the first stage, α -ketoglutaric acid is produced by one type of microbe while another microbe converts α-ketoglutaric acid is converted into glutamic acid). A controlled low biotin level and supply of fatty acid derivatives are responsible for increasing membrane permeability and thus glutamic acid is excreted from the cells. One stage fermentation process is in common practice. Bacterium prefers glucose and starch hydrolysates as carbon source. However, fructose, sucrose, some pentose sugars may also be used. Urea, ammonium salts, ammonia, peptone, corn-steep, liquor meat extract, hydrolyzed casein and fish-meal are different nitrogen sources and any one of these may be used in media. Temperature is maintained between 27-30°C and pH 7-8.5. As earlier discussed in two-stage process, two types of bacteria are required. *E.coli, Aerobacter cloacae* and some species of *Kluyvera* (*K. citrophila, K. noncitrophila*) form α - ketoglutaric acid. A number of genera are capable of converting α -ketoglutaric acid into L-Glutamic acid. These genera are *Pseudomonas, Xanthomonas, Erwinia* and *Bacillus.* Some fungal genera like *Penicllium, Aspergillus* and *Rhizopus* may also be used for this purpose.

PRODUCTION OF L-LYSINE

Lysine is used to supplement cereals and breads. A two-stage process is used for the commercial production of lysine. In this method, first diaminopimelic acid is produced which is later converted to L-Lysine. *Eshcherichia coli* grown in medium containing glycerol, corn steep liqour, and ammonium compounds to produce diaminopimelic acid. *Enterobacter aerogenes* chemically converts this compound into lysine. Now, one stage process is carried out with homoserine auxotrophs of *Cornebacterium glutamicum.* Synthesis of homoserine is blocked and lysine accumulates. Bacterium utilizes glucose as carbon source. L-aspartic acid is required for better yield of lysine. Biotin stimulates activity of enzyme, phosphoenolpyruvate carboxylase and thus, more carbon dioxide was fixed for lysine accumulation. It is therefore, recommended biotin as a supplement in production medium. A mutant of *Bacillus licheniformis* has been developed. This bacterium prefers cane-molasses in production medium.

PRODUCTION OF ANTIBIOTICS

Alexander Fleming (1929) discovered penicillin from *Penicillum chrysogenum* and laid the milestone in the field of production of antibiotics by microorganisms. A large number of antibiotics are produced by microbes, especially by actinomycetes. Generally, submerged culture method is used for the production of antibiotics. Antibiotics (often comes under secondary metabolite produced in idiophase, i.e., after active growth phase) produced by microbes accumulates in culture media. This needs processing and purification of antibiotics.

PRODUCTION OF PENICILLIN

Penicillin is produced by a fungi *Penicillum chrysogenum.* Penicillin may be produced by surface-culture or by submerged culture method. However, commercially it is produced by submerged culture process in rotary drums and deep tanks with proper agitation and aeration. Fungus is cultured in the medium proposed by Moyer and Coghill to achieve sporulation. In surface culture process, dried spores mixed with wheat flour or finely ground oat hulls is used to inoculate the production medium. In submerged culture process, production medium is inoculated with dry spores, or by spore suspension or by pellets. Spore suspension is prepared with ungerminated spores mixed with 0.1% soap solutions. For a good yield of penicillin, production medium should have proper carbon, nitrogen sources with minerals etc. 6% lactose is used as carbon source. 3% lactose with cornstarch and corn dextrin may also be used. Sodium nitrate, ammonium sulphate, ammonium acetate, corn-steep liquor are used as nitrogen source. Potassium dihydrogen phosphate, magnesium sulphate and sulphates of iron and copper are added as minerals. 24°C temperature and pH 5 to 7.5 favours maximum penicillin production. Stirrers are used for proper agitation of medium and as the fungus is aerobic in nature, hence aeration is essential. During agitation and aeration, foam is produced, so antifoam agent like tributyl citrate, octadecanol (3% dissolved in lard oil) alkaterge C are used. Fermentation is completed within six to seven days. Extraction and purification of penicillin is carried in few steps. The first step, fermented production medium is filtered in rotary vacuum filters to remove mycelium. Penicillin is extracted from the medium by countercurrent extraction process. pH of mycelium free production medium is adjusted between 2-2.5. Organic solvents like amyl actate, ether, chloroform, acetone, etc. are used for extraction of penicillin. Very soon, penicillin is extracted back into an aqueous buffer at pH 7-7.5. Organic solvents are again added to aqueous solution. This is a crude preparation and penicillin is extracted as salts. Aqueous solution of sodium hydroxide is added followed by charcoal treatment to the crude preparation. The product gained after filtration, absorption, precipitation and crystallization is the final product. This product is modified by chemical procedures to yield a wide variety of semisynthetic penicillins.

PRODUCTION OF STREPTOMYCIN

Streptomycin is produced by *Streptomyces griseus*. It was discovered by Waksman and his collegues in 1943. Submerged culture method is adopted for streptomycin production. For the production of this antibiotic, fermentation is carried out with soybean-based medium supplemented with glucose as carbon source. Nitrogen source is already present in soybean extracts. However, other carbon source (fructose, xylose, etc) and nitrogen source (peptone, meat extract) may be used. Fats, oils and fatty acids favours antibiotic production. Initially, during fermentation, first stage is production of mycelium followed by antibiotic production. Mycelium growth is checked at this stage. Under controlled nitrogen supply and growth of mycelium is checked, there is production of antibiotic. By the end of fermentation, mycelium undergoes autolysis. After fermentation, streptomycin is adsorbed in charcoal. It is further purified with acidic treatment and concentrating at 60°C under vacuum. Counter-current extraction method may be applied for purification.

PRODUCTION OF TETRACYCLINE

Tetracyclines is a broad spectrum antibiotic. This is a group of compounds, which includes chlortetracycline, oxytetracycline and tetracycline. It is produced by some species of *Streptomycin* (*S.aureofaciens, S.rimosus*). Fermentation is carried out by submerged culture process. Production medium consists of molasses, starch as carbon source, ammonium salts, urea, groundnut flour and corn-steep liquor as nitrogen source. Temperature is maintained at 28°C and pH 6.0. Chlorotetracycline and Oxytetracycline is produced after 2 and 5 days of incubation at 27 and 28°C, respectively. The fermented production medium is centrifuged. The filtrate is treated with charcoal so that antibiotic is adsorbed on it. At this step, pH is kept at a low scale. Methanol, ethanol or acetone is used for elution. Another method of extraction of antibiotic from fermented culture filtrate is with the use of suitable organic solvents.

PRODUCTION OF VITAMINS

Vitamins are produced by synthetic processes. However, production of vitamin B2 (riboflavin) and vitamin B12 (cyanocobalamin) by microbes is more economical.

PRODUCTION OF VITAMIN B12

Vitamin B12 is mainly produced by submerged culture method. The microbes producing vitamin B12 are bacteria (*Bacillus, Pseudomonas, Propionibacterium*) and actinomycetes (*Streptomyces, Nocardia*). *Streptomyces olivaceus* NRRL B-1125 strain is used in the commercial production of vitamin B12. Spores are collected by growing these actinomycetes on Bennett's media. Usually, two or three successive transfers are made to obtain the required amount of inoculum. It is then transferred to the production medium. The composition of production medium consists dextrose, corn-steep liquor or soybean meal (distillers' soluble), Calcium carbonate and cobalt chloride. pH is adjusted to 7.0. Aeration and agitation of medium is essential. Sterile air is passed through at a rate of about 0.25-0.5 volume per volume, of medium per minute. During fermentation, formation of foam is a serious problem. Thus, antifoam agents like soybean oil, corn oil and silicones are used. To recover cobalamine, first fermented medium along with mycelium is boiled and pH is maintained below 5. The vitamin associated with mycelium is liberated. Following this, production medium is filtered and mycelium is removed. The broth is treated with cyanide so the cobalamine is converted to cyanocon-balamine. The broth is passed through a column packed with adsorbing agents (activated charcoal, bentonite, fuller's earth and ion exchange resin). The cyanocobalamine adsorbed in these agents is eluted. For elution, aqueous solution of materials raging from organic bases to hydrochloric acid is used. After this, extraction is carried out with countercurrent distribution between cresol, amylphenol, or benzyl alcohol and water or a single extraction into an organic solvent. Vitamin is precipitated as copper or zinc cyanide cyanoco-balamin. In aqueous concentrates, consists of dissolving a zinc salt in the slight acid solution and then raising the pH to bring about precipitation of zinc hydroxide. In the last step, chromatography on alumina and final crystallization from methanol-acetone, or acetone-water, is carried out. To obtain cobalamin as food supplement, the broth is further dried. For drying broth is first subjected to vacuum drying followed by drum-drying or spray-drying. Bacterium *Propionibacterium freudenreichii* is used for the production of vitamin B12. The production medium consists of glucose, corn-steep, betaine and cobalt. Fermentation is completed by anaerobic and aerobic cycles carried out in two separate tanks. The anaerobic cycle completes in 70 hours and cobinamide is formed. During fermentation, pH falls in the starting but later it rises. 5, 6-dimethylbenziminazole is added to the production medium. In the other cycle of fermentation, nucleotide is formed. It combines with the cobinamide and ultimately cobalamin is formed.

PRODUCTION OF RIBOFLAVIN (B2)

Molds (*Ereomothecium ashbyii, Ashbya gossypii*) produce riboflavin. The production medium contains glucose, collagen, soya oil and glycine. Inoculum or spores of *Ashbya gossypii* are collected and inoculated in fermentation tank. During fermentation, agitation and aeration are essential. Recovery of vitamin depends upon the type of product i.e., for animal feed or as a fine chemical for food and drugs. For animal feed, the broth is evaporated and dried. To prepare fine chemical, vitamin is prepared as pure crystalline form. For this purpose, broth is first heated at 120°C for one hour and filtered. Riboflavin is precipitated with dithionite and it is further purified and crystallized.

PRODUCTION OF MCIROBIAL INSECTICIDES

Bassi (1838), first proposed the use of microbes in controlling insect pests. He experimented with muscardine or calcino disease of silkworm. Insect's pathogens (microbes) are used in controlling insects is known as biological control. Pathogens, which are used in controlling insects, should have properties like: it should be a host specific, fast in action, ability to resist environmental conditions. Vegetative cells are relatively less effective while the spores have some protein crystals, which are toxic to insects. *Bacillus thuringiensis, B.popillae, Pseudomonas fluorescens,* and upto certain extent, *Pseudomonas aeruginosa* are bacteria used for the purpose of insecticide. About 500 fungal species are pathogenic on insects. However, species belonging to four genera are widely used as insecticides. These are *Beauveria bassiana, Metarhizium anisopliae, Verticillium lecanii* and *Entomophthora* spp. Three viruses, viz. Nuclear polyhedrosis virus (NPV), granulosis virus (GV) and cytoplasmic polyhedrosis virus (CPV) are used as insecticide.

Huge mass of *Bacillus thuringiensis* can be produced by fermentation technology. When the cell lyse, the spores and crystals are released into the medium. The medium is then centrifuged and made up as a dust or wettable powder. Submerged culture method as well as semisolid fermentation methods are used in mass production of this bacterium. In *submerged culture* process, the inoculum of bacteria is prepared in a separate medium and inoculum is transferred into production medium. The composition of production medium is as following

Beet molasses	1.86%
Cotton seed flour	1.40%
Corn-steep liquor solids	1.70%
$CaCO_3$	0.10%

Dulmage (1970) has proposed three different production media, these are Tryptone medium, Proflo medium and Soya bean meal medium. Out of these three media, tyrptone medium and soya bean meal medium is economical to use. After the fermentation process is over, pH is adjusted to 7.0 with HCl and it is centrifuged. Sediments are suspended in 4-6% lactose and stirred. During stirring, acetone is added. After stirring for 30 minutes, it is allowed to stand for 10 minutes and filtered with suction. The residue is again collected and stirred with small volume of acetone. It is again filtered and residue is allowed to dry overnight at room temperature. In semi-solid fermentation method, bacteria is grown in flask and than transferred to seed tank, the medium for the two have following composition

Medium for flask		Medium for seed tank
Dextrose	15.0 gm/lt	10.0 gm/lt
Corn-steep liquor	5.0 gm/lt	4.5 gm/lt
Yeast autolysate	5.0 gm/lt	6.0 gm/lt
K_2HPO_4	4.0 gm/lt.	3.5 gm/lt
NaOH	-	0.43 gm/lt
$CaCl_2$	-	0.10 gm/lt

Temperature is maintained at 30°C and pH between 7.0-7.2. Bacterium is grown in flask for 8 hours and than transferred to seed tank for 16 hours. To produce spores and toxins, culture is incubated in production medium. The latter has following composition

Wheat bran	545 gm
Expanded perlite	380 gm
Soya bean meal	62 gm
Dextrose	36 gm
Lime	3.6 gm
NaCl	0.9 gm
$CaCl_2$	0.29 gm
Water	160 ml.

The microbe is grown by semisolid culture method. The bin is perforated at the bottom and dry air is passed through it. After incubation of about 36 hours, the bran cakes are dried and grinded. A powder or dust formulation containing spores, crystals and exotoxins are produced.

Insecticides are also produced by fungi. *Baeuveria bassiana, Meturrhizum amisopliae* and *Acrostalagmus aphidum* produce insecticide. Fungi are grown in semi-solid medium. Media is rich in corn meal. After the full growth of mycelium, culture is dried to obtain sporulation. The conidia have a short life and they start germinating.

The cultural product has to be formulated (in the form of liquid or dust). Formulation is essential so that the insecticide remains adhered to the plant especially in rainy season and in other adverse conditions. For liquid formulation, water is used as diluent. Use of oil is also recommended. However, in case of oil, it should have some characteristic like: adhesiveness, solubility, cheapness and much more oil should not be phytotoxic. Use of microbial insecticide in dust or wettable form is a very old practice and still recommended for many insecticides. For dusty formulation, some additives (like spreaders, stickers and protectants) are used. Dried milk, powdered casein, gelatin and detergents etc. are used as spreaders, while flour, dextrin, gums and polyvinyl chloride are used as stickers.

SECTION – V

CHAPTER - 30

APPARATUS AND TECHNIQUES

CENTRIFUGE

Principle Centrifugation is a technique of separating the particles of various size and densities by applying centrifugal force. If any liquid is rotated horizontally around a central axis, gravitational field higher than the gravitational pull of earth is created. The gravitational field developed due to centrifugal force separates the particles of a rotating liquid. The sedimentation of a particle depends upon the size and density of that particle as well as the RPM of the centrifuge.

Isaac Newton defined centrifugal force as a tendency of any rotating object to move away from its center of rotation. It is expressed in the same terms as the gravitational force or a unit of earth.

In vacuum, the centrifugal force = 2rm

Where is angular velocity, r is the radius of rotation and m is the mass of the object.

The centrifugal force is generally expressed as relative Centrifugal force, "RCT" in "g" units. It is expressed by the following formula.

S2r g units

F = 89500

Where F is centrifugal force, S is rotation (resolution per minute, rpm) and r is the radius in centimeters.

Sedimentation coefficient is calculated by measuring the velocity of the particle at a fixed centrifuge speed (radians/sec) and the log of distance moved by the particle against time taken in seconds (t). Sedimentation coefficient (s) is expressed by following formula.

In

$$S = 2t = \frac{2.303 \log 10}{2t}$$

Sedimentation coefficient of a particle is characteristic. For biological molecules, it is very small. It is expressed in terms of Svedberg constant (S) the basic unit is taken as 1*10 –3. To explain this, let us take the example of ribosome (70S) unit and its two subunits (30S and 50S). The sedimentation constant of these molecules is 70*10-3; 50*10-3; 30 X10-3, respectively.

APPARATUS

Centrifuge is an instrument, which generates the centrifugal force. It has a rotor chamber. In this chamber rotor is rotated around a fixed central axis with the help of electric motors. Rotors have to bear tremendous spin and strains during centrifugation. There are three types of rotors: (1) Fixed angle rotors, the centrifuge tubes containing the sample are kept at an angle 30 to the horizontal. The particle rotate by lateral force and strike on the sides of tubes, which results in driving downwards of particles and thus cause sedimentation (2) The zero angle rotors, the tubes are placed vertically and the particles are separated as vertical zone. When the centrifuge is stopped, the separated particles reorient in horizontal zone. (3) Swinging bucket rotors, the tubes spin horizontally. There is also a device to control speed. The liquid to be centrifuged is kept in the tubes and run in a particular speed. It is necessary to balance each tube and metal holder against opposed tube, and holder. Two tubes are filled to approximately equal levels of suspension or one tube with suspension and other with water. Centrifuges are of three types: (1) Low speed centrifuges, such as clinical centrifuge, with maximum rotation of 5000 rpm. (2) High speed centrifuges with maximum speed of 20,000 rpm and (3) Ultra centrifuges with maximum speed of 1,00,000 rpm. These centrifuges have cooling device (refrigeration) and can work in vacuum. The low speed centrifuges are started and stopped by switch on and off the power supply. In high speed centrifuge, there is a timer and a calibrated speed control. The analytical ultracentrifuge uses an optical system to photograph the sedimentation of particles during centrifugation.

Types of centrifugation

Centrifugation is of two types: (1) Differential centrifugation and (2) Density gradient centrifugation.

In differential centrifugation or pelleting, there is repeated centrifugation and a gradual increase in speed or rpm at each step. The pellets are collected from homogenous suspension after each centrifugation, hence the name differential centrifugation. A homogenous suspension containing particles to be centrifuged is operated at a required speed and time. After centrifugation the pellets or sediments settles in the bottom and remaining is the supernatant. The supernatant is decanted into a beaker or another tube if it is to be recentrifuged. For recentrifugation, centrifuge is operated at a higher speed and again the pellets are collected. Even the pellets may be suspended in a different suspension and recentrifuged to collect different particles present in it. This method is applied to fractionate different organelles of the cells.

Density gradient centrifugation is of two types: (1) Zonal centrifugation and (2) Isopyenic centrifugation. In zonal centrifugation, sedimentation coefficient of a particle plays its role in sedimentation. Separation is carried out on different densities (layers) of sucrose or glycerol. These solutions are not capable to form density gradient and thus, before centrifugation layers of different gradient are prepared. In isopyenic centrifugation, salt solution (like cesium chloride) and sample solutions are thoroughly mixed and centrifuged. Salt solution forms gradient (from top to bottom) by itself during centrifugation.

Sucrose Density Gradient Centrifugation of Enzymes (SDGC)

Enzymes have large size and high molecular weight. This character may be used for separation and purification of enzyme. Enzymes are filtered by dialysis, thus small molecules are filtered off. Svedberg developed analytical ultracentrifugation for enzyme separation. Enzymes in solution sediment at high centrifugal field or zonal centrifugation or high density gradient is a widely used and versatile procedure for separating the enzymes. The enzymes, which have globular protein structure, may be separated by SDGC. In most common procedure a continuous density gradient of sucrose is first prepared in a plastic centrifuge tube. A device is used that mixes concentrated sucrose solution and water in decreasing ratio so that the density of the medium is greatest at the bottom. Enzyme standard and unknown layer on a gradient of buffered 5-20% sucrose prepared by repeated freeze thawing of 20% sucrose. Tubes are centrifuged in a horizontal position in a rotor at high speed. This causes enzyme to sediment down the density gradient at its own rate. The separated bands or zones formed are due to particle weight, density and shape. Following overnight centrifugation 10(5) xg, the tube contents are removed drop-wise and analyzed for enzyme. Alternatively, the plastic tube can be frozen solid and then cut into thin sclices for analysis. The mobility of the unknown enzyme expressed relative to standard of known molecular weight is than computed and calculated.

pH METER

Principle

An acid is a proton donor and base is a proton acceptor. On dissociation, acid yields H^+ ions and base yield $\overline{OH}$ ion. A strong acid or base ionizes or dissociates completely while a weak acid or base shows incomplete dissociation and thus the concentration of free H^+ or OH^- ions depends on the value of their dissociation constants. Ampholytes are substances, which behave both, like an acid and a base depending upon the pH of the solution, e.g., amino acids.

Dissociation of water is following.

$H_2O - H^+ + OH^-$

Thus, the ionization constant k is

$$K = \frac{(H^+)\ \overline{OH}}{(H_2O)} \quad - (1)$$

= 1.81 * 10-6 at 25°C and the concentration of H_2O in pure water is 55.5 moles/litre.

Equation 1 may be expressed as

$$K(H_2O) = (H^+)\ (OH^-) \quad - \quad (2)$$

The degree of dissociation of water is very less and is actually constant for all practical purposes.

$$K(H_2O) = (H^+) (OH^-) = kw$$

$$\text{Or,} \quad kw - (H^+) (OH^-) - 1.8*10\text{-}16*55.5 \text{ moles/ liter.}$$

$$= 1*10^{-14} \text{ M.}$$

The concentrations of H^+ and OH^- have been determined experimentally. It is as totaling 10^{-4}N. As water is neutral and concentrations of H^+ and OH^- is equal, hence this may be expressed as

$$Kw = 10^{-7}*10^{-7} = 10^{-14}$$

pH is defined as the negative logarithm of hydrogen ion activity, i.e.

$$pH = -10g\ aH^+$$

For H+ concentration the formula is

$$pH = -10g\ (II^+)$$

Phenolphthalein, methyl red, methyl orange, bromothymol blue etc are used for the measurement of pH. In effect, the constitution of the dye is changed by the H^+ ion concentration resulting in a colour change. However, this method is not accurate and there are many difficulties in measuring pH for biological experiments. Variation in temperature may cause colour variations. The presence of pigments, salt, colloids etc in experimental solutions may alter the colour of dye. Therefore, for accurate measurement of pH electrometeric apparatus is used in biological labs.

Apparatus

pH meter is a potentiometer, which measures the voltage between two electrodes placed in a solution. When a piece of metal is immersed in a solution of its ions, and electrical potential is established between the two. If a chemically inert metal such as gold or platinum is immersed in a solution of H^+ and if, hydrogen gas is bubbled around it, the electrode will behave as if it were made of metallic hydrogen. A potential is created between electrode and solution. This potential is reproducible and its magnitude depends upon the concentration of hydrogen ions, W. Nernst derived an equation to express the voltage existed between the electrode and hydrogen ion containing solution (E, electromotive force, emf).

$$E = \frac{Eo - RT \text{ in } (H^+)}{F}$$

Where, Eo = the voltage which exists between the electrode and a solution containing hydrogen-ions at unit concentration.

E = voltage at any other concentration.

R = the gas law constant

F = Faraday constant

The two electrodes used in pH meter are a calomel electrode and a glass electrode. The calomel electrode is the external reference electrode whose electric potential is always constant. The glass electrode is the standard test electrode whose electrical potential depends on the pH of the test solution.

The emf (E) and its relation with potentials of calomel electrode (Eref) and glass electrode (E glass) can be expressed as

$$E - Eref = E\ glass$$

The potential of calomel electrode is + 0.250 V at normal temperature. The potential of glass electrode depends upon the pH of the test solution. Thus, the pH of the solution can be determined as

$$pH = \frac{Eglass - Eref}{0.0591} \text{ at } 25°C$$

Substituting the value for Ecalomel

$$pH = \frac{Eglass - 0.250\ V}{0.0591} \text{ at } 25°C$$

Colorimetric Analysis is of 2 types :

(1) Visual colorimetry: This is the type of analysis in which natural or artificial white light is used as a light source and determination made with a colorimeter or colour compurator.

(2) Photoelectric colorimetery – In this type of analysis, eye is replaced by a photoelectric cell. The latter eliminates the error due to the personal characteristics of observer. Photoelectric colorimeter is usually emplyed with light contained within a comparative narrow range of wavelength obtained by passing white light through filters photoelectric colorimeters containing filters are also known as filter photometer.

The calomel electrode is made up of thick glass that is impermeable to H^+ ions. It is filled with mercury, mercury chloride and saturated solution of potassium chloride. These solutions exist in ionized forms and the degree of ionization varies. The calomel electrode is dipped in concentrated KCI solution. The KCI salt bridge forms a contact between the calomel electrode and the test solution. It is created through a fine capillary in the glass casing known as porous plug. The potential of calomel electrode is independent of pH. The glass electrode contains silver, silver chloride and 0.1M HCI solution. This electrode is dipped in 0.1M HCI solution. The tip of the glass electrode is made of special thin (0.05 to 0.1mm) borosilicate glass, which is permeable to H^+ ions only but not to other cations or anions. When the electrode is placed in strong acid (low pH), the calomel electrode will not respond as it in not permeable to H^+, the H^+ ions pass through the glass membrane and neutralizes the electrons of the electrodal reaction. The electrons will flow from calomel to glass electrode because electrons moves from more positive electrode to the other. When the electrodes are placed in strong alkaline solution (high pH), the Hi ions moves out of glass bulb rendering a momentary negative charge and hence the electrons flow from glass to calomel electrode. The potential developed between the two electrodes, results in flow of current between the glass and calomel electrodes. The $H^+/0H^-$ion concentration governs the magnitude of this current, which can be read by calibrated dial directly.

There are certain steps in the operation of pH meter. Firstly, the electrode is to activated by keeping it overnight in a buffer of pH 6-7 or in 0.1 M HCI and then place it in distilled water. The temperature is adjusted by temperature compensation knob. The electrode is removed from beaker, washed with water and soaked with tissue paper. Now the electrode is placed in a standard buffer (of known pH). The pH is accordingly adjusted with pH adjust knob. Again, the electrode is removed from this buffer solution, washed with distilled water and wiped with tissue paper. Now finally the electrode is dipped in the solution whose pH is to be noted. The calibrated dial will show the pH of the solution. The reading is noted is noted and the electrode is again placed in distilled water.

COLORIMETER

Principle

Colorimeter is an instrument, which measures the intensity of colour absorbed by a substance. The basic principle of this instrument is that when a beam of incident light passes through a coloured solution, it absorbs some light so that the intensity of transmitted light is always lcss than that of the incident light. As the solution goes darker, the intensity of emitted light decreases exponentially and vice-versa.

The visible light is made up of seven colours. The distance between the crest of one wave and the next variation of colour of a system with change in concentration of some companent forms the basis of colorimetric analysis and it is concerned with the determination of the concentration of a substance by measurement of the relative absorption of light with respect to a known concentration of susstance varies in much the same way as the ripples seen when a stone is dropped into water.

Lamber's law when a beam of monochromatic light passes through an absorbing medium, the amount of light that is absorbed is directly proportional to the length of the medium. It is expressed as

$$\text{Log} = \frac{Io}{I} = klb$$

Where, Io = the original intensity of the light beam. I = the intensity of the beam after passing through the homogenous substance, b = the thickness of the layer of solution through which the light has passed. KI = the proportionality constant the value of which depends on the absorption characteristics of each compound, units of thickness, temperature, and wavelength.

Beer's law When a beam of monochromatic light passes through an absorbing medium, the amount of light that is, absorbed is directly proportional to the number of light absorbing molecules in that medium. It is expressed as

$$\text{Log} = \frac{Io}{I} = k2c$$

Advantage of Colorimetric and Spectrophotometric analysis. Both the methods are easy simple and they provide a significant method of determining minute quantities of substances.

Where Io = the original intensity of the light beam. I = intensity of the beam passing through solution C = the concentration of the solution in moles per liter. K2 = the proportionality constant the value of which depends on the absorption characteristics of each compound, units of thickness, temperature and wavelength.

Combining the two expressions

$$\text{Log}\ \frac{Io}{I} = kcb$$

Id/I is called the optical density (its reverse I/Io is transmittance). K is called the extinction coefficient and may be written as E for convenience, or as. It is specific for a certain wavelength and is larger when a particular wavelength is absorbed efficiently by a compound. It is possible to use *I* instead of *b* to obtain the following equation.

$$\text{O.D.} = \text{E cI.}$$

Thus, the quantity and quality of light absorbed or transmitted by a solution depends on the solvent and is directly proportional to the concentration of the solution and the thickness of the layer used.

Apparatus Colorimeter is an instrument to measure the intensities of the light before and after it has passed through a coloured solution. It's basic parts are: (1) A light source. (2) A condensing lens to render the light rays parallel. (2) A filter to generate monochromatic light. (4) A sample holder. (5) A photocell to convert light energy (photons) into electrical energy. (6) A galvanometer to measure the electrical energy (current).

The light source is generally a 100-watt double-filament tungsten bulb. The light emitted from this lamp is composed of wavelengths between 400 and 900 nm. It passes through a slit, a condenser lens, and a filter and finally e merges as a parallel beam of monochromatic light. The choice of proper filter is essential; as for each analytical experiment filters used may be different. Three filters blue, green and red filters generally are used. The monochromatic light passes through the sample solution and the transmitted light falls on the photocell. The photocell converts the transmitted light energy into electrical energy, which is amplified and measured by the galvanometer. The galvano-meter is calibrated to read the Absorbance/Transmittance directly.

Operation of Klett-Summerson photoelectric colorimeter Fix the proper filter in the space provided. Pointer is adjusted by the small knob. The colorimeter tube containing distilled water is placed in the instrument. Now the large knob is used to turn the scale and the reading is brought to zero. The current is switched on and at this moment, pointer moves somewhat away from the line on the pointer scale. Now with the help of zero knob scale is set to zero. The apparatus is left on for sometime to settle. To note down the reading of an unknown solution, water is removed from the tube and it is filled with the solution whose reading is to be noted. The tube is placed inside the colorimeter, this brings about a deflection in pointer. Scale knob is turned until the pointer has been brought back exactly to the zero line. The reading at this point is the reading of unknown solution.

SPECTORPHOTOMETER

Spectrophotometer is a sensitive colorimeter, giving better precision and resolution. It has multiple uses in a biological lab. It is essential to understand some concepts and definition before going into the details of this apparatus.

Absorbance Logarithm to the base 10 of the reciprocal of the transmittance A = log 10 (/T).

Transmittance The ratio of the radiant power transmitted by a sample to the radiant power incident on the sample.

In spectrophotometric analysis, light of definite wavelength extending to the UV region of the spectrum constitutes the source of light and thus necessitates the use of a spectrophotometer. The latter may be regarded as a combination of spectrometer and a photometer. A photometer is an instrument which is used for the production of coloured light to selected colour or wavelength when it is used as a part of spectrophotometer it is usually known as monochromator. A photometer is a device, which is used, for measuring the entenrity of chromatography may be regarded as an analytical technique employed for the purification and separation of organic and inorganic substances. It is also found useful for the fractionation of complex mixture, separation of closely related compounds, such as isomers and in the isolation of unstable substances.

It is used for measuring the – intensity of light of when incorporated in a spectrophotometer it is used as a part of a spectrohophotometer it is used to measure the intensity of the monochromatic light produced by the monochromatic.

Apparatus

The basic components of spectrophotometer are light source, condensing lens, monochromator, sample holder, detector and recorder. Spectrophotometer can be operated in visible and uv light. For visible light, tungsten bulb is used, while for uv light deuterium or hydrogen lamp is used as light source. Light emitted by the lamp is reflected by a plane mirror so that it passes through a slit. Light source gives a light of different wave length, thus a monochromator is used. The later is a prism, which can select out light of a particular wavelength. Samples are placed in a cuvette, made up of glass or plastic or silica or quartz. Light passing through the sample has to be converted into converted into electrical energy, so that it may be recorded. Photocell or photomultiplier tube is used for this purpose.

CHROMATOGRAPHY

Chromatography (Gr. Khroma = colour, graphein = to write), technique was introduced by M.S. Tswett (1906) in this technique, compounds are partitioned in two immiscible solvents. The relative amount of compounds distributed in the two solvent systems is expressed by 'distribution coefficient' or 'partition coefficient'. It is the ratio of concentration of the compound in two different immiscible solvents.

There are two phases in chromatographic systems, i.e., the stationary phase and mobile phase. The mixture of compounds is separated due to their different affinities for a stationary phase (solid or liquid) and their differential solubility in a moving phase (liquid or gas). The compounds separate due to continually changing equilibria. The shuttling of molecules between the stationary and moving phase. The mobile phase carries molecules to new areas of the adsorbent where they are adsorbed. The molecules, which are very soluble in the moving phase and have little affinity for the adsorbent will move faster and a greater distance than those having, less affinity for the solvent and a grater affinity for the adsorbent. Different type of chromatography has different stationary and mobile phase. Based on this system, chromatography is divided into seven types: adsorption, partition, gas-solid, ion-exchange, gel permeation and affinity chromatography.

Adsorption chromatography

The stationary phase is solid and mobile is either a gas or a liquid. Solid particles (of alumina, silicic acid, sugar, activated charcoal, cellulose) are used as adsorbent. These particles are able to hold molecules on its surface mainly by electrostatic interactions or even by van der Waals forces, dipole interactions, hydrogen bonding, steric factors etc. The adsorbent should not react to solvent or to compounds to be separated. Adsorption is maximal in nonpolar solvents and decreases as the polarity of the solvent increases.

The distribution coefficient of a substance between the for phases of a system is the most important factors in any form of chromatography. The distribution coefficient (k) is given by

$$K = \frac{\text{Amount of solute per unit of stationary phase}}{\text{Amount of solute per unit of mobile phase}}$$

In the case of adsorption chromatography the distribution coefficient depends mainly upon the temperature and the concentration of the substance.

Applications

1. In the purification suddan red.
2. Separation of the urinary –17 – ketosteroids
3. Separation of 17- ketosteroids glucuronides .
4. Separation of plasma Cortisol etc.

Partition chromatography

Both stationary and mobile phases arc liquid. Paper chromatography is the most common and simple type of partition chromatography. The partitioning may be liquid-liquid partition (water acts as stationary phase and organic solvent form mobile phase) or solid-liquid partition (adsorption of solutes to cellulose fibers and desorption or elution of the solutes by the same mobile organic phase during the separation process). Most of the biomolecules are separated by thin layer chromatography (TLC). In this technique a thin slurry of adsorbent (silica get, aluminium oxide, kieselghur, cellulosic powder) is applied on glass plate. This thin layer of adsorbent is stationary phase, and acts as a filter paper in paper chromatography.

In paper and thin layer chromatography, compound (to be separated) and solvent move either on a filter paper or on a thin layer of adsorbent. The relationship between the distance traveled by the solute and solvent system is known as resolution front (Rf) value. Rf value is the ratio of the linear rate of movement of the solute zone to the linear rate of movement of the solvent. It may be identify compounds.

$$Rf = \frac{\text{distance from origin traveled by compound}}{\text{distance of solvent front from origin}}$$

Ion exchange chromatography (IEC)

Stationary phase is a matrix with charged groups and mobile phase in salt in liquid. The biomolecules are electrically charged. The pH of the medium has a role in maintaining the charge on these molecules and the difference in their charge helps in their separation in IEC. In this technique, ion exchangers are used. Based on the charge they are anion and cation exchanger. Anion exchanger carry positively charged groups on the matrix. While, cation exchangers carry negative charges

On them, positively charged biomolecules will bind to negatively charged ion exchange matrix while, negatively charged biomonlecules will bind to positively charged ion exchanger. Synthetic resin (synthesized by copolymersing styrene with divinyl benzene) is used. Now other ion exchangers like cellulose, Sephadexes and Sepharoses are also used.

Gel permeation chromatography

Stationary phase is gel with porous structure, and mobile phase is liquid Molecules are separated according to their size and shape.

ELECTROPHORESIS

Electrophoresis is separation of electrically charged biomolecules (amino acid, protein, and nucleic acid) due to their migration under the influence of an electrical field. Tiselius, was the pioneer to develop this technique. He demonstrated the movement of biomolecules under electrical field, and he carried out an experiment to separate the samples in free solution (under electrical field) without using solid support. Now electrophoresis is performed on a solid supporting medium with buffers. Molecules form zone on this support. The zones are formed due to the similar size, shape and electrical charge. Thus, this technique is also named as 'zonal electrophoresis'. For supporting medium, Whatsman filter paper or cellulose acetate papers are used. Gel (as supporting medium) is prepared with starch, agar, polyacrylamide and agarose. The factors governing the movement of biomolecules under the electrical field are: electrical field strength, sample, buffers and supporting medium. Voltage or current or resistance control the movement of a charged molecule under the electrical field.

Mobility is directly proportional to voltage and current and inversely proportional to resistance. The electrical charge, shape and size of the molecule has a direct influence on mobility. The rate of migration increases with an increase in the net charge on the molecule. Rate of migration is directly proportional to the size of the molecule. However, molecules of the same size but different shape show difference in mobility. Composition, concentration and pH of buffer have a great role in migration of samples. Supporting medium can influence migration by three means: adsorption, electroendoosmosis and molecular sieving. Adsorption reduces the rate as well as resolution of separation. This results in tailing of bands. During electrophoresis, biomolecules generally move towards anode, but due to electroendoosmosis, the dissociatable cations of the supporting medium and their associated water molecules move towards the cathode. This disrupts the separation process. The inter-wining of molecules of supporting medium results in a formation of sieve like structure. There are definitc pore size, which may be increased or decreased by changing the concentration of gel. The pore size will control the movement of molecules according to their size and shape.

Polyacrylamide gel electrophoresis(PAGE): this technique is applied for the separation of peptides, proteins, and small molecular weight nucleic acids. It is also used for DNA sequencing. Proteins are amphoteric in nature, pH of the medium, governs the electric charge of proteins by ionizing side chains such as carboxyl and amino groups. Every protein is influenced by pH, i. e., at certain pH proteins are highly active (having maximum electrical charge) and at certain pH, it is totally inactive (with no electrical charge), pH value at which, proteins have no electrical charge is known as isoelectric point Proteins show negative charge at pH above this point and move towards anode and vice versa. In PAGE, pH of the medium is kept constant so that protein moves in electric field. Polyacrylamide gels are synthetic gels and are used as supporting medium. It is prepared by synthetic chemicals. The components of polyacrylamide gel are acrylamide, N, N-Methylemide-bisacrylamide, initiators and catalyst. Acrylamide forms the linear backbone of the gel, N, N-Methylene-bis-acrylamide forms cross linking, ammonium per sulphate or riboflavin are used as initiators, these compounds are responsible for initiating polymerization reaction, TEMED (N, N, N', N'-Tetramethylenediamine) is used as catalyst. Ammonium per sulphate is a homogenous polymerizer, while riboflavin is a non-homogenous polymcrizcr. Ammonium sulphatic when mixed with water, dissociates in two persulphate radicals, which immediately combine with TEMED. From TEMED, the free radicals are transferred to acrylamide monomers. These monomer complexes polymerize in the presence of cross linking reagents. For electrophoresis, some buffers are used. These are separating gel buffers, stacking gel buffers, sample buffer and tank buffer.

CHAPTER - 31

MICROSCOPY

Microscopy is the use of a microscope. A microscope is an instrument that magnifies the size of the image of an object it helps in viewing objects too small to be visible with the naked eye. Microscopes are of two categories, light (or optical) and electron, depending upon the principle on which, magnification is based. Light microscopy in which magnification is obtained by a system of optical lenses using waves, including (a) bright field, (b) dark field, (c) fluorescence and (d) phase contrast microscopy. The electron microscope uses a beam of electrons in place of light waves to produce the image. Specimens can be examined by either transmission or scanning electron microscope.

TYPES OF MICROSCOPES

Bright field Microscope (Maximum useful magnification – 1,500 x; Resolution power 100-200 nm), mostly used for the study of microorganisms. For a proper visualization staining is necessary.

Dark field Microscope (Magnification - 1,500 x; Resolution power 100-200 nm) used for viewing live microorganisms. Specimens are not stained and are examined in dark background hence they appear bright. It is helpful in studying external morphology.

Ultra violet Microscope (Magnification 2,500 x; Resolution power 100 nm) It has an improved resolution over normal light microscope. Electron microscope is used in place of this microscope.

Fluorescence microscope (Magnification 1,500 x, Resolution power 100-200nm) Fluorescent stains are used for the study of microorganisms. It is useful in diagnostic procedures for identifying microorganisms.

Phase contrast Microscope (Magnification 1,500 x, Resolution power 100-200 nm) It is used to examine structures of living microorganisms. Staining is not required for the study of microorganisms.

Nomarski differential interference (Magnification 1,500x, Resolution power 100-200nm) A sharp multi-coloured image having three dimension appearance, may be studied with the help of this microscope.

Confocal scanning (Magnification 1,500 x, Resolution power 100-200 nm) A single microorganism may be studied in the mixture of various types microorganisms, fluorescent stains are used which produces blue-free image. It produces three dimension images.

TEM (Magnification 500,000 – 1,000,000 x, Resolution power 1-2 nm) used to view ultra structure of microorganism including viruses.

SEM (Magnification 10,000-1,000,000 x, Resolution power 1-2 nm) used for showing detailed surface structures of microorganisms, produces a three dimensional image.

LIGHT MICROSCOPY

Visible light or ultra violet light is used a source to illuminate objects. The light passes through several glass lenses that alter the path of the light and produce a magnified image of the object. The quality of the image and the magnification that can be achieved depend on the properties of visible light and those of the glass lenses of the microscope.

Properties of light There are two important aspects in light microscopy. They are, the physical properties of light and how light interacts with objects. In electromagnetic radiation, the range of wavelengths (400-700 nm) forms the visible light. When light passes through a substance, it is known as transmission and if it bounces off an object, it is reflection. When light is broken up into differing wavelengths it is diffracted. In microscopy, when light passes through the aperture of a microscope lens, it is diffracted and causes blurring of an image. When light moves through medium of one density into a medium of another density, such as from air into a glass microscope lens, it is refracted. The refractive index is the ratio of the speed of light in a given medium to the speed of light in a vaccum. Light is refracted when it passes through a glass microscope lens.

Distortion of microscope lenses In microscope distorted image is a basic problem. It is mainly due to the fact that parallel rays entering a microscope lens do not focus (forms image) exactly at the same point. Distortion based on the shape of the lens is called spherical aberration. As the visible light is made up of different wavelengths hence it will focus at slightly differing focal points. This leads to distortion based on the colour of light called chromatic aberration. To overcome these distortions, compound lenses are constructed. These lenses are of different shapes and glass composition.

Magnification It is enlargement of the image of an object. A compound light microscope uses multiple lenses to refract light to achieve magnification. Such magnification permits the structure(s) of the specimen to be seen. Total magnification is the product of the magnifying powers of the individual lenses.

Resolution Resolution is the degree to which, the details in the specimen is retained in the magnified image. The resolving power (R) of a microscope is the ability to distinguish two adjacent points as distinct and separate. The resolving power of a light microscope depends on the wavelength of light (λ) and a property of the objective lens, called the numerical aperture (NA). The formula for calculating the approximate resolving power of a light microscope is

$$dR = \frac{0.5\lambda}{N \sin \alpha}$$

where λ = Wavelength of light

α = half angle of the objective lens

N = refrective index

Numerical Aperture (NA) It is a property of lens that defines the amount of light that can enter it. The formula for calculating the numerical aperture is

$$NA = n \sin \theta$$

Where, n is refractive index of the medium filling the space between the specimen and the front of the lens

θ is the angle of the most oblique rays of light that can enter the objective lens.

TYPES OF LIGHT MICROSCOPE

In microbiology different types of light microscopes are used. The choice of a microscope, depends upon the size of the object, the degree of detail that must be viewed, the nature of the specimen.

BRIGHT FIELD MICROSCOPES

In bright field microscope visible light is transmitted through the specimen. In this type of microscopy, the microscopic field is brightly lighted and the microorganisms appear dark because they absorb some of the light. Staining the microorganisms enable them to absorb much light resulting in greater contrast and colour differentiation. Generally, a condenser is used which focuses the light on the specimen and two sets of lenses i.e., objective and ocular gives a magnified image of the specimen. A typical microscope used in bacteriology has objective lenses that magnify 10 x,40 x and 100 x and ocular lens of 10 x. This results in magnification of 100, 400 and 1000 times (of the specimen), respectively. If the various lenses are adjusted so that once the specimen is focused with one lens it remains in focus when one switches to other objective lenses, the microscope is said to be parfocal. The resolution of the typical bright-field microscope for viewing bacteria is 200 nm. Bacteria almost always are stained before viewing with the bright field microscope.

DARK-FIELD MICROSCOPES

In dark field microscopy the effect produced by the dark-field technique is that of a dark background against which objects are brilliantly illuminated. A special type of condenser is used that transmits a hollow cone of light from the source of illumination. The condenser lens of the dark-field microscope does not permit light to be transmitted directly through the specimen and into the objective lens. Most of the light directed through the condenser does not enter the objective. The microscopic field is kept dark. The dark-field condenser lens focuses light on the specimen at an oblique angle, such that light that does not reflect off an object does not enter the objective lens. Therefore, only the light that reflects off the specimen will be seen, and the light simply passing through the slide will not enter the objective. Staining is not required. The microorganisms are suspended in fluid wet mount and hanging drop preparation.

PHASE CONTRAST MICROSCOPY

Phase contrast microscopy is based on the fact that the rate at which light travels through objects is inversely related to their refractive indices. The frequency of light waves is independent of the medium through which they travel. When the phase of light ray traveling through an object of higher refractive index it will be relatively retarded. Light that passes through a cell or a cell structure is slowed down relative to the light that passes directly through the less dense surrounding medium. The greater the refractive index of the cell or cellular structure, the greater the retardation of the light wave. In phase contrast microscopy there is a special type of condenser having system of rings. The objective separates the diffracted and undiffrected rays from specimen. The microscope is equipped with a glass ring, which introduces an additional phase difference. When the light travels through the glass ring the two sets of rays recombine. The total phase shift is equal to one-half of a wavelength. Hence, when the rays are recombined in the image plane, destructive interference greatly increases the contrast of cells or intracellular structures that differ slightly in refractive index from their surrounding. Staining is not required and living microorganisms can be studied. This microscope optically changes differences in phase into differences in intensity that produce differences in contrast.

ELECTRON MICROSCOPY

The electron microscope has sufficient magnification and resolution to view viruses and the internal structures of all microorganisms. In electron microscope a beam of electron is used. It has a great resolution and much higher magnification. An electromagnetic field acts on a beam of electron in a way analogous to the action of a glass lens on a beam of photons. When electron beam is accelerated through an electric field, its wavelength is inversely proportional to the square root of the accelerating voltage. When electron beam is generated at a high accelerating voltage, the wavelength is much shorter than visible light. This provides a much higher magnification. In electron microscope the accelerating wavelength used is 60,000 volts. At this voltage, the emitted electron has a very short wave length approximately 0.005 nm, which gives resolution of approximately 0.2 nm. A wavelength of only 0.04 nm, about 10,000 times shorter than that of visible light is routinely obtainable with accelerating voltage of about 100kv.

Preparation of specimen In the preparation of specimen one has to be very careful for problems like artifact. An artifact is the appearance of something in an image that is not a true representation of the features of the specimen on view. The fixed specimen is dehydrated. These processes must be carried out carefully. Fixation means to preserve intact specimen and to maintain their natural orientation. Fixation needs precautions so as to avoid shrinking, stretching, or otherwise distorting of the microorganisms and alter the image. A thin section is prepared with the help of microtome. In electron microscopy, staining is carried out with electron dense heavy metal salts. The heavy metal stains scatter the electron beam and the stained areas thus appear black. The stains contain heavy metal compounds, such as phosphotungstic acid. It has heavy metal tungsten.

Freeze Etching Techique This technique is used to reveal biochemically defined layers of a microorganisms, including organelle structures. The specimen is first frozen at –196°C and is fractured by striking it with a knife blade (freeze-fractured). The fractured specimen is then etched. Such etching raises part of the surface layer of the specimen. After this, the specimen is exposed to vapors of heavy metal. The specimen is kept at an angle to produce a shadow effect, it is then rotated at 90° angle. At this angle it is exposed to vaporized carbon to produce a replica of the surface. Biological material if adhered on the surface is removed. Carbon replica can be examined under electron microscope. By this technique there are no chances of artifacts. The external and internal surface structures can be studied.

TRANSMISSION ELECTRON MICROSCOPY

In transmission electron microscopy (TEM) an electron beam passes through the specimen. In this type of microscope, an electron gun is used. The electron beam is ejected from a hot tungsten filament. The electrons are accelerated as a fine electron beam past an anode by a high voltage established between filament and the anode. This beam is focused on the specimen with an electromagnetic condenser lens. For a clear passage for electron beam, the air should be removed as to avoid collision between the electron beam and gas molecules. For this purpose high efficiency vaccum system is used. The gas molecules may scatter the electron beam and make it impossible to resolve high magnification image. The image is rendered visible by allowing it to impinge on a phosphorescent screen. This is similar to the front of a cathode ray tube in a television set. The specimen must be completely dehydrated before examination. A very thin section (100 nm or less) can be observed in the conventional electron microscope. The electron beam cannot penetrate through matter hence, new electron microscopes have been developed which have accelerating voltages of a million KV. The electron beam in such microscopes can penetrate through intact cells of more than a micrometer thickness. Fine details, even upto the level of molecular arrangements can be seen with the transmission electron microscope.

SCANNING ELECTRON MICROSCOPY

Scanning electron microscopes (SEM) uses an electron beam that is scanned across the surface of a specimen. The primary electron beam ejected from electron gun hits electron out of the specimen surface. This produces secondary electrons, which are transmitted to a collector. Few numbers of electrons are reflected or back scattered from the specimen surface. The secondary electrons produce an intense signal on the collector. Electrons reaching the collector are transmitted to a detector consisting of a substance that emits light when struck by electrons. The emitted light is converted to an electrical current that is used to control the brightness of an image on a cathode ray tube screen. The secondary electrons emitted from each point on the specimen are characteristic of the surface at that point. The intensity of the image reflects the composition and topography of the specimen surface. The surface topography controls the number of secondary electron creating a contrast. The shadowed image gives a three-dimensional appearance. It high lights the topography of the specimen surface, as seen with the Scanning electron microscope.

Preparation of specimens for SEM The specimen is fixed and dehydrated carefully, so that the natural shape of the specimen is retained. Artifacts may appear while dehydrating specimens. To overcome from this problem critical-point drying technique is applied. This technique is based on a critical point in terms of temperature and pressure for a solvent in a closed container where the density of the liquid is equal to the density of the vapour. At this point, the liquid in a specimen and the gaseous phase over it become indistinguishable. This permits conversion of the liquid to a gas without any surface tension damage to the specimen. When all the gas is removed, the specimen is dry. In this technique, the specimen is kept in ethanol or acetone, so that water is removed. After this, specimen is transferred in a sealed chamber by using pressurized liquid carbon di oxide. Liquid CO_2 act as a transitional liquid. The temperature is raised above 32°C, the critical point. The fluid vaporizes without any surface tension. The pressure and gas are released slowly, leaving a dry, undistorted specimen. After dehydration, specimen is coated with a layer of gold or fold-palladium by a sputter coater. This improves the characteristics of the specimen. Proper coating is essential so that there is no surface charging. Biological specimens have nonconducting surfaces, which results in surface charging. After coating specimens are mounted for examination. To observe the internal surface structures of a specimen, it is subjected to cryofracturing. The specimen is frozen in liquid nitrogen and then fractured with a sharp blade.

CHAPTER - 32

CULTURAL TECHNIQUES

Microorganisms are grown in laboratory under artificial conditions in or on nutrient media is known as culture media. A particular species of microbe is grown on culture media. This process is isolation. Microbial cell or spore or any such part of the body used for cultivation of microbe is known as inoculum.

Brefeld was the pioneer worker to grow fungi. He introduced the practice of isolating single cells, as well as the cultivation of fungi on solid media. He used gelatin as solidifying agent. J. Schroeter (1872) observed that isolated bacterial growth can be grown on solid substrates such as potato, bread and egg albumen. The colonies arose differed from one another but within each colony the bacteria were of one type. Koch inoculated bacteria on cut surface of potatoes placed in sterile glass vessels. There are certain disadvantages as the surface is moist which allows motile bacteria to spread freely over it, secondly, the surface is opaque and hence it is difficult to observe the bacterial colonies. Potato is not an appropriate medium for bacteria and hence it is not suitable for culturing bacteria. Previously gelatin was used as solidifying agent but as it is a protein and hence highly susceptible for microbial digestion and liquefaction. Another disadvantage is that it changes from gel to a liquid at temperature above 28°C. Now agar is used as solidifying agent. It is a polysaccharide, extracted from red algae. It is a complex polysaccharide, that is attacked by relatively few bacteria so the problems of its liquification, is rare. It requires 100°C for agar to melt and once it is melted, it remains a liquid until the temperature falls above 44°C.

CULTURE MEDIA

To grow microorganism in a laboratory, it needs proper nutrients and favourable conditions for growth. Carbon and nitrogen are the major nutrients. A microbe needs phosphate, sulphate, iron, magnesium, sodium, potassium etc in traces. All these ingredients are mixed in a proper proportion to enable microbe for a proper growth of microbe. The combination of such nutrients is called nutrient media/culture media. There are three types of culture media.

(a) Natural or emperical media, (b) Synthetic media and (c) living culture media.

Natural or Empirical media This type of media is prepared from the extract of plants or animal tissues. The exact composition of media is not known. Extract of any plant or animal, vegetable and fruit juice, meat extract, are mainly used in this media. Peptone is a type of protein and it can only form a media. Blood serum is also used for culturing microorganisms. Blood is mixed with meat infusion. Lactose is added to this infusion broth and this media is known as lactose broth. This media is used to culture microorganism whose picture is not clear as to which of nutrients will bring the best growth of that microbe.

Synthetic media are the artificial media whose exact chemical constituent is known. Media is prepared by the combination of organic and inorganic constituents. Such media are helpful in studying the nutritional requirement of a microbe.

Living culture media Obligate parasites like viruses, rickettsia etc are grown in living media. Living cell of animal and plant tissues are used for culturing such microbe.

Based on utility, culture media are of following types

Purpose media it is generally used to culture bacteria and fungi.

Differential media it is used to study the growth product of a particular bacteria.

Selective media in this media small quantity of a antibiotics, dye, acids, alkalies etc, is added. This restricts some microorganisms to grow while others are encouraged for growth.

BASAL INGREDIENTS OF A CULTURE MEDIA

A culture media is prepared by a combination of nutrients. These nutrients are of plant or animal origin or chemical substances. The basic requirements are carbon and nitrogen source, trace nutrients and in some cases certain growth factors like vitamins, amino acids, hormones etc are also required.

Carbon source: Carbon is said to be energy source for the growth of a microbe. In a culture media, carbohydrates are the main carbon source. Following carbohydrate may be used for the growth of a microbe.

1. Pentose: arabinose, xylose, rhabinose.
2. Hexose: glucose, fructose, galactose.
3. Diasaccharides: Sucrose, maltose, lactose etc.
4. Trisaccharides: Raffinose.
5. Polysaccharides: Starch, insulin, dextrin, glycogen.
6. Alcohol: Glycerol, mannitol, sorbitol.

While preperaing media, precautions on heating should be observed. A high temperature and low quantity of water may hydrolyse or even decompose carbohydrates specially, at pH above 7.0. Carbohydrates may also react with many inorganic salts. The products of these reactions may inhibit the growth of microorganisms. As precautions following care should be adopted

1. Sugar and peptone must be dissolved completely before autoclaving.

2. High concentration media (specially containing peptone-dextrose) may not be autoclaved.
3. Autoclaving temperature and time should be maintained.
4. Medium with pH 8.0 should be avoided and autoclaving should be for a brief time. Each flask should contain small amount of media.
5. Nitrogen source: Nitrogen is provided in organic or inorganic forms. Peptone is a common nitrogen source in the culture medium for bacteria. Peptones are hydrolyzed proteins of animals or plants. It is rich in proteins, peptides, polypeptides, proteoses, amino acids, carbohydrates, various inorganic and organic micronutrients and essential minerals. Yeast extract is another important nitrogen source. It is prepared by autolysis or plasmolysis of cells of *Saccharomyces*. It is rich in amino acids, peptides, water-soluble vitamins, carbohydrates and minerals. Malt extract is prepared from malted berry. It is good source of both carbohydrate and nitrogen. It also contains varying quantities of lipids, fatty and organic acids, phosphorous, sulphur compounds and vitamins. Meat meals, casein, soya meals, ground meal are other protein hydrolycates used in culture media.

Minerals Inorganic mineral salt, are also added in the medium. Salts containing K^+, Mg^{++}, Fe^{++}, PO_4^{---} and SO_4^- are essential for the growth and metabolsim of microbes. Copper, cobalt, zinc, iron, magnesium, molybdenum, manganese and calcium are required by some microorganisms as trace elements.

Chelates Chelating agents are organic compounds which form complex substances with heavy metals and keep them in solution in culture media. Some carboxylic acids like citrate, succinate, acetate are used as chelating agents for calcium, magnesium and iron etc. Some amino acids like cystine, histidine and glycine are also used as chelating agents. Ethylenediaminetetratic acid (EDTA) is now a popular chelating agent.

STERILIZATION TECHNIQUES

In a microbiology laboratory, it is important to control unwanted microorganisms. Sterilization and disinfection are the common processes. There are some terms, which are to be clarified to understand the concepts.

Sterilization The destruction of all forms of microbial life, including heat resistant spore.

Disinfection The destruction of harmful microbes. This term is generally applied for inanimate objects (non-spore forming pathogen). The chemicals used are known as disinfectants. In practice, this term is used for a chemical treatment of an inert surface or substance.

Degerming It is a simple method of disinfection. It is to degerminate and not to kill microbes mechanically in a limited area by swabbing it with alcohol.

Sanitization It is washing the articles (glassware, table top) with hot water or chemicals. It lowers the microbial count and controls their transmission. To reduce the number is judged by public health standards.

Antisepsis Chemical treatment of living tissues so as to control pathogen. The chemical used is known as antiseptic.

Sterilization is a process of destroying all forms of life by physical or chemical agents. Removal of microorganisms is also considered as sterilization. Filteration or centrifugation techniques are adopted for keeping away. While culturing a microorganism, number of unwanted microorganisms may grow alongwith the needed microorganisms. These unwanted microorganisms are to be destroyed. It is possible to control such microorganisms by creating conditions, which they cannot tolerate. The antimicrobial agents may be microbicidal (cide = kill, i.e., these agents kill the microorganisms) or microbistatic (static = stand still, i.e., these agents stop the growth and reproduction of microorganisms). Various methods are adopted for sterilization practices. They are classified under following heads.

MOIST HEATING

Moist heating means, use of boiling water or its vapours to kill microorganisms. Wet heating results in coagulation of structural proteins. This technique is applied for sterilization of liquid media, clothes and gloves etc. There are two measurements of heating. Thermal death time (TDT)-It is the length of time to kill all bacteria in a liquid culture at a given temperature. Thermal death point (TDP)-is the lowest temperature required to kill all bacteria in a liquid culture in 10 minutes. Decimal reduction time or D value (DRT) is the time (in minutes) in which 90% of a population of bacteria is killed at a given temperature. Methods used for moist heating are (1) autoclaving, (b) Tyndallization, (c) Boiling and (d) heating.

Autoclaving is a process conducted in a specially designed vessel. The principle of autoclave is that the steam is allowed to form in the inner cylinder by heating water. The pressure increased due to heating is measured by pressure gauge. The steam pressure increases the temperature inside as per needed. In general, 15 pounds per square inch in excess of the atmospheric pressure is used. This temperature is equivalent to a temperature of 121°C (250°F) at sea level. This temperature and pressure is applied for 15 minutes. However, the time of operation at a temperature or pressure level to achieve complete sterilization depends upon the nature of material being sterilized.

Tyndallization In some culture media, carbohydrate and other components may be destroyed by heating in autoclave. To overcome this problem a technique is developed in which media is exposed to steam at atmospheric pressure. This process is known as Tyndallization or fractional sterilization. Water is boiled at 100°C and steam is collected. The media is exposed for 30 minutes for three successive days. Vegetative cells are killed after first exposure. Dormant and resistant spores become active after the media is cooled. The cells developed from these spores are killed by exposure to steam. No apparatus is required. Media is kept in a loosely covered container so that no pressure is developed. Media is steamed after an interval of 24 hours for three successive days. It is a time consuming process. Non-nutrient media is not sterilized in this method, as the spores will not germinate in the absence of media. They will remain dormant and hence it will again contaminate.

Steaming: In this method media is sterilized in steam at 100°C for 90 minutes. Steaming is applied for those substances which cannot be placed or immersed in water and which cannot be sterilized by steam pressure. A vessel with closed lid and having a steam outlet is used for this purpose. It contains a perforated shelf. The media to be sterilized is placed on this perforated shelf. Water is kept boiling at the bottom of the vessel. Steam generated by boiling water passes through this perforated shelf sterilizing the media.

DRY STERILIZATION

Dry heating is carried out in a hot air oven or by flaming. Glass wares, metal instruments and articles are sterilized by dry heating. It is not that effective as moist heating. Much more, it is time consuming. Articles to be sterilized in an oven are wrapped in a paper (kraft paper/ brown paper) and kept in an oven under hot air. Sterilization is carried out at 160°C for one hour. Microbes and their spores are killed due to destructive oxidation of the cell content.

Inoculating needles are sterilized by flaming. Inoculating needle and loops are kept in a flame until it is red hot. It is cooled in air.

CHEMICAL STERILIZATION

Chemical sterilization is carried out by some chemicals known as disinfectants. The priciple of using disinfectant is to control microorganisms by denaturating or inactivating their proteins and enzymes. These chemicals are as following

Halogen and halogen compounds

These compounds are useful against most microorganisms. Chlorine and iodine are most effective antimicrobial agents. Chlorine and chlorine compounds oxidize essential proteins or enzymes. The germicidal activity of chlorine is caused by the hypochlorous acid, it is formed when chlorine combines with water. A liquid form of compressed chlorine gas is used for disinfecting municipal water. Iodine is available as a tincture. Iodine inactivates proteins. Iodine combines with the amino acid tyrosine. The latter is a common component of many enzymes and other cellular proteins. Proteins are oxidized due to oxidation of sulfhydryl group of amino acids.

Compounds of heavy metals

Mercuric chloride is vastly used. It is useful for disinfecting desk tops and inanimate objects. Mercuric chloride (1:1000) and silver nitrate (1:1000) are used for disinfecting the surface of test materials. Proteins are inactivated and denaturalized by mercuric chloride and silver nitrate, respectively.

Phenols

Phenols and its derivatives (lysol, cresol etc.) are active on the surface. They disrupt cell membrane, inactivate enzymes and denaturing proteins. Phenolics (derivatives of phenol) are often used as disinfectants because they remain active in the presence of organic compounds. They are stable and persist for long periods after applications.

Alcohol

Ethyl alcohol and isopropyl alcohol denatures or inactivates proteins. Ethyl alcohol (70% by weight) is used for disinfecting laboratory desk-tops. It has low toxicity. It is applied on skin as disinfectant. Formaldehyde inactivates enzyme. It is a microbistatic.

Detergent Soap

Roccal, Zephiran disrupt cell membrane, denature proteins and inactivates enzymes. It is used as skin disinfectant.

GASEOUS STERILIZATION

Gaseous sterilization is applied where sterilization by heat or other agents is not possible. Ethylene oxide (C_2H_4O) is used for this purpose. It is toxic to viruses, bacteria and fungal cells. It even kills heat resistant spores and endospores. This gas is also harmful to man and hence it should be handled with care. Vapour of ethylene oxide is used under pressure in special equipment. Ethylene oxide is used in disinfecting chemical respirators. Heart-lung machines, opthalomoscopes, blankets, bedding etc.

RADIATION

Radiation is used to kill microbes. The wavelength, intensity and duration are important factors, which controls microbial growth. There are two types of radiation used for killing microbes. They are ionizing radiation and non-ionizing radiation. Ionizing radiation is of short wavelengths (less than 1 nm). Gamma rays, X-rays or high energy, electron radiation are ionizing radiation. Ionizing particles or energy packets pass through the vital portions of the cell. This is called as hits. Few hits may cause non-lethal mutation, but a high dose of such hits causes lethal mutations. Non-ionizing has wavelength of about 1 nm. UV light damages the DNA of exposed cells by causing bonds to form between adjacent thymines in DNA chains. These thyamine dimers inhibit correct replication of the DNA during reproduction.

ISOLATION OF PURE CULTURE

Natural environment has a mixed population of microorganisms. Hence, it becomes necessary to isolate the species of microbe to be studied. This is known as isolation. There are different techniques to isolate microorganism from different sources.

Serial dilution In a suspension or liquid source, there is a mixed population of microbes. It can be purified by serial dilutions. A mixed culture is serially diluted in test tubes of sterile medium. Dilutions are made up to the point of where, the number of cell are extinct. In this way, the last tube will contain only a single organism.

Streak plate method Sterile petri dishes are plated with solid culture media. A small amount of mixed culture is streaked on the surface of this culture media. To prepare a streak, inoculting needle with platinum or nichrome wire loop is used. The sample is streaked in such a way as to provide successive dilution in first stroke. Wire loop is not reinoculated while preparing a streak. The streak thins out the culture sufficiently. It separates the aggregated cells of the sample. The plates are incubated overnight. Cells spread by the streak starts developing colonies. These colonies are pure as they are developed by a single cell. Inoculum from each colony is reinoculated on a separate petri dish.

Pour plate method In this method serially diluted sample (mixed with culture medium) is plated on a sterile petri plate. An aliquot of diluted inoculum is added in tubes of melted and cooled agar medium. The temperature of medium is kept at about 45 to 50°C. The melted medium with inoculum is spread on a sterile petri plate and allowed to solidify. These plates are incubated for overnight. The colonies develoed after this incubation period are reinoculated on a separate petri plate.

Enrichment culture This technique is adopted for isolation of specific microbes. The mcirobes present is relatively small number is a mixed population. Much more, the growth of such microbes is slow as compared to the other microbes of the population. For this purpose, culture media is specially designed so that it permits the growth of a specific microbe and inhibits the growth of other types. Beijerinck and Winogradsky isolated *Nitrobacter* by this method.

Single cell isolation This is an accurate method of isolation. A single cell of the desired microbe is picked up and placed in a watch glass or a perti plate. It is watched under the microscope and the unwanted mycelium etc is washed out. A single cell or spore is picked up with a capillary pipette and inoculated on a culture medium. Micromanipulator is a device used with microscopes. It has micrometer adjustment. Micropipettes or other instruments can be moved forward and backward, right and left, and up and down. A drop of serially diluted medium containing microbe is placed on a sterile cover slip. This cover slip is inverted on a slide. This technique is known as hanging drop technique. The drop is observed under the microscope and a cell is gently sucked up by a micropipette. It is transferred to a large drop of sterile broth and incubated.

MAINTENANCE OF CULTURE

Maintenance of isolated cultures is an important task in microbiology laboratory. To maintain a culture care is taken so that the microbe's viability is not lost and it does not undergoes mutation. There are following methods for maintenance of culture.

Periodic transfer of culture A microbe is grown on agar slants and after it completes its period of logarthmic growth, it is stored in refrigerator. Low temperature prevents the accumulation of metabolic products in the media. This enables microbes to survive for a longer period. Frequent transfer of microbes results in mutation and hence this practice is avoided.

Use of mineral oil Microbial culture may be stored by using the mineral oil. Slants are covered by ½ inches above the tip of the slant. Such storage does not affect viability for few years.

Maintenance of soil microorganisms Soil microorganism are maintained in soil medium. For the preparation of soil medium, 1% dextrose is mixed and stirred in fertile soil. Soil used is filtered form sieved in 2mm sieve. It is kept in test tubes at least 50 cm deep. Test tubes containing soil is first kept in boiling water for 15 minutes and plugged. It is further autoclaved for two successive days. The culture is inoculated in this medium and preserved for long months.

Maintenance of plate culture A petri plate containing proper medium is inoculated with culture. When colonies develop covering whole plate, a drop of formaldehyde is placed on the lid. This will fix the colonies. The plate is sealed with vaseline and stored at 4°C. Sealing of plate stops gaseous exchange inside and outside environment of the plate. This is to further stop of the growth of the microorganism. When the growth is completely inhibited, clean embedding resin is spread over the surface of the plate so that there is safe displaying in the glass.

Maintenance by lyophilization (Freeze drying) This is an important method and microbes may be stored for years in their inactive state. This technique completes in few steps, culture is frozen and then water is removed by sublimation from the ice as vapour. Thus, the water content of cell will not convert into ice crystals and so it does not damages the cellular structure. The culture is stored in ampoules. Ampoules are of about 6 mm internal diameter and are made up of neutral glass. The ampoules are first treated with 2% HCl and rinsed with sterile distilled water. Number is placed in the ampoule, which shows the index number. Dry ampoules are plugged with absorbent cotton wool and placed for sterilization within a metal box made of monel. It is autoclaved and after autoclaving, ampoules are stored at 60°C. First step of this technique is pre-drying. It is the treatment of heat at almost the same temperature as required for growth. The media may also be the same of different for pre-drying and for maintenance of culture (e.g., in case of *Bacillus subtilis*, bacterium is grown on nutrient agar but for pre-drying, soil-extract-agar medium is used). There may be atmospheric gases inside the plates or carbon dioxide and air, only carbon di oxide, atmospheric methane, nitrogen etc. After the microbe has completed the logarithmic growth phase, suspension fluid is spread over the agar slope. Culture is rubbed with a Pasteur pipette and this suspension is transferred to ampoules immediately so that, spores do not start germinating. The next step is primary drying. Filled ampoules are placed on a perforated plate, the cuet stand. Cotton plug is removed from the neck and latter is flamed. Now this cuet stand is fixed in previously flamed canister. The canister is covered and fixed on centrifuge head. This is now subjected to drying chamber of a machine. This is heated in vacuum and centrifuge is operated at 500 rpm for 2-4 hours so that free water is evaporated. After this chamber is gently ventilated. The ampoules are removed and again plugged with pre-sterilized non-absorbent cotton wool and tube is constricted. Next step is secondary drying. The primary dried ampoules are fixed on the nipples of the secondary dryer. There is a plate at the bottom. Phosphorus pentaoxide is kept in this tray. Vacuum pump is turned on and after secondary drying about only 1% of moisture is left and the culture takes a powdered form. This process takes about 18 to 20 hours. Ampoules are finally sealed and stored in dark at 4°C temperature. Whenever, the ampoules are to be reopened and the culture is to be reactivated, first the neck along with cotton plug is removed. 0.5 ml of sterile nutrient broth is poured in the ampoules and mixed with culture. The suspension is sub-cultured in a suitable medium. This will again activate the freeze-dried microorganism.

Maintenance by Sordelli's method In this technique, microbe is inoculated on a suitable medium. When the colonies are fully grown, it is emulsified in a loop-ful of horse serum. This suspension is kept in a small tube, which is kept in a large tube. The sizes of these tubes are 8 X 60 mm and 10 x 15 mm, respectively. Before keeping small tube inside the large tube, some phosphorus pentaoxide is kept in the large tube very carefully so that it doesn't adheres on the walls of the large tube. The small tube with culture is properly plugged and kept in this large tube. Now the large tube is constricted near the rim and joined to the vacuum pump. The tube is then sealed and stored. Whenever, required the tube are gently broken and microbe is subcultured. Thus it is again ready for use.

CHAPTER - 33

STAINING TECHNIQUE

Most of the microorganisms are colourless and transparent. Their cells are stained by using dyes. Dye is synthetic products having both chromophores and auxochrome groups. Dyes are of acid or basic types depending upon the electrical charge on the dye portion of the molecule. Acid dyes are salts of colour acids and ionize to give the dye portion of the molecules a negative charge. Basic dyes are salts of colour bases and their ionization yield a positive charge. Stains are applied in aqueous medium. Solutions are prepared in water or alcohol. Certain structures are not stained directly. They need mordant. A mordant is not a dye but it forms insoluble compounds with a stain thus staining the structure. In microbiological studies, tannic acid, salts of ammonium, iron, tin, zinc, copper and chromium are used as mordant.

SIMPLE STAINING

In staining process when only one dye is used it is called simple staining. Simple staining helps only in examining size and shape of bacteria. It is not helpful in differentiating structures. Simple staining of bacterial cell is performed by using basic dyes like methylene blue, crystal violet, basic fuchsin, eosin-Y, safranin-O etc. Acid dyes are used for negative staining. In negative staining, cells do not retain colour. Sodium eosinate and nigrosine are acid dye. While staining bacterial cell with these dyes, the cell wall being acidic in nature does not combines with these stains. These dyes simply accumulate around the cell leaving it colourless. Thus the outline of the cell may be studied.

DIFFERENTIAL STAINING

In differential staining two or more dyes are used. Gram staining and acid fast staining are the two differential staining techniques common in micobiological studies.

GRAM STAINING

In 1884, a Danish bacteriologist Hans Christian Gram developed a staining method for bacteria, which has significance in classifying bacteria and even for clinical aspect. It is a differential method of staining. After staining, those bacteria which retain stain, are known as Gram positive while those fail to retain stain are called Gram negative. A thin film of material containing bacteria is spread over the surface of the slide. This film is air dried or gently passed through the flame of a Bunsen burner. This is known as smear. A smear stained with crystal violet. This is the primary staining. After a minute it is washed and few drops of iodine solution is applied on this smear. Iodine solution is taken as mordant. After some time, it is washed with organic solvents like alcohol or alcohol-acetone solution. It is the decolourizing agent. If, after washing the bacterial cells retain stain, they are Gram positive and if they fail, it is Gram negative. Gram negative bacterial smear is counter stained with basic dye, safranin. The Gram positive bacteria have some physico-chemical properties of cell wall. The cell wall of Gram positive bacteria have more peptidoglycan while, the cell wall of Gram negative bacteria have more lipopolysaccharide. The crystal violet and iodine solution enters in the cells of both types of bacteria and forms a complex known as crystal violet-iodine (CV-I) complex. When the smear is washed with organic solvents, the CV-I complex becomes intact with peptidoglycan. Thus, cells retain colour. In Gram-negative bacteria, the organic solvent disrupts the lipopolysaccharide layer and washes CV-I complex and the cells fail to retain colour.

ACID FAST STAINING

This technique is applied for *Mycobacterium.* It is a Gram positive bacteria but fails to retain stain. Cell of this bacterium, is covered by a waxy material. It does not take counter stain. It is stained with red coloured, carbol-fuchsin dye. This dye is more soluble in phenol than in water or acid-alcohol. In this method, carbolduchsin is applied to a smear. This is gently heated so that the dye can penetrate inside the cells and it is retained there only. After sometime the slide is cooled and washed with water and then by decolourizer (acid-alcohol). The dye is more soluble in cell wall waxes and thus it is retained.

SPECIAL STAINS

Special staining technique is applied for staining capsule, flagella and endospore if present.

CAPSULE STAINING

Some bacteria have gelatinous capsule covering their cell. Components of capsule are soluble in water and may be dislodged or removed during washing. Negative staining is applied using Indian ink or nigrosin. The bacteria is simple stained with safranin. It appears as halos surrounding each stained bacterial cell.

ENDOSPORE STAINING

Endospores are formed in seven genera of bacteria. It provides resistance to bacterial cell under adverse conditions. It is stained by Schaeffer-Fulton endospore stain. Malachite green, is applied on the smear. It is a primary stain. The stain treated smear is heated to steaming for about five minutes. This enables the stain to penetrate in endospore wall. After cooling it is washed with water for about thirty seconds. The smear is counter stained with safranin. Endospores appear green within red or pink cells.

FLAGELLA STAINING

Flagella staining is a delicate process. A mordant and carbolfuchsin stain is used in flagella staining. Mordants act as a surface coating. It is a mixture of tannic acids and some metallic salts. Mordant and stain may be applied as a combined solution or may be applied separately.

REFERENCE

A very, O.T., Macleod, C.M. and McCarty.M. (1944) Studies on the chemical nature of the substance inducing transformation of pneumococcal types J.Exp. Med. **79**:137.

Abbott, L.K. and Robson, A.D. (1985). Formation of external hyphae in soil and four species of vesicular-arbuscular mycorrhizal fungi. New Phytol., 245-255.

Agrios, G.N. (1988) Plant - Pathology, Academic Press, inc, Newyork.

Alexander, C; Alexander, I.J. and Hadley, G (1984). Phosphate uptake by *Goodyera repens* in relation to mycorrhizal infection. New Phytol. **97**, 401-411.

Alexander, M. (1961). Introduction to Soil Microbiology. John Wiley and Sons, Inc. New York and Lond.

Alexander, M. (1977) Introduction to Soil Microbiology 2nd Edu. John Wiley and Sons, Inc. New York and London.

Alexopoulous, C.J. and Mims, C.W., Introductery Mycolgy 3rd Edu. John Wiley & Sons, 1962.

Ananthanarayan, Round Paniker, C.K.J. (2000) Text Book of Microbiology, 6th Edu. Orient Longman, Hyderabad.

Atleinson, B. and Mavituna, F. (1983), Biochemial engineering and biotechnology handbook, Macmillan Publisher Ltd., New York.

Bagyaraj, D.J. (1984). Biological interaction with VAmycorrhizal fungi. In: "VA Mycorrhiza" (Powell, C.LI. and Bagyaraj, D.J. eds.) CRC Press. Boca Ration, FL, pp. 131-153.

Bailey, J.E. and Ollis, D.F. (1977). Biochemical engineering fundamentals, McGraw-Hill book company, London.

Bakshi, B.K. 1976. Forest Pathology-Principle and Practices. Controllers of Publication.

Beckwith, T.D. (1911). Phytopathology, **1**, 169-176.

Bergey's Manual of Det. Bact. "7th Ed." Williams and Wilkins, Baltimore.

Birge, E.A. Bacterial and Bacteriophage Genetics, Newyork Springer - Verlag, 1981.

Borkar, P.S. (Editor) (1982), Hindustan Antibiotics bulletin Vol. 24, H.A. Ltd.s, India.

Brbank, W.D. (1942). Physiol. Zool. **15**, 342-362.

Brock, T.D. Milestones in Microbiology Reprint (1975) Washington, D.C., American Souety for Microbiology.

Broda, P. (1979) Plasmids. W.H. Freeman.

Burges, A.; Hurst, H.M. and Walkden, B. (1964). Geochim. Cosmochim. Acta. **28**, 1547-1554.

Burges, N.A. and Fenton, E. (1953). Tran. Br. Mycol. Soc. **36**, 104.

Burnett, J.H. Mycogenesis : An Introduction to the Genral Genelies of Fungi, New York, John Wiley & Sons, 1975.

Bushell, M.E. and Slater, J.H. (Editors) (1981), Mixed Culture fermentations, Academic Press, London and New York.

Casida, L.E. (1968), Industrial microbiology, John Wiley & Sons Ltd., New York.

Clarkson, D. (1985). Factors affecting mineral nutrition acquisition by plants. Annu. Rev. Plant Physiol., **36**, 77-115.

Clifton, C.E. (1957), Introduction to Bacterial Physiology, McGraw-Hill Book Company Inc., New York.

Davis, R.M. (1980). Influence of Glomus fasciculatus on *Thielaviopsis basicola* root rot of citrus. Plant Dis., **64**, 839-840.

Fedorov, M.V. (1952). "Biological fixation of atmospheric nitrogen". 2nd ed. Gosudarstv. Izdatel. Sel' skokhoz. Let. Moscow (Russian).

Fox, Brian A. (1970), Food Science - a chemcial approach, Hodder and Stoughton, London.

Frazier, W.C. (1967), Food Microbiology, Mc Graw-Hill Book Company, London & New York.

Gabriel, B.L. Biological Electron Microscopy, Van Nostromy Reinhold, New York, 1982.

Gardner, Eldon, J. (1972), History of Biology, Wiley Eastern Limited, New Delhi, India.

Garrett, R.H. and Grisham, C.M. (1999) Biochemistry, 2nd edu. New York Saunders.

George A. Wistreich & Max D. Lechtman, Minobiolon (1984) 4th Edu. Macmillan Publishing Company pp 916 pp. Ainsworth, G.C. Introduction to the History of Plant Pathology Cambridge, England, 1981.

Gilman, J.C. (1957). "A Manual of Soil Fungi". Jowa State College Press, Ames, Jowa.

Hairston, N.G. (1965). In "Progress in Protozoology", London (1965), pp. 114.

Hammer, B.W. (1938), Dairy bacteriology, Wiley, London.

Han. Y.M. (1978), Adv. appl. microbiol. **23**, 119-149.

Harley, J.L. and Smith, S.E. (1983). Mycorrhizal symbiosis. Academic Press, London.

Harley, J.L. and Waid J.S. (1955). Plant. Soil, **7**, 96-112.

Henderson, V.E. and Katznelson, H. (1962). Phytopathology, **52**, 423-427.

Hewitt, W. (1977), Microbiological assay, Academic Press, London.

Hughes, S.S. The Virous, it History of the Corcept, New York, Science History Publications, 1977.

Jacob, F., and J. Monod, (1961) Genetic regulatory mechanisms in the synthestis of proteins. J, Mol. Biol, 3:318.

James D. Watson, Tania A. Baker, Stephen P. Bell, Alexender Gann, Michael Levine, Richard Losick, Molecular Biology of the Gene Pearson Education, V Edu. 2004.

Kendrew, J.C. (1961) Three Dimensional Structure qa protien Molecule, Sci Amer. 205-96-110.

Ketchum, Paul A, Microbiology Concept a application. John Wiley and Sons, 1988.

Khan, .M. and Mehrotra, R.M.L. (1973) Microbiology for Medical Students, Vol. I., Luchwow Publishing House, Lucknow

Kornberg, A. and Baker .A. (1992) DNA Replication, 2nd edition, W.H. Preeman, New York.

Kornberg, A. and T. Baker, (1992) DNA replication, 2nd Edu. W.H. Freeman.

Lehninger, A (1971) Bioenergetics, 3rd edition, Benjamin Cummings, Menlo Pank, Californika.

Luria, S.E. and M. Delburk (1943). Mutations of bacteria from virus sensitivity to virus resistance, Genetics, **28**, 491.

Matthews, R.E.F. Plant Virology, Acodemic Press, Inc. 1981.

Mims, C.A. (1987) The Pathogenesis of Infections Disease, 3rd Edu. London Academic Press.

Mishra, R.R. and Srivastava V.B. (1970). Leaf surface fungi of *Oryza sativa* Linn. Mycopath. Mycol. Appl. 44, 289-294.

Mishra, R.R. and Srivastava V.B. (1970). Leaf surface microflora of Triticum aestivum. Bull. Torr. Bot. Club. **97**, 364-367.

Parkinson, D. (1965), In "Plant Microbes Relationship", pp. 69-75. Publishing House of the Czechoslovak Academy of Sciences, Prague.

Paul, P. and Clark, F.E. (1989) Soil Microbiology and Biotechnology. Academic Press, New York.

Pauling L, (1960) The nature of the chemical bond, 3rd edu. Cornell, University Press, Ithaca, New York.

Pelezar, M.J., Jr. and Reid, R.D. (1985). "Microbiology", McGrow Hill Book Co., Inc., N. York.

Pres cott & Cunn's - Industrial Microbiology 4th Edn. edited by Gerald Reed, CSBS Publishers & Distributors, New Delhi.

Reed, G. (Editor), (1982), Prescott & Dunn's Industrial Microbiology, Macmillan Publishers, Ltd., U.K.

Robinson, R.K. (Editor), (1981), Dairy Microbiology; Vol. 2, Applied Science Publishers, New Jersey and London.

Ronald M. Alas, Basic and Practical Microbiology, Macmillan Publishing Company, 1986.

Rose, A.H. (1961), Industrial Microbiology, Butterworth & Co. (Publishers) Ltd., London.

S chiegel, H.G. and Bowien, B. editors (1989) Autotrophic bacteria, Madison, Wis. Science Tech. Publishers.

Schlegel, H.G. (1993) General Microbiology, 7th Ethni Cambridge University Press Cambridge.

Singh, R.N. (1961). "Role of Blue-green Algae in Nitrogen Economy of Indian Agriculture". Indian Coun. Agric. Res., New Delhi.

Smith, K.M. and D.A. Ritchie, Introduction to Virology, Chapman & Hall, 1980.

Smith, K.M. Plant Viruses, 6th edu. London, Chapman & Hall, 1977.

Smith, S.E., St. John. B.J.. Smith, F.A. and J.D. Nicholas (1985). Activity of glutamine synthetase and glutamate dehydrogenase in Trifolium subterraneum L. and Allium Cepa L.. Effect of mycorrhizal infection and phosphate nutrition. New Phytol., **99**, 211-227.

Stanier, R.Y, J. Jugraham, M.L. Wheelis and P.R. Painter, The Microbial Ward 5th Edu. Prentice - Hall, Englewood Chiefts, N.J., 1986.

Stanl F.W. (1987) Genetic recombination. Sci AM. 256 (2) : 91-101

Subba Rao, N.S. (1999) Soil Microbiology, Oxford & IBH Publishing Co. Pvt. Ltd. New Delhi.

Tortora, G.J., Funke, B.R. and Case, C.L. (1998) Microbiology : An Introduction, 6th Edu. The Benjamin - Cummings Publishing Co., Redwood City, CA.

Umbreti, W.W. (1962) Modern Microbiology, Freeman & Co. London.

Waksman, S.A. (1927). "Principles of Soil Microbiology." Bailliere, Tindall and Cox, London.

Waksman, S.A. (1959). The Actinomycetes. Vol. 1. Nature occurrence and activities. The Williams and Wilkins Co., Baltimore.

Watson, J.D. and Crick F.H.C. (1953), Molecular Structure of nucleic acids : A structure for deoxyribonycleic acids. Natre, 171 : 737-738.

Weir, DM and Stewart, J. (1997) Immunology, 8th Edu. Edinburgh, Churchill - Livingstore.

Wolfgang K. Joklik, Hilda P. Willett (1968) Zinsser Microbiology, Appleton - Century Crofts, 16th Edu. 1165pp.

Zurig S.E., James C. Darnell, Jr. David Balbmore, Allan Campbell. John. Wiley & Sons, 1978